Section I
Purines and Analogs

[1] 9-Benzyladenine, 9-(ω-Chloroalkyl)adenines, and 3,9-(Oligomethylene)adeninium Chloride

Synthesis of 9-Alkyladenines from the Sodium Salt of Adenine

KERMIT L. CARRAWAY, PING CHEONG HUANG, and T. GORDON SCOTT

DEPARTMENT OF CHEMISTRY AND CHEMICAL ENGINEERING, UNIVERSITY OF ILLINOIS, URBANA, ILLINOIS 61801

1 (135.1) $\xrightarrow{NaH}$ 2 (157.1) Na^+ $\xrightarrow{RX}$

3, R = $C_6H_5CH_2$ (225.2)

3a, R = $(CH_2)_nCl$ (169.6 + 14.0*n*) (*n* = 2–10)

3a ⟶ 4 Cl^- (169.6 + 14.0*n*) (*n* = 3–5)

INTRODUCTION

9-Benzyladenine[1] (**3**) is a useful intermediate in the synthesis of 1-substituted adenines,[2] as the 9-benzyl group is a removable protecting group[2,3] that directs an incoming substituent to the 1-position. The 9-(ω-chloroalkyl)adenines (**3a**) provide versatility in the synthesis of related 9-(ω-substituted alkyl)adenines, and an interesting application is their cyclization to 3,9-(oligomethylene)adeninium chlorides (**4**). The title compounds **3** and **3a** are prepared by alkylation of the sodium salt of adenine, prepared from adenine and sodium hydride in *N,N*-dimethylformamide. This procedure increases the selectivity of the alkylation[4] to give an increased ratio of the 9- to the 3-isomer as compared with earlier methods,[5] and permits crystallization of the desired product from the reaction mixture.

PROCEDURE

Sodium Salt of Adenine[4] (2)

A suspension of 25 g. (0.19 mole) of adenine (**1**) and 5 g. (0.21 mole) of sodium hydride (53.9%, dispersed in mineral oil; Metal Hydrides, Inc.) in 300 ml. of anhydrous *N,N*-dimethylformamide is stirred at room temperature for 1 hr. A white suspension of the sodium salt is formed; the suspension is used for the alkylation step.

9-Benzyladenine[4] (3)

To the stirred suspension of sodium adeninide (**2**) is added, dropwise, 25 ml. (0.21 mole) of benzyl bromide, and the stirring is continued overnight. The product (**3**) is precipitated when the reaction mixture is cooled, and is recrystallized from absolute ethanol as colorless needles, m.p. 233.5–235.5°; yield 11.7 g. (27%), homogeneous by thin-layer chromatography on silica gel.

9-(ω-Chloroalkyl)adenines[6] (3a)

To a stirred suspension of the sodium salt of adenine (**2**) is added 0.24 mole of a 1-bromo-ω-chloroalkane, and stirring is continued at room temperature for 10 hr. The resulting clear, light-yellow suspension is filtered, and the solvent is removed under diminished pressure. The residue is washed with a small volume of ether, and is recrystallized from aqueous methanol. The yields and melting points are tabulated below:

n	*Compound* **3a**	*Yield, %*	*m.p., degrees*
2	9-(2-Chloroethyl)adenine[7]	38	204–205
3	9-(3-Chloropropyl)adenine[7]	32	189–190
4	9-(4-Chlorobutyl)adenine	28	138–140
5	9-(5-Chloropentyl)adenine	31	155–157
6	9-(6-Chlorohexyl)adenine	27	128–130
7	9-(7-Chloroheptyl)adenine	40	153–154
8	9-(8-Chlorooctyl)adenine	36	149–150
9	9-(9-Chlorononyl)adenine	30	132–134
10	9-(10-Chlorodecyl)adenine	30	120–121

Extended heating is to be avoided, especially where $n = 3$–5, as intramolecular cyclization then occurs. Intentional cyclization is described next.

3,9-(Oligomethylene)adeninium Chlorides[6] (4)

The cyclic compound (**4**, $n = 3$–5) is obtained (a) by heating the corresponding 9-(ω-chloroalkyl)adenine (**3a**) at a temperature 10° above its melting point until resolidification takes place, or (b) by heating compound **3a** in a small proportion of *N,N*-dimethylformamide at 150° for 1 hr. The products (**4**) are recrystallized from aqueous ethanol, the yields are 90–95%, and the melting points are all >320° for 10-amino-5,6-dihydro-4*H*-pyrimido[1,2,3-*cd*]purin-3(or 7)-ium chloride (3,9-trimethyleneadeninium chloride), 11-amino-4,5,6,7-tetrahydro[1,3]diazepino-[1,2,3-*cd*]purin-3(or 8)-ium chloride (3,9-tetramethyleneadeninium chloride), and 12-amino-5,6,7,8-tetrahydro-4*H*-[1,3]diazocino[1,2,3-*cd*]-purin-3(or 9)-ium chloride (3,9-pentamethyleneadeninium chloride).

REFERENCES

(1) J. W. Daly and B. E. Christensen, *J. Org. Chem.*, **21**, 177 (1956); J. A. Montgomery and C. Temple, Jr., *J. Am. Chem. Soc.*, **83**, 630 (1961).

(2) N. J. Leonard and T. Fujii, *Proc. Natl. Acad. Sci. U. S.*, **51**, 73 (1964).

(3) J. A. Montgomery and H. J. Thomas, *J. Am. Chem. Soc.*, **85**, 2672 (1963); N. J. Leonard and K. L. Carraway, *J. Heterocyclic Chem.*, **3**, 485 (1966).

(4) K. L. Carraway, Ph.D. Thesis, University of Illinois, 1966.

(5) G. Thoiss, *Z. Physiol. Chem.*, **13**, 395 (1889); M. Krüger, *ibid.*, **18**, 423 (1894).

(6) P. C. Huang, Ph.D. Thesis, University of Illinois, 1966.

(7) J. H. Lister and G. M. Timmis, *J. Chem. Soc.*, **1960**, 327.

[2] 9-(1-Hydroxy-2-octyl)adenine

Preparation of a 9-Substituted Purine by Cyclization of a Pyrimidine with Triethyl Orthoformate

HOWARD J. SCHAEFFER and CHARLES F. SCHWENDER

DEPARTMENT OF MEDICINAL CHEMISTRY, STATE UNIVERSITY OF NEW YORK AT BUFFALO, BUFFALO, NEW YORK 14214

1 (164.0) + **2** (145.2) → **3** (272.2) → **4**, R = Cl (282.8); **5**, R = NH_2 (263.3)

INTRODUCTION

9-(1-Hydroxy-2-octyl)adenine (**5**) is a potent inhibitor of the adenosine deaminase of calf-intestinal mucosa.[1] Its synthesis is readily accomplished by a modification of a general procedure for the preparation of 9-substituted adenines.[2,3]

PROCEDURE

5-Amino-4-chloro-6-(1-hydroxy-2-octylamino)pyrimidine (3)

A mixture of 12.0 g. (73.0 mmoles) of 5-amino-4,6-dichloropyrimidine (**1**) (Sec. II [27]), 11.6 g. (80.3 mmoles) of 2-amino-1-octanol (**2**),[4] 8.59 g. (85.0 mmoles) of triethylamine, and 125 ml. of butyl alcohol is refluxed for 21 hr. under a nitrogen atmosphere. After the reaction mixture has been evaporated *in vacuo* to a sirup, the crude product is dissolved in methanol. Addition of water to the cooled, methanolic solution gives 16.1 g. of crude, crystalline product, m.p. 94–102°. It is recrystallized from methanol–water; yield 14.9 g. (74.8%), m.p. 102–105°.

6-Chloro-9-(1-hydroxy-2-octyl)purine (4)

A suspension of 6.23 g. (22.8 mmoles) of compound **3**, 74.9 mg. (0.682 mmole) of ethanesulfonic acid, and 20 ml. of triethyl orthoformate is stirred at room temperature for 3 hr. Cooling the mixture and addition of hexane gives 3.57 g. of a colorless solid. The filtrate is evaporated *in vacuo* to a sirup, which is crystallized from chloroform and hexane to give an additional 2.25 g. The crops are combined, and recrystallized from hexane to give pure **4**; yield 5.58 g. (86.9%), m.p. 72–74°.

9-(1-Hydroxy-2-octyl)adenine (5)

A mixture of 1.00 g. (3.53 mmoles) of 6-chloro-9-(1-hydroxy-2-octyl)-purine (**4**) and 25 ml. of methanolic ammonia (20%) is heated at 90° for 19 hr. in a steel bomb. The reaction mixture is evaporated *in vacuo* to a solid which, after recrystallization from ethanol–water, gives 770 mg. of a colorless, crystalline solid, m.p. 147–150°. Further recrystallization of the crude product from ethanol–water gives pure **5**; yield 459 mg. (49.3%), m.p. 150–152°.

REFERENCES

(1) H. J. Schaeffer and C. F. Schwender, to be published.
(2) J. A. Montgomery and C. Temple, Jr., *J. Am. Chem. Soc.*, **79**, 5238 (1957).
(3) C. Temple, Jr., C. L. Kussner, and J. A. Montgomery, *J. Med. Pharm. Chem.*, **5**, 866 (1962).
(4) O. Vogl and M. Pöhm, *Monatsh.*, **84**, 1097 (1953).

[3] *N*-(3-Methyl-2-butenyl)adenine

[*N*-(Isopentenyl)adenine]

Combination of 6-Chloropurine with a Primary Amine

SIDNEY M. HECHT, JOHN P. HELGESON, and TOZO FUJII[a]

DEPARTMENT OF CHEMISTRY AND CHEMICAL ENGINEERING, UNIVERSITY OF ILLINOIS, URBANA, ILLINOIS 61801

$$\text{NK} + \text{ClCH}_2\text{CH{=}C(CH}_3)_2 \longrightarrow \text{NCH}_2\text{CH{=}C(CH}_3)_2$$

1 (185.2) **2** (104.6) **3** (215.2)

1. N_2H_4
2. HCl

$$\text{(CH}_3)_2\text{C{=}CHCH}_2\text{NH}_2\cdot\text{HCl} \xrightarrow[\text{Et}_3\text{N}]{\mathbf{5}\ (154.6)} \text{HNCH}_2\text{CH{=}C(CH}_3)_2$$

4 (121.6) **5** (154.6) **6** (203.2)

INTRODUCTION

N-(3-Methyl-2-butenyl)adenine (**6**) (Sec. I [4]), which was initially obtained synthetically,[1,2] is a highly active cytokinin.[3–6] In the tobacco assay, compound **6** is about ten times as active as kinetin in the region of highest sensitivity, and in high concentrations, it represses growth less than does kinetin (*N*-furfuryladenine).[5] Compound **6** occurs in Nature,[7–10] and the synthetic route to it described here is similar to the

[a] Present address: Department of Pharmacy, Kanazawa University, Kanazawa-Shi, Ishikawa-Ken, Japan.

generally applicable method for producing N^6-substituted adenines from 6-chloropurine (**5**) and primary amines,[11] except that, in the present instance, the amine is generated in the reaction mixture [as it is more convenient to store 3-methyl-2-butenylamine as the hydrochloride (**4**)].

PROCEDURE

N-(3-Methyl-2-butenyl)phthalimide (3)

To a stirred suspension of 65 g. (0.35 mole) of potassium phthalimide (**1**) in 1 l. of *N,N*-dimethylformamide is added dropwise 30 g. (0.29 mole) of 1-chloro-3-methyl-2-butene (**2**). The mixture is heated at 70° for 2 hr. with stirring, allowed to cool, and poured into 3 l. of ice water. After several hours, the resulting fine, white precipitate is collected by filtration and dried, yielding 54.4 g. (88%) of crude **3**, m.p. 90–92°.

3-Methyl-2-butenylamine Hydrochloride (4)

To a suspension of 50.0 g. (0.23 mole) of crude compound **3** in 150 ml. of methanol is added 13.0 ml. of an 85% solution of hydrazine hydrate in water. The solution is refluxed, with stirring, for 2 hr., and allowed to cool. To the solution is added 100 ml. of water, the solution is evaporated to dryness under diminished pressure, and the residue is dissolved in 500 ml. of water. The pH of the solution is adjusted to about 1 with concentrated hydrochloric acid, and the resulting white precipitate is filtered off, well washed with water, and discarded. The aqueous solutions are combined and evaporated to dryness, and the solid product is recrystallized, with decolorization (carbon), from absolute ethanol; yield 14.2 g. (50.5%) of pure **4**, m.p. 190–192°.

N-(3-Methyl-2-butenyl)adenine (6)

A mixture of 20 g. (0.13 mole) of 6-chloropurine[12] (**5**), 20 g. (0.16 mole) of compound **4**, and 90 ml. of triethylamine in 1 l. of butyl alcohol is refluxed for about 1 hr., and allowed to cool. The solvent is completely removed under diminished pressure, and 750 ml. of water is added to the residual solid. The pH is adjusted to about 8, the suspension is refrigerated for a few hours, and the resulting crystals are filtered off, dried, and recrystallized several times (with decolorization) from absolute ethanol; yield 16.0 g. (60.6%) of pure **6**, m.p. 216–217°.

REFERENCES

(1) A. Cavé, Dr. Natl. Sci. Thesis, University of Paris, 1962.
(2) N. J. Leonard and T. Fujii, *Proc. Natl. Acad. Sci. U. S.*, **51**, 73 (1964).

(3) A. Cavé, J. A. Deyrup, R. Goutarel, N. J. Leonard, and X. G. Monseur, *Ann. Pharm. Franc.*, **20**, 285 (1962).

(4) N. J. Leonard, lecture at Fairleigh–Dickinson University, Madison, New Jersey, October 21, 1963; see *Trans. Morris County Res. Council*, **1**, 11 (1965).

(5) J. H. Rogozinska, J. P. Helgeson, and F. Skoog, *Physiol. Plantarum*, **17**, 165 (1964).

(6) G. Beauchesne and R. Goutarel, *Physiol. Plantarum*, **16**, 630 (1963).

(7) D. Klämbt, G. Thies, and F. Skoog, *Proc. Natl. Acad. Sci. U. S.*, **56**, 52 (1966).

(8) J. P. Helgeson and N. J. Leonard, *Proc. Natl. Acad. Sci. U. S.*, **56**, 60 (1966).

(9) As the 9-ribonucleotide in serine transfer ribonucleic acid: H. G. Zachau, D. Dütting, and H. Feldmann, *Angew. Chem.*, **78**, 392 (1966).

(10) As the 9-ribonucleotide in (yeast) soluble ribonucleic acid (sRNA) and calf-liver sRNA: R. H. Hall, L. Stasiuk, M. J. Robins, and R. Thedford, *J. Am. Chem. Soc.*, **88**, 2614 (1966).

(11) G. B. Elion, E. Burgi, and G. H. Hitchings, *J. Am. Chem. Soc.*, **74**, 411 (1952).

(12) (a) A. Bendich, P. J. Russell, Jr., and J. J. Fox, *J. Am. Chem. Soc.*, **76**, 6073 (1954); (b) A. G. Beaman and R. K. Robins, *J. Appl. Chem.* **(London)**, **12**, 432 (1962).

[4] *N*-(3-Methyl-2-butenyl)adenine

N-*Alkylation of Adenine*

ROSS H. HALL[a] and MORRIS J. ROBINS[b]

DEPARTMENT OF EXPERIMENTAL THERAPEUTICS, ROSWELL PARK MEMORIAL INSTITUTE, BUFFALO, NEW YORK 14203

$Me_2C{=}CHCN$ (**1**, 81.1) → $Me_2C{=}CHCH_2NH_2$ (**2**, 85.2) —[6-chloropurine **3** (154.6)]→ 6-($HNCH_2CH{=}CMe_2$)purine (**4**, 203.2)

INTRODUCTION

N-(3-Methyl-2-butenyl)adenine (**4**) (Sec. I [3]) may be prepared by the condensation of 3-methyl-2-butenylamine (**2**) with 6-chloropurine (**3**), as described herein. The base (**4**) cannot be obtained by acid hydrolysis of the corresponding D-ribonucleoside, namely, *N*-(3-methyl-2-butenyl)-adenosine [a component of soluble D-ribonucleic acid (s-RNA)[1]], because a series of degradation reactions occurs.[1,2]

PROCEDURE

3-Methyl-2-butenylamine[c] (2)

A solution of 8.1 g. (0.1 mole) of senecionitrile (3,3-dimethylacrylonitrile, **1**) (K & K Chemical Co., Plainview, N. Y.) in 50 ml. of absolute ether is slowly added to a stirred suspension of 3.42 g. (0.09 mole) of lithium aluminum hydride in 200 ml. of absolute ether at room temperature. The solution is stirred at room temperature overnight, and a solution of 3.6 g. (0.2 mole) of water in 50 ml. of tetrahydrofuran is added dropwise, with cooling. The mixture is filtered, the filtrate is dried with sodium sulfate, and filtered, and the filtrate is evaporated to a sirup, which is distilled at 109–110°/760 mm. Hg; yield 3.08 g. (36%) of product (**2**).

[a] Present address: Department of Biochemistry, McMaster University, Hamilton, Ontario, Canada.

[b] Present address: Department of Chemistry, University of Utah, Salt Lake City, Utah 84112.

[c] Semenow and coworkers[3] prepared this compound from 1-chloro-3-methyl-2-butene.

N-(3-Methyl-2-butenyl)adenine (4)

A solution of 1.6 g. (0.01 mole) of 6-chloropurine (**3**) and 1.7 g. (0.02 mole) of 3-methyl-2-butenylamine (**2**) in 100 ml. of 2-methoxyethanol is refluxed for 3.5 hr., the solution is cooled, and 10.4 ml. of *N* sodium hydroxide is added. The mixture is evaporated to dryness under diminished pressure, and the product (**4**) is recrystallized once from water and once from 1:25 (v/v) ethanol–water; yield 1.6 g. (70%), m.p. 212°.

REFERENCES

(1) R. H. Hall, L. Stasiuk, M. J. Robins, and R. Thedford, *J. Am. Chem. Soc.*, **88**, 2614 (1966).
(2) M. J. Robins, R. H. Hall, and R. Thedford, *Biochemistry*, **6**, 1837 (1967).
(3) D. Semenow, C.-H. Shih, and W. G. Young, *J. Am. Chem. Soc.*, **80**, 5472 (1958).

[5] 3-(3-Methyl-2-butenyl)adenine (Triacanthine)

Direct Alkylation of Adenine at the 3-Position

TOZO FUJII[a] **and NELSON J. LEONARD**

DEPARTMENT OF CHEMISTRY AND CHEMICAL ENGINEERING, UNIVERSITY OF ILLINOIS, URBANA, ILLINOIS 61801

NH_2 N N N N H + $(CH_3)_2C{=}CHCH_2Br$ ⟶ NH_2 N N $\overset{+}{N}$ N H Br^- $CH_2CH{=}C(CH_3)_2$

1 (135.1) **2** (149.0) **3** (284.2)

NH_4OH

NH_2 N N N N $CH_2CH{=}C(CH_3)_2$

4 (203.3)

INTRODUCTION

Triacanthine, the structure of which has been established as 3-(3-methyl-2-butenyl)adenine (**4**), has been isolated from a wide variety of natural sources.[1–6] Its synthesis *via* the direct alkylation of adenine[7] illustrates a convenient, general method for the preparation of 3-substituted adenines.[5,7–11]

PROCEDURE

3-(3-Methyl-2-butenyl)adenine Hydrobromide (3)

A mixture of 4.05 g. (30 mmoles) of adenine (**1**), 4.57 g. (31 mmoles) of 1-bromo-3-methyl-2-butene (**2**) (Sec. V [156]), and 30 ml. of *N*,*N*-dimethylacetamide is stirred at about 30° for 7 hr. The clear, pale-yellow

[a] Present address: Department of Pharmacy, Kanazawa University, Kanazawa-Shi, Ishikawa-Ken, Japan.

solution is evaporated to dryness under diminished pressure, and the resulting, almost colorless solid is washed three times with 80-ml. portions of ether. The solid is then triturated with 15 ml. of hot absolute ethanol, and the mixture is kept in a refrigerator for one day, giving crystals (7.06 g.) of crude hydrobromide; paper chromatography and paper electrophoresis indicate the presence of three components in the crystalline mixture. Two recrystallizations from absolute ethanol afford 4.72 g. (55%) of colorless pillars, m.p. 218–219° (dec.), shown to be pure **3** by a single spot at R_f 0.70 on a paper chromatogram, by use of 15:4:1 butyl alcohol–water–acetic acid.

3-(3-Methyl-2-butenyl)adenine (Triacanthine) (4)

To a solution of 3.10 g. (10.9 mmoles) of compound **3** in 20 ml. of warm water is added, dropwise, concentrated aqueous ammonium hydroxide until the pH of the solution is 7–8. The mixture is kept in a refrigerator for 5 hr., and the resulting crystals are collected by filtration, washed with a small amount of water, dried, and recrystallized from absolute ethanol, to give colorless prisms of **4**; yield 1.80 g. (82%), m.p. 231–232°.

REFERENCES

(1) A. S. Belikov, A. I. Bankowsky, and M. V. Tsarev, *Zh. Obshch. Khim.*, **24**, 919 (1954).
(2) M.-M. Janot, A. Cavé, and R. Goutarel, *Bull. Soc. Chim. France*, **1959,** 896.
(3) X. G. Monseur and E. L. Adriaens, *J. Pharm. Belg.*, **1960**, 279.
(4) N. J. Leonard and J. A. Deyrup, *J. Am. Chem. Soc.*, **82**, 6202 (1960).
(5) N. J. Leonard and J. A. Deyrup, *J. Am. Chem. Soc.*, **84**, 2148 (1962).
(6) A. Cavé, J. A. Deyrup, R. Goutarel, N. J. Leonard, and X. G. Monseur, *Ann. Pharm. Franc.*, **20**, 285 (1962).
(7) N. J. Leonard and T. Fujii, *J. Am. Chem. Soc.*, **85**, 3719 (1963).
(8) J. W. Jones and R. K. Robins, *J. Am. Chem. Soc.*, **84**, 1914 (1962).
(9) B. C. Pal, *Biochemistry*, **1**, 558 (1962).
(10) J. A. Montgomery and H. J. Thomas, *J. Am. Chem. Soc.*, **85**, 2672 (1963).
(11) C. J. Abshire and L. Berlinguet, *Can. J. Chem.*, **42**, 1599 (1964).

[6] 7-Benzylguanine

Synthesis of a 7-Substituted Purine from a Pyrimidine Precursor; Cyclization using Diethoxymethyl Acetate

G. T. ROGERS and T. L. V. ULBRICHT

TWYFORD LABORATORIES, ELVEDEN ROAD, LONDON, N.W. 10, ENGLAND

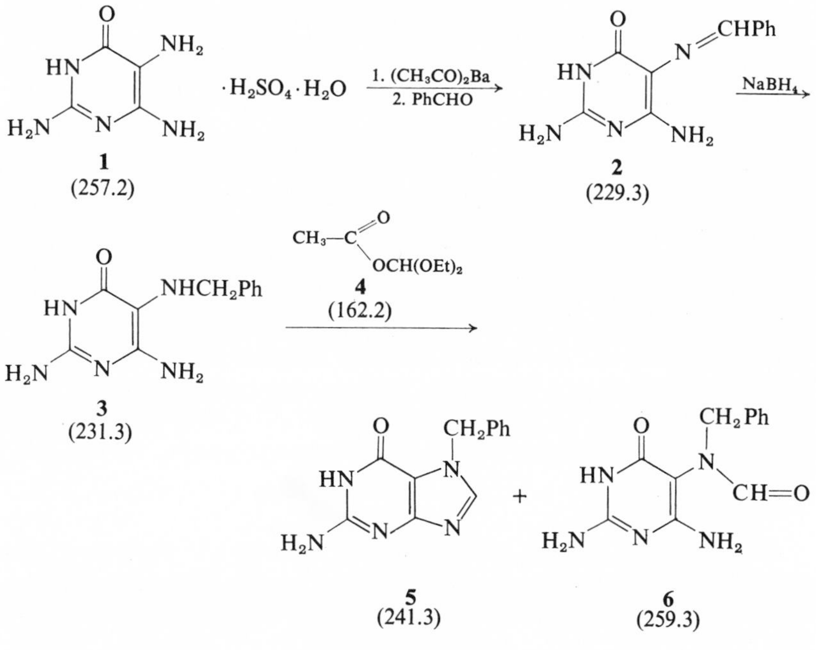

1 (257.2) $\cdot H_2SO_4 \cdot H_2O$ —1. $(CH_3CO)_2Ba$, 2. PhCHO→ 2 (229.3) —$NaBH_4$→

3 (231.3) —$CH_3C(=O)OCH(OEt)_2$ 4 (162.2)→ 5 (241.3) + 6 (259.3)

INTRODUCTION

Although guanine may be alkylated at N-7, it cannot be benzylated[1] at N-7. However, 7-benzylguanine can be synthesized from the appropriate pyrimidine by the procedure described here. Neither the immediate precursor (**3**), nor its formyl derivative, is cyclized by the cyclizing reagents commonly used, but cyclization proceeds readily with diethoxymethyl acetate (anhydride of acetic acid with diethyl orthoformate) (**4**). This compound[2] is a very effective cyclizing agent for 5,6-diaminopyrimidines.[3]

PROCEDURE

2,6-Diamino-5-(benzylideneamino)-4-pyrimidinol[4] (2)

To a hot solution of barium acetate (5 g., 0.02 mole) in 100 ml. of water containing a little sodium hydrosulfite is added 2,5,6-triamino-4-pyrimidinol sulfate (**1**) (5 g., 0.02 mole), the mixture is heated, with stirring, on a steam bath for 30 min., and the precipitate of barium sulfate is filtered off. To the filtrate is added a solution of benzaldehyde (2.5 g, 0.024 mole) in 20 ml. of ethanol, and the mixture is vigorously shaken, and kept at 0° overnight. The resulting suspension is filtered, and the crystals are successively washed with water, cold ethanol, and ether, and dried; yield 3.75 g. (82%) of **2**, a yellow solid, m.p. 276–278° (dec.), λ_{max} at pH 7.0, 233, 288, and 364 nm.; at pH 2.0, 251 nm.

2,6-Diamino-5-(benzylamino)-4-pyrimidinol (3)

A stirred suspension of compound **2** (2.3 g., 0.01 mole) in 40 ml. of 95% ethanol is dissolved by slowly adding 50 ml. of 3:1 water–ethanol containing fresh sodium borohydride (0.9 g., 0.024 mole). The mixture is slowly warmed to 40°, and the reaction is followed spectrophotometrically by observing the disappearance of the ultraviolet absorbance of the starting material at 364 nm. at pH 7.0. When the reaction is complete, the solution is rendered neutral with glacial acetic acid, and cooled to 5°, and the resulting solid is filtered off, washed with water, and dried *in vacuo*; yield 1.45 g. of yellow needles of **3**, m.p. 248–250° (dec.), raised to 253–254° by recrystallization from aqueous ethanol. Concentration of the filtrate, followed by cooling at 5° overnight, gives an additional 330 mg. of **3**; total yield 1.78 g. (77%); $\lambda_{max}^{50\%\ EtOH}$ 236 and 280 nm.; R_f 0.46 in water (thin-layer chromatography on silica gel GF_{254}, Merck).

7-Benzylguanine (5)

A mixture of compound **3** (0.5 g., 2.2 mmoles) and diethoxymethyl acetate (**4**) (30 ml.) is stirred at 100–120° for 2 hr.; the solid dissolves soon after heating has begun. The solvents are then evaporated off under diminished pressure, and the residue is digested with 2 *N* sodium hydroxide (25 ml.) on a steam bath for 30 min. Water (75 ml.) is added, and the solution is cooled and rendered neutral, giving compound **5** (100 mg.). The filtrate is evaporated to 45 ml., and kept at room temperature, giving an additional 70 mg. of **5**. The crops are combined, washed successively with water, ethanol, and ether, and dried; total yield 0.17 g. (33%), m.p. >300°; λ_{max} at pH 7.0, 247–8 and 285 nm.; at pH 13.0, 282 nm. (ϵ 8,200); at pH 1.0, 254 nm. (ϵ 9,600) and 272 nm. (inflection).

The above filtrate is concentrated to about 20 ml., and kept at 5° overnight, giving colorless crystals of 2,6-diamino-5-(benzylformylamino)-4-pyrimidinol (**6**); yield 320 mg. (57%), m.p. 300°, λ_{max} pH 7.0, 272 nm.; pH 1.0, 269 nm.; pH 13.0, 264 nm.; R_f 0.56 in 1:1 ethanol–chloroform on silica gel GF_{254} (Merck). The product is identical with a sample prepared by refluxing **3** with formic acid. The formyl derivative (**6**) is unchanged when treated with diethoxymethyl acetate; it is, therefore, not an intermediate in the cyclization of **3**.

REFERENCES

(1) G. T. Rogers and T. L. V. Ulbricht, unpublished results; J. A. Montgomery, personal communication.
(2) H. W. Post and E. R. Erickson, *J. Org. Chem.*, **2**, 260 (1937).
(3) J. A. Montgomery and L. B. Holum, *J. Am. Chem. Soc.*, **80**, 404, 409 (1958).
(4) W. Traube and W. Nithack, *Ber.*, **39**, 227 (1906).

[7] 3-Methylguanine

Preparation of 3-Methylpurines by the Ring Closure of a 5,6-Diaminopyrimidine

LEROY B. TOWNSEND and ROLAND K. ROBINS

DEPARTMENT OF CHEMISTRY AND DEPARTMENT OF BIOPHARMACEUTICAL SCIENCES, UNIVERSITY OF UTAH, SALT LAKE CITY, UTAH 84112

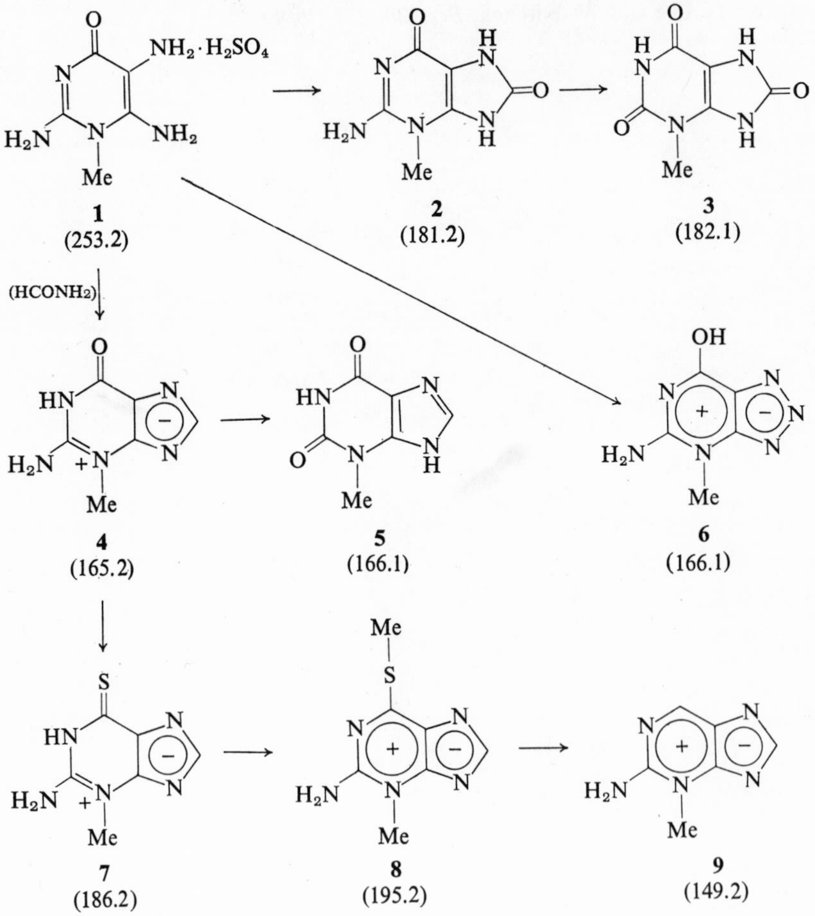

INTRODUCTION

Ring closure of 2,5,6-triamino-1-methyl-4(1*H*)-pyrimidinone (**1**) proceeds readily in boiling formamide, to provide[1] the only hitherto-undescribed isomer of *N*-methylguanine, *viz.*, 3-methylguanine (**4**). The chemical reactivity of **4** is unusual, as the standard, acidic conditions for converting guanines into xanthines are unsuccessful, and refluxing 2 *N* sodium hydroxide converts **4** into 3-methylxanthine (**5**) in good yield. Ring closure of **1** by fusion with urea furnishes 2-amino-3-methyl-3*H*-purine-6,8(1*H*,9*H*)-dione (**2**) which, on treatment with refluxing 2 *N* sodium hydroxide, gives 3-methyluric acid (**3**). Treatment of **1** with nitrous acid provides another example of ring closure,[2] and gives the interesting compound 5-amino-4-methyl-*v*-triazolo[4,5-*d*]pyrimidin-7-ol (**6**, 3-methyl-8-azaguanine). Thiation of **4** with phosphorus pentasulfide and pyridine affords 2-amino-3-methylpurine-6(3*H*)-thione (**7**, 3-methyl-6-thioguanine) which, on subsequent methylation, yields 2-amino-3-methyl-6-(methylthio)purine (**8**). Substantiation for the actual site of methylation is provided by the fact that treatment of **8** with Raney nickel[3] gives 2-amino-3-methylpurine (**9**). The use of 3-methylguanine (**4**) as a model compound for determination of the actual site of glycosylation is of increasing importance, in view of the recent reports[4] that direct glycosylation of preformed guanine analogs has produced isomeric mixtures.

PROCEDURE

2-Amino-3-methyl-3*H*-purine-6,8(1*H*,9*H*)-dione Hemihydrate (2)

An intimate mixture of 2,5,6-triamino-1-methyl-4(1*H*)-pyrimidinone (**1**) (4.0 g., 15.8 mmoles) with urea (10.0 g., 16.7 mmoles) is carefully heated to 130°, giving a melt; the temperature is then raised to 150°, and kept at 150° for *ca.* 20 min., whereupon the melt partly solidifies. The temperature is then kept at 160° for *ca.* 5 min., and the mixture is allowed to cool to room temperature. The resulting solid is dissolved in warm *N* potassium hydroxide, the mixture is treated with activated carbon, the suspension is filtered, and the pH of the filtrate is adjusted to 6. The solution is kept at 5° for 18 hr., and the solid that separates is collected by filtration, and dried at 110°, giving 2.8 g. (48%) of a tan-colored product. For purification, a sample is dissolved in ammonium hydroxide, and the pH is adjusted to 6, giving **2** as the hemihydrate, m.p. >300°.

Anal. Calcd. for $C_6H_7N_5O_2 \cdot 0.5\ H_2O$: C, 37.9; H, 4.2; N, 36.8. Found: C, 37.6; H, 4.6; N, 37.0.

3-Methyluric acid (3)

A solution of **2** (0.1 g., 0.16 mmole) in 35 ml. of 2 *N* sodium hydroxide is refluxed for 4 hr., and evaporated to dryness under diminished pressure. The resulting residue is dissolved in 25 ml. of water at 95°, the solution is treated with activated carbon, the suspension is filtered, and the filtrate is brought to pH 6 with glacial acetic acid. The solution is kept at 5° for 18 hr., and the resulting solid is filtered off, washed with acetone (200 ml.), and dried at 110°, giving 0.07 g. (14%) of **3**, which is recrystallized from water.

3-Methylguanine (4)

A mixture of 2,5,6-triamino-1-methyl-4(1*H*)-pyrimidinone sulfate (**1**) (50 g., 0.195 mole) and 250 ml. of formamide is refluxed for 1.5 hr., and cooled, and the precipitate is filtered off, washed successively with water and acetone, and dried at 110°; yield 27 g. (80%) of crude **4**; this is recrystallized from water; m.p. >300°; λ_{max}^{pH1} 264 nm. (ϵ 11,200) and 244 (infl.) nm. (ϵ 8,300); λ_{max}^{pH11} 273 nm. (ϵ 13,700).

3-Methylxanthine (5)

A solution of **4** (1 g., 6 mmoles) in 20 ml. of 2 *N* sodium hydroxide is refluxed for 4 hr., and then evaporated to dryness under diminished pressure on a steam bath. Water (25 ml.) is added, the solution is re-evaporated to dryness, and the residue is dissolved in boiling water, the solution treated with activated carbon, and the suspension filtered. The filtrate is kept at 5° overnight, and the resulting precipitate is filtered off; yield 0.85 g. (84%) of product, m.p. >300°; it may be recrystallized from water.

5-Amino-1,6-dihydro-4-methyl-7*H*-*v*-triazolo[4,5-*d*]pyrimidin-7-one (6)

To a solution of 6 ml. of glacial acetic acid in 100 ml. of water is added 12 g. (0.047 mole) of compound **1** at room temperature. A solution of 7 g. of sodium nitrite in 15 ml. of water is then slowly added during 10–15 min., by which time the temperature is *ca.* 40°. The mixture is stirred for a further 15 min., and then cooled in an ice bath for 2 hr. The resulting solid is filtered off, washed with 100 ml. of water at 5°, and suspended in 150 ml. of hot water. Sufficient *N* sodium hydroxide is added to afford a clear solution, activated carbon is added, the suspension is filtered, and the hot filtrate is brought to pH 6 by addition of glacial acetic acid. The solution is evaporated to dryness under diminished pressure, and the resulting solid is successively washed with 400 ml. of boiling water and 400 ml. of boiling acetone, and dried at 110°; yield 4.5 g. (57%), m.p. >300°.

Anal. Calcd. for $C_5H_6N_6O$: C, 36.1; H, 3.6; N, 50.6. Found: C, 35.6; H, 3.5; N, 50.3.

2-Amino-3-methylpurine-6(3*H*)-thione (7)

A suspension of 20 g. (0.12 mole) of compound **4** and 100 g. of phosphorus pentasulfide in 1.4 l. of pyridine is refluxed for 16 hr., and then evaporated to dryness under diminished pressure on a steam bath. To the residue is added 1 l. of water, and the mixture is heated on a steam bath for 3 hr., and cooled overnight. The resulting precipitate is filtered off, washed with water, and dissolved in hot, dilute, aqueous ammonia, and the solution is treated with activated carbon, and the suspension filtered. The pH of the filtrate is adjusted to 6 with glacial acetic acid, and the solution is allowed to cool. The resulting precipitate is filtered off, washed with water, and dried; yield 19 g. (87%) of pale-yellow solid, m.p. >300°, which may be recrystallized from water. λ_{max}^{pH1} 340 nm. (ϵ 33,600), 256 nm. (ϵ 6,260), and 237 nm. (ϵ 4,360); λ_{max}^{pH11} 330 nm. (ϵ 33,600), 279 nm. (ϵ 12,000), and 235 nm. (ϵ 8,550).

2-Amino-3-methyl-6-(methylthio)purine (8)

A solution of **7** (1.0 g., 5.5 mmoles) in 25 ml. of *N* potassium hydroxide containing 1.5 g. of methyl iodide is stirred and heated at 35–40° for 4 hr., and kept at 5° for 4 hr., and the resulting solid is filtered off; yield 0.8 g. (74%) of **8**; after recrystallization from water, it affords the hemihydrate, m.p. 289–292°; λ_{max}^{pH1} 318 nm. (ϵ 15,900) and 275 nm. (ϵ 14,200); λ_{max}^{pH11} 319 nm. (ϵ 16,300), 264 nm. (ϵ 11,400), and 232 nm. (ϵ 10,600).

2-Amino-3-methylpurine (9)

To a solution of **8** (0.5 g., 0.28 mmole) in absolute ethanol (50 ml.) at 70° is added 2.0 g. (wet wt.) of Raney nickel prewashed with absolute ethanol, and the suspension is refluxed for 1 hr.; 5.0 g. (wet wt.) of prewashed Raney nickel is added, and the suspension is refluxed for a further 6 hr., and cooled. The catalyst is filtered off, and washed with 30 ml. of boiling water, and the filtrate and washings are combined, and treated with activated carbon; the suspension is filtered, and the filtrate is evaporated to dryness under diminished pressure, giving 0.15 g. (39%) of **9**. This is dissolved in 35 ml. of absolute ethanol, and the solution is treated with activated carbon; the suspension is filtered, and the filtrate is concentrated to 10 ml. and kept at 5° for 18 hr.; m.p. 260–270° (gradual decomposition); λ_{max}^{pH1} 317 nm. (ϵ 13,600) and 256 nm. (ϵ 3,600); λ_{max}^{pH11} 311 nm. (ϵ 13,900).

Anal. Calcd. for $C_6H_7N_5$: C, 48.4; H, 4.70. Found: C, 48.5; H, 4.70.

REFERENCES

(1) L. B. Townsend and R. K. Robins, *J. Am. Chem. Soc.*, **84**, 3008 (1962).
(2) R. Weiss, R. K. Robins, and C. W. Noell, *J. Org. Chem.*, **25**, 765 (1960).
(3) L. B. Townsend and R. K. Robins, *J. Org. Chem.*, **27**, 990 (1962); *J. Heterocyclic Chem.*, **3**, 241 (1966).
(4) K. Imar, A. Nohara, and M. Honjo, *Chem. Pharm. Bull.* (Tokyo), **14**, 1377 (1966); S. R. Jenkins, F. W. Holly, and E. Walton, *J. Org. Chem.*, **30**, 2851 (1965); Z. A. Shabarova, Z. P. Polyakova, and M. A. Prokofev, *Zh. Obshch. Khim.*, **29**, 215 (1959).

[8] 1-Methylhypoxanthine

N-*Substituted Purines*

H. JEANETTE THOMAS, CARROLL TEMPLE, JR.,
and JOHN A. MONTGOMERY

KETTERING-MEYER LABORATORY, SOUTHERN RESEARCH INSTITUTE,
BIRMINGHAM, ALABAMA 35205

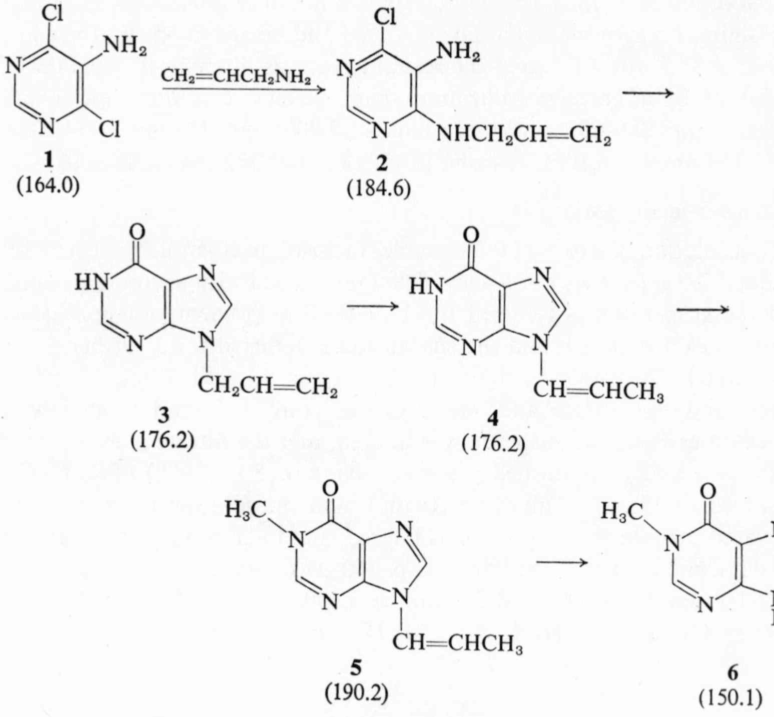

$CH_2{=}CHCH_2NH_2$

1 (164.0) → 2 (184.6) → 3 (176.2) → 4 (176.2) → 5 (190.2) → 6 (150.1)

INTRODUCTION

1-Methylpurines occur in Nature[1,2] and exhibit biological activity.[3] Directing a methyl or other alkyl group to N-1 of a purine can be accomplished by substituting N-9 with a removable protecting group.[4–6] The procedure described here has, with appropriate modification, also been applied to adenine.[7]

PROCEDURE

6-(Allylamino)-5-amino-4-chloropyrimidine (2)

A solution of 50.0 g. (0.305 mole) of 5-amino-4,6-dichloropyrimidine (**1**) (Sec. II [27]) in 200 ml. of 1:1 ethanol–allylamine is heated in a stainless-steel bomb at 130° for 3.5 hr. The bomb is then cooled, and the mixture is evaporated to dryness under diminished pressure to an oil. Extraction of the oil with six 600-ml. portions of hot benzene, and evaporation of the combined extracts *in vacuo*, gives a solid which is recrystallized from 12 l. of petroleum ether (b.p. 65–85°); yield of **2**, 44.3 g. (79%), m.p. 113°, λ_{max}^{pH7} 260, 290 nm. (log ϵ 3.952, 3.959).

9-Allylhypoxanthine (3)

A solution of 7.4 g. of **2** in 100 ml. of anhydrous formic acid is refluxed for 3 hr., and evaporated to dryness *in vacuo*. The resulting residue is dissolved in 100 ml. of water, and the solution is neutralized with concentrated ammonium hydroxide and evaporated to dryness *in vacuo*. The residue is recrystallized from 150 ml. of acetonitrile; yield 4.81 g. (68%) of **3**, m.p. 248–251°, λ_{max}^{pH7} 250 nm. (log ϵ 4.129).

9-Propenylhypoxanthine (4)

A mixture of 63 mg. (1.16 mmoles) of sodium methoxide and 204 mg. (1.16 mmoles) of 9-allylhypoxanthine (**3**) in 4 ml. of anhydrous methyl sulfoxide is placed in a 10-ml., round-bottomed flask (fitted with a drying tube) and heated in an oil bath at 100° for 20 min. The solution is cooled to room temperature and diluted with 15 ml. of water, and the pH is lowered to 8 with carbon dioxide. The precipitate that forms is collected by filtration, and washed with cold water; yield 150 mg. (74%), m.p. 301–303° (dec.).

1-Methyl-9-propenylhypoxanthine (5)

To a solution of 352 mg. (2.00 mmoles) of 9-propenylhypoxanthine (**4**) and 0.35 ml. (2.00 mmoles) of methyl *p*-toluenesulfonate in 30 ml. of *N,N*-dimethylacetamide is added 324 mg. (2.35 mmoles) of anhydrous potassium carbonate, and the suspension is heated and stirred at 100° for 2 hr., cooled, and filtered. The filtrate is evaporated to dryness under diminished pressure, giving a white residue which is partitioned between chloroform and water (25 ml. each). The chloroform layer is dried with magnesium sulfate, and evaporated to dryness under diminished pressure. The crystalline residue is washed with ethanol, and dried at 78°/0.07 mm. Hg over phosphorus pentaoxide; yield 357 mg. (94%), m.p. 220°.

1-Methylhypoxanthine (6)

To a cold, stirred solution of 357 mg. (1.88 mmoles) of 1-methyl-9-propenylhypoxanthine (**5**) in 15 ml. of 0.5 *M* methanolic sodium hydroxide is slowly added a 4% aqueous solution of potassium permanganate. When the dark-brown precipitate that forms becomes very thick, it is removed by filtration. Addition of the permanganate solution is continued until no more brown precipitate is formed. A final filtration gives a pale yellow filtrate which is adjusted to pH 4–5 with *N* hydrochloric acid, and evaporated to dryness under diminished pressure. The residue is crystallized from 15 ml. of water; yield 164 mg. (58%), m.p. 315–320°, λ_{max} in nm. (log ϵ): pH 1, 249 (3.97), pH 7, 250 (3.95), and pH 13, 260 (3.98).

REFERENCES

(1) D. Ackerman and P. H. List, *Z. Physiol. Chem.*, **323**, 192 (1961).
(2) R. W. Park, J. F. Holland, and A. Jenkins, *Cancer Res.*, **22**, 469 (1962).
(3) L. L. Bennett, Jr., M. H. Vail, S. Chumley, and J. A. Montgomery, *Biochem. Pharmacol.*, **15**, 1719 (1966).
(4) E. Shaw, *J. Am. Chem. Soc.*, **80**, 3899 (1958).
(5) J. A. Montgomery and H. J. Thomas, *J. Org. Chem.*, **28**, 2304 (1963).
(6) R. W. Balsiger, A. L. Fikes, T. P. Johnston, and J. A. Montgomery, *J. Org. Chem.*, **26**, 3446 (1961).
(7) J. A. Montgomery and H. J. Thomas, *J. Org. Chem.*, **30**, 3235 (1965).

[9] The Mercury Salt of 2-Acetamido-6-chloropurine

A Purine Intermediate for the Mercuri Procedure

EDWARD M. ACTON and ROBERT H. IWAMOTO

LIFE SCIENCES RESEARCH, STANFORD RESEARCH INSTITUTE, MENLO PARK, CALIFORNIA 94025

1 (169.6) $\xrightarrow{Ac_2O}$ 2, R = H (253.6); 3, R = Ac (295.7)

NH_4OH, CH_3OH → 4 (211.6)

4 $\xrightarrow{HgCl_2,\ NaOH}$ 5 (443.2) [Ac . . . . (Hg, O_2)]

INTRODUCTION

2-Amino-6-chloropurine (**1**) is useful for the synthesis of purine nucleosides having different functional groups at C-2 and C-6, but only if the 2-amino group is first protected.[1] This is done by polyacetylation of **1**, followed by mild cleavage of all but the N^2-acetyl group of **4**, because attempts at direct monoacetylation generally give the N^9-acetyl derivative. The mercury salt (**5**) of **4** is of undetermined composition, and apparently involves complexing with the N^2-acetyl group, as evidenced by the absence of infrared carbonyl absorption in **5**; by the mercuri procedure, it nevertheless forms N^9-nucleosides derived from **4** (Sec. III [72] and [87]).

PROCEDURE

Di- and Tri-acetylation of 2-Amino-6-chloropurine (1) to give 2 and 3

A mixture of 16.0 g. (0.094 mole) of 2-amino-6-chloropurine[2] (**1**) and 500 ml. of acetic anhydride, in a 1-l., round-bottomed flask (fitted with a reflux condenser having a drying tube to exclude moisture), is heated to reflux in an oil bath. Two drops of phosphoric acid (reagent grade, 85%) are added, and complete dissolution is then attained after refluxing for 5–10 min.[a] Refluxing is continued for 40 min., and the acetic anhydride is removed *in vacuo* at 70°. The residual, yellow solid is stirred with 600 g. of ice–water for 1 hr. (to hydrolyze any remaining acetic anhydride), and is collected on a filter and washed successively with water and ether, until the odor of acetic acid has been removed. The solid (19.8 g.) is ground, digested with 1 l. of boiling toluene, and the resultant hot suspension filtered from 1.5 g. of brown solid. The filtrate is concentrated *in vacuo* to 400 ml., and kept at 5° for 1 hr. The resultant precipitate is collected, and washed with cold toluene; yield 17.8 g. Melting points of this mixture of **2** and **3** vary, in different experiments, from 172–179° to 172–202°, as the composition varies.

The N^9-acetyl group of these purines is readily lost by hydrolysis; also the 2-(*N*,*N*-diacetylamino) group is easily hydrolyzed to the 2-acetamido group, which is relatively stable. It is important that no 9-acetyl-2-amino-6-chloro-9*H*-purine be present, and this is best determined by paper chromatography. With Whatman No. 1 paper by a descending technique, in 5% disodium hydrogen phosphate, the product (**2** and/or **3**) exhibits two ultraviolet-absorbing spots, of R_f 0.68 (attributed to **2**) and R_f 0.41 [the same as that of 2-acetamido-6-chloropurine (**4**) from the next step, apparently formed here by deacetylation during chromatography], compared to one spot of R_f 0.26 for both 2-amino-6-chloropurine (**1**) and its N^9-acetyl derivative (*i.e.*, also deacylated during chromatography). In water-saturated 1-butanol, the mixture (**2** and/or **3**) shows one spot, R_f 0.70 (attributed to **2**) as compared with R_f 0.40 for **1** [in this system, both 2-acetamido-6-chloropurine (**4**) and 9-acetyl-2-amino-6-chloro-9*H*-purine are stable and indistinguishable, R_f 0.30].

2-Acetamido-6-chloropurine (4)

Partial deacylation of the mixture of **2** and **3** is carried out by dissolving 17.8 g. of the mixture in 850 ml. of methanol containing 85 ml. of concentrated ammonium hydroxide. After 15 min. at room temperature, the solution is evaporated, giving a white solid which is washed onto a filter with cold water and dried. The yield of **4** is 12.7 g. (64%); R_f as described in the foregoing, $\lambda_{max}^{0.01\,N\,NaOH}$ 236 nm. (ϵ 20,100), 285 nm. (ϵ 8,040). The compound does not melt below 280°. It dissolves in *N* sodium hydroxide,

[a] On a 5-g. scale, 1 drop is sufficient; on a 50-g. scale, 5 drops are used. If dissolution is not complete at this point, additional phosphoric acid is added dropwise. We are indebted to Mr. O. P. Crews of these laboratories for these details. Mr. Fred Keller of Riker Laboratories, Northridge, California, suggested that phosphoric acid is more effective when added to the boiling mixture under reflux than at room temperature.

and is reprecipitated by neutralizing the solution with *N* hydrochloric acid; this may be used as a method of further purification.

Mercury Salt (5) of Compound 4

To a solution of 28.65 g. (0.106 mole) of mercuric chloride in 1.5 l. of 50% aqueous ethanol is added 11.16 g. (0.053 mole) of pulverized **4**. The suspension is stirred, and treated with 105.7 ml. of *N* sodium hydroxide, added dropwise at such a rate that the resultant, yellow-orange color of mercuric oxide is completely discharged before the next drop is added[b]; 50–70 min. is required. The insoluble product weighs 22.09 g. (94%); it is best collected after adding 25 g. of Celite to aid filtration, and is washed successively with water and ethanol, and dried.

REFERENCES

(1) R. H. Iwamoto, E. M. Acton, and L. Goodman, *J. Med. Chem.*, **6**, 684 (1963).

(2) G. D. Daves, Jr., C. W. Noell, R. K. Robins, H. C. Koppel, and A. G. Beaman, *J. Am. Chem. Soc.*, **82**, 2633 (1960); R. W. Balsiger and J. A. Montgomery, *J. Org. Chem.*, **25**, 1573 (1960); also, commercially available.

[b] A permanent yellow color may be obtained when the addition is complete, but it is not essential.

[10] 3-Benzylpurines

N-*Substituted Purines*

H. JEANETTE THOMAS and JOHN A. MONTGOMERY

KETTERING-MEYER LABORATORY, SOUTHERN RESEARCH INSTITUTE, BIRMINGHAM, ALABAMA 35205

NH_2 ... PhCH$_2$Cl / CH$_3$CONMe$_2$ → ... NOCl ... P$_2$S$_5$ →

1 (135.1) 2 (225.3) 3 (226.2) 4 (242.3)

INTRODUCTION

Some 3-substituted purines[1,2] occur in Nature, and others show interesting biological properties.[3,4] Still other 3-substituted purines[5,6] have shown synthetic utility, for example in the synthesis of 7-glycosylpurines, *e.g.*, the nucleoside from pseudovitamin B_{12}, namely, 7-α-D-ribofuranosyladenine.[7]

PROCEDURE

3-Benzyladenine[6,8,9] (2)

A suspension of 5.00 g. (37.0 mmoles) of adenine (**1**) and 13.55 g. (108 mmoles) of benzyl chloride in 100 ml. of *N,N*-dimethylacetamide in a 250-ml., round-bottomed flask fitted with a drying tube is stirred magnetically, with heating in an oil bath at 115°, for 18 hr. The resulting solution is evaporated to dryness under diminished pressure, and the residue is crystallized from 700 ml. of ethanol as the colorless hydrochloride; yield 5.82 g. (60%), m.p. 265°.

To obtain the free base, 3.02 g. (11.5 mmoles) of the hydrochloride is dissolved in 50 ml. of hot water, and the solution is cooled to room temperature and neutralized with 0.9 ml. of concentrated ammonium hydroxide. The precipitate that forms is collected by filtration; yield 2.62 g. (99% from the hydrochloride), m.p. 275–277°; λ_{max} in nm. (log ϵ): pH 1, 275 (4.24); pH 7 and 13, 272 (4.10).

3-Benzylhypoxanthine[5,10] (3)

A solution of 2.64 g. (11.7 mmoles) of 3-benzyladenine (**2**) in a mixture of 13.5 ml. of dry pyridine, 7.8 ml. of acetic anhydride, and 18.2 ml. of glacial acetic acid in a 50-ml., three-necked, round-bottomed flask [fitted with a reflux condenser (drying tube) and dropping funnel] is stirred magnetically and cooled in an ice bath while a solution of 3.40 g. of nitrosyl chloride in 26 ml. of acetic anhydride is slowly added through the dropping funnel. The reaction solution is then stirred for 1.5 hr. at room temperature. When, at the end of this time, a positive potassium iodide–starch test is obtained, the solution is evaporated to a sirup under diminished pressure below 40°. A solution of the residue in 200 ml. of water is extracted with three 100-ml. portions of chloroform. After being dried with magnesium sulfate, the combined chloroform extracts are evaporated to dryness under diminished pressure. The orange, crystalline residue is recrystallized from 100 ml. of ethanol (with charcoal treatment); yield of **3**, 1.61 g. (61%), m.p. 245–247°; λ_{max} in nm. (log ϵ): pH 1, 254 (4.05); pH 7, 265 (4.14); pH 13, 264 (4.02) and 277 (sh) (3.97).

3-Benzylpurine-6(3*H*)-thione[5] (4)

A mixture of 950 mg. (4.20 mmoles) of 3-benzylhypoxanthine (**3**) and 3.16 g. (14.2 mmoles) of phosphorus pentasulfide in 50 ml. of dry pyridine, in a 100-ml., round-bottomed flask fitted with a reflux condenser (drying tube), is stirred magnetically while refluxing for 5 hr. The dark mixture is evaporated to about 25 ml. under diminished pressure, and slowly poured into one liter of boiling water. After the mixture has been boiled for 30 min., it is chilled in an ice bath, and the resulting dark-brown solid is collected by filtration; yield of **4**, 770 mg. The solid is dissolved in 50 ml. of hot *N*,*N*-dimethylformamide, and is reprecipitated as a light-yellow solid by the addition of 200 ml. of ethanol; yield 553 mg. Another 72 mg. is obtained by triturating the residue from the filtrate of the first crop with 50 ml. of 1:5 *N*,*N*-dimethylformamide–ethanol; total yield of **4**, 625 mg. (61%), m.p. 147–148°; λ_{max} in nm. (log ϵ): pH 1, 234 (3.98), 275 (3.78), and 321 (4.47); pH 7, 239 (4.17) and 316 (4.35).

REFERENCES

(1) N. J. Leonard and J. A. Deyrup, *J. Am. Chem. Soc.*, **80**, 2751 (1958).
(2) H. S. Forest, D. Hatfield, and J. M. Lagowski, *J. Chem. Soc.*, **1961**, 963.
(3) K. Gerzon, I. S. Johnson, G. B. Boder, J. C. Cline, P. J. Simpson, and C. Speth, *Biochim. Biophys. Acta*, **119**, 445 (1966).
(4) A. M. Michelson, C. Monny, R. A. Laursen, and N. J. Leonard, *Biochim. Biophys. Acta*, **119**, 258 (1966).
(5) J. A. Montgomery and H. J. Thomas, *J. Org. Chem.*, **28**, 2304 (1963).
(6) J. A. Montgomery and H. J. Thomas, *J. Heterocyclic Chem.*, **1**, 115 (1964).
(7) J. A. Montgomery and H. J. Thomas, *J. Am. Chem. Soc.*, **87**, 5442 (1965).
(8) R. Denayer, *Bull. Soc. Chim. France*, **1962**, 1358.
(9) N. J. Leonard and T. Fujii, *J. Am. Chem. Soc.*, **85**, 3719 (1963).
(10) H. J. Thomas and J. A. Montgomery, *J. Org. Chem.*, **31**, 1413 (1966).

[11] 6-(Benzylthio)-7-ethylpurine

An Illustration of a General Method for the Preparation of 7-Substituted Purines

JOHN A. MONTGOMERY and KATHLEEN HEWSON

KETTERING-MEYER LABORATORY, SOUTHERN RESEARCH INSTITUTE, BIRMINGHAM, ALABAMA 35205

$PhCH_2S$ NH_2 N N NH_2 **1** (232.3) —HCOOH→ $PhCH_2S$ NH CHO N N NH_2 **2** (260.3) —EtI, K_2CO_3, $HCONMe_2$→

$PhCH_2S$ Et N CHO N N NH_2 **3** (288.4) → $PhCH_2S$ Et N N N N **4** (270.4)

INTRODUCTION

The preparation of 6-(benzylthio)-7-ethylpurine[1] (**4**) by the ethylation of *N*-[4-amino-6-(benzylthio)-5-pyrimidinyl]formamide (**2**) in *N,N*-dimethylformamide (DMF) to give **3**, followed by ring closure, illustrates a general procedure[1] for the preparation of 7-substituted purines that are also substituted at the 1-, 2-, 3-, or 6-position.[1–3]

PROCEDURE

_N_-[4-Amino-6-(benzylthio)-5-pyrimidinyl]formamide[1] (2)

A solution of 4,5-diamino-6-(benzylthio)pyrimidine (**1**) (5.0 g., 22 mmoles) in 98–100% formic acid (90 ml.), in a 200-ml., round-bottomed flask equipped with a thermometer and a calcium chloride drying tube, is heated for 2 hr. at 80°. After being cooled to room temperature, the solution is filtered through dry Celite, and the filtrate is evaporated to dryness *in vacuo*. The solid residue (**2**) is triturated with ethanol–ether (20 ml.), collected by filtration, washed with ether, and dried at 78°/0.05 mm. Hg; yield, 5.1 g. (91%), m.p. 227°. λ_{max} in nm. (log ε): pH 1, 245 (4.16); pH 7, 226 (4.36); and pH 13, 230 (4.28).

6-(Benzylthio)-7-ethylpurine[1] (4)

To a mixture of *N*-[4-amino-6-(benzylthio)-5-pyrimidinyl]formamide (**2**) (5.5 g., 21 mmoles) and anhydrous potassium carbonate (2.9 g., 21 mmoles) in DMF (spectrograde; 50 ml.) is rapidly added, with stirring, ethyl iodide (3.4 ml., 42 mmoles). After being stirred at room temperature for 24 hr.,[a] the reaction solution is refluxed for 1 hr., cooled, and poured with stirring onto crushed ice (200–300 ml.). The resulting crystals are collected by filtration, washed with cold water, air-dried, and recrystallized from ethanol; yield of **4**, 1.6 g. (28%), m.p. 99°. λ_{max} in nm. (log ε): pH 1, 295 (sh), 302 (4.17), 315 (sh); pH 7 and 13, 295 (4.19), 300 (sh).

REFERENCES

(1) J. A. Montgomery and K. Hewson, *J. Org. Chem.*, **26**, 4469 (1961).
(2) J. A. Montgomery, K. Hewson, and C. Temple, Jr., *J. Med. Pharm. Chem.*, **5**, 15 (1962).
(3) R. Denayer, *Bull. Soc. Chim. France*, 1358 (1962).

[a] Alkylation is completed at room temperature, and the alkylated pyrimidine may be isolated if desired. The solution is evaporated to dryness *in vacuo* at 40°, and the residue triturated with water until solidification is complete. The white solid is collected by filtration, washed with fresh water, and recrystallized from ethanol to give the pure *N*-[4-amino-6-(benzylthio)-5-pyrimidinyl]-*N*-ethylformamide (**3**); m.p. 180°; λ_{max} in nm. (log ε): pH 1, 245 (4.17), 300.5 (4.08); pH 7 and 13, 228 (4.35), 282 (3.90).

[12] 1-Methylpurine

Preparation of 1-Methylpurine and Derivatives via *Displacements of an Alkylthio or Thio Group*

LEROY B. TOWNSEND and ROLAND K. ROBINS

DEPARTMENT OF CHEMISTRY, UNIVERSITY OF UTAH, SALT LAKE CITY, UTAH 84112

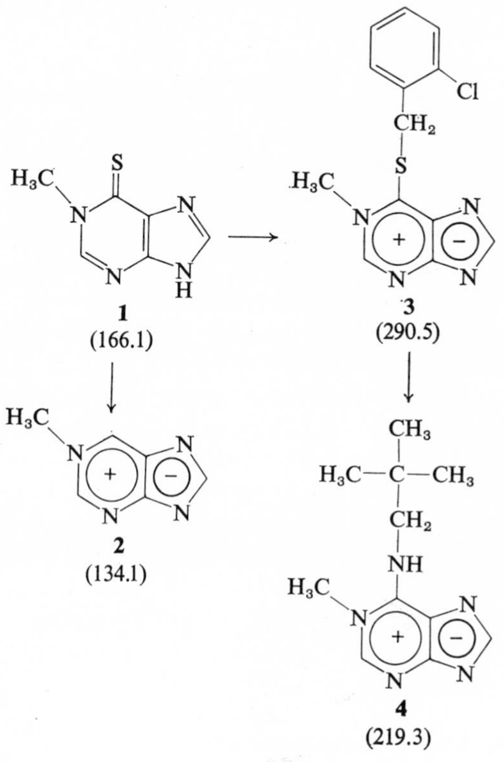

1 (166.1) → **3** (290.5)

1 (166.1) → **2** (134.1)

3 (290.5) → **4** (219.3)

INTRODUCTION

1-Methylpurine (**2**) has been prepared[1] by the dethiation of 1-methylpurine-6(1*H*)-thione[1] (**1**, 1-methyl-6-mercaptopurine) with Raney nickel. Treatment of **1** with *o*,α-dichlorotoluene in *p*-dioxane under basic conditions affords the versatile intermediate 6-(*o*-chlorobenzylthio)-1-methylpurine (**3**). The increase in ease of nucleophilic displacement of an alkylthio group in 1- and 3-methylpurine derivatives as compared with the corresponding

7- and 9-methylpurine derivatives has been well documented.[1,2] The preparation of 1-methyl-*N*-neopentyladenine (**4**) provides an example of the facile nucleophilic displacement of the 6-alkylthio group if N-1 of a 6-alkylthiopurine is substituted.

It has been demonstrated[3] that, after direct glycosylation of a preformed purine derivative, the *N*-methyl derivatives (as model compounds) are indeed necessary in determining the actual site of glycosylation. The preparation of 1-methylpurine also affords a model compound that is essential for study before an unequivocal assignment[4] for the actual site of glycosylation or other alkylation of preformed purine, *per se*, can be achieved.

PROCEDURE

1-Methylpurine (2)

To a solution of 1-methylpurine-6(1*H*)-thione[1] (**1**) (4 g., 0.024 mole) in water at 80–90° is added 48 g. of activated, Davison[a] sponge nickel catalyst, and the mixture is refluxed for 20 min. The catalyst is filtered off, and the filtrate is rapidly evaporated to dryness under diminished pressure (rotary flash evaporator). The very dark residue is triturated with 800 ml. of acetone at room temperature for 12 hr. The suspension is filtered, and the filtrate is concentrated to 180 ml. on a steam bath; petroleum ether (b.p. 60–110°) is then slowly added to the boiling solution until a solid begins to separate. The solution is kept at 5° for 18 hr., and the resulting solid is filtered off; yield 1.8 g. (56%) of a light-yellow compound that is *ca.* 90% pure (judged by comparison of its ultraviolet absorption spectrum with that of an analytically pure sample). This product is extracted with 500 ml. of boiling ethyl acetate, and the extract is concentrated on a steam bath until deposition of solid begins. The solution is kept at 5° for 18 hr., and the resulting solid is filtered off; when recrystallized from acetone–petroleum ether (b.p. 60–110°), compound **2** has m.p. 234–235°; $\lambda_{max}^{pH\ 1}$ 268 nm. (ϵ 6,600); $\lambda_{max}^{pH\ 11}$ 275 nm. (ϵ 6,050).

6-(*o*-Chlorobenzylthio)-1-methylpurine (3)

To a stirred solution of 1-methylpurine-6(1*H*)-thione (**1**) (2.5 g., 0.015 mole) and 1.5 g. of potassium hydroxide in 25 ml. of water at 40–45° is added a solution of *o*,α-dichlorotoluene (5 g.) in 10 ml. of *p*-dioxane in small portions during 2 hr., and the mixture is then stirred at 40–45° for 1 hr. The mixture is kept at 5° for 2 hr., and the gummy mass at the bottom of the reaction vessel is filtered off, and triturated with 50 ml. of petroleum ether (b.p. 60–110°) at room temperature for 1 hr. The resulting, colorless

[a] Davison Chemical Co., Baltimore, Md. 21203.

crystals of **3** are filtered off; yield 2.8 g. (64%); recrystallized from heptane–ethyl acetate, m.p. 152°; $\lambda_{max}^{pH\,1}$ 312 nm. (ϵ 13,700).

1-Methyl-*N*-neopentyladenine (4)

To a solution of 1.4 g. of neopentylamine in 20 ml. of anhydrous methanol at room temperature is added 2.0 g. (6.9 mmoles) of **3**. The clear solution is stirred at room temperature for 6 hr., and then evaporated to dryness under diminished pressure (rotary flash evaporator) at room temperature. The resulting residue is extracted with 50 ml. of boiling heptane, and the extract is discarded. The solid remaining is recrystallized from heptane–ethanol, giving 1.1 g. (73%) of colorless crystals of **4**; recrystallized from ethyl acetate, m.p. 275–277°; $\lambda_{max}^{pH\,1}$ 265 nm. (ϵ 13,100); $\lambda_{max}^{pH\,11}$ 278.5 nm. (ϵ 13,400) and 228 nm. (ϵ 16,000).

Anal. Calcd. for $C_{11}H_{17}N_5$: C, 60.30; H, 7.76; N, 32.0. Found: C, 60.39; H, 7.90; N, 31.8.

REFERENCES

(1) L. B. Townsend and R. K. Robins, *J. Org. Chem.*, **27**, 990 (1962).

(2) L. B. Townsend and R. K. Robins, *J. Am. Chem. Soc.*, **84**, 3008 (1962); *J. Heterocyclic Chem.*, **3**, 241 (1966); F. Bergmann, G. Levin, A. Kalmus, and H. Kwietny-Govrin, *J. Org. Chem.*, **26**, 1504 (1961); J. W. Jones and R. K. Robins, *J. Am. Chem. Soc.*, **84**, 1914 (1962).

(3) L. B. Townsend, R. K. Robins, R. N. Loeppky, and N. J. Leonard, *J. Am. Chem. Soc.*, **86**, 5320 (1964).

(4) T. Hashizume and H. Iwamura, *Tetrahedron Letters*, **1966**, 643.

[13] 3-Methylpurine

*Preparation of 3-Methylpurine and 3-Methyl-3*H*-purine Derivatives* via *Displacements of the 6-Alkylthio Group*

LEROY B. TOWNSEND, ARTHUR F. LEWIS,
JESSE W. JONES, and ROLAND K. ROBINS

DEPARTMENT OF CHEMISTRY, UNIVERSITY OF UTAH, SALT LAKE CITY,
UTAH 84112

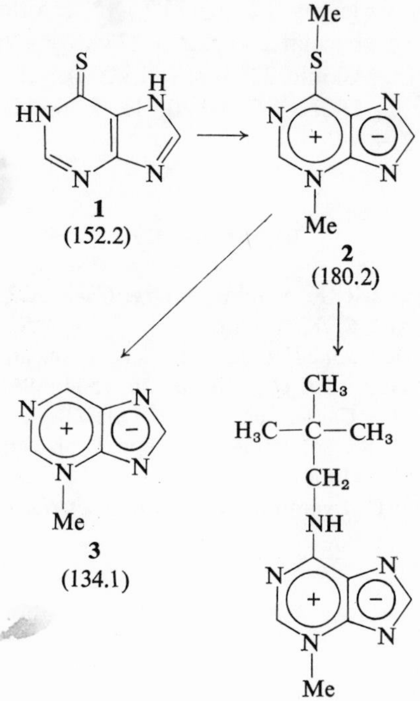

INTRODUCTION

3-Methylpurine (**3**) has been prepared[1] from the versatile intermediate 3-methyl-6-(methylthio)-3*H*-purine[2,3](**2**) after several unsuccessful attempts by dethiation and dehalogenation had been reported.[3,4] The increase in ease of nucleophilic displacement of an alkylthio group on 1- and 3-methylpurine derivatives as compared with the corresponding 7- and 9-methylpurine derivatives has been well documented.[3,5] It is of interest that several 3-substituted purine derivatives have demonstrated[6] the ability to rearrange (usually by heating) to the corresponding 9-substituted purine derivatives. The one-step preparation[2] of **2**, as compared with the original[3] 9-step preparation, has been found to afford more reproducible results simply by lowering the temperature of the reaction mixture to 95–100°. This preparation of 3-methylpurine (**3**) gives the only remaining *N*-methylpurine which, as a model compound, is required before an unequivocal assignment[7] for the actual site of glycosylation of preformed purine can be achieved.

PROCEDURE

3-Methyl-6-(methylthio)-3*H*-purine (2)

A mixture of 20 ml. of *N,N*-dimethylacetamide and 50 g. of methyl *p*-toluenesulfonate is heated to 95°. With stirring, four 5-g. portions (0.128 mole) of purine-6(1*H*)-thione (**1**) monohydrate are added,[a] and the mixture is stirred and heated at 95° ± 2° for 1 hr., allowed to cool to room temperature, and extracted with five 100-ml. portions of anhydrous ether.[b] The ether extracts are discarded, and the residue is dissolved in 250 ml. of boiling acetone. The solution is kept at 5° for 24 hr., and the resulting solid is filtered off, washed with acetone (100 ml.), and air-dried to yield 12.9 g. of the *p*-toluenesulfonate of **2**. The filtrate and washings are combined, and kept at −20° for 48 hr., to yield another 0.9 g. of the salt; total yield, 13.8 g.

The *p*-toluenesulfonate of **2** (13.8 g.) is mixed with 100 ml. of water, and the mixture is stirred and gradually heated until all of the solid has dissolved. The solution is then cooled to room temperature, and the pH is adjusted to 9 by the careful addition of ammonium hydroxide (28%). A thick, colorless solid separates, and the suspension is kept at 5° for 24 hr. The solid (**2**) is collected by filtration, washed successively with a small volume of cold water and a small volume of acetone, and air-dried; yield, 4.5 g. (21%), m.p. 154–157°. A small sample, recrystallized once from water, and once from acetone, has m.p. 163–165°; $\lambda_{max}^{pH\ 1}$ 235 nm. (ϵ 9,200), 274 nm. (ϵ 4,860), and 317 nm. (ϵ 25,200); $\lambda_{max}^{pH\ 11}$ 237 nm. (ϵ 10,250), and 311 nm. (ϵ 17,200).

3-Methyl-3*H*-purine (3)

3-Methyl-6-(methylthio)-3*H*-purine (**2**) (0.5 g., 2.78 mmoles) is dissolved in anhydrous methanol at 60°. Raney nickel (7.5 g., wet wt.), prewashed with anhydrous methanol (6 × 50 ml.), is added, and the mixture is refluxed. The rate of dethiation is followed by observation of the ultraviolet absorption of the mixture; when the peak at 312 nm. (methylthio

[a] The temperature rises on addition of **1**, and is allowed to fall to 95° before the next addition is made.

[b] The residue is a very sticky, partly solid mass at this point.

group) has disappeared (*ca.* 10–20 min.), with concomitant appearance of a peak at 277 nm., the catalyst is filtered off, and washed with three 25-ml. portions of boiling, anhydrous methanol. The filtrate and washings are combined, and rapidly evaporated to dryness under diminished pressure at room temperature. The resulting residue is triturated for 15 min. with acetone (50 ml.) at room temperature, the suspension is filtered, and the filtrate is evaporated to dryness under diminished pressure. The resulting residue is recrystallized from heptane–benzene, to afford 0.2 g. (54%) of **3**, m.p. 184–185° (dec.); $\lambda_{max}^{pH\,1}$ 275 nm. (ϵ 9,780); λ_{max}^{MeOH} 277 nm. (ϵ 6,950); $\lambda_{max}^{pH\,11}$ 276 nm. (ϵ 6,980); $\lambda_{min}^{pH\,1}$ 231 nm. (ϵ 1,200); λ_{min}^{MeOH} 238 nm. (ϵ 670); $\lambda_{min}^{pH\,11}$ 237 nm. (ϵ 1,070).

3-Methyl-*N*-neopentyladenine (4)

A solution of **2** (1.0 g., 5.5 mmoles) in anhydrous methanol (10 ml.) containing neopentylamine (1.0 g.) is refluxed on a steam bath for 12 hr., and evaporated to dryness under diminished pressure (on a steam bath). The resulting residue is extracted with 50 ml. of boiling heptane, and the extract is discarded. The solid is recrystallized twice from ethanol–heptane, to give 0.5 g. (41%) of **4**. A sample recrystallized from ethyl acetate and dried at 60° gives colorless crystals, m.p. 231°; $\lambda_{max}^{pH\,1}$ 284 nm. (ϵ 20,800); λ_{max}^{MeOH} 290.5 nm. (ϵ 14,500); $\lambda_{max}^{pH\,11}$ 287.5 nm. (ϵ 15,600).

Anal. Calcd. for $C_{11}H_{17}N_5$: C, 60.30; H, 7.76; N, 32.00. Found: C, 60.22; H, 7.89; N, 31.82.

REFERENCES

(1) L. B. Townsend and R. K. Robins, *J. Heterocyclic Chem.*, **3**, 241 (1966).
(2) J. W. Jones and R. K. Robins, *J. Am. Chem. Soc.*, **84**, 1914 (1962).
(3) F. Bergmann, G. Levin, A. Kalmus, and H. Kwietny-Govrin, *J. Org. Chem.*, **26**, 1504 (1961).
(4) F. Bergmann, Z. Neiman, and M. Kleiner, *J. Chem. Soc.*, **1966,** 10; Z. Neiman, M. Kleiner, and F. Bergmann, *Proc. 34th Meeting Israel Chem. Soc.*, **2**, 243 (1964).
(5) L. B. Townsend and R. K. Robins, *J. Org. Chem.*, **27**, 990 (1962); *J. Am. Chem. Soc.*, **84**, 3008 (1962).
(6) B. Shimizu and M. Miyaki, *Chem. Ind.* (London), **1966,** 664 and references cited therein.
(7) T. Hashizume and H. Iwamura, *Tetrahedron Letters*, **1966,** 643.

[14] Trimethylpurin-6-ylammonium Chloride

A Purine Containing Quaternary Nitrogen

JEROME P. HORWITZ

DETROIT INSTITUTE OF CANCER RESEARCH DIVISION OF THE MICHIGAN CANCER FOUNDATION, DETROIT, MICHIGAN 48201

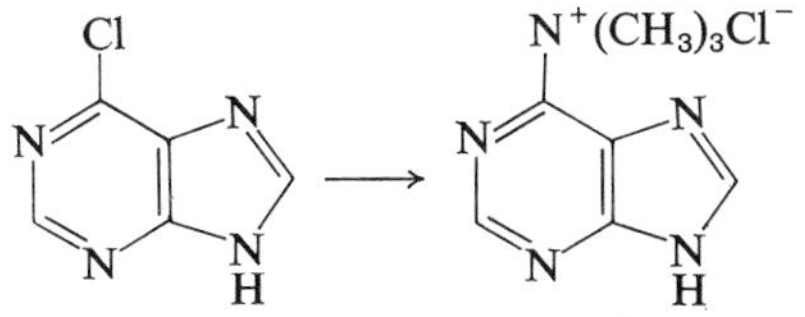

Cl — N — N — N — N H → $N^+(CH_3)_3Cl^-$ — N — N — N — N H

1 (154.6) **2** (213.7)

INTRODUCTION

Quaternary nitrogenous purines of the type illustrated by the title compound (**2**) are readily obtained by the action of trimethylamine on the corresponding 6-chloropurine.[1–3] Trimethylpurin-6-ylammonium chloride (**2**) inhibits Ehrlich ascites tumor, an epidermoid carcinoma (DC-5), and adenocarcinoma 755 in mice.[1] In humans, objective tumor regressions have also been observed with this agent.[4]

PROCEDURE

Trimethylpurin-6-ylammonium Chloride (2)

To 50 ml. of dry *N,N*-dimethylformamide in a 100-ml., 2-necked flask is added 6.18 g. (0.04 mole) of 6-chloropurine (**1**), and the suspension, protected from moisture by a soda-lime tube, is warmed until dissolution has been effected. The flask is then fitted with a gas-inlet tube, and a stream of anhydrous trimethylamine is bubbled into the reaction mixture, at 0°, until 8.0 g. (0.136 mole) has been absorbed. The mixture is allowed to warm to room temperature, and is kept at room temperature overnight; the product is deposited in the form of a dense, crystalline mass. The precipitate is collected, washed successively with *N,N*-dimethylformamide and ether, and dried in a vacuum oven at 45° for 18 hr.; yield 7.50 g. (80%), m.p. 166–170° (dec.). Purification[a] is achieved by dissolving the crude material in 15 ml. of water at room temperature, and precipitating

[a] Recrystallization of the crude product from methanol yields a product that is a 1:1 mixture of quaternary chloride and the corresponding methoxide; see Ref. 3.

the product (**2**) by the addition of 200 ml. of acetone; yield 5.45 g. (64%), m.p. 189–191° (dec.[b]); $\lambda_{max}^{0.1\,N\,HCl}$ 265.7 nm. (ϵ 8,272), $\lambda_{min}^{0.1\,N\,HCl}$ 220 nm. (ϵ 732).

REFERENCES

(1) J. P. Horwitz and V. K. Vaitkevicius, U. S. Patent 3,274,193 (1966); *Experientia*, **17**, 552 (1961).

(2) L. R. Lewis, F. H. Schneider, and R. K. Robins, *J. Org. Chem.*, **26**, 3837 (1961).

(3) E. J. Reist, A. Benitez, L. Goodman, B. R. Baker, and W. W. Lee, *J. Org. Chem.*, **27**, 3274 (1962).

(4) V. K. Vaitkevicius, M. L. Reed, R. L. Fox, and R. W. Talley, *Cancer Chemotherapy Rept.*, No. **27**, 55 (1963); V. K. Vaitkevicius and M. L. Reed, *Proc. Am. Assoc. Cancer Res.*, **7**, 72 (1966).

[b] The decomposition point is dependent on the rate of heating, and this may, in turn, account for the occasional observation of higher values (193–195°; 203–206°) for material that has essentially the same elementary analysis and spectral properties.

[15] 2,8-Diaminopurine-6-thiol

Direct Replacement of Oxygen with Sulfur in a Purine Derivative

ALDEN G. BEAMAN,[a] JESSE W. JONES,[b] and ROLAND K. ROBINS

DEPARTMENT OF CHEMISTRY, UNIVERSITY OF UTAH, SALT LAKE CITY, UTAH 84112

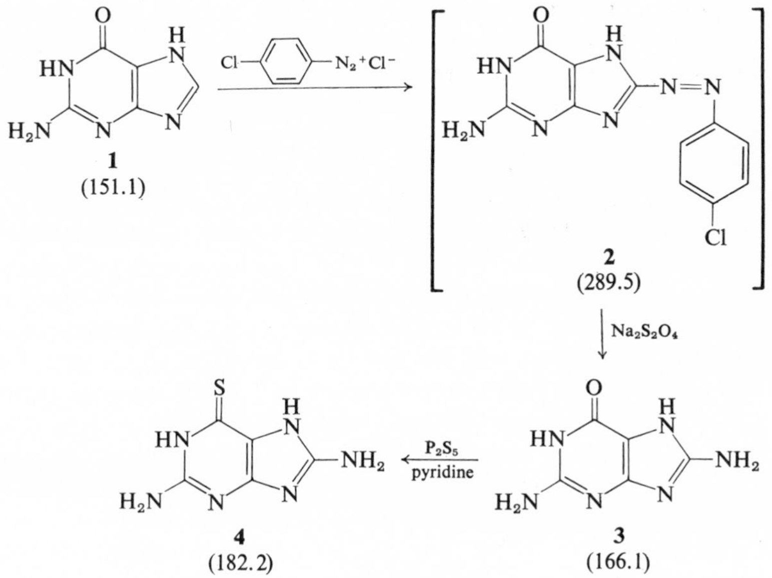

1 (151.1) → [with $Cl\text{-}C_6H_4\text{-}N_2^+Cl^-$] → 2 (289.5) → [$Na_2S_2O_4$] → 3 (166.1) → [$P_2S_5$, pyridine] → 4 (182.2)

INTRODUCTION

H. Fischer[1] prepared 2,8-diaminohypoxanthine (**3**) by coupling diazotized 3,4-dichloroaniline with guanine, followed by reduction with sodium hydrosulfite to yield **3**. The present procedure is a modification of Fischer's procedure,[1] as very limited directions are given in the original description. Treatment of compound **3** with phosphorus pentasulfide in pyridine yields 2,8-diaminopurine-6-thiol[2] (**4**); this is a toxic purine derivative that has shown some inhibition against adenocarcinoma 755 in mice.[3] Compound **4** is a useful intermediate in the synthesis of 2,8-diaminopurine.[4]

[a] Present address: Hoffmann-La Roche, Inc., Research Division, Nutley 10, New Jersey.

[b] Department of Chemistry, Texas College, Tyler, Texas.

PROCEDURE

2,8-Diaminohypoxanthine (3)

p-Chloroaniline (92 g., 0.72 mole) is ground to a very fine powder in a mortar, and added, with magnetic stirring, to 400 ml. of concentrated hydrochloric acid in a 1-l. beaker.[c] The resulting solution is cooled to 0° in an ice–salt bath, and, with continuous stirring, a solution of 65 g. (0.94 mole) of sodium nitrite in 90 ml. of water is dropped in slowly at such a rate that the reaction temperature is kept at 0–5°. A small amount of supercooled, crushed ice (from a freezer) is added occasionally, so as to permit addition of the sodium nitrite solution at a reasonable rate. Meanwhile, 170 g. of potassium hydroxide is dissolved in 1 l. of water, 100 g. (0.66 mole) of guanine (**1**) is added, and the clear solution is stirred, and cooled to 0–7° in an ice bath. Into this cooled solution (in a 4-l. beaker equipped with a powerful, mechanical stirrer) is run slowly from an addition funnel, with stirring, about half of the diazo solution. Crushed, supercooled ice (to a total of about 1 kg.) is added in small portions, so that the diazo solution may be run in at a moderate rate while the reaction temperature is kept at 0–7°. After about half of the diazo solution has been added, an ice-cold solution of 170 g. of potassium hydroxide in 170 ml. of water (pre-prepared) is added. Then the other half of the diazo solution is added from the funnel, as before, the reaction temperature being kept at 0–7° during the addition, and the mixture is stirred, in an ice-bath, for 2.5 hr., during which time the product sets to a thick, brick-red paste. This paste is transferred to a 16-l., stainless-steel pail, and 25 g. of thiourea is added (to decompose the excess nitrite). A cooled solution of 150 g. of potassium hydroxide in 200 ml. of water is added, and the mixture is thoroughly stirred.[d] With frequent stirring by hand, the mixture is heated to 85–90° on a heavy-duty hot-plate. The hot mixture is removed from the hot-plate, and, with good hand-stirring, 550 g. of sodium hydrosulfite ($Na_2S_2O_4$) is added during about 5 min.; the color of the solution changes to a hazy yellow, and a black sludge is formed on top.[e] Decolorizing carbon (60 g.) is added, the pail is re-placed on the hot-plate, and the suspension is heated to boiling, boiled for 10 min. with frequent stirring, and filtered hot through a thoroughly preheated, 24-cm. Büchner funnel. The carbon is washed with 1 l. of boiling water and the deep-yellow filtrate and the washings (about 5.5–6 l.) are combined, reheated to boiling, and the pH

[c] The powdered *p*-chloroaniline may be dissolved in hydrochloric acid at about 80°, and the finely divided hydrochloride precipitated by cooling the solution to 0° with magnetic stirring.

[d] Because of frothing and spurting on subsequent heating, use of an extra-large vessel is desirable.

[e] If necessary, enough concentrated potassium hydroxide solution is added to bring the pH to 10.

adjusted to 5–6 by slow addition, with stirring, of 50% acetic acid. The mixture is kept for 15 min., the suspension is then filtered hot on a 24-cm. Büchner funnel, and the solid is thoroughly washed by suspension in water and filtration, followed by suspension in acetone and filtration, and dried; yield 90–95 g. (82–86%). If this product is desired highly pure, it may be converted into the sulfate in 10% sulfuric acid solution.

2,8-Diamino-6-purinethiol[2] (4)

In a 5-l., 3-necked flask, fitted with a sealed, Hershberg mechanical stirrer, and a large reflux-condenser, and having a large (preferably at least ₮ 55–50) stoppered neck, is placed 3.3 l. of refined pyridine. The pyridine is heated to 80°, heating is stopped, and, with stirring, 1 kg. of phosphorus pentasulfide is added in 4 or 5 portions; the mixture is then refluxed, but all of the phosphorus pentasulfide does not dissolve. With continued stirring, 333 g. of compound **3** is added in portions (*ca.* 30 g. each), and the mixture is refluxed, with vigorous stirring, for 5.25 hr. After 4.5 hr., the reflux condenser is replaced with a Claisen head and distillation assembly, and some of the pyridine is permitted to distil off at atmospheric pressure. The hot, cloudy mixture is then poured slowly, with vigorous stirring, into 11 l. of boiling, distilled water in a 10-gallon, stainless-steel receptacle in the hood. The flask is rinsed with 6 l. of boiling, distilled water (in several portions), and the rinsings are added. The resulting mixture is stirred for 15 min. (not longer), and is then promptly filtered hot, on two 32-cm. Büchner funnels. The solid is washed with 1–2 l. of distilled water, and is then suspended in 5 l. of distilled water, and the suspension is added to 6.5 l. of boiling, distilled water and heated to boiling, with occasional stirring. Just enough concentrated potassium hydroxide solution is added to the boiling suspension to dissolve the solid, affording a dark solution. The pH of the boiling solution is adjusted to 6 by slow addition of glacial acetic acid, with vigorous stirring, causing precipitation of a light-tan solid. The suspension is kept for 10 min., and is then promptly filtered hot on a 24-cm. Büchner funnel. The solid is thoroughly washed by suspension in distilled water and filtration, and by suspension in acetone and filtration, and dried; yield of light-tan, crude **4** is 240 g. (66%). The product is approximately 95% pure, and may be further purified by reprecipitation.

REFERENCES

(1) H. Fischer, *Z. Physiol. Chem.*, **60**, 69 (1909).
(2) J. W. Jones and R. K. Robins, *J. Am. Chem. Soc.*, **82**, 3773 (1960).
(3) R. K. Robins, *J. Med. Chem.*, **7**, 186 (1964).
(4) A. F. Lewis, A. G. Beaman, and R. K. Robins, *Can. J. Chem.*, **41**, 1807 (1963).

[16] 2-Amino-1-methylpurine-6(1*H*)-thione and 5-Amino-1,6-dihydro-6-methyl-7*H*-*v*-triazolo[4,5-*d*]-pyrimidin-7-one

Ring Closures of 5,6-Diaminopyrimidines

C. WAYNE NOELL,[a] LEROY B. TOWNSEND, and ROLAND K. ROBINS

DEPARTMENT OF CHEMISTRY, UNIVERSITY OF UTAH, SALT LAKE CITY, UTAH 84112

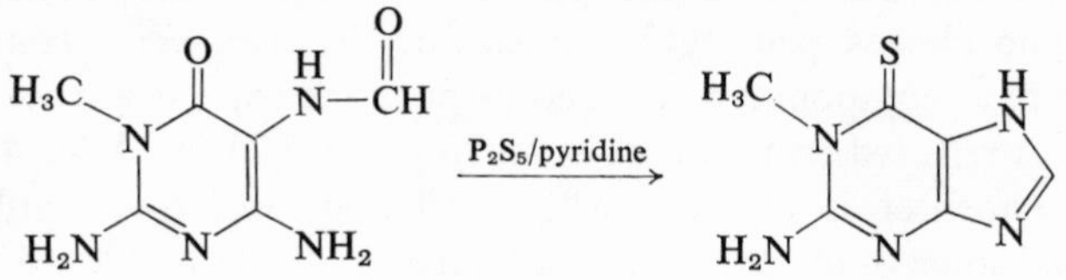

P_2S_5/pyridine

1 (183.2) → **2** (181.2)

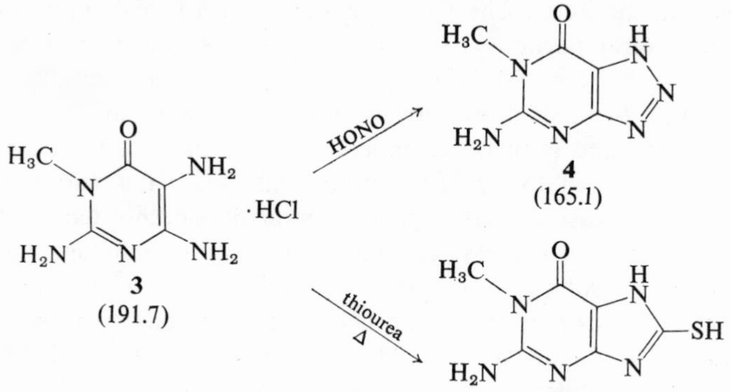

HONO

thiourea, Δ

3 (191.7) · HCl → **4** (165.1); **3** → **5** (197.2)

INTRODUCTION

2-Amino-1-methylpurine-6(1*H*)-thione (**2**) is an agent having a wide spectrum of antitumor activity, and is active against a variety of neoplasms.[1,2] The preparation of **2** has been accomplished by the direct thiation of 1-methylguanine,[1] and by simultaneous ring-closure and thiation of 2,6-diamino-5-formamido-3-methyl-4(3*H*)-pyrimidinone (**1**) with phosphorus pentasulfide in pyridine.[1–3] Acid hydrolysis of **1** readily

[a] Present address: Midwest Research Institute, 425 Volker Blvd., Kansas City, Missouri.

affords the 2,5,6-triaminopyrimidine (**3**), which may be ring-closed to "1-methyl-8-azaguanine" (**4**) with nitrous acid. Fusion[6] of **3** with thiourea gives 8-mercapto-1-methylguanine (**5**).

PROCEDURE

2-Amino-1-methylpurine-6(1*H*)-thione (2)

2,6-Diamino-5-formamido-3-methyl-4(3*H*)-pyrimidinone[4] (**1**) (50 g., 0.271 mole) and 150 g. of phosphorus pentasulfide are added, with stirring, to 1.8 l. of pyridine, and the mixture is refluxed, with stirring, for 30 hr. and filtered. The precipitate is mixed with 1 l. of water, and the suspension (*A*) is kept for 1 hr. In the meantime, the pyridine filtrate is evaporated under diminished pressure to a gummy residue; this is added to suspension *A*, and the mixture is kept at room temperature for 3 hr., and then on a steam bath for 3 hr., and allowed to cool. The precipitate is filtered off, washed with water, and dissolved in 1 l. of boiling, 2% potassium hydroxide solution. The hot solution is treated with decolorizing carbon, the suspension is filtered, and the hot filtrate is acidified with glacial acetic acid, and allowed to cool to room temperature. The resulting precipitate is filtered off, and recrystallized from 2 l. of water; a second recrystallization from water gives 9 g. (18%) of pure **2**, which is dried at 95°, m.p. 340–342°, $\lambda_{max}^{pH\,1}$ 256 and 345 nm. (ϵ 9,400 and 21,000); $\lambda_{max}^{pH\,11}$ 234 and 336 nm. (ϵ 16,800 and 21,000).

5-Amino-1,6-dihydro-6-methyl-7*H*-*v*-triazolo[4,5-*d*]pyrimidin-7-one (1-Methyl-8-azaguanine, 4)

2,5,6-Triamino-3-methyl-4(3*H*)-pyrimidinone[5] hydrochloride (**3**) (40 g., 0.21 mole) is added to dilute acetic acid (20 ml. of glacial acetic acid diluted to 300 ml. with water), and to the mixture is slowly added (during 10–15 min.) a solution of sodium nitrite (22.0 g.) in 40 ml. of water. The temperature of the mixture is then 45–50°; the mixture is stirred for 10 min., kept at 0.5° for 2 hr., and filtered, and the solid is washed with 250 ml. of water at 5°. The solid is added to 100 ml. of water at 90–95°, and sufficient *N* sodium hydroxide is added to effect dissolution; the solution is treated with decolorizing carbon, and filtered, and the pH of the filtrate is adjusted to 6 with glacial acetic acid. The suspension is stirred at room temperature for 30 min., and filtered, and the solid is successively washed with 250 ml. of water at 5° and acetone (500 ml.), and dried at 110°; yield 30.0 g. (87%) of the hemihydrate of **4**, m.p. $>300°$; $\lambda_{max}^{pH\,1}$ 252 nm. (ϵ 5,650) and 264 nm. (sh) (ϵ 4,160).

8-Mercapto-1-methylguanine (5)

2,5,6-Triamino-3-methyl-4(3*H*)-pyrimidinone[5] hydrochloride (**3**) (3.0 g., 15.7 mmoles) and thiourea (9.0 g.) are thoroughly mixed, and the mixture is heated on a hot-plate to a melt (at approximately 200°) until the melt partly solidifies. The mass is cooled, and dissolved in 200 ml. of *N* potassium hydroxide, and the solution is heated to 90–95°, treated with decolorizing carbon, and filtered, and the pH of the filtrate is adjusted to 6 with glacial acetic acid. A solid that separates from the hot solution is filtered off, washed with water at 5°, dissolved in hot *N* potassium hydroxide (200 ml.) at 90°, and the pH of the solution adjusted to 6 with glacial acetic acid. The solid that separates from the hot solution is filtered off, successively washed with 200 ml. of water at 5° and 500 ml. of acetone, and dried at 110°; yield 3.4 g. (97%) of **5**, m.p. >300°; $\lambda_{max}^{pH\,1}$ 301 nm. (ϵ 19,700), 268 nm. (ϵ 17,300), and 229.5 nm. (ϵ 13,200); $\lambda_{max}^{pH\,11}$ 302 nm. (ϵ 17,000) and 244.5 nm. (ϵ 24,002).

REFERENCES

(1) C. W. Noell, D. W. Smith, and R. K. Robins, *J. Med. Pharm. Chem.*, **5**, 996 (1962).
(2) R. K. Robins, *J. Med. Chem.*, **7**, 186 (1964).
(3) G. D. Daves, Jr., C. W. Noell, R. K. Robins, H. C. Koppel, and A. G. Beaman, *J. Am. Chem. Soc.*, **82**, 2633 (1960).
(4) F. G. Mann and J. W. G. Porter, *J. Chem. Soc.*, **1945**, 751.
(5) W. Pfleiderer, E. Liedek, R. Lohrmann, and M. Rukweid, *Chem. Ber.*, **93**, 2010 (1960).
(6) R. K. Robins, *J. Am. Chem. Soc.*, **80**, 6671 (1958).

[17] 9-Benzyl-9*H*-purine-6(1*H*)-thione

Synthesis of 6,9-Disubstituted Purines

JOHN A. MONTGOMERY and CARROLL TEMPLE, JR.

KETTERING-MEYER LABORATORY, SOUTHERN RESEARCH INSTITUTE, BIRMINGHAM, ALABAMA 35205

1 (164.0) $\xrightarrow[\Delta]{PhCH_2NH_2}$ **2** (234.7)

2 $\xrightarrow{(EtO)_3CH/HCl}$ **4**

3 (154.6) $\xrightarrow[Me_2SO,\ K_2CO_3]{PhCH_2Cl}$ **4** (244.7) + **5** (244.7)

4 $\xrightarrow{(H_2N)_2CS,\ \text{propyl alcohol}/\Delta}$ **6** (242.3)

INTRODUCTION

A number of 9-alkylpurines have been found to inhibit animal[1] and human[2] neoplasms, and to inhibit the growth of cells resistant to purine-6-(1*H*)-thione (6-mercaptopurine).[3]

The preparation of 9-benzyl-6-chloro-9*H*-purine[4] (**4**) by the acid-catalyzed cyclization of 5-amino-4-(benzylamino)-6-chloropyrimidine (**2**), prepared from 5-amino-4,6-dichloropyrimidine (**1**) with ethyl orthoformate[5] or by the alkylation of 6-chloropurine (**3**) with benzyl chloride,[6] illustrates two versatile methods for the synthesis of 9-alkyl-6-chloropurines,[5,6] which are valuable intermediates for the preparation of potential purine antimetabolites, such as 9-benzyl-9*H*-purine-6(1*H*)-thione (**6**).

PROCEDURE

9-Benzyl-6-chloro-9*H*-purine (4)

A. Cyclization.—A stirred mixture of 82.0 g. (0.500 mole) of 5-amino-4,6-dichloropyrimidine (**1**) (Sec. II [27]) and 268 g. (2.50 moles) of benzylamine is heated in an oil bath at 100–110°. A brown solution is formed as the internal temperature rises to 150°. At this point, the solution is cooled to 100°, and then heated at 100–110° for 2 hr. The mixture is cooled, and the solid is collected by filtration and washed with water, giving 97.6 g. of 5-amino-4-(benzylamino)-6-chloropyrimidine (**2**), m.p. 202–204° (lit.[4] m.p. 207–209°). The reaction filtrate is evaporated to dryness *in vacuo*, and the resulting residue is washed with water to give an additional 11.7 g. of **2**, m.p. 197–202°. The total yield of crude **2** is 109 g. (93%).

To a suspension of 109 g. (0.466 mole) of **2** in 1.75 l. of triethyl orthoformate at room temperature is slowly added 43.4 ml. of concentrated hydrochloric acid.[a] The mixture is stirred at room temperature for 20 hr.[b] and then evaporated to dryness under diminished pressure. The resulting residue is recrystallized from hot petroleum ether (b.p. 85–105°); yield 108 g. (in two crops), m.p. 94–95°. A second recrystallization of the combined crops, from petroleum ether, gives 98.5 g. (81% from **1**) of **4**, m.p.[c] 95–96° (lit.[4] m.p. 84–85.5°),[c] $\lambda_{max}^{pH\,7}$ 266 nm. (log ϵ 3.992).

B. Alkylation.—A mixture of 100 g. (0.648 mole) of 6-chloropurine (**3**), 164 g. (1.30 moles) of benzyl chloride, and 89.6 g. (0.648 mole) of anhydrous potassium carbonate in 850 ml. of methyl sulfoxide is stirred at room temperature for 5 hr. After the reaction mixture has been diluted with 3.8 l. of water, the resulting oily and aqueous layers are separated. The aqueous layer is extracted with five 1,250-ml. portions of ether, and the first three extracts are added, with stirring, to the oily layer. The solid that forms is collected by filtration, washed successively with the two remaining ether extracts, and air dried, to give 41.4 g. of crude 7-benzyl-6-chloro-7*H*-purine (**5**), m.p. about 140° (with presoftening).[d] Concentration

[a] If the mixture becomes hot, hydrolysis of the chloro group of **4**, to give 9-benzylhypoxanthine, can occur.

[b] Completion of the reaction can be ascertained by determining the ultraviolet spectrum of an aliquot portion.

[c] Some samples melt at 87°, solidify, and remelt at 96°.

[d] Only low yields of the 7-isomer were obtained with other alkyl halides.[6]

of the combined filtrate and ether washings to a volume of about 2.0 l. causes deposition of an additional 6.7 g. of **5**, m.p. 148°. The ether filtrate (*A*) is saved. The combined crops of **5** are extracted with three 300-ml. portions of hot petroleum ether (b.p. 85–105°), and the combined extracts (*B*) are also saved. The residual solid is recrystallized from a large volume of water to give 22.2 g. (14%) of **5**, m.p. 151–153°, $\lambda_{max}^{pH\ 7}$ 269 nm. (log ϵ 3.887).

The ether filtrate (*A*) is concentrated to a slurry, and 67.7 g. of 9-benzyl-6-chloro-9*H*-purine (**4**), m.p. 92°, is collected by filtration. An additional 34.6 g. of **4**, m.p. 91°, is obtained in the following manner. Evaporation of the filtrate gives a residue that is mixed with the petroleum ether extract (*B*), and the suspension is heated to boiling; the solution is decanted through a fluted filter-paper, and the filtrate is cooled, whereupon it deposits **4**. Extraction of the residue with the filtrate is repeated until no additional **4** crystallizes from the cooled filtrate. Then the combined crops of **4** are recrystallized from a large volume of petroleum ether (b.p. 85–105°); yield 85.9 g. (54%), m.p. 96°.

9-Benzyl-9*H*-purine-6(1*H*)-thione (6)

A solution of 56.0 g. (0.229 mole) of 9-benzyl-6-chloro-9*H*-purine (**4**) in 1.5 l. of propyl alcohol containing 21.0 g. (0.276 mole) of thiourea is refluxed for 2 hr. The resulting mixture is cooled, and the solid is collected by filtration, washed with propyl alcohol, and dried *in vacuo* at 100° over phosphorus pentaoxide; yield 52.0 g. (94%) of pure **6**, m.p. >264°, $\lambda_{max}^{pH\ 13}$ 311 nm. (log ϵ 4.366).[e]

REFERENCES

(1) H. E. Skipper, J. A. Montgomery, J. R. Thomson, and F. M. Schabel, Jr., *Cancer Res.*, **19**, 425 (1959).
(2) "On the Chemotherapy of Cancer," J. A. Montgomery, *Progr. Drug Res.*, **8**, 431 (1965).
(3) G. G. Kelly, G. P. Wheeler, and J. A. Montgomery, *Cancer Res.*, **22**, 329 (1962).
(4) S. M. Greenberg, L. O. Ross, and R. K. Robins, *J. Org. Chem.*, **24**, 1314 (1959).
(5) J. A. Montgomery and C. Temple, Jr., *J. Am. Chem. Soc.*, **83**, 630 (1961).
(6) C. Temple, Jr., C. L. Kussner, and J. A. Montgomery, *J. Med. Pharm. Chem.*, **5**, 866 (1962).

[e] If further purification is desired, this material may be reprecipitated, from solution in sodium hydroxide, with glacial acetic acid.

[18] 4*H*-Pyrazolo[3,4-*d*]pyrimidin-4-one

ROLAND K. ROBINS

DEPARTMENT OF CHEMISTRY, UNIVERSITY OF UTAH, SALT LAKE CITY, UTAH 84112

1 (126.1) + $HC(=O)-NH_2$ $\xrightarrow[180°]{\Delta}$ **2** (136.1)

INTRODUCTION

4*H*-Pyrazolo[3,4-*d*]pyrimidin-4-one (**2**), an analog of hypoxanthine, was first prepared by Robins[1] from 3-aminopyrazole-4-carboxamide (**1**). Schmidt and Druey[2] later reported the synthesis of **2** from ethyl 3-aminopyrazole-4-carboxylate.[2] The 4*H*-pyrazolo[3,4-*d*]pyrimidine derivatives were first shown by Feigelson, Davidson, and Robins[3] to be good inhibitors of xanthine oxidase. 4*H*-Pyrazolo[3,4-*d*]pyrimidin-4-one (**2**), known as "allopurinol" and "zyloprim," is clinically active against hyperuracemia[4] and gouty arthritis.[5] Compound **2** is also an important intermediate in the synthesis of 4-aminopyrazolo[3,4-*d*]pyrimidine[1] and its 4-substituted derivatives,[1] which have shown significant antitumor activity[6] in experimental animals and in clinical trial.[7]

PROCEDURE

4*H*-Pyrazolo[3,4-*d*]pyrimidin-4-one (2)

3-Aminopyrazole-4-carboxamide (**1**)* hemisulfate[1] (25 g., 0.14 mole) is heated with 100 g. of formamide (reagent grade) in an open flask (inside temperature, 180°) for 40 min. The solution is then cooled, and 100 ml. of water is added. The product is filtered off, and washed with 1 l. of cold water. The light-tan product is dried in an oven at 110°; yield 17.5 g. (90%). A pure, colorless product is obtained by treating a hot, ammoniacal solution with activated carbon, followed by filtration, and reprecipitation from the hot filtrate by addition of acetic acid.

REFERENCES

(1) R. K. Robins, *J. Am. Chem. Soc.*, **78**, 784 (1956).
(2) P. Schmidt and J. Druey, *Helv. Chim. Acta*, **39**, 987 (1956).

(3) P. Feigelson, J. D. Davidson, and R. K. Robins, *J. Biol. Chem.*, **226**, 993 (1957).

(4) R. W. Rundles, *Trans. Assoc. Am. Physicians*, **76**, 126 (1963); J. R. Kleinberg, *Arthritis Rheumat.*, **6**, 779 (1963).

(5) J. B. Wyngaarden, *Arthritis Rheumat.*, **8**, 883 (1965); J. B. Houpt, *ibid.*, **8**, 899 (1965).

(6) H. E. Skipper, R. K. Robins, J. R. Thomson, C. C. Cheng, R. W. Brockman, and F. M. Schabel, *Cancer Res.*, **17**, 579 (1957); E. Y. Sutcliffe, K.-Y. Zee-Cheng, C. C. Cheng, and R. K. Robins, *J. Med. Pharm. Chem.*, **5**, 588 (1962); F. R. White, *Cancer Chemotherapy Rept.*, **3**, 26 (1959); R. K. Robins, *J. Med. Chem.*, **7**, 186 (1964).

(7) J. L. Scott and L. V. Foye, Jr., *Cancer Chemotherapy Rept.*, **20**, 73 (1962).

Section II
Pyrimidines and Analogs

[19] Cytosine

The Use of Pyrimidinethiols as Intermediates

B. W. ARANTZ and D. J. BROWN

JOHN CURTIN SCHOOL OF MEDICAL RESEARCH, CANBERRA, AUSTRALIA

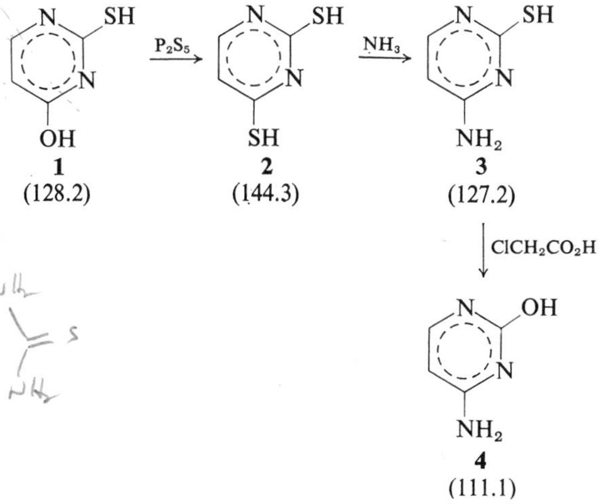

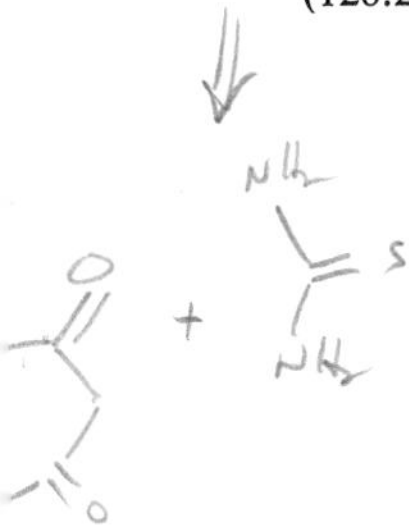

INTRODUCTION

Cytosine (**4**) has been prepared by three practical routes.[1–4] The best one begins with thiation of commercially available 2-thiouracil[5] (**1**), and this stage can now be completed within 15 min. by using phosphorus pentasulfide in the relatively inexpensive 2-picoline. The resulting 2,4-dithiouracil (**2**) is selectively aminated[2,3] with aqueous ammonia to give 4-amino-2-pyrimidinethiol (**3**), which is converted into cytosine (**4**) by boiling in aqueous chloroacetic acid.[2,3]

PROCEDURE

2,4-Dithiouracil (2)

A suspension of 70 g. (0.55 mole) of 2-thiouracil (**1**) in 1 l. of 2-picoline[a] in a 2-l. flask fitted with a reflux condenser is set in a steam bath and mechanically stirred. The internal temperature is raised to 40°, and 245 g. (1.1 moles) of phosphorus pentasulfide is added as quickly as possible, with vigorous stirring. Full steam-heating and stirring are continued for

[a] If necessary, the 2-picoline should be dehydrated with potassium hydroxide pellets for 3–5 days.

exactly 15 min. The steam bath is replaced immediately by an ice bath, and, when the temperature of the mixture falls to 10–15°, the liquid layer is decanted and evaporated to dryness in a rotary evaporator under diminished pressure. The residual solid (*ca.* 4 g.) is added to the bulk of the crude dithiol (see below).

The black residue remaining after decantation is decomposed by gradually adding 1 l. of water, the mixture is refrigerated, and the brown solid is filtered off. The crude product is combined with the first batch, and is recrystallized from 10 l. of boiling water (with decolorization with carbon), giving 44 g. (56%) of yellow, crystalline dithiol (**2**), decomposing at about 265°. Paper chromatography in 3% aqueous ammonium chloride solution shows the product to be free from 2-thiouracil and other impurities.

4-Amino-2-pyrimidinethiol (3)

To 500 ml. of 10 *N* ammonium hydroxide is added 34 g. of **2**, and a stream of ammonia, at the rate of one bubble per second, is passed[b] into the solution for 24 hr., with heating under reflux on a steam bath. The reaction mixture is kept at room temperature for 24 hr., and the product (23.7 g.) is filtered off and recrystallized from 900 ml. of boiling water,[c] with the use of decolorizing carbon if necessary. The yield is 20.3 g. (66%) of colorless **3**, m.p. 275°; the product is chromatographically homogeneous.

Cytosine (4)

A solution of 20 g. of pure **3** and 16 g. of chloroacetic acid in 160 ml. of water is refluxed for 45 min.; 160 ml. of concentrated hydrochloric acid is then added, and heating is continued for 2 hr. The suspension is filtered, and the filtrate is evaporated to dryness under diminished pressure (rotary evaporator). The resulting, crude hydrochloride of **4** is dissolved in 70 ml. of warm water, the pH of the solution is adjusted to 9 with concentrated ammonium hydroxide, and the solution is refrigerated. The colorless solid (17.7 g.) that separates is filtered off, and washed successively with a little cold water and ethanol. Recrystallization from 150 ml. of boiling water gives 16.8 g. (95%) of pure **4**, m.p. *ca.* 306° (dec.). A sodium-fusion test for sulfur is negative, and the product is chromatographically homogeneous. Its molar absorbancies are as specified.[6]

[b] The inlet tube should be wide and straight to facilitate clearance of blockages that occur during the first five hr.; use of a safety valve in the inlet tube for ammonia is advisable.

[c] A steam-heated funnel is needed, because of the steep solubility gradient of **3** in water.

REFERENCES

(1) P. J. Tarsio and L. Nicholl, *J. Org. Chem.*, **22**, 192 (1957).

(2) G. H. Hitchings, G. B. Elion, E. A. Faclo, and P. B. Russell, *J. Biol. Chem.*, **177**, 357 (1949).

(3) D. J. Brown, *J. Soc. Chem. Ind.* (*London*), **69**, 353 (1950).

(4) R. S. Karlinskaya and N. V. Khromov-Borisov, *Zh. Obshch. Khim.*, **27**, 2113 (1957).

(5) G. B. Elion and G. H. Hitchings, *J. Am. Chem. Soc.*, **69**, 2138 (1947).

(6) "Specifications and Criteria for Biochemical Compounds," R. S. Tipson (ed.), National Academy of Sciences–National Research Council, Publication No. 1344, Washington, D. C. (1967), N-19, p. 413.

[20] *N*-Alkylcytosines

Aminolysis of Alkoxypyrimidines

W. SZER and D. SHUGAR

INSTITUTE OF BIOCHEMISTRY AND BIOPHYSICS, POLISH ACADEMY OF SCIENCES, AND DEPARTMENT OF BIOPHYSICS, UNIVERSITY OF WARSAW, WARSAW, POLAND

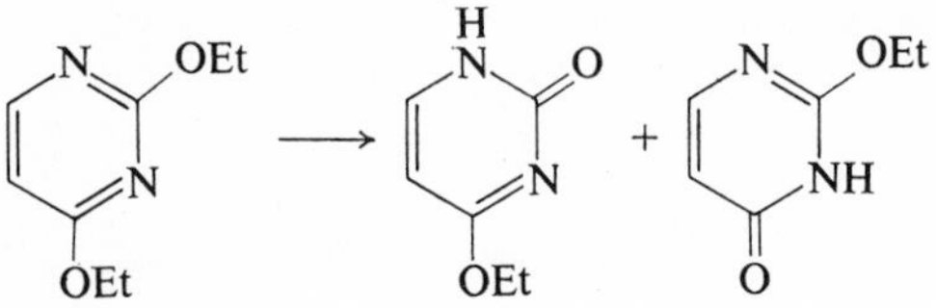

1 (168.2) ⟶ **2** (140.1) + **2a** (140.1)

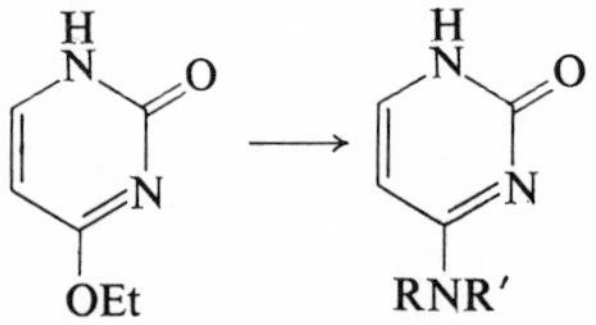

2 (140.1) ⟶ **3**, R = H, R′ = CH_3 (125.7); **3a**, R = R′ = CH_3 (139.2)

INTRODUCTION

N-Methylcytosine[1] (**3**) has recently been found as a minor component of ribosomal ribonucleic acid.[2] Its synthesis is achieved *via* the useful intermediate, 4-ethoxy-2-pyrimidinol[3] (**2**), having a readily exchangeable 4-ethoxy group. The preparation of **2** is based on the original procedure of Hilbert and Jansen,[3] and includes the simultaneous isolation of the isomeric 2-ethoxy-4-pyrimidinol (**2a**). Some improvements have been introduced, and paper chromatography is now employed for following the course of the reaction and estimating the purity of the products.

PROCEDURE

4-Ethoxy-2-pyrimidinol (2)

To a solution prepared by adding, portionwise, 21 g. (0.9 mole) of sodium to 500 ml. of 90% ethanol, in a round-bottomed flask fitted with a reflux condenser, is added 100 g. (0.6 mole) of 2,4-diethoxypyrimidine (**1**)

(Sec. II [30]). The clear solution is refluxed for 70 hr. in an oil bath, until the characteristic odor of **1** has disappeared. The reaction is interrupted after 14 and 32 hr., and the crude sodium salt of 4-ethoxy-2-pyrimidinol (**2**), which separates on cooling to room temperature (successively, 19 g. and 8 g.) is collected on a sintered-glass funnel; after 52 hr., the solution is chilled for several hours to induce crystallization (affording a further 3 g.). After 70 hr., the mixture is concentrated to one-third of its original volume, and another 3.5 g. of **2** is removed after cooling, to bring the total yield of the crude sodium salt of **2** to 33.5 g. (35%); after recrystallization from water, m.p. 305° (dec.). Recrystallization is not essential for use of the product in the subsequent aminolysis. Paper chromatography (Whatman No. 1, ascending) of the crude salt with solvent A (upper phase of 169:45:15 benzene–ethanol–water) shows one major product, R_f 0.55, and a trace of uracil, R_f 0.02. The crude sodium salt of **2** (33.5 g.) is dissolved in 100 ml. of water at 60°, and 1:4 (v/v) acetic acid–water is added dropwise, with stirring, to pH 6.5; compound **2** separates on slow cooling. The mixture is refrigerated overnight, and then filtered, to yield 21.3 g. of **2**; an additional 3.0 g. is obtained on concentrating the filtrate to half its original volume; total yield 24.3 g. (29%). This product is chromatographically homogeneous, m.p. 163–166°; on recrystallization from 17:3 (v/v) ethyl acetate–ethanol, m.p. 167–167.5°.

If the sodium salt of the isomeric 2-ethoxy-4-pyrimidinol[4] (**2a**) is desired, it may be isolated from the mother liquor remaining after removal of the crude sodium salt of **2**. The solution is evaporated to dryness, and the residue is recrystallized twice from 4:1 (v/v) benzene–ethanol, to give 35 g. (36%) of the sodium salt of **2a**; this is dissolved in 50 ml. of water at 60°, and 2-ethoxy-4-pyrimidinol (**2a**) is isolated as described for **2**; yield 19.2 g. (23%). After two recrystallizations from water, m.p. 127–129°; R_f 0.83 in solvent A, showing a trace of **2**.

N-Methylcytosine (3)

4-Ethoxy-2-pyrimidinol (**2**) (1 g., 0.071 mole) is treated with 25 ml. of a solution of methylamine (30–40% by weight) in anhydrous methanol in a sealed tube at 100° for 18 hr.; the mixture is then cooled, giving a colorless, crystalline precipitate of **3** (0.43 g.). The tube is opened, and the product is filtered off. The filtrate is evaporated to dryness, re-evaporated thrice with methanol, and the residue dissolved in the minimal volume of hot methanol. The solution is cooled, affording 0.38 g. of **3**. The two crops are combined, and recrystallized from hot methanol to which several drops of ethyl acetate are added; yield 0.68 g. (75%), m.p. 271° (dec.); in 0.01 *N* HCl λ_{max} 276 nm. (ϵ_{max} 10.6 × 10^3), λ_{min} 240 nm.; pK_1 (for protonation of N-3) 4.40; pK_2 (for dissociation of proton number 1) 12.7. R_f in solvent A,

0.06; R_f in solvent B (water-saturated butyl alcohol having ammonia in the vapor phase), 0.45.

N,N-Dimethylcytosine[4] (3a)

Compound **3a** is obtained from **2** (1 g., 0.071 mole) by treatment with a solution of dimethylamine (30–40% by weight) in anhydrous methanol, as described above for **3**. The product is recrystallized from methanol–ethyl acetate; yield 0.71 g. (72%); m.p. 254–255° (dec.); in 0.01 *N* HCl λ_{max} 282 nm. (ϵ_{max} 13.6 × 10^3), λ_{min} 252 nm.; pK_1 4.15, pK_2 12.8; R_f in solvent A, 0.15, R_f in solvent B, 0.48.

REFERENCES

(1) D. J. Brown, *J. Appl. Chem.* (London), **9**, 203 (1959); W. Szer and D. Shugar, *Acta Biochim. Polon.*, **13**, 177 (1966).
(2) J. L. Nichols and B. G. Lane, *Biochim. Biophys. Acta*, **114**, 469 (1966).
(3) G. E. Hilbert and E. P. Jansen, *J. Am. Chem. Soc.*, **57**, 552 (1935).
(4) I. Wempen, R. Duschinsky, L. Kaplan, and J. J. Fox, *J. Am. Chem. Soc.*, **83**, 4755 (1961).

[21] *N*-Alkyl-1-methylcytosines

Aminolysis of Alkoxy-1-methylpyrimidines

W. SZER and D. SHUGAR

INSTITUTE OF BIOCHEMISTRY AND BIOPHYSICS, POLISH ACADEMY OF SCIENCES,
AND DEPARTMENT OF BIOPHYSICS, UNIVERSITY OF WARSAW,
WARSAW, POLAND

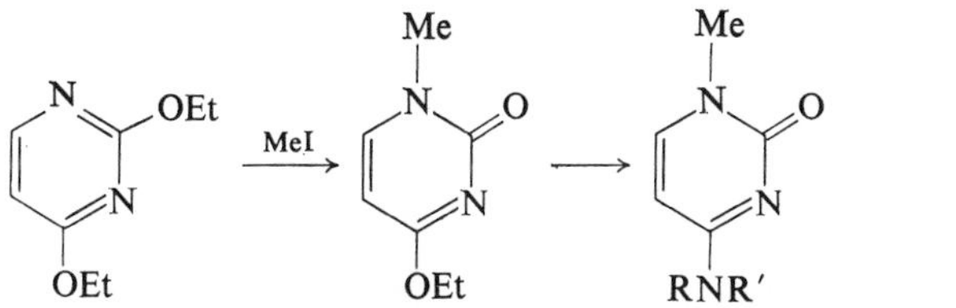

1 (168.2) $\xrightarrow{\text{MeI}}$ **2** (154.2) $\longrightarrow$ **3**, R = R′ = H (125.1); **3a**, R = H, R′ = Me (139.2); **3b**, R = R′ = Me (153.2)

INTRODUCTION

1-Methylcytosine derivatives have proved to be extremely useful model compounds in studies on the physicochemical properties of, and the nature of the reactions undergone by, the corresponding cytosine nucleosides.[1] They are most readily obtained *via* the important intermediate 4-ethoxy-1-methyl-2-pyrimidinol[2] (**2**), the 4-ethoxy group of which undergoes facile exchange.

PROCEDURE

4-Ethoxy-1-methyl-2-pyrimidinol (2)

A solution of 25 g. (0.15 mole) of 2,4-diethoxypyrimidine (**1**) (Sec. II [30]) in 25 ml. of freshly distilled methyl iodide is kept in the dark at room temperature for about 20 hr. The large, colorless crystals deposited from the solution are collected by filtration, and washed with ether. Recrystallization from ethanol–ether gives **2** in essentially quantitative yield, m.p. 136°; R_f 0.89 and 0.80 in solvents[a] A and B, respectively.

[a] For solvent systems A and B, see Sec. II [20].

1-Methylcytosine (3)

A solution of 1 g. (0.065 mole) of **2** in 20 ml. of dry methanol presaturated with ammonia is kept in a sealed tube at 100° for 18 hr., and cooled; compound **3** separates in large crystals. Recrystallization from ethanol affords 0.68 g. (84%), m.p. 303°; in 0.01 *N* HCl λ_{max} 283 nm. (ϵ_{max} 12.3 × 10^3), λ_{min} 242.5 nm.; pK 4.55; R_f in solvent A, 0.06; R_f in solvent B, 0.34.

***N*,1-Dimethylcytosine (3a)**

Compound **3a** is obtained from **2** (1 g., 0.065 mole) by treatment with a solution of methylamine (30–40% by weight) in anhydrous methanol as described for **3**. After recrystallization from methanol–ether, m.p. 180–181°; yield 0.65 g. (72%); in 0.01 *N* HCl λ_{max} 285 nm. (ϵ_{max} 12.4 × 10^3), λ_{min} 244 nm.; pK 4.40; R_f in solvent A, 0.13; R_f in solvent B, 0.58.

***N*,*N*,1-Trimethylcytosine (3b)**

Compound **3b** is obtained from **2** (1 g., 0.065 mole) by treatment with a solution of dimethylamine (30–40% by weight) in anhydrous methanol as described for **3**. After recrystallization from methanol–ether, m.p. 178–179°; yield 0.62 g. (62.5%); in 0.01 *N* HCl λ_{max} 288 nm. (ϵ_{max} 14.3 × 10^3), λ_{min} 247 nm.; pK 4.20; R_f in solvent A, 0.46; R_f in solvent B, 0.63.

REFERENCES

(1) J. J. Fox and D. Shugar, *Biochim. Biophys. Acta*, **9**, 369 (1952); G. W. Kenner, C. B. Reese, and A. R. Todd, *J. Chem. Soc.*, **1955,** 855; W. Szer and D. Shugar, *Acta Biochim. Polon.*, **13**, 177 (1966).
(2) G. E. Hilbert and T. B. Johnson, *J. Am. Chem. Soc.*, **52**, 2001 (1930).

[22] 1-Substituted 5,6-Dihydrocytosines

Cyclization of 1-Substituted 1-(2-cyanoethyl)ureas with Sodium Ethoxide

C. C. CHENG and LELAND R. LEWIS

MIDWEST RESEARCH INSTITUTE, KANSAS CITY, MISSOURI 64110

$N{\equiv}C{-}CH{=}CH_2$ + $H_2N{-}R$ ⟶ $N{\equiv}C{-}CH_2{-}CH_2{-}NH{-}R$ $\xrightarrow[H^+]{KCNO}$

1 (53.1)

2, $R = CH_3$ (31.1)

2a, $R = (CH_2)_3CH_3$ (73.1)

3, $R = CH_3$ (84.1)

3a, $R = (CH_2)_3CH_3$ (126.2)

$N{\equiv}C{-}CH_2{-}CH_2{-}N(R){-}C(=O){-}NH_2$ $\xrightarrow{NaOCH_3}$ 5,6-dihydrocytosine (1-R)

4, $R = CH_3$ (127.2)

4a, $R = (CH_2)_3CH_3$ (169.2)

5, $R = CH_3$ (127.2)

5a, $R = (CH_2)_3CH_3$ (169.2)

INTRODUCTION

The importance of 5,6-dihydrocytosines and related compounds is demonstrated by the fact that 5,6-dihydro-5′-cytidylic acid has been isolated from liver.[1] This led Cohen and co-workers[2] to postulate that some 5,6-dihydrocytosines might be precursors of cytidine derivatives. 5,6-Dihydrocytosines have been prepared by catalytic reduction of cytosine and related compounds.[2,3] However, the scope of this reaction is limited, and the reduction cannot readily be stopped at the dihydro stage. The present method[4] offers a new and unambiguous synthesis of 5,6-dihydrocytosines substituted at the 1-position—the site of attachment of the β-D-ribofuranosyl group in the related nucleosides.

PROCEDURE

3-(Methylamino)propionitrile (3)

To 800 g. (10 moles) of 40% aqueous methylamine (**2**), cooled in an ice bath, is rapidly added 212 g. (4 moles) of acrylonitrile (**1**). After the addition is complete, the mixture is stirred at room temperature for 4 hr., and heated on a steam bath for 1 hr. To the clear solution is added 400 g. of anhydrous potassium carbonate, and the mixture is stirred until all of the solid dissolves, whereupon the solution separates into two layers. The organic layer is separated, and distilled under diminished pressure. The product[5,6] is collected at 67–68°/10 mm. Hg, and weighs 302 g. (93%).

3-(Butylamino)propionitrile (3a)

Acrylonitrile (**1**) (49 g., 0.92 mole) is added dropwise to 50 g. (0.69 mole) of butylamine (**2a**) at 20–30°. The resulting solution is gently refluxed on a steam bath for 2 hr., and then distilled under diminished pressure. The product[6] is collected at 103°/12.5 mm. Hg, and weighs 85 g. (98%).

1-(2-Cyanoethyl)-1-methylurea (4)

To a cool (20°) solution of 55 g. (0.68 mole) of potassium cyanate in 150 ml. of water is added, with stirring, 42 g. (0.50 mole) of **3** dissolved in a solution of 50 ml. of concentrated hydrochloric acid in 90 ml. of water. The resulting, clear solution is stirred at room temperature for 12 hr., and then evaporated to dryness *in vacuo*. The solid residue is extracted with three 200-ml. portions of absolute ethanol, and the extract is evaporated to dryness. The crude product is recrystallized from propyl alcohol to give 48.2 g. (76%) of **4** as colorless, fine needles, m.p. 119–121°.

1-Butyl-1-(2-cyanoethyl)urea (4a)

A solution of 80 g. (0.63 mole) of **3a** in 150 ml. of water is mixed at 0° with 70 ml. of concentrated hydrochloric acid dissolved in 50 ml. of water. The resulting solution is added dropwise to a solution of 70 g. (0.87 mole) of potassium cyanate in 200 ml. of water. The reaction mixture, which becomes opalescent after the addition, is stirred overnight, and then evaporated to dryness *in vacuo*. The white residue is extracted with three 350-ml. portions of methanol, the extract is evaporated to dryness, the resulting glass is dissolved in boiling butyl alcohol, the mixture is filtered, and the filtrate is evaporated. The resulting semi-solid is crystallized from 1:6 ethanol–isopropyl ether to give 73.5 g. (79%) of **4a** as colorless crystals, m.p. 69–71°.

5,6-Dihydro-1-methylcytosine (5)

A mixture of 6.3 g. (0.05 mole) of **4** and 1 g. of sodium methoxide is dissolved in 60 ml. of boiling, absolute ethanol. The solution is refluxed, with stirring, for 30 min., during which time a white precipitate gradually forms. The reaction mixture is cooled, and the solid is filtered off, and washed with cold ethanol. It is recrystallized from ethanol to give 5.3 g. (84%) of analytically pure product, which melts at 223–225° (dec.), $\lambda_{max}^{pH\ 11}$ 245 nm. (ϵ 7,600).

1-Butyl-5,6-dihydrocytosine (5a)

A mixture of 8.5 g. (0.05 mole) of **4a**, 1 g. of sodium methoxide, and 80 ml. of absolute ethanol is refluxed for 2 hr., the clear solution is concentrated to *ca.* 20 ml. under diminished pressure, and the resulting solid is filtered off and washed with cold ethanol. Recrystallization of the product from ethanol gives 7.1 g. (84%) of colorless needles, m.p. 188–189° (dec.), $\lambda_{max}^{pH\ 11}$ 239 nm. (ϵ 3,800).

REFERENCES

(1) L. Grossman and D. W. Visser, *J. Biol. Chem.*, **216**, 775 (1955).
(2) M. Green and S. S. Cohen, *J. Biol. Chem.*, **228**, 601 (1957); S. S. Cohen, J. Lichtenstein, H. D. Barner, and M. Green, *ibid.*, **228**, 611 (1957); M. Green, H. D. Barner, and S. S. Cohen, *ibid.*, **228**, 621 (1957).
(3) C. Janion and D. Shugar, *Acta Biochim. Polon.*, **7**, 309 (1960).
(4) C. C. Cheng and L. R. Lewis, *J. Heterocyclic Chem.*, **1**, 260 (1964).
(5) A. H. Cook and K. J. Reed, *J. Chem. Soc.*, **1945,** 399.
(6) D. S. Tarbell, N. Shakespeare, C. J. Claus, and J. F. Bunnett, *J. Am. Chem. Soc.*, **68**, 1217 (1946).

[23] 5-(Hydroxymethyl)cytosine

Removal of an Alkylthio Group by Hydrolysis

T. L. V. ULBRICHT

TWYFORD LABORATORIES, ELVEDEN ROAD, LONDON, N.W. 10, ENGLAND

SMe, N, N, HOH_2C, NH_2 — dil. HCl → H, N, O, N, HOH_2C, NH_2

1 (171.2) **2** (141.1)

INTRODUCTION

5-(Hydroxymethyl)cytosine occurs in place of cytosine in the 2′-deoxyribonucleic acid of T-even phages of *Escherichia coli.*[1]

It has been synthesized, in low yield, by the reduction of ethyl 4-amino-2-hydroxy-5-pyrimidinecarboxylate,[2] and, in better yield, by the acid hydrolysis of 4-amino-2-(ethylthio)-5-pyrimidinemethanol[2,3] or of the 2-(methylthio) compound.[4] The latter is readily available (Sec. II [32]).

PROCEDURE

5-(Hydroxymethyl)cytosine (2)

4-Amino-2-(methylthio)-5-pyrimidinemethanol (**1**) (Sec. II [32]) (17.1 g., 0.1 mole) in 1.2 *N* hydrochloric acid (340 ml.) is refluxed for 6 hr. The solution is cooled, the pH is adjusted to 6–7 with 6 *N* sodium hydroxide, and the solution is kept at 0° overnight. The crystalline product is filtered off, and the mother-liquors are concentrated to give additional material. The product is recrystallized from water (decolorizing carbon). Each crop of crystals is checked for purity by paper chromatography on Whatman No. 1 paper, descending, R_f 0.2 (butyl alcohol–water), 0.015 (ethyl acetate), and recrystallized again if necessary; yield of purified product, 11.0 g. (78%). It does not melt; λ_{max} 270 nm. (ϵ 5,380) at pH 6.85, and 280 nm. (ϵ 9,670) at pH 1.0.

REFERENCES

(1) T. L. V. Ulbricht, *Progr. Nucleic Acid Res.*, **4**, 189 (1965).
(2) C. S. Miller, *J. Am. Chem. Soc.*, **77**, 752 (1955).
(3) A. Dornow and G. Petsch, *Ann.*, **588**, 45 (1954).
(4) T. L. V. Ulbricht and C. C. Price, *J. Org. Chem.*, **21**, 567 (1956).

[24] Hydrouracils

Ring Closure of α,β-Unsaturated Acids with Urea

KWANG-YUEN ZEE-CHENG, ROLAND K. ROBINS, and C. C. CHENG

MIDWEST RESEARCH INSTITUTE, KANSAS CITY, MISSOURI 64110

$(CH_2OH)_2$, Δ

1, R = H (72.1) + **3** (60.1) → **4**, R = H (114.1)

2, $R = CH_3$ (86.1) → **5**, $R = CH_3$ (128.1)

INTRODUCTION

5,6-Dihydropyrimidines are important intermediates in the biosynthesis and degradation of pyrimidines.[1] Existing methods for the preparation of 5,6-dihydropyrimidines include the catalytic hydrogenation of pyrimidines,[2] and the condensation of α,β-unsaturated carbonyl compounds with compounds of the amidine type.[3,4] The first method is limited, in that the yields are usually low, the reactions are difficult to control, and, quite often, the structures of the products are uncertain. Usually, the second method has been conducted by fusion and, as a result, the isolation procedures have been quite involved, the products rather impure, and the yields often low. The present procedure[5] is essentially the second method, modified by use of an inert solvent as the reaction medium.

PROCEDURE

Hydrouracil (4)

A mixture of 504 g. (7 moles) of acrylic acid (**1**), 841 g. (14 moles) of urea (**3**), 6 g. of hydroquinone, and 200 ml. of ethylene glycol in an open beaker is carefully heated to 130°, with vigorous stirring. Heating is discontinued, and the temperature of the stirred mixture spontaneously rises to 180°. After the exothermic reaction subsides, the mixture is heated at 200–210° for 1 hr., and cooled to 150°; it is then cautiously added to 1.5 l.

of water, and the solution is boiled, decolorized with activated carbon, and filtered. On cooling the filtrate, 192 g. (24%) of colorless crystals of **4**, m.p. 268–270°, are obtained. Recrystallization from ethanol raises the m.p. to 276–278° (lit.,[6] m.p. 275°).

6-Methylhydrouracil (5)

A mixture of 86.1 g. (1 mole) of crotonic acid (**2**) and 150 g. (2.5 moles) of urea (**3**) in 300 ml. of ethylene glycol is heated slowly, with stirring, to 190°. The temperature is maintained at 185–195° for 1 hr., and, on cooling, colorless crystals slowly separate from the viscous solution. The mixture is refrigerated overnight, and the product is filtered off, washed with cold ethanol and a small volume of ice–water, and dried at 120° to give 58 g. (45%) of **5**, m.p. 216–218°. Recrystallization from ethanol raises its m.p. to 217–218°. The product is identical with a sample prepared according to the procedure of Fischer and Roeder.[3]

REFERENCES

(1) I. Lieberman and A. Kornberg, *J. Biol. Chem.*, **207**, 911 (1954); **212**, 909 (1955); C. Cooper, R. Wu, and D. W. Wilson, *ibid.*, **216**, 37 (1955); R. A. Yates and A. B. Pardee, *ibid.*, **221**, 743 (1956); J. L. Fairley, R. L. Herrmann, and J. M. Boyd, *ibid.*, **234**, 3229 (1959); E. S. Canellakis, *ibid.*, **221**, 315 (1956); P. Fritzson and A. Pihl, *ibid.*, **226**, 229 (1957); L. L. Campbell, Jr., *ibid.*, **227**, 693 (1957); L. K. Mokrash and S. Grisolia, *Biochim. Biophys. Acta*, **27**, 226 (1958); **33**, 444 (1959); **34**, 165 (1959); **39**, 361 (1960); K. Fink, A. J. Merrit, and A. Ehrlich, *Arch. Biochem. Biophys.*, **197**, 441 (1952); L. L. Campbell, Jr., *J. Bacteriol.*, **73**, 225 (1957).

(2) P. A. Levene and F. B. LaForge, *Ber.*, **45**, 608 (1912); E. B. Brown and T. B. Johnson, *J. Am. Chem. Soc.*, **45**, 2702 (1923); **46**, 702 (1924); J. C. Ambelang and T. B. Johnson, *ibid.*, **61**, 74 (1939); **63**, 1934 (1941); J. J. Fox and D. V. Praag, *ibid.*, **82**, 486 (1960).

(3) E. Fischer and G. Roeder, *Ber.*, **34**, 3751 (1901).

(4) J. Evans and T. B. Johnson, *J. Am. Chem. Soc.*, **52**, 4993 (1930); T. B. Johnson and J. E. Livak, *ibid.*, **58**, 299 (1936); A. P. Phillips and J. Mentha, *ibid.*, **76**, 574 (1954); P. Biginelli, *Ber.*, **24**, 1317 (1891); W. Traube and R. Schwarz, *ibid.*, **32**, 3163 (1899); M. Bachstez and G. Cavallini, *ibid.*, **66**, 681 (1933); L. Birkofer and I. Storch, *ibid.*, **86**, 529 (1953).

(5) K.-Y. Zee-Cheng, R. K. Robins, and C. C. Cheng, *J. Org. Chem.*, **26**, 1877 (1961).

(6) H. Weidel and E. Roithner, *Monatsh.*, **17**, 174 (1896).

[25] Orotaldehyde
(6-Uracilcarboxaldehyde)

Oxidation of a Methyl Group with Selenium Dioxide

KWANG-YUEN ZEE-CHENG and C. C. CHENG

MIDWEST RESEARCH INSTITUTE, KANSAS CITY, MISSOURI 64110

H₃C–(6-methyluracil) $\xrightarrow{SeO_2}$ HC(=O)–(uracil)

1 (126.1) → **2** (140.1)

INTRODUCTION

Existing methods for the preparation of orotaldehyde (6-uracilcarboxaldehyde), a compound closely related to the biologically important orotic acid, involves the condensation of ethyl 4,4-diethoxyacetoacetate with thiourea, followed by hydrolysis of the resulting diethyl acetal with acid, and conversion of the thioxo group of 2-thioorotaldehyde to an oxo group with hydrogen peroxide in alkali.[1] The intermediate ethyl 4,4-diethoxyacetoacetate is prepared by the method of Johnson and Mikeska[2] from very pure[3] dichloroacetic acid *via* ethyl diethoxyacetate.[1,4] The present, simple method is not only more convenient, but may also be applied to the preparation of related classes of compounds.[5]

PROCEDURE

Orotaldehyde (2)

A mixture of 63 g. (0.5 mole) of 6-methyluracil[6] (**1**), 66.6 g. (0.6 mole) of selenium dioxide, and 1.5 l. of glacial acetic acid is refluxed, with mechanical stirring, for 6 hr. During this time, the white suspension of selenium dioxide is gradually replaced by gray, elemental selenium. The hot reaction mixture is filtered, and the selenium cake is extracted with two 250-ml. portions of boiling acetic acid. The yellow filtrate and extracts are combined and evaporated to dryness under diminished pressure, giving 60 g. of a yellow solid, which gives a positive test with (2,4-dinitrophenyl)-hydrazine. The crude orotaldehyde (**2**), which still contains some selenium and excess selenium dioxide, is purified as follows. The solid is dissolved n 600 ml. of warm water, and an aqueous solution of sodium hydrogen ulfite (30 g./60 ml. of water) is cautiously added, in small portions, to the tirred mixture; it is then boiled with activated charcoal and Celite for 0 min., and filtered. The pH of the filtrate is adjusted to 1 with concentrated hydrochloric acid, and the solution is cooled, giving 25 g. of pure rotaldehyde (**2**), m.p. 273–275° (dec.) (slow heating causes carbonization at 273–275° without melting). An additional 16 g. of product is btained by concentrating the mother liquor, bringing the total yield of **2** o 58%. An analytical sample is prepared by recrystallization from water, nd the off-white solid, m.p. 273–275° (dec.) [lit.[1b] m.p. 273–275° (dec.)], s dried *in vacuo* at 120°. $\lambda_{max}^{pH\,1,7}$ 261 nm. (ϵ 13,300); $\lambda_{max}^{pH\,11}$ 225 nm. ϵ 16,100) and 291 nm. (ϵ 9,200).

REFERENCES

1) (a) T. B. Johnson and L. H. Cretcher, Jr., *J. Am. Chem. Soc.*, **37**, 2144 (1915). (b) T. B. Johnson and E. F. Schroeder, *ibid.*, **53**, 1989 (1931). (c) E. Ludolphy, *Chem. Ber.*, **84**, 385 (1951). (d) M. Claesen and H. Vanderhaege, *Bull. Soc. Chim. Belges*, **66**, 292 (1957). (e) C. Piantadosi, V. G. Skulason, J. L. Irvin, J. M. Powell, and L. Hall, *J. Med. Chem.*, **7**, 337 (1964).

2) T. B. Johnson and L. A. Mikeska, *J. Am. Chem. Soc.*, **41**, 810 (1919).

3) Commercial dichloroacetic acid failed to give a satisfactory yield of ethyl diethoxyacetate. Pure dichloroacetic acid is prepared from chloral hydrate according to the procedure of G. W. Pucher, *J. Am. Chem. Soc.*, **42**, 2251 (1920). See the discussion in Ref. 1b.

4) T. B. Johnson and L. H. Cretcher, Jr., *J. Biol. Chem.*, **26**, 99 (1916).

5) K.-Y. Zee-Cheng and C. C. Cheng, *J. Heterocyclic Chem.*, **4**, 163 (1967).

6) J. J. Donleavy and M. A. Kise, *Org. Syn.*, *Coll. Vol. II*, 422 (1943).

[26] Orotic Acid

(6-Uracilcarboxylic acid)

Two Alternative Syntheses via *2-(Methylthio)orotic acid*

1. *Acid Hydrolysis of a Methylthio Group*

G. DOYLE DAVES, JR., FRED BAIOCCHI, ROLAND K. ROBINS, and C. C. CHENG

MIDWEST RESEARCH INSTITUTE, KANSAS CITY, MISSOURI 64110

EtOOC–CH=C(ONa)–COEt(=O) + H_2N–C(=NH)–SCH_3 · 0.5 H_2SO_4 $\xrightarrow[NaOH]{H_2O}$

1 (210.2) **2** (139.2)

3 (186.2) $\xrightarrow{HCl}$ **4** (156.1)

2. *Oxidative Hydrolysis of a Methylthio Group*

WAYNE H. NYBERG and C. C. CHENG

MIDWEST RESEARCH INSTITUTE, KANSAS CITY, MISSOURI 64110

3 (186.2) $\xrightarrow[H_2SO_4]{HNO_3}$ **4** (156.1)

INTRODUCTION

Orotic acid (**4**) occupies a significant and unique position in the biosynthesis of pyrimidines.[1] Previous synthetic routes include (*1*) the oxidation of 6-methyluracil,[2] orotaldehyde,[3] or 6-uracilacetic acid,[4] (*2*) the ring expansion of a hydantoin intermediate,[5] (*3*) dethiation of 2-thioorotic acid,[6] and (*4*) a procedure involving carbonation of 2,4-diethoxy-6-pyrimidinyllithium.[7] The two synthetic methods described here serve as convenient and improved preparations for this biologically important compound.

PROCEDURE

2-(Methylthio)orotic acid (3)

To 6 l. of water containing 400 g. of sodium hydroxide are added 1,050 g. (5 moles) of the sodium salt of diethyl oxalacetate (**1**) and 695 g. (5 moles) of *S*-methylisothiouronium sulfate (**2**). The resulting solution is stirred at room temperature for 8 hr., boiled briefly with charcoal, and then filtered. The hot filtrate, while being stirred vigorously, is acidified to pH 1 with hydrochloric acid. After 30 min., the precipitated product is filtered off, washed with water, and suspended in a large volume of acetone. The acetone mixture is stirred for 1 hr., and the product is filtered off, and dried. The yield of **3** is 563 g. (60%), m.p. 252–254°; this material is sufficiently pure for synthetic purposes. Recrystallized from water, compound **3** melts at 253–254° (lit.[8] m.p. 250–252°), $\lambda_{max}^{pH\,1}$ 240 nm. (ϵ 8,700) and 310 nm. (ϵ 5,100); $\lambda_{max}^{pH\,11}$ 250 nm. (ϵ 3,900).

Orotic Acid (4)

A. By Acid Hydrolysis

Compound **3** (600 g., 3.22 moles) in 9 l. of 2 *N* hydrochloric acid is refluxed, with stirring, for 3 hr. The solid product is then filtered off, and suspended in 6 l. of hot water. Solid potassium hydroxide is carefully added until complete dissolution is attained, charcoal is added, and the suspension is boiled for 15 min. and then filtered. The pH of the hot filtrate is adjusted to 1 with hydrochloric acid, and the precipitated product is filtered off, well washed with water and acetone, and dried; yield of orotic acid (**4**) monohydrate, 447 g. (79%), m.p. 339–341°.

B. By Oxidative Hydrolysis

To 40 ml. of concentrated sulfuric acid is carefully added 10 g. (0.053 mole) of **3** ($<40°$). The resulting solution is cooled to 15°, and 8 ml. of white, fuming nitric acid is added dropwise, while maintaining the temperature below 20°. After the addition is complete, the solution is stirred in an ice bath for 1 hr., and then poured onto ice. The solid product is filtered off, and washed with water and acetone, to yield 8 g. (87%) of **4** as the monohydrate. The product is identical with that prepared by the first method.

REFERENCES

(1) G. D. Daves, Jr., F. Baiocchi, R. K. Robins, and C. C. Cheng, *J. Org. Chem.*, **26**, 2755 (1961), and references cited therein.
(2) R. Behrend and K. Strave, *Ann.*, **378**, 153 (1911).
(3) T. B. Johnson and E. F. Schroeder, *J. Am. Chem. Soc.*, **53**, 1989 (1931); **54**, 2941 (1932); C. Heidelberger and R. B. Hurlbert, *ibid.*, **72**, 4704 (1950); F. Korte, W. Paulus, and K. Störiko, *Ann.*, **619**, 63 (1958).
(4) G. E. Hilbert, *J. Am. Chem. Soc.*, **54**, 2076 (1932).
(5) H. L. Wheeler, *Am. Chem. J.*, **38**, 358 (1907); M. Bachstez, *Ber.*, **63**, 1000 (1930); *Brit. Pat.* 800,709 (1958); *Chem. Abstracts*, **53**, 16172 (1959); J. F. Nyc and H. K. Mitchell, *J. Am. Chem. Soc.*, **69**, 1382 (1947); S. Takeyama, T. Noguchi, Y. Miura, and M. Ishidate, *Chem. Pharm. Bull.* (Tokyo), **4**, 492 (1956).
(6) R. Kitamura, *Yakugaku Zasshi*, **57**, 209 (1937).
(7) B. W. Langley, *J. Am. Chem. Soc.*, **78**, 2136 (1956).
(8) H. Vanderhaeghe, *Bull. Soc. Chim. Belges*, **62**, 611 (1953).

[27] 5-Amino-4,6-dichloropyrimidine

The Preparation of an Intermediate in the Synthesis of Purines

JOHN A. MONTGOMERY, WILLIAM E. FITZGIBBON, JR., VLADIMIR MINIC, and CHARLES A. KRAUTH[a]

KETTERING-MEYER LABORATORY, SOUTHERN RESEARCH INSTITUTE, BIRMINGHAM, ALABAMA 35205

HO, N, N, HO — **1** (112.1) $\xrightarrow[CH_3CO_2H]{HNO_3}$ HO, N, N, O_2N, HO — **2** (157.1) $\xrightarrow{POCl_3}$ Cl, N, N, O_2N, Cl — **3** (194.0) $\xrightarrow[Ni]{H_2}$ Cl, N, N, H_2N, Cl — **4** (164.0)

INTRODUCTION

5-Amino-4,6-dichloropyrimidine[1] (**4**) is used extensively as a key intermediate in the synthesis of 9-substituted purines (Sec. I [2], [8], and [17]). The procedures described here may readily be used for quantities much larger than those indicated.

PROCEDURE

5-Nitro-4,6-pyrimidinediol[2] (2)

A solution of 350 ml. of red, fuming nitric acid (sp. gr. 1.59–1.60) in 1 l. of glacial acetic acid is cooled in an ice bath to below 20°. Then, with constant stirring, 250 g. (2.23 moles) of 4,6-pyrimidinediol[3] (**1**) is added in small portions at such a rate that the temperature does not exceed 20°. When the addition is complete, the ice bath is removed, stirring of the reaction mixture is continued for 1 hr., and the mixture is poured onto 3 kg. of crushed ice, with efficient stirring. The precipitated product is collected by filtration, washed successively with 2 l. of water and 1 l. of acetone, and

[a] Present Address: Taejon Presbyterian College, Department of Chemistry, 133 Ojung-Dong, Taejon, Korea.

dried at 60° under vacuum to constant weight; the product (**2**) weighs 280 g. (80%), m.p. >300°. It is used for the next step without further purification.

4,6-Dichloro-5-nitropyrimidine[2] (3)

In a 5-l., round-bottomed flask (fitted with a mechanical stirrer, an addition funnel, and a reflux condenser bearing a drying tube) is placed a solution of 280 g. (1.98 moles) of 5-nitro-4,6-pyrimidinediol (**2**) in 1.57 l. of freshly distilled phosphoryl chloride, and the solution is stirred and heated to gentle reflux. Then, with continued stirring and heating, a solution of 396 ml. (2.44 moles) of *N,N*-diethylaniline in 670 ml. of freshly distilled phosphoryl chloride is slowly added during 1.5 to 2 hr., stirring and refluxing are continued for a further 5 hr., and the mixture is allowed to cool overnight. The excess phosphoryl chloride (about 800 ml.) is removed at 65°/20 mm. Hg. The residual material, which should not be too viscous to be poured readily, is poured cautiously over 10 l. of crushed ice with very efficient stirring. The resulting mixture is immediately extracted with 9 lbs. of ether and then with three successive 5-lb. portions of ether. The ether extracts are combined, washed with two 2-l. portions of water, dried with anhydrous magnesium sulfate, filtered, and evaporated to dryness under diminished pressure. Recrystallization of the crude residue from petroleum ether (b.p. 85–105°) gives a light-tan solid (**3**); yield 315 g. (91%), m.p. 101°.

5-Amino-4,6-dichloropyrimidine[1] (4)

A portion (100 g., wet) of Raney nickel catalyst is washed four times with dry methanol by decantation. The catalyst and a solution of 315 g. (1.62 moles) of 4,6-dichloro-5-nitropyrimidine (**3**) in 4 l. of dry methanol are placed in a 12-l., round-bottomed flask equipped with a hydrogen inlet-tube and an efficient agitator powered by a totally enclosed motor (*e.g.*, a Vibro-mix). The hydrogenation is conducted at room temperature and atmospheric pressure, with introduction of hydrogen (from a large, laboratory gas-holder) over the surface of the agitated mixture. After about 5 hr., approximately 90% of the calculated amount of hydrogen has been taken up and there is no further uptake. The mixture is filtered, and the catalyst is washed with three 300-ml. portions of anhydrous methanol. The filtrate and washings are combined, and evaporated to dryness under diminished pressure, and the resulting solid is crystallized from petroleum ether (b.p. 85–105°). The colorless needles obtained weigh 221 g. (83%), m.p. 145–146°; $\lambda_{max}^{H_2O}$ 247 nm. (log ϵ 3.98), 308 nm. (log ϵ 3.79).

REFERENCES

(1) R. K. Robins, K. J. Dille, and B. E. Christensen, *J. Org. Chem.*, **19**, 930 (1954).
(2) W. R. Boone, W. G. M. Jones, and G. R. Ramage, *J. Chem. Soc.*, **1951**, 99.
(3) G. W. Kenner, B. Lythgoe, A. R. Todd, and A. Topham, *J. Chem. Soc.*, **1943**, 388. A mesomeric betaine structure has been proposed for this compound.[4]
(4) A. R. Katritsky, F. D. Popp, and A. J. Waring, *J. Chem. Soc.*, **1966**, 565.

[28] 2,4-Diamino-5-nitropyrimidine and 2,4-Diamino-6-chloro-5-nitropyrimidine

Direct Nitration of 2,4-Diaminopyrimidines

DARRELL E. O'BRIEN, C. C. CHENG, and W. PFLEIDERER

MIDWEST RESEARCH INSTITUTE, KANSAS CITY, MISSOURI 64110

$\xrightarrow[\text{excess } H_2SO_4]{HNO_3}$

1, X = H·H_2SO_4 (208.2) → **2**, X = H (155.1)

1a, X = Cl (144.6) → **2a**, X = Cl (189.6)

INTRODUCTION

Failure to accomplish direct nitration of 2,4-diaminopyrimidine (**1**) has been recorded.[1] Nitration of 2,4-diamino-6-chloropyrimidine under conventional conditions results in the formation of 2-amino-6-chloro-4-nitraminopyrimidine.[2] The conditions of nitration described here, in which a large excess of concentrated sulfuric acid, with respect to nitric acid is employed,[3] not only give the desired nitration products in satisfactory yields, but may also be used for the nitration of similar compounds that are not otherwise amenable to nitration.

PROCEDURE

2,4-Diamino-5-nitropyrimidine (2)

To a solution of 5 g. (0.024 mole)[4] of **1** (hydrogen sulfate) in 25 ml. of concentrated sulfuric acid is added dropwise, with stirring, 5 ml. of fuming nitric acid (s.g. 1.50), the temperature being maintained at 30–35°. After the addition is complete, the solution is stirred at 30–35° for 30 min., and then slowly added to 300 g. of flaked ice with vigorous stirring. The pH of the resulting mixture is adjusted to 8–9 by dropwise addition of concentrated ammonium hydroxide, the temperature being kept below 15°. The light-yellow precipitate is filtered off, washed successively with water and ethanol, and dried. The product, 0.8 g. (22%), m.p. 245–250° (dec.), is identical with a sample prepared from 2,4-dichloro-5-nitropyrimidine and ammonia.[5]

2,4-Diamino-6-chloro-5-nitropyrimidine (2a)

To a solution of 30 ml. of fuming nitric acid (s.g. 1.50) in 150 ml. of concentrated sulfuric acid is added, at 30–35° (with external cooling), 28.9 g. (0.2 mole) of compound[6] (**1a**). The resulting solution is stirred at 30–35° for 30 min., and then slowly poured, with vigorous stirring, over 1 kg. of flaked ice. The pH of the mixture is carefully adjusted to 8–9 by dropwise addition of concentrated ammonium hydroxide, while the temperature is kept below 10°. It is then filtered, and the yellow, solid product is washed with four 150-ml. portions of water and dried *in vacuo* over calcium chloride. The dry product is recrystallized from absolute ethanol to give 23.5 g. (62%) of **2a** as yellow crystals, melting at 220–222° (melting-point apparatus preheated to 215°), followed by resolidification upon rapid heating. No definite melting point is observed upon slower heating. $\lambda_{max}^{pH\,1}$ 216 nm. (ϵ 7,000) and 338 nm. (ϵ 13,400); $\lambda_{max}^{pH\,11}$ 235 nm. (ϵ 14,600), 267 nm. (ϵ 5,100), and 340 nm. (ϵ 22,300).

REFERENCES

(1) D. J. Brown, *The Pyrimidines*, A. Weissberger, Ed., Interscience Publishers, New York, N. Y., 1962, pp. 10 and 140.

(2) D. E. O'Brien, F. Baiocchi, R. K. Robins, and C. C. Cheng, *J. Med. Chem.*, **6**, 467 (1963).

(3) D. E. O'Brien, C. C. Cheng, and W. Pfleiderer, *J. Med. Chem.*, **9**, 573 (1966).

(4) (a) E. Büttner, *Ber.*, **36**, 2227 (1903); (b) T. B. Johnson and C. O. Johns, *Am. Chem. J.*, **34**, 175 (1905).

(5) (a) O. Isay, *Ber.*, **39**, 250 (1906); (b) A. Albert, D. J. Brown, and G. Cheeseman, *J. Chem. Soc.*, **1951**, 474; (c) E. C. Taylor and M. J. Thompson, *J. Org. Chem.*, **26**, 5224 (1961).

(6) S. Gabriel, *Ber.*, **34**, 3362 (1901).

[29] 2-Substituted 4-Pyrimidinecarboxylic Acids

Oxidation of a Methyl Group with Nitrous Acid

G. DOYLE DAVES, JR., DARRELL E. O'BRIEN, LELAND R. LEWIS, and C. C. CHENG

MIDWEST RESEARCH INSTITUTE, KANSAS CITY, MISSOURI 64110

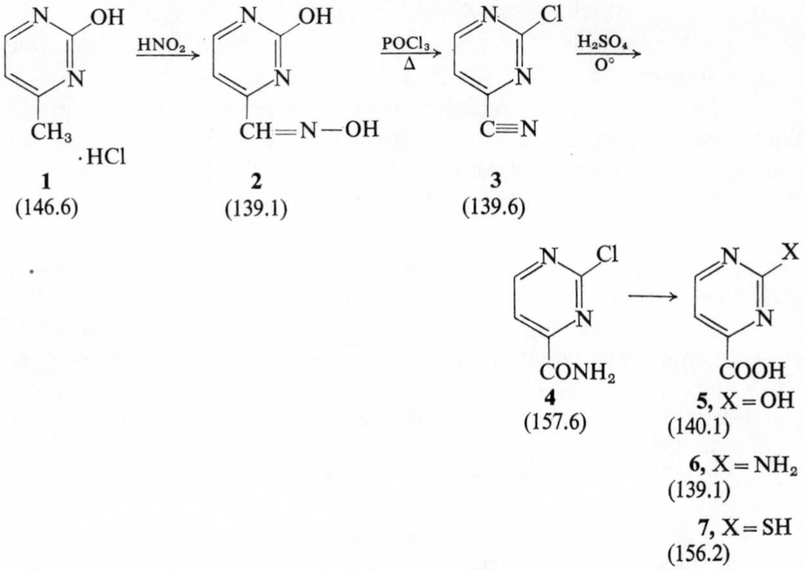

CH_3 ·HCl
1
(146.6)

$\xrightarrow{HNO_2}$

CH=N—OH
2
(139.1)

$\xrightarrow[\Delta]{POCl_3}$

C≡N
3
(139.6)

$\xrightarrow[0°]{H_2SO_4}$

$CONH_2$
4
(157.6)

⟶

COOH
5, X = OH
(140.1)
6, X = NH_2
(139.1)
7, X = SH
(156.2)

INTRODUCTION

A procedure frequently employed for the preparation of pyrimidinecarboxylic acids is the direct oxidation of alkyl-substituted pyrimidines with potassium permanganate.[1,2] Because permanganate oxidation often gives variable results and low yields, the present method[3] offers an alternative approach from the same starting material (the methylpyrimidines). In addition, other functional groups may be readily introduced by treating the intermediate chloropyrimidines with appropriate nucleophilic agents.

PROCEDURE

2-Hydroxy-4-pyrimidinecarboxaldehyde Oxime (2)

To a solution of 14.7 g. (0.1 mole) of 4-methyl-2-pyrimidinol hydrochloride[4] (**1**) in 100 ml. of 50% acetic acid cooled at 15° is rapidly added, with vigorous stirring, 10.4 g. (0.15 mole) of sodium nitrite. An exothermic reaction occurs (the temperature rises to *ca.* 40°) and a pale-yellow solid is precipitated. The reaction mixture is stirred at room temperature for 3 hr., and the solid product is collected by filtration, washed with ice–water, and dried at 80°. Recrystallization from water gives 11.0 g. (79%) of analytically pure product (**2**), which melts with rapid decomposition at 226°; $\lambda_{max}^{pH\,1}$ 220 nm. (ϵ 9,900), 251 nm. (ϵ 8,500), 330 nm. (ϵ 5,800); $\lambda_{max}^{pH\,11}$ 239 nm. (ϵ 7,400), 274 nm. (ϵ 11,000), 324 nm. (ϵ 8,600).

2-Chloro-4-pyrimidinecarbonitrile (3)

A mixture of 50 g. (0.36 mole) of **2** with 200 ml. of cold phosphoryl chloride is *carefully* warmed until a vigorous reaction begins. Heating is immediately discontinued and, shortly, complete dissolution occurs. To the solution is added 25 ml. of *N*,*N*-dimethylaniline, and the resulting dark solution is refluxed for 30 min., cooled, and slowly added to 1.5 kg. of crushed ice. The aqueous solution is extracted with ether (4 × 300 ml.) and the combined extracts are successively washed with water (2 × 150 ml.), saturated aqueous sodium hydrogen carbonate (2 × 100 ml.), and water (2 × 100 ml.), and dried with sodium sulfate. Evaporation of the ethereal solution gives 36 g. (77%) of crude product. Recrystallization from ethyl acetate–heptane affords 30 g. (60%) of analytically pure **3**, m.p. 50–51°, λ_{max}^{EtOH} 269 nm. (ϵ 4,200). The product fails to show a characteristic nitrile absorption band[5] at 2,250 $cm.^{-1}$.

2-Chloro-4-pyrimidinecarboxamide (4)

To 50 ml. of concentrated sulfuric acid cooled to 15° is slowly added 14 g. (0.1 mole) of **3**. After addition is complete, the mixture is stirred at room temperature for 4 hr. The resulting solution is then added, with vigorous stirring, to 300 g. of crushed ice, and the white precipitate is filtered off and well washed with ice–water. Recrystallization of the product from water gives 7.5 g. (47%) of white needles, m.p. 151–153°, $\lambda_{max}^{pH\,1}$ 267 nm. (ϵ 4,300); $\lambda_{max}^{pH\,11}$ 268 nm. (ϵ 4,300).

2-Hydroxy-4-pyrimidinecarboxylic Acid (5)

A mixture of 15.8 g. (0.1 mole) of **4** and 500 ml. of 3% aqueous sodium hydroxide is refluxed for 3 hr. The clear solution is treated with charcoal and filtered, and the pH of the filtrate is adjusted to 1 with dilute hydrochloric acid. When the suspension is cool, the precipitated product is separated by filtration; recrystallization of the crude product from dilute, aqueous ethanol gives 7.3 g. (52%) of analytically pure **5**, m.p. 225° (dec.), $\lambda_{max}^{pH\,1}$ 324 nm. (ϵ 5,750); $\lambda_{max}^{pH\,11}$ 306 nm. (ϵ 4,300).

2-Amino-4-pyrimidinecarboxylic Acid (6)

A mixture of 15.8 g. (0.1 mole) of **4** and 200 ml. of concentrated ammonium hydroxide is heated in an autoclave at 130° for 5 hr. The cooled reaction mixture is evaporated to dryness, and the residue is dissolved in 100 ml. of 10% aqueous sodium hydroxide. The solution is then refluxed for 1 hr., decolorized with charcoal, and filtered. The pH of the filtrate is adjusted to 4 with dilute hydrochloric acid, and the acid solution is cooled, giving 9.8 g. (70%) of analytically pure **6**, m.p. 275° (subl.); $\lambda_{max}^{pH\,1}$ 320 nm. (ϵ 4,200); $\lambda_{max}^{pH\,11}$ 306 nm. (ϵ 3,600). The product is identical with that prepared by the oxidation method.[2]

2-Thio-4-pyrimidinecarboxylic Acid (7)

A mixture of 7.9 g. (5 mmoles) of **4**, 7.9 g. (0.1 mole) of thiourea, and 250 ml. of absolute ethanol is refluxed for 3 hr. The resulting solution is evaporated to dryness, and the solid residue is dissolved in 100 ml. of 10% aqueous sodium hydroxide. The solution is boiled for 1 hr., treated with charcoal, and filtered, and the pH of the filtrate is adjusted to 1 with dilute hydrochloric acid. On cooling the solution, an orange solid separates; this is filtered off, and recrystallized from water to give 2.9 g. (37%) of **7**, m.p. 201–202°; $\lambda_{max}^{pH\,1}$ 284 nm. (ϵ 17,000); $\lambda_{max}^{pH\,11}$ 230 nm. (ϵ 5,800), and 272 nm. (ϵ 15,800).

REFERENCES

(1) *Cf.* D. J. Brown, *The Pyrimidines*, A. Weissberger, Ed., Interscience Publishers, New York, N. Y., 1962.
(2) T. Matsukawa and K. Shirakawa, *Yakugaku Zasshi*, **72**, 909 (1952).
(3) G. D. Daves, Jr., D. E. O'Brien, L. R. Lewis, and C. C. Cheng, *J. Heterocyclic Chem.*, **1**, 130 (1964).
(4) D. M. Burness, *J. Org. Chem.*, **21**, 97 (1956).
(5) The fact that 2-chloro-4-pyrimidinecarbonitrile (**3**) fails to show the characteristic nitrile absorption at 2,250 cm.$^{-1}$ is not without precedent. P. Sensi and G. G. Gallo [*Gazz. Chim. Ital.*, **85**, 224 (1955)] have reported that the presence of electron-withdrawing groups *ortho* or *para* to an aromatic nitrile group tends to diminish the intensity of the nitrile bands.

[30] The Diethyl and Dimethyl Ethers of 2,4-Pyrimidinediol and of 5-Methyl-2,4-pyrimidinediol

Pyrimidine Intermediates in the Hilbert–Johnson Synthesis

CLARITA C. BHAT and H. RANDALL MUNSON

DEPARTMENT OF CHEMISTRY, GEORGETOWN UNIVERSITY,
WASHINGTON, D.C. 20007

$\xrightarrow{POCl_3}$ $\xrightarrow{NaOR'}$

1, R = H (112.1)

2, R = CH_3 (126.1)

3, R = H (149.1)

4, R = CH_3 (163.1)

5, R = H, R′ = CH_3 (140.1)

5a, R = H, R′ = C_2H_5 (168.2)

6, R = CH_3, R′ = CH_3 (154.2)

6a, R = CH_3, R′ = C_2H_5 (182.2)

INTRODUCTION

2,4-Dimethoxypyrimidine[1] (**5**), 2,4-diethoxypyrimidine[1] (**5a**), 2,4-di-ıethoxy-5-methylpyrimidine[2] (**6**), and 2,4-diethoxy-5-methylpyrimidine[2] **6a**) are widely used as key intermediates in the Hilbert–Johnson proce-ure[3] for the synthesis of pyrimidine nucleosides. Through the use of the ubject derivatives, nucleosides of uracil (**1**), thymine (**2**), cytosine, and -methylcytosine may be prepared. Substantial improvements in the ields have been made through certain modifications in the original reparations[1,2] of **5**, **5a**, **6**, and **6a**.

PROCEDURE

,4-Dichloropyrimidine[4] (3)

A suspension of 40 g. (0.36 mole) of uracil (**1**) in 160 ml. of phosphoryl hloride, in a 250-ml., round-bottomed flask fitted with a reflux condenser drying tube), is heated in an oil bath at 100–110° for 2.5 hr. The brown olution is cooled to room temperature, and the excess phosphoryl hloride is removed at 30°/20 mm. Hg by use of a Buchler Model PF-9G rotary evaporator.[a] The viscous mixture is very slowly poured, wit stirring, onto 350 g. of crushed ice.[b] During the addition, 50 ml. of ethe is added to facilitate decomposition of the complex. After removal of th ether layer, the aqueous solution is extracted with five 100-ml. portions c ether. The combined ether extracts are washed with two 125-ml. portion of saturated aqueous sodium carbonate, and dried with calcium chlorid The ether is removed at 30° under diminished pressure, and the crud crystalline residue is recrystallized from pentane, giving 42.7 g. (80%) c 2,4-dichloropyrimidine (**3**), m.p. 61–61.5°.

2,4-Dichloro-5-methylpyrimidine[2] (4)

Details for the reaction of 20 g. (0.16 mole) of thymine (**2**) and 80 ml. c phosphoryl chloride, and subsequent processing of the reaction product are the same as given in the preceding preparation of **3**. After evaporatio of the dried ether extract, the residue is purified by distillation at 108 109°/11 mm. Hg; yield 21.5 g. (83%), m.p. 25–26°.

2,4-Dimethoxypyrimidine (5)

A solution of 50 g. (0.34 mole) of 2,4-dichloropyrimidine (**3**) in 250 m of dry methanol is added slowly, with stirring, to a solution of 16 g. c sodium in 300 ml. of dry methanol. The mixture is refluxed for 10 min cooled, and filtered by suction. The separated sodium chloride is rinse with ether, and the combined filtrate and washings are evaporated und diminished pressure at 35°. The oily residue is dissolved in 200 ml. c ether, and the solution is washed with two 100-ml. portions of 30% aqueo sodium hydroxide, and then with 100 ml. of water. The ether layer dried with sodium sulfate, filtered, and evaporated under diminishe pressure at 30°. The residue is purified by distillation at 37°/0.1 mm. Hg yield 44 g. (93%), b.p. 202°.

2,4-Diethoxypyrimidine (5a)

A solution of 43 g. (0.29 mole) of **3** in 200 ml. of absolute ethanol added dropwise, with stirring, to a solution of 13 g. of sodium in 200 m of absolute ethanol. The solution is refluxed for 10 min. and filtered, an the separated sodium chloride is washed with ether. The combined filtra and washings are evaporated under diminished pressure, and the residue redissolved in 200 ml. of ether. The ether solution is washed with tw 100-ml. portions of 30% aqueous sodium hydroxide, and then with 50 m of water. After being dried with sodium sulfate, the solution is evaporate

[a] The phosphoryl chloride collected in the distillate amounts to 64 ml. and can used in subsequent preparations.

[b] At no time should the temperature of the ice mixture exceed 20°.

under diminished pressure, leaving an oily residue which is distilled at 55°/0.3 mm. Hg; yield 45 g. (93%), m.p. 19–20°.

2,4-Dimethoxy-5-methylpyrimidine (6)

A solution of 45 g. (0.28 mole) of 2,4-dichloro-5-methylpyrimidine (**4**) in 200 ml. of dry methanol is treated with sodium methoxide (prepared from 13.8 g. of sodium and 250 ml. of dry methanol) according to the directions given for the preparation of **5**. The reaction products are processed in a similar manner, giving a solid residue which, after recrystallization from pentane, yields 38 g. (88%) of **6**, m.p. 60–61°.

2,4-Diethoxy-5-methylpyrimidine (6a)

A solution of 20 g. (0.12 mole) of 2,4-dichloro-5-methylpyrimidine (**4**) in 100 ml. of absolute ethanol is treated with sodium ethoxide (5.7 g. of sodium in 100 ml. of dry ethanol) according to the directions given for the preparation of **5a**. Evaporation of the dried ether extract gives a crude, crystalline residue, which is recrystallized from pentane; yield 19.6 g. (89%), m.p. 35.5–36.5°.

REFERENCES

(1) G. E. Hilbert and T. B. Johnson, *J. Am. Chem. Soc.*, **52**, 2001 (1930).
(2) W. Schmidt-Nickels and T. B. Johnson, *J. Am. Chem. Soc.*, **52**, 4511 (1930).
(3) G. E. Hilbert and T. B. Johnson, *J. Am. Chem. Soc.*, **52**, 4489 (1930).
(4) G. E. Hilbert and T. B. Johnson, *J. Am. Chem. Soc.*, **52**, 1152 (1930).

[31] 4-Amino-2-methoxy-5-pyrimidinemethanol (Bacimethrin)

Treatment of Ethyl 4-Amino-2-(ethylsulfonyl)-5-pyrimidine-carboxylate with Sodium Methoxide, Followed by Reduction

HENRY C. KOPPEL, ROBERT H. SPRINGER,
ROLAND K. ROBINS, and C. C. CHENG

MIDWEST RESEARCH INSTITUTE, KANSAS CITY, MISSOURI 64110

1 (259.3) $\xrightarrow{NaOMe}$ 2 (197.2) $\xrightarrow{LiAlH_4}$ 3 (155.2)

INTRODUCTION

Bacimethrin (**3**), an interesting antibiotic substance that has activity against the growth of various yeasts and bacteria,[1] possesses a structure strikingly similar to those of the biologically active 5-(hydroxymethyl)-cytosine (HMC),[2] toxopyrimidine (4-amino-2-methyl-5-pyrimidinemeth-anol),[3] methioprim [4-amino-2-(methylthio)-5-pyrimidinemethanol],[3d,e] and Bayer DG-428 [2-(*o*-chlorobenzylthio)-4-(dimethylamino)-5-methyl-pyrimidine].[4] Two synthetic procedures for the preparation of the title compound have been devised in our Laboratory.[5] The following method generally gives the higher yield.

PROCEDURE

Ethyl 4-Amino-2-methoxy-5-pyrimidinecarboxylate (2)

To a stirred suspension of 28 g. (0.108 mole) of powdered ethyl 4-amino-2-(ethylsulfonyl)-5-pyrimidinecarboxylate[6] (**1**) in 200 ml. of anhydrous methanol cooled to 0° is added, dropwise, a solution of sodium methoxide prepared by adding 3.7 g. (0.16 g.-atom) of sodium to 50 ml. of methanol. The temperature during the addition is kept below 10°, and the mixture is then stirred for 1 hr. The resulting, colorless solid is filtered off, washed with a small volume of cold benzene, and recrystallized from benzene to yield 16 g. (76%) of **2** as long, colorless needles, m.p. 151–153°; $\lambda_{max}^{pH\,1}$ 239 nm. (ϵ 12,200) and 273 nm. (ϵ 6,500); $\lambda_{max}^{pH\,11}$ 241 nm. (ϵ 11,400) and 286 nm. (ϵ 8,300).

Bacimethrin (3)

To a stirred mixture of 36.9 g. (1 mole) of lithium aluminum hydride in 1.5 l. of anhydrous tetrahydrofuran is added dropwise a solution of 100 g. (0.49 mole) of[a] **2** in 500 ml. of anhydrous tetrahydrofuran,[b] and the mixture is refluxed, with stirring, for 4 hr. To the reaction mixture is carefully added 300 ml. of water; the resulting slurry is centrifuged, and separated from the liquid, and extracted with three 500-ml. portions of hot tetrahydrofuran. The extracts are combined with the original solution, dried with sodium sulfate, and filtered, and the filtrate is evaporated to dryness. The residue, after recrystallization from ethanol–ether, followed by recrystallization from ethyl acetate, yields 52 g. (69%) of colorless prisms, m.p. 173–174°, $\lambda_{max}^{H_2O}$ 227 nm. (ϵ 7,600) and 271 nm. (ϵ 7,300); $\lambda_{max}^{0.1\,N\,HCl}$ 229 nm. (ϵ 8,400) and 261 nm. (ϵ 9,500); $\lambda_{max}^{0.1\,N\,NaOH}$ 231 nm. (ϵ 6,200) and 271 nm. (ϵ 7,600).

REFERENCES

(1) F. Tanaka, S. Takeuchi, N. Tanaka, H. Yonehara, H. Umezama, and Y. Sumiki, *J. Antibiotics, Ser. A.*, **14**, 161 (1961).

(2) (a) G. R. Wyatt and S. S. Cohen, *Nature*, **170**, 1072 (1952); (b) G. R. Wyatt and S. S. Cohen, *Biochem. J.*, **55**, 774 (1953); (c) A. Dornow and G. Petsch, *Ann.*, **588**, 45 (1954); (d) C. S. Miller, *J. Am. Chem. Soc.*, **77**, 752 (1955).

(3) (a) R. Abderhalden, *Arch. Ges. Physiol.*, **240**, 647 (1938); **242**, 199 (1939); (b) A. Watanabe, *Yakugaku Zasshi*, **59**, 133 (1939); M. Kawashima, *ibid.*, **77**, 758 (1957); (c) K. Makino, T. Kinoshita, T. Sasaki, and T. Shioi, *Nature*, **173**, 34 (1954); **174**, 275, 1056 (1954); (d) T. L. V. Ulbricht and C. C. Price, *J. Org. Chem.*, **21**, 567 (1956); T. Okuda and C. C. Price, *ibid.*, **23**, 1738 (1958), and references cited therein; (e) R. Guthrie, M. E. Loebeck, and M. J. Hillman, *Proc. Soc. Exptl. Biol. Med.*, **94**, 792 (1957); (f) A. Schellenberger and K. Winter, *Z. Physiol. Chem.*, **322**, 173 (1960).

(4) K. Westphal and R. Bierling, *Naturwissenschaften*, **46**, 230 (1959).

(5) H. C. Koppel, R. H. Springer, R. K. Robins, and C. C. Cheng, *J. Org. Chem.*, **27**, 3614 (1962).

(6) J. M. Sprague and T. B. Johnson, *J. Am. Chem. Soc.*, **57**, 2252 (1935).

[a] Use of the 2-(methylsulfonyl) homolog instead of **1** gives, after the replacement reaction, a very low yield of **3**.

[b] This compound may also be prepared, in 60% yield, by using anhydrous ether as the solvent and introducing the (ethoxycarbonyl)pyrimidine by continuous extraction in a Soxhlet apparatus.

[32] 4-Amino-2-(methylthio)-5-pyrimidinemethanol (Methioprim)

A Biologically Active[1] *Intermediate in the Synthesis of 5-(Hydroxymethyl)cytosine*[2]

T. L. V. ULBRICHT

TWYFORD LABORATORIES, ELVEDEN ROAD, LONDON, N.W. 10, ENGLAND

1 (199.2) $\xrightarrow[KOH]{Me_2SO_4}$ 2 (213.3) $\xrightarrow{LiAlH_4}$ 3 (171.2)

INTRODUCTION

Methioprim [4-amino-2-(methylthio)-5-pyrimidinemethanol, **3**] is a biologically active compound,[1] and is also the key intermediate in the only practicable synthesis of 5-(hydroxymethyl)cytosine[2] yet devised (Sec. II [23]). The original procedure[2] is satisfactory on a small scale, but it required modification for large-scale preparations.

PROCEDURE

Ethyl 4-Amino-2-(methylthio)-5-pyrimidinecarboxylate (2)

A mixture of finely powdered ethyl 4-amino-2-mercapto-5-pyrimidinecarboxylate[3] (**1**) (150 g., 0.75 mole) and a solution of 45 g. (0.80 mole) of potassium hydroxide in 1.5 l. of 95% ethanol, in a 3-l., round-bottomed flask fitted with a reflux condenser, stirrer, thermometer, and dropping funnel, is warmed on a water-bath to 60°. Methyl sulfate (100 g., 75 ml., 0.76 mole) is added during 10 min., with vigorous stirring, the temperature being kept at 65–70°. The mixture is stirred at 70° for 2 hr., is then made alkaline with a solution of potassium hydroxide (5 g.) in 95% ethanol (170 ml.), and is stirred vigorously for 10 min. Methyl sulfate (11 g., 8.2 ml.) is added, any material adhering to the inside of the flask is freed, and the mixture is stirred and refluxed for 2 hr. The reflux condenser is now replaced by a condenser and distillation adapter fitted for distillation, and the ethanol is evaporated off under diminished pressure (with continuous stirring to avoid bumping). Water (1.3 l.) is added to the residue, and the ıixture is warmed to 70°, stirred vigorously for 1 hr., and filtered. The olid product is washed thoroughly with water, and recrystallized from 1 l. f boiling ethanol, unreacted starting-material being filtered off from the ot suspension. Concentration of the mother liquors yields additional roduct; total yield 123 g. (77%), m.p. 128–130°.

-Amino-2-(methylthio)-5-pyrimidinemethanol (3)

A 3-l., three-necked, round-bottomed flask is equipped in the middle ıeck with a sealed stirrer, in a second neck with a stopper, and in the third ıeck with a Soxhlet extractor in which is a thimble containing 63.9 g. 0.3 mole) of compound **2**. The extractor is surmounted by an efficient eflux condenser, connected *via* a calcium chloride tube to a nitrogen upply and an escape valve. After the apparatus has been flushed with ıitrogen, 1 l. of anhydrous ether is introduced, 17.1 g. (0.45 mole) of ithium aluminum hydride is added, and the suspension is stirred until lissolution occurs. The solution is gently heated, so that the ether is condensed and falls into the extractor at a moderate rate (excessive heating eads to violent evolution of heat when the pyrimidine solution flows down nto the flask). After the mixture has been refluxed for 7 hr., it is cooled to oom temperature, and excess lithium aluminum hydride is decomposed y the cautious addition of 40 ml. of ethyl acetate and not more than 20 ml. f water. The solid is filtered off, and extracted with six 400-ml. portions f boiling acetone. The filtrate and extracts are combined, and evaporated o dryness, and the residue is recrystallized from benzene (decolorizing arbon) to give colorless needles; yield 42 g. (82%), m.p. 126–128°.

REFERENCES

1) T. L. V. Ulbricht, *Progr. Nucleic Acid Res.*, **4**, 189 (1965).
2) T. L. V. Ulbricht and C. C. Price, *J. Org. Chem.*, **21**, 567 (1956).
3) T. L. V. Ulbricht, T. Okuda, and C. C. Price, *Org. Syn.*, **39**, 34 (1959).

[33] 4-Pyrimidinethiol
(4-Mercaptopyrimidine)

Direct Thiation of 4-Pyrimidinol with Purified Phosphorus Pentasulfide

HENRY C. KOPPEL, ROBERT H. SPRINGER, ROLAND K. ROBINS, and C. C. CHENG

MIDWEST RESEARCH INSTITUTE, KANSAS CITY, MISSOURI 64110

P_2S_5 (pyridine)

OH **1** (96.1) → SH **2** (112.2)

INTRODUCTION

Direct conversion of the hydroxy (oxo) derivatives of purines an pyrimidines into the corresponding mercapto (thioxo) derivatives by mean of phosphorus pentasulfide usually gives the desired products in low yield A modified method, originally developed in Professor J. F. W. McOmie' laboratory at Bristol University, England, was found satisfactory for th thiation described here.[1]

PROCEDURE

4-Pyrimidinethiol (2)

A mixture of 52 g. (0.54 mole) of 4-pyrimidinol[2] (**1**) and 125 g. of *purifie* phosphorus pentasulfide[a] in 1 l. of pyridine is refluxed, with vigorou mechanical stirring, for 1 hr. The hot, dark-brown solution is heated on a steam bath for 3 hr., and a small amount of insoluble material tha separates is filtered off. The filtrate is concentrated under diminishe pressure to *ca.* 200 ml., and cooled in a refrigerator. A brown, crystallin solid is deposited; this is collected by filtration, washed with ice–water and recrystallized three times from 500-ml. portions of boiling water, t give pale-yellow, long, silky needles, m.p. 190–192° (dec.); yield 45 g (74%). The product is identical with that prepared *via* the 4-chloropyr imidine intermediate.[3]

[a] Purified phosphorus pentasulfide is prepared by extracting the commercia material with carbon disulfide in a Soxhlet apparatus. Pure phosphorus pentasulfid separates as yellow crystals in the extraction flask.

REFERENCES

(1) H. C. Koppel, R. H. Springer, R. K. Robins, and C. C. Cheng, *J. Org. Chem.*, **26**, 792 (1961).
(2) D. J. Brown, *J. Soc. Chem. Ind.* (London), **69**, 353 (1950).
(3) M. P. V. Boarland and J. F. W. McOmie, *J. Chem. Soc.*, **1951**, 1218.

[34] 2-Chloro-5-fluoro-4-pyrimidinol

Preferential Replacement of a Halogen Atom

G. J. DURR

DEPARTMENT OF CHEMISTRY, LE MOYNE COLLEGE, SYRACUSE, NEW YORK 13214

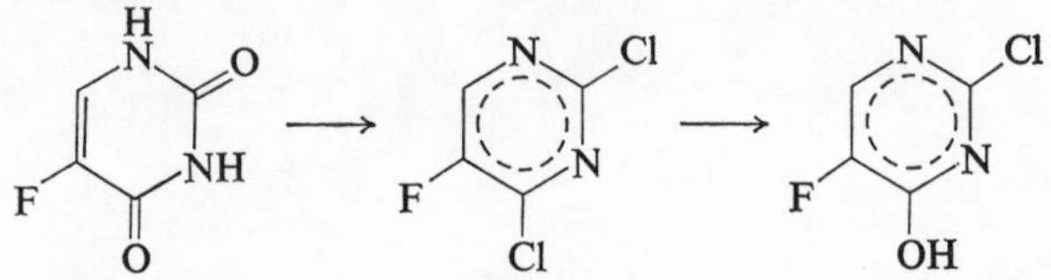

1 (130.1) → 2 (167.0) → 3 (148.5)

INTRODUCTION

2-Chloro-5-fluoro-4-pyrimidinol (**3**) shows antitumor activity[1] against P1798 Lymphosarcoma and Sarcoma 180. Its synthesis is representative of nucleophilic substitution reactions of 2,4-dichloropyrimidines.[2] Whereas replacement of a chlorine atom from other 2,4-dichloropyrimidines has been shown to give a mixture of the C-2 and C-4 substitution products,[3] **3** is formed from 2,4-dichloro-5-fluoropyrimidine (**2**)[4–6] without formation of the isomeric 4-chloro-5-fluoro-2-pyrimidinol.[1] Reaction of **2** with an excess of aqueous sodium hydroxide gives **3**.

PROCEDURE

2,4-Dichloro-5-fluoropyrimidine[4] (2)

A mixture of 10.4 g. (0.08 mole) of 5-fluorouracil[7] (**1**), 16.6 g. (0.14 mole) of *N*,*N*-dimethylaniline, and 90 ml. of phosphoryl chloride is stirred magnetically and refluxed gently for 2 hr., giving a clear, dark solution. The excess phosphoryl chloride is removed, at 20 mm. Hg, on a steam bath.[a] After removal of the solvent, the residue is cautiously poured, with vigorous stirring, into a mixture of 50 g. of ice and 50 g. of ether. The separated aqueous layer is extracted with 25 ml. of ether, the combined ether extracts are dried with anhydrous sodium sulfate, and filtered, and the ether is removed under diminished pressure. The yellow, solid residue weighs 13.5 g., and may be used directly in the following reaction. Purification by vacuum distillation yields 12.5 g. (93%) of **2**, b.p. 67–68°/6 mm. Hg. On cooling, the product solidifies to needles, m.p. 35–36°.

[a] Introduction of a boiling stick prevents bumping.

2-Chloro-5-fluoro-4-pyrimidinol[1] **(3)**

A mixture of 1.96 g. (11.7 mmoles) of **2** and 6.2 ml. (11.8 mmoles) of 1.9 *N* aqueous sodium hydroxide in a 25-ml. flask is heated to 45° in an oil bath, and stirred magnetically for 45 min. At the end of this time, the reaction mixture consists of an oily, lower phase and a neutral, aqueous upper phase. To the heterogeneous mixture an additional 6.2 ml. of 1.9 *N* sodium hydroxide is added, and stirring is continued for 15 min.; at this point, the oily phase is no longer present. After being cooled, the reaction mixture is treated with 1 ml. of concentrated hydrochloric acid, causing immediate precipitation of the desired product **3**; yield 1.38 g. (80%), m.p. 170–171°. Recrystallization is effected from absolute ethanol.[b] The purified product melts at 176–177°, then crystallizes, and remelts at 228–243°.

REFERENCES

(1) G. J. Durr, *J. Med. Chem.*, **8**, 253 (1965).
(2) D. J. Brown, *The Pyrimidines*, Interscience Publishers, New York, N.Y., 1962, pp. 183–208.
(3) E. C. Taylor and M. J. Thompson, *J. Org. Chem.*, **26**, 5224 (1961).
(4) F. Hoffmann–La Roche and Co., *Brit. Pat.* 877,318 (1960) [*U. S. Pat.* 3,040,026 (1962)]; *Chem. Abstracts*, **56**, 8724 (1962).
(5) M. G. Biressi, M. Carrissimi, and F. Ravenna, *Gazz. Chim. Ital.*, **93**, 1268 (1963).
(6) L. D. Protsenko and Yu. I. Bogodist, *Zh. Obshch. Khim.*, **33**, 537 (1963).
(7) R. Duschinsky, E. Pleven, and C. Heidelberger, *J. Am. Chem. Soc.*, **79**, 4559 (1957).

[b] Prolonged heating is to be avoided during the recrystallization, since autocatalysis converts **3** into **1**.

[35] 2-Amino-6-chloro-5-nitro-4(3*H*)-pyrimidinone

Pyrimidine Intermediate in the Synthesis of 9-Substituted Guanines

DOUGLAS T. BROWNE

DEPARTMENT OF CHEMISTRY AND CHEMICAL ENGINEERING, UNIVERSITY OF ILLINOIS, URBANA, ILLINOIS 61801

1 (145.6) $\xrightarrow{HNO_3}$ **2** (208.6) $\xrightarrow[H_2SO_4]{conc.}$ **3** (190.6)

INTRODUCTION

2-Amino-6-chloro-5-nitro-4(3*H*)-pyrimidinone[1–5] (**3**) may be condensed with a wide variety of primary amines to yield precursors of 9-substituted guanines.[1,6] It has been found that nitration of 2-amino-6-chloro-4(3*H*)-pyrimidinone (**1**) is conveniently achieved by dissolving its nitric salt (**2**) in concentrated sulfuric acid. This method offers greater simplicity and reproducibility of yield than published procedures[1–5] involving nitration of **1** by mixtures of fuming nitric acid and concentrated sulfuric acid.

PROCEDURE

2-Amino-6-chloro-5-nitro-4(3*H*)-pyrimidinone[1–5] (3)

Most of a 31.4-g. (0.216 mole) portion of 2-amino-6-chloro-4(3*H*)-pyrimidinone[7] (**1**) is dissolved in 6 l. of anhydrous methanol at room temperature. The remainder of **1** dissolves upon addition of 72 ml. (1.1 moles) of concentrated nitric acid. The pale-yellow solution is concentrated under diminished pressure to 1 l., cooled to $-20°$, and filtered, giving 40.3 g. (90%) of the nitric salt (**2**) as a white, amorphous solid, m.p. 141–142° (*violent* dec.). The salt is added to 150 ml. of vigorously stirred, concentrated sulfuric acid kept at 0° with an ice bath, and the suspension is stirred for 10 min. at 0°. The ice bath is then removed, and the mixture is stirred at room temperature for 2 hr. The pale-yellow solution is poured over ice, and the resulting suspension is filtered while cold. The pale-yellow precipitate is successively washed with cold water and anhydrous acetone, and treated with 3.5 l. of boiling, absolute methanol. The suspension is filtered while hot, and the filtrate, kept at −20°, deposits 25.4 g. (69%) of 2-amino-6-chloro-5-nitro-4(3*H*)-pyrimidinone (**3**) as yellow plates, m.p. 278° (*violent* dec.). The product is stable for several months, at least, if stored at −20°, but undergoes decomposition (presumably by self-condensation) if stored in a desiccator at room temperature.

REFERENCES

(1) J. Davoll and D. D. Evans, *J. Chem. Soc.*, **1960**, 5041.
(2) W. Pfleiderer and H. Walter, *Ann.*, **677**, 113 (1964).
(3) A. Stuart and H. C. S. Wood, *J. Chem. Soc.*, **1963**, 4186.
(4) A. Stuart, D. W. West, and H. C. S. Wood, *J. Chem. Soc.*, **1964**, 4769.
(5) R. Lohrmann and H. S. Forrest, *J. Chem. Soc.*, **1964**, 460.
(6) D. T. Browne, unpublished results.
(7) H. S. Forrest, R. Hall, H. J. Rodda, and A. R. Todd, *J. Chem. Soc.*, **1951**, 3.

[36] 1-(3-Bromopropyl)thymine and *N*-Acetyl-1-(3-bromopropyl)cytosine

Direct Alkylation of Trimethylsilyl Derivatives of Pyrimidines with Unactivated Alkyl Halides

DOUGLAS T. BROWNE

DEPARTMENT OF CHEMISTRY AND CHEMICAL ENGINEERING, UNIVERSITY OF ILLINOIS, URBANA, ILLINOIS 61801

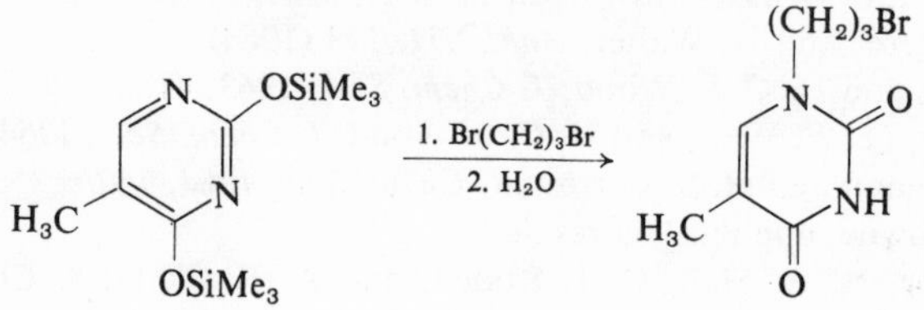

$OSiMe_3$, H_3C, $OSiMe_3$ → 1. $Br(CH_2)_3Br$; 2. H_2O → $(CH_2)_3Br$, O, NH, H_3C, O

1 (270.2) **2** (247.1)

$OSiMe_3$, $H_3C—C$, O, $SiMe_3$ → 1. $Br(CH_2)_3Br$; 2. H_2O → $(CH_2)_3Br$, O, H, $C—CH_3$, O

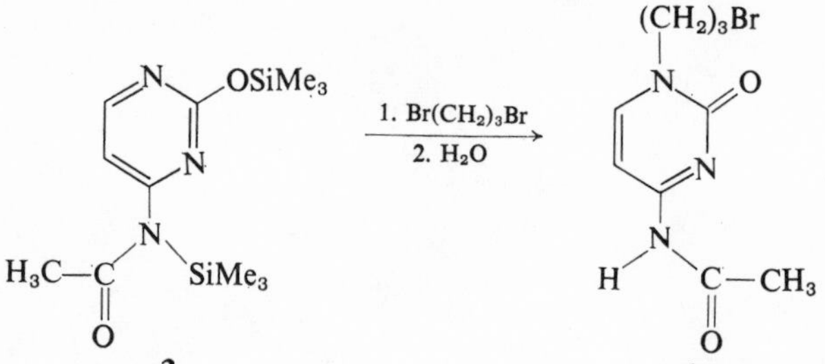

3 (328.2) **4** (274.1)

INTRODUCTION

The bis(trimethylsilyl) derivatives of *N*-acetylcytosine and thymine have been employed in the synthesis of pyrimidine nucleosides.[1] The synthesis of 1-methylthymine by treatment of 5-methyl-2,4-bis(trimethylsiloxy)pyrimidine (**1**) with methyl iodide has also been reported.[2] It has been found that treatment of the trimethylsilyl derivatives **1** and **3** with an excess of unactivated alkyl bromide at room temperature affords the corresponding, 1-substituted pyrimidine in good yield. The syntheses illustrated require mild reaction conditions, because of the possibility of intramolecular cyclization.

PROCEDURE

1-(3-Bromopropyl)thymine (2)

A 53-g. (0.20 mole) portion of 5-methyl-2,4-bis(trimethylsiloxy)-pyrimidine[3] (**1**) is dissolved, with slight heating, in 600 g. (3.0 moles) of 1,3-dibromopropane under an atmosphere of dry nitrogen. The colorless solution is kept at room temperature for 10 days. The pale-yellow solution is poured into 1 l. of water, and the aqueous layer is extracted with four 800-ml. portions of chloroform. The chloroform layers are combined and dried (anhydrous sodium sulfate). Removal of the chloroform under diminished pressure gives a pale-yellow solid. Recrystallization from 120 ml. of hot absolute ethanol yields 39.5 g. (82%) of 1-(3-bromopropyl)-thymine (**2**) as colorless platelets, m.p. 136–138°.

N-Acetyl-1-(3-bromopropyl)cytosine (4)

A 33.7-g. (0.124 mole) portion of 2-(trimethylsiloxy)-4-[*N*-(trimethyl-silyl)acetamido]pyrimidine[3] (**3**) is dissolved in 600 g. (3.0 moles) of 1,3-dibromopropane under an atmosphere of dry nitrogen. The initially colorless solution is kept at room temperature for 10–20 days. The dark-red solution and yellow precipitate are poured into 500 ml. of water, and the aqueous layer is extracted with six 500-ml. portions of chloroform. The chloroform extracts are combined and dried (anhydrous sodium sulfate), and the chloroform is removed under diminished pressure. The resulting dark-red solution in the excess 1,3-dibromopropane is applied directly to a column of 700 g. of silica gel (Davison, No. 923; 100–200 mesh) packed with the aid of chloroform. *N*-Acetyl-1-(3-bromopropyl)cytosine (**4**) is eluted with 2.5:97.5 (v/v) methanol–chloroform as pale-yellow needles, yield 12.0 g. (43%); m.p. 139–142°.

REFERENCES

(1) T. Nishimura and I. Iwai, *Chem. Pharm. Bull.* (Tokyo), **12**, 357 (1964).
(2) E. Wittenburg, *Chem. Ber.*, **99**, 2380 (1966).
(3) T. Nishimura and I. Iwai, *Chem. Pharm. Bull.* (Tokyo), **12**, 352 (1964).

[37] 1-Propylthymine and 1-Propylcytosine

Synthesis of 1-Substituted Derivatives of Thymine and Cytosine by Direct Alkylation

DOUGLAS T. BROWNE

DEPARTMENT OF CHEMISTRY AND CHEMICAL ENGINEERING, UNIVERSITY OF ILLINOIS, URBANA, ILLINOIS 61801

$$\mathbf{1}\ (126.1) + CH_3CH_2CH_2Br\ (123.0) \xrightarrow[Me_2SO]{K_2CO_3} \mathbf{2}\ (168.2)$$

$$\mathbf{3}\ (153.1) + CH_3CH_2CH_2Br\ (123.0) \xrightarrow[Me_2SO]{K_2CO_3} \text{1-}(CH_2)_2CH_3\text{-}N^4\text{-acetylcytosine} \xrightarrow{NH_3} \mathbf{4}\ (153.2)$$

INTRODUCTION

Because of the greater simplicity, synthesis of 1-substituted pyrimidines by sequences involving direct alkylation of pyrimidine precursors is often more desirable than by routes involving ring-closure steps. The course of direct alkylation reactions is, however, frequently difficult to predict, and mixtures of products are often formed. It has been found that direct alkylation of thymine (**1**) or *N*-acetylcytosine (**3**) with an alkyl bromide under conditions similar to those employed[1] for the synthesis of 1-alkyluracils produces the corresponding 1-alkylpyrimidine in about 50% yield.[a]

[a] No attempt has been made to optimize the yield at this stage, but rather to avoid dialkylation.

PROCEDURE

1-Propylthymine[2] (2)

A 5.0-g. (0.040 mole) portion of thymine (**1**) is dissolved in 135 ml. of dry methyl sulfoxide, and 1.6 g. (0.013 mole)[b] of 1-bromopropane and 5.5 g. (0.044 mole) of anhydrous potassium carbonate are added. The resulting suspension is stirred at room temperature for 11 hr., and filtered, and the methyl sulfoxide is removed under diminished pressure, giving a colorless semi-solid. This is mixed with 500 ml. of water, and the suspension is extracted with three 300-ml. portions of chloroform. The chloroform extracts are combined and dried (anhydrous sodium sulfate). Removal of the chloroform under diminished pressure affords a pale-yellow solid; two recrystallizations from absolute ethanol give 1-propylthymine (**2**) as colorless prisms, m.p. 135–137°, yield 0.98 g. (44%, based on 1-bromo-propane).

1-Propylcytosine (4)

A 5.0-g. (0.033 mole) portion of *N*-acetylcytosine[3] (**3**) is dissolved, with heating, in 800 ml. of dry methyl sulfoxide. The solution is cooled to room temperature, and 4.0 g. (0.033 mole) of 1-bromopropane and 4.95 g. (0.036 mole) of anhydrous potassium carbonate are added with stirring. The suspension is stirred at room temperature for 14 hr., and filtered, and the solvent is removed under diminished pressure, giving a pale-yellow sirup. This is mixed with 500 ml. of water, and the suspension is extracted with four 300-ml. portions of chloroform. The chloroform extracts are combined, and dried (anhydrous sodium sulfate). Removal of the solvent under diminished pressure gives a pale-yellow solid,[c] which is dissolved in a mixture of 100 ml. of methanol and 75 ml. of methanolic ammonia (presaturated at room temperature). The solution is kept at room temperature for 7.5 hr., and white platelets separate out. The suspension is evaporated to dryness *in vacuo*, giving a colorless solid which is dissolved in 100 ml. of hot 95% ethanol. The solution is cooled, and deposits 2.76 g. (55%) of 1-propylcytosine (**4**) as colorless platelets, m.p. 256.5–259.5°.

REFERENCES

(1) B. R. Baker and G. B. Chheda, *J. Pharm. Sci.*, **54**, 25 (1965).
(2) G. Shaw and R. N. Warrener, *J. Chem. Soc.*, **1959**, 50.
(3) D. M. Brown, A. R. Todd, and S. Varadarajan, *J. Chem. Soc.*, **1956**, 2384.

[b] Addition of a full equivalent of the alkyl bromide leads to formation of some 1,3-dialkylated product.
[c] The solid contains about 20% of *O*-alkylated material; this is removed in the recrystallization step.

[38] 6-Methyl-*as*-triazine-3,5(2*H*,4*H*)-dione (6-Azathymine)

Use of Cyclization of α-Keto Acid Thiosemicarbazones in the Preparation of as-*Triazine Derivatives*

M. BOBEK and J. GUT

INSTITUTE OF ORGANIC CHEMISTRY AND BIOCHEMISTRY, CZECHOSLOVAK ACADEMY OF SCIENCES, PRAGUE, CZECHOSLOVAKIA

Me, O, C, HOOC — **1** (88.1) —[H_2N–NH–C(=S)–H_2N]→ Me, N, NH, HOOC, C, S, H_2N — **2** (161.1) —MeI→

[Me, N, NH, HOOC, N, H, SMe] HI — **3** (303.0) → Me, N, NH, O, N, SMe — **4** (157.2) —HI→ Me, N, NH, O, N, H, O — **5** (127.1)

INTRODUCTION

6-Azathymine [6-methyl-*as*-triazine-3,5(2*H*,4*H*)-dione, **5**] is an antimetabolite, and the mechanism of its action has been studied in detail.[1] Earlier syntheses[2–5] use alkaline media to effect the cyclization that affords the *as*-triazine system. The present procedure[6] is an example of the utility of cyclization of *S*-alkylisothiosemicarbazones (**3**), which proceeds by refluxing in an aqueous medium. In view of the uniform reaction-course and the satisfactory yields in the individual steps, the synthesis may be performed in a continuous series of operations, without isolation of intermediates. The present procedure may be used in the synthesis[6] of "6-azauracil," and "5-(fluoromethyl)-6-azauracil," and some other 5-substituted derivatives of "6-azauracil."

PROCEDURE

Pyruvic Acid Thiosemicarbazone[4] (2)

A solution of 18.2 g. (0.2 mole) of pure thiosemicarbazide in 100 ml. of boiling water is treated with 17.6 g. (0.2 mole) of freshly distilled (*in vacuo*) pyruvic acid (**1**), the resulting mixture is kept at room temperature for 12 hr., and the first crop (28.0 g.) of crystals, m.p. 189°, is collected with suction. A second crop, having the same m.p., is obtained by concentrating the mother liquor to about one quarter of the original volume, and cooling; total yield 29.7 g. (92%).

6-Azathymine [6-Methyl-*as*-triazine-3,5(2*H*,4*H*)-dione,[6] 5]

A magnetically stirred mixture of 150 ml. of water, 16.1 g. (0.1 mole) of compound **2**, and 10 ml. of methyl iodide is heated in a bath at 45–50° for 3–5 hr., during which time the thiosemicarbazone (**2**) should dissolve. The excess methyl iodide is removed by evaporation at atmospheric pressure, and the residual solution is refluxed for 90 min. in a hood (evolution of methanethiol), and kept at room temperature for 12 hr. The crystals of **5** deposited (7 g.) are collected with suction, and successively washed at 0° with 10 ml. of water, 20 ml. of ethanol, and 20 ml. of ether. The product melts at 212–213° and does not require recrystallization. The mother liquors and washings are combined, and concentrated under diminished pressure to 25 ml., and the concentrate is kept at 0° for 12 hr. The crystals (4 g.) that separate (and which melt at the same temperature as the main portion) are collected with suction, and washed successively with 5 ml. of water, 10 ml. of ethanol, and 10 ml. of ether. Further concentration of the mother liquors, to 10 ml., and treatment of the concentrate with 10 ml. of ethanol affords 0.6 g. of a less-pure product. Crops 2 and 3 are purified by recrystallization from water; total yield 10.56 g. (82%).

REFERENCES

(1) J. Škoda, *Progr. Nucleic Acid Res.*, **2**, 201 (1963).
(2) J. Thiele and J. Bailey, *Ann.*, **303**, 75 (1898).
(3) J. Bougault and L. Daniel, *Compt. Rend.*, **186**, 1216 (1928).
(4) J. Gut, *Collection Czech. Chem. Commun.*, **23**, 1588 (1958).
(5) P. K. Chang, *J. Org. Chem.*, **23**, 1951 (1958).
(6) M. Bobek, J. Farkaš, and J. Gut, *Collection Czech. Chem. Commun.*, **32**, 1295 (1967).

[39] 5-Amino-*as*-triazin-3(2*H*)-one (6-Azacytosine)

Selective Substitution of as-*Triazine-3,5*(2H,4H)-*dione at C-5*

J. PLIML, A. PÍSKALA, and J. GUT

INSTITUTE OF ORGANIC CHEMISTRY AND BIOCHEMISTRY, CZECHOSLOVAK ACADEMY OF SCIENCES, PRAGUE, CZECHOSLOVAKIA

1 (113.1) $\xrightarrow{P_2S_5}$ **2** (129.1) $\xrightarrow[NaOMe]{MeI}$ **3** (143.2) $\xrightarrow{NH_3}$ **4** (112.1)

INTRODUCTION

6-Azacytosine [5-amino-*as*-triazin-3(2*H*)-one, **4**] has been studied as a potential carcinostatic agent, and has been shown[1] to possess inhibitory properties against adenocarcinoma 755. Extensive attention has also been devoted to the 1-D-ribofuranosyl-6-azacytosine derivatives.[2] 6-Azacytosine (**4**) was originally prepared[3] by amination of 4-thio-6-azauracil [5-thio-*as*-triazine-3,5(2*H*,4*H*)-dione, **2**] under pressure. The present procedure, involving the more-reactive methylthio derivative (**3**), obviates the use of pressure and may also be applied to the preparation of *N*-alkyl derivatives of 6-azacytosine.[4]

PROCEDURE

4-Thio-6-azauracil [5-Thio-*as*-triazine-3,5(2*H*,4*H*)-dione, 2][5]

A solution of 22.6 g. (0.2 mole) of 6-azauracil [*as*-triazine-3,5(2*H*,4*H*)-dione, **1**] in 200 ml. of boiling, anhydrous pyridine is rapidly treated with a solution of 26.7 g. (0.12 mole) of phosphorus pentasulfide in 400 ml. of boiling, anhydrous pyridine. The resulting mixture is immediately immersed in a bath preheated to 130°, and is refluxed for 2 hr. The bath is then removed, and the dark-red solution is cooled, kept at room temperature for 12 hr., decanted from the heavy oil at the bottom of the flask, and evaporated to dryness under diminished pressure. The crystalline residue is triturated with 500 ml. of water, and the suspension is made alkaline, at 20°, with 250 ml. of 2 *N* sodium hydroxide, and filtered with suction through a bed of moist, activated carbon to remove a small amount of tan crystals. The filtrate is washed with three 200-ml. portions of ether, and acidified (to Congo Red paper) with about 250 ml. of 2 *N* hydrochloric acid. On scratching the inner walls of the flask with a sharp-edged glass rod, the solution begins to deposit bright-red crystals of 4-thio-6-azauracil (**2**). After 10 hr. at room temperature, the crystals (21 g.) are collected, and the mother liquor is extracted with seven 300-ml. portions of ether. Evaporation of the combined ethereal extracts affords an additional crop (about 6 g.) of the product. The crops are combined, and dissolved in 500 ml. of hot water, the solution is treated with activated carbon, and filtered, and the filtrate is cooled; yield 20 g. (75%) of bright-red crystals, m.p. 227°.

5-(Methylthio)-*as*-triazin-3(2*H*)-one[5] (3)

To a solution of sodium methoxide, prepared from 2.3 g. (0.1 mole) of sodium and 150 ml. of absolute methanol, is added 12.9 g. (0.1 mole) of 4-thio-6-azauracil (**2**). When the latter has dissolved, 10 ml. of methyl iodide is added, and the mixture is kept at room temperature for 4 hr., and evaporated under diminished pressure to a sirup which is triturated with 50 ml. of water. The mixture is kept at 0° for 2 hr., and the crystalline product is collected with suction, washed with 25 ml. of water precooled to 0°, and dried in a vacuum desiccator over concentrated sulfuric acid. The dried product, m.p. 165–168°, weighs 10 g. (66%), and may be used directly in the following reaction. On recrystallization from ethanol, the compound melts at 176–178° (Kofler block).

6-Azacytosine [5-Amino-*as*-triazin-3(2*H*)-one, 4]

A mixture of 14.3 g. (0.1 mole) of 5-(methylthio)-*as*-triazin-3(2*H*)-one (**3**) and 150 ml. of methanol presaturated at room temperature with dry ammonia gas is kept, with occasional shaking, in a stoppered vessel at room temperature for 24 hr. The solution is evaporated to dryness under diminished pressure, and the residue is recrystallized from 1.0 l. of boiling water to give 8.3 g. of 6-azacytosine (**4**). An additional crop of the product is obtained by concentrating the mother liquor to 200 ml. Total yield, 9.7 g. (86%) of pure, colorless product which does not melt up to 350°, $\lambda_{max}^{H_2O}$ 258 nm. (log ϵ 3.78).

REFERENCES

(1) G. B. Elion, S. Bieber, H. Nathan, and G. H. Hitchings, *Cancer Res.*, **18**, 802 (1958).
(2) J. Škoda, *Progr. Nucleic Acid Res.*, **2**, 211 (1963).
(3) E. A. Falco, E. Pappas, and G. H. Hitchings, *J. Am. Chem. Soc.*, **78**, 1938 (1956).
(4) J. Gut, J. Jonáš, and J. Pitha, *Collection Czech. Chem. Commun.*, **29**, 1394 (1964).
(5) J. Gut, M. Prystaš, and J. Jonáš, *Collection Czech. Chem. Commun.*, **26**, 986 (1961).

[40] s-Triazine-2,4(1H,3H)-dione (5-Azauracil)

Preparation of 5-Azapyrimidines by Condensation of Urea Derivatives with Ethyl Orthoformate

A. PÍSKALA and J. GUT

INSTITUTE OF ORGANIC CHEMISTRY AND BIOCHEMISTRY, CZECHOSLOVAK ACADEMY OF SCIENCES, PRAGUE, CZECHOSLOVAKIA

3 $H_2NC(O)NH_2$ (**1**) $\xrightarrow[Ac_2O]{HC(OEt)_3}$ **2** (173.1) $\xrightarrow{HCl}$ **3** (113.1)

INTRODUCTION

5-Azauracil [*s*-triazine-2,4(1*H*,3*H*)-dione, **3**] is an antimetabolite, and the mechanism of its action has been studied in detail.[1] The present procedure exemplifies the general method of synthesis of 5-azapyrimidines.[2–5] The literature records other syntheses of 5-azauracil[3,5] (**3**) which are, however, less suitable.

PROCEDURE

Adduct (2) of Urea (1) with 5-Azauracil [*s*-Triazine-2,4(1*H*,3*H*)-dione, 3]

A mixture of 30 g. (0.5 mole) of dry, powdered urea (**1**), 80 ml. of ethyl orthoformate, and 30 ml. of acetic anhydride is placed in a 750-ml., round-bottomed flask, and refluxed in an oil bath (140°), with stirring, until a vigorous reaction sets in (5–10 min.); the oil bath is then removed. At the end of the exothermic reaction (approximately 10 min.), the resulting slurry is refluxed for 1 hr., cooled, and kept at +3° for 2 hr. Adduct **2** is collected with suction, washed with 60 ml. of ethanol, and dried at 20°/0.3 mm. Hg; yield 22 g. (76%), m.p. 234–235° (dec.). This product is sufficiently pure to be used in the next step.

5-Azauracil [*s*-Triazine-2,4(1*H*,3*H*)-dione, 3]

A mixture of 17.3 g. (0.1 mole) of adduct **2** and 20 ml. of concentrated hydrochloric acid is heated in a 250-ml., round-bottomed flask until **2** dissolves. The solution is then immediately cooled, and diluted with 100 ml. of ethanol; 5-azauracil (**3**) is gradually deposited in the form of an unstable adduct with ethanol.[6] The mixture is kept at +3° for 5 min.,[a] and the adduct is collected with suction, washed with ethanol, and recrystallized from 60 ml. of water. The resulting 5-azauracil (**3**) monohydrate is dried at 100°/0.3 mm. Hg; m.p. 281–282° (dec.); yield 9.5 g. (84%).

REFERENCES

(1) J. Škoda, *Progr. Nucleic Acid Res.*, **2**, 197 (1963).
(2) A. Pískala and J. Gut, *Collection Czech. Chem. Commun.*, **28**, 2376 (1963).
(3) J. Gut, *Advan. Heterocyclic Chem.*, **1**, 189 (1963); A. Pískala and J. Gut, *Collection Czech. Chem. Commun.*, **28**, 1681 (1963).
(4) A. Pískala and F. Šorm, *Collection Czech. Chem. Commun.*, **29**, 2060 (1964).
(5) H. Bredereck, F. Effenberger, and A. Hofmann, *Chem. Ber.*, **97**, 61 (1964).
(6) P. Pithová, A. Pískala, J. Pitha, and F. Šorm, *Collection Czech. Chem. Commun.*, **30**, 90 (1965).

[a] Prolonged storage is not suitable, as compound **2** is re-formed.

[41] 4-Amino-*s*-triazin-2(1*H*)-one (5-Azacytosine)

Preparation of 5-Azacytosines by Condensation of Amidinourea with N,N-*Dimethylformamide Dimethyl Acetal*

A. PÍSKALA and F. ŠORM

INSTITUTE OF ORGANIC CHEMISTRY AND BIOCHEMISTRY, CZECHOSLOVAK ACADEMY OF SCIENCES, PRAGUE, CZECHOSLOVAKIA

HN=C(NH$_2$)–NH–CN (**1**, 83.1) ⟶ H_2N–C(=O)–NH–C(=NH)–NH$_2$·HCl (**2**, 138.6) —$Me_2NCH(OMe)_2$⟶ 5-azacytosine (**3**, 112.1)

1	**2**	**3**
(83.1)	(138.6)	(112.1)

INTRODUCTION

5-Azacytosine (**3**) inhibits the growth of *Escherichia coli*,[1] and 4-amino-1-β-D-ribofuranosyl-*s*-triazin-2(1*H*)-one (5-azacytidine) shows a marked bacteriostatic and cytostatic activity.[2] The method[3] given here is of a general character, and may also be used in the preparation of substituted 5-azacytosines. The other known syntheses[4] of compound **3** are, in our experience, less suitable for laboratory-scale preparations.

PROCEDURE

Amidinourea Hydrochloride (2)

In a 750-ml., round-bottomed flask is placed an approximately 4 *M* solution of hydrogen chloride in absolute methanol (200 ml.) and, at 0–5°, 33.6 g. (0.4 mole) of cyanoguanidine (**1**) is added, with stirring. The mixture is kept (drying tube) in an ice bath until almost all of the cyanoguanidine has dissolved; then the ice bath is removed, the flask is equipped with a reflux condenser, and the mixture is kept at room temperature. In the course of several minutes, a vigorous evolution of methyl chloride sets in, and the temperature rises to about 46°. When the reaction is complete (after 10–20 min.), the flask is kept in an ice-bath for 1 hr. The product (**2**) that separates is collected with suction, washed with three 50-ml. portions of chilled methanol, and dried in a vacuum desiccator over concentrated sulfuric acid and potassium hydroxide; yield 38.5 g. (70%), m.p. 173–175°

(dec.). This material may be directly used in the subsequent step. An additional crop (7.8 g., 14%) of the same quality is obtained by evaporating the mother liquors to a thick slurry, filtering, washing with methanol, and crystallizing from 250 ml. of ethanol.

5-Azacytosine 4-[Amino-*s*-triazin-2(1*H*)-one, 3]

A solution of 13.86 g. (0.1 mole) of amidinourea hydrochloride (**2**) in 150 ml. of hot, absolute methanol is cooled in an ice bath, and treated with a solution prepared from 2.3 g. (0.1 g.-atom) of sodium in 50 ml. of absolute methanol. The sodium chloride that separates is immediately filtered off, and the filtrate is treated with 11.9 g. (0.1 mole) of *N*,*N*-dimethylformamide dimethyl acetal. The resulting solution is kept (drying tube) at room temperature for 24 hr. The solid that separates is collected with suction and saved, and the filtrate is evaporated to dryness under diminished pressure. The residue is triturated with 50 ml. of methanol, and the crystals are collected with suction, to obtain an additional crop of the crude product. The two crops are combined, and recrystallized from 800 ml. of water; yield 6.7 g. (60%) of 5-azacytosine (**3**),$\lambda_{max}^{0.1\,N\,HCl}$ 248 nm. (log ϵ 3.80). The product does not melt up to 350°. An additional crop (12.3 g., 11%) is obtained by concentrating the mother liquors under diminished pressure to 50 ml. and recrystallizing from water.

REFERENCES

(1) J. Škoda, A. Čihák, J. Gut, M. Prystaš, A. Pískala, C. Párkányi, and F. Šorm, *Collection Czech. Chem. Commun.*, **27**, 1735 (1962).
(2) F. Šorm, A. Pískala, A. Čihák, and J. Veselý, *Experientia*, **20**, 202 (1964); F. Šorm and J. Veselý, *Neoplasma*, **11**, 123 (1964).
(3) A. Pískala, *Collection Czech. Chem. Commun.*, **32**, 3966 (1967).
(4) J. Gut, *Advan. Heterocyclic Chem.*, **1**, 189 (1963); A. Pískala and J. Gut, *Collection Czech. Chem. Commun.*, **28**, 1681 (1963); H. Bredereck, F. Effenberger, and A. Hofmann, *Chem. Ber.*, **97**, 61 (1964); I. Flament, R. Promel, and R. H. Martin, *Bull. Soc. Chim. Belges*, **73**, 585 (1964).

[42] 1-Substituted Uracils

Cyclization of 1-Substituted 1-(2-Cyanoethyl)ureas in Acid, Followed by Bromination and Dehydrobromination

C. C. CHENG and LELAND R. LEWIS

MIDWEST RESEARCH INSTITUTE, KANSAS CITY, MISSOURI 64110

$\xrightarrow{H^+}$ $\xrightarrow[-HBr]{Br_2,}$

1, R = CH_3 (127.2)	**2**, R = CH_3 (128.1)	**3**, R = CH_3 (126.1)
1a, R = $(CH_2)_3CH_3$ (169.2)	**2a**, R = $(CH_2)_3CH_3$ (170.2)	**3a**, R = $(CH_2)_3CH_3$ (168.2)
1b, R = C_6H_{11} (195.3)	**2b**, R = C_6H_{11} (196.3)	**3b**, R = C_6H_{11} (194.2)
1c, R = $CH_2C_6H_5$ (203.3)	**2c**, R = $CH_2C_6H_5$ (204.2)	**3c**, R = $CH_2C_6H_5$ (202.2)

INTRODUCTION

An unambiguous method for synthesizing 1-substituted uracils involves the preparation of a substituted β-alanine, and cyclization to a 1-substituted hydrouracil, followed by bromination and dehydrogenation.[1–3] However, isolation of 1-substituted hydrouracils from the reaction mixture often presents much difficulty. The modified method[4] described herein utilizes 1-substituted 1-(2-cyanoethyl)ureas, instead of the corresponding β-alanines, for the preparation of intermediate hydrouracils that can be readily isolated.

PROCEDURE

1-(2-Cyanoethyl)-1-cyclohexylurea (1b)

Compound **1b** is prepared in a manner similar to that for 1-(2-cyanoethyl)-1-methylurea (**1**) or 1-butyl-1-(2-cyanoethyl)urea (**1a**) (Sec. II [22]); yield 55% (recrystallized from benzene), m.p. 108–109°.

1-Benzyl-1-(2-cyanoethyl)urea (1c)

The procedure for the preparation of **1c** is similar to that for **1b**; yield 92% (recrystallized from ethyl acetate), m.p. 75–78°.

1-Substituted Hydrouracils (2-2c)

A mixture of 10 g. of a 1-substituted 1-(2-cyanoethyl)urea, 50 ml. of methanol, and 100 ml. of 10% hydrochloric acid is refluxed for 2 hr. The resulting solution is evaporated to dryness under diminished pressure, and the crude products are purified individually in the following ways.

1-Methylhydrouracil (2)

Compound **2** is obtained from 1-(2-cyanoethyl)-1-methylurea (**1**) (Sec. II [22]) as long needles (after recrystallization from isopropyl alcohol); yield 5.4 g. (55%), m.p.[2,5,6] 173°.

1-Butylhydrouracil (2a)

Compound **2a** is obtained from 1-butyl-1-(2-cyanoethyl)urea (**1a**) (Sec. II [22]) as large, hexagonal plates (after recrystallization from methanol–ethyl acetate); yield 8.2 g. (83%), m.p. 78–79°.

1-Cyclohexylhydrouracil (2b)

Compound **2b** is obtained from compound **1b**, and is recrystallized from methanol–water; yield 6.6 g. (67%), m.p. 180–182°.

1-Benzylhydrouracil (2c)

Compound **2c** is obtained from compound **1c**, and is recrystallized from benzene; yield 6.2 g. (63%), m.p.[2] 125–127°.

1-Methyluracil (3)

In a three-necked, round-bottomed flask (equipped with a mechanical stirrer, condenser, and dropping funnel) is placed 63 g. (0.49 mole) of compound **2**, followed by 250 ml. of glacial acetic acid. The mixture is heated until refluxing, and to the hot, stirred solution is added dropwise 80 g. (1 mole) of bromine dissolved in 120 ml. of glacial acetic acid.[a] The mixture is then refluxed for 3 hr., and the pale-yellow solution is concentrated almost to dryness in a rotary evaporator; water is added to the residual sirup, and the product gradually solidifies on standing. The pH of the resulting mixture is adjusted to 5 with 5% aqueous sodium hydroxide, and the solid is filtered off, and washed with water. Recrystallization from ethanol gives 5-bromo-1-methylhydrouracil as long, colorless needles; yield 62.0 g. (100%), m.p. 153–155°, resolidifying at 159° and remelting at 206–213° (dec.) (lit.[2,6] m.p. 132–135°).

All of the 5-bromo-1-methylhydrouracil is added, in portions, to 500 ml. of boiling *N*,*N*-dimethylformamide, and the resulting, light-yellow solution

[a] The color of the bromine disappears rapidly, and the hydrogen bromide formed escapes through the condenser.

is boiled for 4 hr. and evaporated to dryness *in vacuo*. The crude, solid product is recrystallized from absolute ethanol to give 57 g. (92%) of **3**, m.p.[2,6,7] 230–231°, $\lambda_{max}^{pH\,1}$ 267 nm. (ϵ 8,900) and $\lambda_{max}^{pH\,11}$ 265 nm. (ϵ 7,000).

1-Butyluracil (3a)

This compound is prepared from compound **2a**, in a manner similar to that for **3**. From 26.7 g. (0.15 mole) of **2a**, 25 g. (0.31 mole) of bromine, and 200 ml. of glacial acetic acid, 23.2 g. (88%) of **3a** [m.p. 100.5–102.5°, $\lambda_{max}^{pH\,1}$ 267 nm. (ϵ 10,700) and $\lambda_{max}^{pH\,11}$ 265 nm. (ϵ 7,600)] is obtained. The intermediate 5-bromo-1-butylhydrouracil melts at 165–167°.

1-Cyclohexyluracil (3b)

This compound is obtained from 40 g. (0.2 mole) of compound **2b**, giving 29.2 g. (74%) of **3b**, m.p. 217–218°; $\lambda_{max}^{pH\,1}$ 268 nm. (ϵ 10,500); $\lambda_{max}^{pH\,11}$ 266 nm. (ϵ 7,200). The intermediate 5-bromo-1-cyclohexylhydrouracil melts at 167–169°.

1-Benzyluracil (3c)

The preparation of compound[2,8] **3c** from 20.4 g (0.1 mole) of compound **2c** gives 15.9 g. (81%) of **3c**, m.p. 173–174°, $\lambda_{max}^{pH\,1}$ 265 nm. (ϵ 9,700) and $\lambda_{max}^{pH\,11}$ 264 nm. (ϵ 7,100).

REFERENCES

(1) E. Fischer and G. Roeder, *Ber.*, **34**, 3751 (1901); T. B. Johnson and J. E. Livak, *J. Am. Chem. Soc.*, **58**, 299 (1936).
(2) J. E. Gearien and S. B. Binkley, *J. Org. Chem.*, **23**, 491 (1958).
(3) R. C. Smith and S. B. Binkley, *J. Org. Chem.*, **24**, 249 (1959).
(4) C. C. Cheng and L. R. Lewis, *J. Heterocyclic Chem.*, **1**, 260 (1964).
(5) D. J. Brown, E. Hoerger, and S. F. Mason, *J. Chem. Soc.*, **1955**, 211.
(6) G. E. Hilbert, *J. Am. Chem. Soc.*, **54**, 2076 (1932).
(7) G. Shaw and R. N. Warrener, *J. Chem. Soc.*, **1958,** 157.
(8) H. L. Wheeler and T. B. Johnson, *Am. Chem. J.*, **37**, 628 (1907); T. B. Johnson and A. W. Joyce, *J. Am. Chem. Soc.*, **38**, 1385 (1916).

[43] 5-(Trifluoromethyl)uracil, 5-(Difluoromethyl)uracil, and 6-(Trifluoromethyl)-*as*-triazine-3,5(2*H*,4*H*)-dione [5-(Trifluoromethyl)-6-azauracil]

The Reaction of Sulfur Tetrafluoride with Carbonyl-substituted Uracil and 6-Azauracil

MATHIAS P. MERTES and SOUHEIL E. SAHEB

DEPARTMENT OF MEDICINAL CHEMISTRY, UNIVERSITY OF KANSAS, LAWRENCE, KANSAS 66044

$\xrightarrow[\text{(HF)}]{SF_4}$ + HF + SOF_2

1a, R = COOH (156.1) → **2a**, R = CF_3 (180.1)

1b, R = CHO (140.1) → **2b**, R = CHF_2 (162.1)

3 (157.1) $\xrightarrow[\text{(HF)}]{SF_4}$ **4** (181.1)

INTRODUCTION

5-(Trifluoromethyl)uracil (**2a**), initially prepared *via* a primary ring-synthesis from 2-(trifluoromethyl)acrylonitrile,[1] is conveniently prepared in one step from 5-uracilcarboxylic acid (1,2,3,4-tetrahydro-2,4-dioxo-5-pyrimidinecarboxylic acid, **1a**) by the action of sulfur tetrafluoride in the presence of hydrofluoric acid.[2] The biological activity of the 2′-deoxynucleotide of **2a** has prompted the synthesis of 5-(difluoromethyl)uracil (**2b**)[3] and 5-(trifluoromethyl)-6-azauracil[4] (**4**).

PROCEDURE

5-(Trifluoromethyl)uracil (2a)

In a reaction vessel (125 ml.; fitted with an inlet valve, copper inlet tube, and a stirrer [a]), in a well ventilated hood, are placed 1.00 g. (0.006 mole) of 5-uracilcarboxylic acid [b] (**1a**) and 0.5 ml. of water. The vessel is sealed and checked for leaks by quickly evacuating the vessel (without removing the water), allowing it to stand, and periodically checking for a change in pressure in the vessel. The evacuated vessel is cooled in a Dry Ice–acetone bath for two hr.; an excess of sulfur tetrafluoride [c] (45 g., 0.41 mole) is then admitted through the copper tubing and condensed in the reaction vessel. After being warmed to room temperature, the contents are agitated and heated at 100° overnight. The cooled reaction vessel is vented, and the exhaust gases are decomposed in 10% aqueous potassium hydroxide. The viscous, green residue in the vessel is recrystallized from water, with use of suitable plastic or other hydrogen fluoride-resistant containers, to give 0.88 g. (77%) of colorless needles, m.p. 247–249° (dec.). Subsequent recrystallization must be conducted in neutral or in acidic media to avoid alkaline solvolysis. $\lambda_{max}^{0.1 N \, HCl}$ (undissociated) 257 nm. (ϵ 8,150), $\lambda_{max}^{pH \, 9.3}$ (monoanion) 281 nm. (ϵ 9,200). Above a pH of 9, slow, alkaline solvolysis to 5-uracilcarboxylic acid (**1a**) occurs.

5-(Difluoromethyl)uracil [2a,3] (2b)

Compound **2b** is prepared, in a manner similar to that described in the preceding experiment, by treating 0.71 g. (0.005 mole) of 5-formyluracil [5] (1,2,3,4-tetrahydro-2,4-dioxo-5-pyrimidinecarboxaldehyde, **1b**) and 0.5 ml. of water with 35 g. of sulfur tetrafluoride, and heating for 10 hr. at 100°. Because the product solvolyzes, even at pH 4, the residue is removed, and is dried in a vacuum desiccator. The product sublimes at 170°/0.1 mm. Hg to yield 0.49 g. (60%) of a colorless powder, decomposing at 285–300°, $\lambda_{max}^{pH \, 1}$ 263 nm. (ϵ 7,450).

6-(Trifluoromethyl)-*as*-triazine-3,5(2*H*,4*H*)-dione (4)

6-Azauracil-5-carboxylic acid (2,3,4,5-tetrahydro-3,5-dioxo-*as*-triazine-6-carboxylic acid) [6] (**3**) (31 g., 0.2 mole) in 14.4 ml. (0.8 mole) of water in a 300-ml. reaction vessel is treated, according to directions given in the preceding experiments, with 190 g. of sulfur tetrafluoride. The mixture is

[a] Magne–Dash high-pressure autoclave, Autoclave Engineers, Inc., Erie, Pa.
[b] Nutritional Biochemicals Corp., Cleveland, Ohio.
[c] Sulfur tetrafluoride (SF_4) can be purchased in 1- and 10-lb. steel cylinders from E. I. duPont de Nemours and Co., Inc., Freon Products Division, Chestnut Run, Wilmington, Delaware. Owing to the relatively high toxicity of sulfur tetrafluoride, a well ventilated hood should be used.

heated at 50° overnight, the exhaust gases are vented, and the residue is dried, and crystallized from chloroform to give 20 g. (56%) of crude product. Analytically pure material (m.p. 153°) is obtained by sublimation, or by repeated recrystallization from chloroform. The product (stable in alkaline media) has $\lambda_{max}^{pH\,1}$ (undissociated) 263 nm. (ϵ 6,500), $\lambda_{max}^{pH\,7.6}$ (monoanion) 258 nm. (ϵ 5,200), and $\lambda_{max}^{pH\,12.6}$ (dianion) 292 nm. (ϵ 6,950).

REFERENCES

(1) (a) C. Heidelberger, D. G. Parsons, and D. C. Remy, *J. Am. Chem. Soc.*, **84**, 3597 (1962); (b) *J. Med. Chem.*, **7**, 1 (1964).
(2) (a) M. P. Mertes and S. E. Saheb, *J. Pharm. Sci.*, **52**, 508 (1963); (b) M. P. Mertes, S. E. Saheb, and D. Miller, *J. Med. Chem.*, **9**, 876 (1966).
(3) M. P. Mertes and S. E. Saheb, *J. Med. Chem.*, **6**, 619 (1963).
(4) (a) M. P. Mertes and S. E. Saheb, *J. Heterocyclic Chem.*, **2**, 491 (1965); (b) T. Y. Shen, W. V. Ruyle, and R. L. Bugianesi, *ibid.*, **2**, 495 (1965); (c) A. Dipple and C. Heidelberger, *J. Med. Chem.*, **9**, 715 (1966).
(5) R. H. Wiley and Y. Yamamoto, *J. Org. Chem.*, **25**, 1906 (1960).
(6) R. B. Barlow and A. D. Welch, *J. Am. Chem. Soc.*, **78**, 1258 (1956).

Section III

Nucleosides

A. Purine Nucleosides and Analogs

[44] 9-β-D-Allofuranosyladenine

Preparation of an Aldohexofuranosyl Nucleoside

PAUL KOHN, RITA H. SAMARITANO, and LEON M. LERNER

DEPARTMENT OF BIOLOGICAL CHEMISTRY, UNIVERSITY OF ILLINOIS AT THE MEDICAL CENTER, CHICAGO, ILLINOIS 60612

HOCH$_2$ / HOCH — lactone, HO, OH
1 (178.2) →
BzOCH$_2$ / BzOCH — lactone, BzO, OBz
2 (594.6)
$(H_3C-CH(CH_3)-CH(CH_3))_2BH$ **3** (154.1) →

BzOCH$_2$ / BzOCH — H, OH, BzO, OBz
4 (596.6) →
BzOCH$_2$ / BzOCH — H, O-*p*-NBz, BzO, OBz
5 (745.8) →

BzOCH$_2$ / BzOCH — Cl, H, BzO, OBz
6
BzNH-purine-N-HgCl **7** (474.3) →
BzOCH$_2$ / BzOCH — BzNH-purine, BzO, OBz
8 (817.8) →
HOCH$_2$ / HOCH — NH$_2$-purine, HO, OH
9 (297.2)

where *p*-NBz is *p*-nitrobenzoyl.

INTRODUCTION

9-β-D-Allofuranosyladenine (**9**) is an example of an aldohexofuranosyl nucleoside that can be prepared from an aldohexono-1,4-lactone. The free hydroxyl groups of the lactone are acylated, and the acylated lactone is reduced with bis(3-methyl-2-butyl)borane (di-*sec*-isoamylborane; disiamylborane)[1] (**3**) to the tetra-*O*-acylaldohexofuranose (**4**). Preparation of the nucleoside then follows established procedures, *viz.*, acylation of the anomeric hydroxyl group, conversion into the *O*-acylglycosyl halide, coupling of this with a suitably protected base, and removal of the protecting groups to yield the nucleoside (**9**). This procedure has been applied to a number of 1,4-lactones, to afford nucleosides of adenine.[2,3]

PROCEDURE

Diborane[1,2]

A distilling flask is equipped with a pressure-equalizing dropping funnel, a magnetic stirring bar, and an inlet tube for nitrogen and an outlet tube for diborane. The mixture of nitrogen and evolved diborane is passed through a trap of sodium borohydride in bis(2-methoxyethyl) ether (to remove contaminating boron trifluoride) and then into a receiving flask containing tetrahydrofuran.

In the distilling flask is placed 24.0 ml. of an ether solution of boron fluoride etherate (32.4 g., 0.228 mole). With magnetic stirring, 114 ml. of *M* sodium borohydride in bis(2-methoxyethyl) ether (freshly distilled under nitrogen from lithium aluminum hydride, at 73°/35 mm. Hg) is slowly added dropwise, and the diborane generated is collected in 60 ml. of tetrahydrofuran (freshly distilled under nitrogen from lithium aluminum hydride, at 65–66°/760 mm. Hg) at 0°. After all of the sodium borohydride has been added (45 min.), the generator is warmed to 70° to distil off the remaining diborane. The resulting solution of diborane in tetrahydrofuran is[a] about 2 *M*.

Bis(3-methyl-2-butyl)borane (disiamylborane)[1,2] (3)

In a flask equipped with a pressure-equalizing dropping funnel and a nitrogen inlet tube is placed 25 ml. (0.24 mole) of 2-methyl-2-butene. The flask is cooled[b] in a refrigerated bath, or an ice–salt mixture, to −5 to −10°. The 2.0 *M* diborane solution (60 ml., 0.12 mole) is added dropwise

[a] Determined by removal of an aliquot (1 ml.), which is pipetted into 10 ml. of acetone. Hydrolysis to boric acid is caused by addition of 10 ml. of water. D-Mannitol (0.7 g.) is added, and the solution is titrated with 0.10 *N* sodium hydroxide to the phenolphthalein end-point.

[b] To prevent formation of trisiamylborane, it is important to keep the temperature below 0°.

to the magnetically stirred solution. After 6 hr. of stirring under a slight static pressure of nitrogen, the solution is diluted to 120 ml. with tetrahydrofuran to give a *M* solution of **3**.

2,3,5,6-Tetra-*O*-benzoyl-D-allono-1,4-lactone[4] (2)

A mixture of 70 ml. of benzoyl chloride and 70 ml. of chloroform is cooled in an ice–salt bath. In a separate vessel, 84 ml. of pyridine and 70 ml. of chloroform are mixed and cooled in an ice–salt bath. The two solutions are then combined in a 1-l. flask in an ice bath, and the solution is stirred with a magnetic stirrer. D-Allono-1,4-lactone[5] (**1**) (20 g., 0.112 mole) is added slowly in small portions, and, after 1 hr., the flask is removed from the ice bath and placed in a refrigerator. After 2 days, chloroform (100–200 ml.) is added to the solution, which is then successively washed with three 100-ml. portions of ice-cold, 5% sulfuric acid, three 100-ml. portions of ice-cold, saturated sodium hydrogen carbonate solution, and three 100-ml. portions of water. The aqueous extracts are extracted with chloroform. The chloroform solutions are combined, dried (anhydrous magnesium sulfate), and evaporated to a sirup which is dissolved in boiling absolute ethanol. The solution is allowed to cool, and is kept at room temperature.[c] The resulting crystals of compound **2** are removed by filtration; yield, 46.3 g. (69.5%). Recrystallization from absolute ethanol affords pure **2**, m.p. 114–115°, $[\alpha]_D^{19}$ $-20.0°$ (*c* 4, chloroform).

2,3,5,6-Tetra-*O*-benzoyl-D-allose[2] (4)

To 100 ml. of tetrahydrofuran containing 0.20 mole of disiamylborane (**3**), under a nitrogen atmosphere, is slowly added 29.7 g. (0.05 mole) of **2** in 85 ml. of tetrahydrofuran. The solution is kept under nitrogen, at room temperature, overnight. Excess of compound **3** is then decomposed by the careful addition of 15 ml. of water. The solution is refluxed for 30 min., and cooled to 0°, and 30 ml. of 30% hydrogen peroxide is slowly added while the pH is maintained between 7 and 8 by addition of 3 *N* sodium hydroxide.[6] The solution is concentrated to a small volume, and extracted several times with chloroform. The chloroform extracts are combined, washed with water, dried (magnesium sulfate), and evaporated to a sirup. The sirup is dissolved in a small volume of chloroform, absolute ethanol is added, and petroleum ether (b.p. 30–60°) is then added to incipient turbidity. On chilling the solution, compound **4** crystallizes; yield, in two crops, 25 g. (83%). Recrystallization from ethanol–chloroform gives pure **4**, m.p. 148–149°, $[\alpha]_D^{20}$ $+83.0°$ (*c* 4, chloroform).

[c] Complete drying and removal of pyridine can be ensured by addition of toluene to the sirup, followed by distillation of the toluene, affording a sirup which is dissolved in ethanol.

2,3,5,6-Tetra-*O*-benzoyl-1-*O*-(*p*-nitrobenzoyl)-D-allose[2] (5)

A solution of 30 g. (0.05 mole) of D-allofuranose tetrabenzoate (**4**) in 450 ml. of pyridine is cooled to 0°, and 30 g. (0.16 mole) of *p*-nitrobenzoyl chloride is added. The solution is stirred at room temperature for 16 hr., concentrated under diminished pressure (25°) to 200 ml., and poured into a mixture of ice and saturated aqueous sodium hydrogen carbonate (1.5 l.), and the mixture is stirred vigorously. When all of the ice has melted, the product (**5**) is filtered off, and washed with water. Recrystallization from absolute methanol–chloroform affords 26.3 g. (68%) of colorless needles of **5**, m.p. 163–164°, $[\alpha]_D^{23}$ −2.5° (*c* 8.14, chloroform).

2,3,5,6-Tetra-*O*-benzoyl-D-allosyl chloride[7] (6)

Compound **5** (21 g., 0.028 mole) is dissolved in 450 ml. of dichloromethane presaturated at 0° with dry hydrogen chloride. Acetyl chloride (30 ml.) is added (to keep the system dry), and the solution is kept at −10° for 4 days and at +4° for 1 day. A precipitate of *p*-nitrobenzoic acid is filtered off, and the filtrate is evaporated to dryness under diminished pressure at 25°. To the resulting sirup, three portions of benzene are added and evaporated off (to remove residual hydrogen chloride and acetic acid). The resulting sirup contains compound **6**.

Chloromercuri-*N*-benzoyladenine[8,9] (7)

Benzoyl chloride (43 ml., 52.4 g., 0.37 mole) is slowly added to a suspension of 20 g. (0.15 mole) of adenine in 50 ml. of dry pyridine. The mixture is refluxed for 2 hr. and the pyridine is removed by evaporation. The residual, gummy mass is triturated with warm, saturated, aqueous sodium hydrogen carbonate solution, and the aqueous solution and gummy residue are transferred to a separatory funnel with the aid of chloroform. The mixture is well shaken, and the chloroform layer is removed and washed once with saturated, aqueous sodium hydrogen carbonate solution. Crystallization of *N*-benzoyladenine begins almost immediately, and is permitted to continue overnight in a refrigerator. The product is filtered off, and the filtrate is concentrated to a gelatinous residue which is dissolved in hot absolute ethanol; the solution is cooled, and further crystallization of product occurs. The crystals are filtered off and the filtrate is carefully concentrated to a volume from which further crystallization occurs.[d] Filtration and concentration are repeated twice more. The crops of crystalline material (obtained from chloroform and from ethanol) are combined, and recrystallized from 2-methoxyethanol (methyl Cellosolve); yield, in three crops, 29.6 g. (84%) of *N*-benzoyl adenine, m.p. 241–242°.

[d] Concentration to too small a volume leads to gel formation.

N-Benzoyladenine (29.6 g., 0.12 mole) is added to 600 ml. of boiling, 50% aqueous ethanol in a 2-l. beaker, and the mixture is stirred with a magnetic stirrer. One equivalent (122 ml.) of *N* sodium hydroxide is slowly added, giving a clear solution. To this is added dropwise a solution of mercuric chloride (1 equiv., 33.4 g.) in the minimal volume of hot, 95% ethanol; a white precipitate forms immediately. After all of the mercuric chloride has been added, the mixture is stirred for a few minutes, 200 ml. of water is added, and the mixture is allowed to cool. The suspension is filtered with suction, and the solid is successively washed thoroughly with water and absolute ethanol; yield 57.1 g. (97%) of compound **7**.

N-Benzoyl-9-(2,3,5,6-tetra-*O*-benzoyl-β-D-allosyl)adenine[3] (8)

To 500 ml. of dry xylene is added 13.4 g. (0.028 mole) of compound **7**, 13.4 g. of Celite 535, and 6 g. of cadmium carbonate. Traces of moisture are removed by distillation of 230 ml. of the xylene. A solution of the halide (**6**) in 260 ml. of dry xylene is added, and the suspension is refluxed for 4 hr., and filtered while hot, giving a filter cake (*A*). The filtrate is evaporated to a solid residue (*B*) under diminished pressure at 50°. The filter cake (*A*) is washed with 250 ml. of boiling chloroform, and the chloroform wash is used to dissolve the solid residue (*B*). The resulting solution is successively washed with four 100-ml. portions of aqueous potassium iodide (30%) and three 100-ml. portions of water, dried (anhydrous magnesium sulfate), and evaporated to dryness under diminished pressure at 40°. The white residue is dissolved in chloroform, and absolute methanol is added to the solution. The resulting crystals of **5** are filtered off, and the filtrate is concentrated to a sirup; this is dissolved in chloroform, and absolute methanol is added, resulting in a second crop of crystals of **5**. In this way, 8 g. of unreacted **5** is obtained. The filtrate contains **8**.

9-β-D-Allofuranosyladenine[3] (9)

The solution of **8** is evaporated to a sirup, which is dissolved in 200 ml. of hot methanol; 14 ml. of *M* sodium methoxide in methanol is added, and the solution is refluxed for 1 hr., rendered neutral with glacial acetic acid, and evaporated to a black residue which is partitioned between 150 ml. of water and 150 ml. of chloroform. The chloroform layer is extracted with two 50-ml. portions of water, and the aqueous extracts are combined, decolorized with decolorizing carbon (Norit A), and filtered through a pad of Celite 535. The nucleoside (**9**) in the aqueous solution is contaminated with a small amount of adenine, and is purified *via* the picrate[10] as follows.

The aqueous solution of **9** is evaporated to a sirup, which is dissolved in 40 ml. of hot absolute methanol. To the methanolic solution is added

150 ml. of a 10% solution of picric acid in methanol; a yellow precipitate is immediately formed. The suspension is kept in an ice bath for several hours, and the product is filtered off; a second crop of the product is obtained by concentrating the filtrate. The crops are combined, washed successively with cold absolute methanol and cold water, and recrystallized from boiling water; yield 1.63 g. (18%) of the picrate, m.p. 195–225° (dec.), $[\alpha]_D^{23}$ −23.4° (*c* 2.77, *N,N*-dimethylformamide).

The picrate is dissolved in 500 ml. of hot water in a beaker, and sufficient Bio-Rad AG-1 X8 (CO_3^{2-}) (200–400 mesh) anion-exchange resin is added to discharge the yellow color. The suspension is stirred for 1 hr., and filtered. The resin is washed with 200 ml. of water, and the filtrate and washings are combined, and evaporated under diminished pressure at 40°. The colorless residue is recrystallized from warm ethanol–water; yield 0.49 g. (5.9% from **5**) of **9**, m.p. 262–264° (dec.), $[\alpha]_D^{22}$ −57.2° (*c* 3.67, *N* hydrochloric acid).

REFERENCES

(1) H. C. Brown, *Hydroboration*, W. A. Benjamin, Inc., New York, N. Y., 1962; G. Zweifel, K. Nagase, and H. C. Brown, *J. Am. Chem. Soc.*, **84**, 183, 190 (1962); H. C. Brown and D. B. Bigley, *ibid.*, **83**, 486 (1961).
(2) P. Kohn, R. H. Samaritano, and L. M. Lerner, *J. Am. Chem. Soc.*, **87**, 5475 (1965).
(3) P. Kohn, R. H. Samaritano, and L. M. Lerner, *J. Org. Chem.*, **31**, 1503 (1966).
(4) P. A. Levene and G. M. Meyer, *J. Biol. Chem.*, **76**, 513 (1928).
(5) J. W. Pratt and N. K. Richtmyer, *J. Am. Chem. Soc.*, **77**, 1906 (1955).
(6) H. C. Brown and B. C. Subba Rao, *J. Am. Chem. Soc.*, **78**, 5694 (1956).
(7) (a) J. Davoll, B. Lythgoe, and A. R. Todd, *J. Chem. Soc.*, **1948**, 967. (b) B. R. Baker and R. E. Schaub, *J. Am. Chem. Soc.*, **77**, 5900 (1955).
(8) M. W. Bullock, J. J. Hand, and E. L. R. Stokstad, *J. Org. Chem.*, **22**, 568 (1957).
(9) J. Davoll and B. A. Lowy, *J. Am. Chem. Soc.*, **73**, 1650 (1951).
(10) B. R. Baker and K. Hewson, *J. Org. Chem.*, **22**, 959 (1957).

[45] 9-(2,3-Anhydro-β-D-lyxofuranosyl)adenine

Synthesis of Anhydronucleosides

WILLIAM W. LEE and ABELARDO P. MARTINEZ

LIFE SCIENCES RESEARCH, STANFORD RESEARCH INSTITUTE, MENLO PARK, CALIFORNIA 94025

1, R = H (307.3) → **3** (345.3) → **4** (249.2)

2, R = $-O_2SCH_3$ (385.4)

INTRODUCTION

Such anhydronucleosides as **4** are versatile, nucleoside intermediates. Substituents are readily introduced into the sugar moiety by attack of various reagents upon the epoxide grouping. The present method for synthesizing **4** proceeds from 9-(3,5-*O*-isopropylidene-β-D-xylofuranosyl)-adenine (**1**) (Sec. III [49]) by the sequence outlined above, and represents improvements in yield and convenience over the original method.[1]

PROCEDURE

9-[3,5-*O*-Isopropylidene-2-*O*-(methylsulfonyl)-β-D-xylosyl]adenine (2)

A solution of 14.0 g. (45.6 mmoles) of 9-(3,5-*O*-isopropylidene-β-D-xylofuranosyl)adenine (**1**) (Sec. III [49]) in 190 ml. of anhydrous pyridine is protected from moisture, stirred, and cooled in an ice bath while 6.8 ml. (88 mmoles) of methanesulfonyl chloride is added. The reaction mixture is kept in the ice bath for 1 hr. and then at room temperature until the reaction is complete[a,b] (3–4 days). The mixture is treated with 10 ml. of water, stirred for 30 min., diluted with 1 l. of water, and extracted with four 200-ml. portions of chloroform. The combined chloroform extracts are washed with 250 ml. of water, dried with anhydrous magnesium sulfate, and evaporated to dryness (rotary, vacuum evaporator) to give 15.2 g. (87%) of[1] **2** as colorless crystals, m.p. 214–215.5°, homogeneous by thin-layer chromatography (tlc).

9-[2-*O*-(Methylsulfonyl)-β-D-xylofuranosyl]adenine (3)

A solution of 15.2 g. (39.5 mmoles) of 9-[3,5-*O*-isopropylidene-2-*O*-(methylsulfonyl)-β-D-xylofuranosyl]adenine (**2**) in 230 ml. of 80% acetic acid (aqueous) is heated on a steam bath until the reaction is complete[c] (about 40 min.[d]). The solution is evaporated to dryness (rotary, vacuum evaporator; bath temperature, 45–50°), the residue is dissolved in 30 ml. of hot methanol, and the solution is diluted with 150 ml. of toluene and evaporated to dryness, giving a solid foam that is crystallized from 50 ml. of methanol. The crystals are collected, washed with 50 ml. of cold methanol (at about 0°), and dried *in vacuo* at 56°, to afford 10.9 g. (80%) of[1] **3**, m.p. 172–173°, homogeneous by tlc.

9-(2,3-Anhydro-β-D-lyxofuranosyl)adenine (4)

A mixture of 10.9 g. (31.6 mmoles) of crystalline[e] 9-[2-*O*-(methylsulfonyl)-β-D-xylofuranosyl]adenine[1] (**3**), 135 g. of Amberlite IR-45 (OH^-)[f] ion-exchange resin, and 1 l. of methanol in a 2-l., round-bottomed flask is stirred (overhead stirrer) in a bath at[g] 35–38° until the reaction is complete[h] (4–5 days). The resin[i] is collected by filtration and thoroughly

[a] Aliquots are removed at intervals and checked by thin-layer chromatography[b] (tlc) in solvent A for completeness of reaction. The R_f values are 0.33 and 0.49 for starting material (**1**) and product (**2**), respectively. It is important that the reaction be complete before proceeding with isolation of **2**.

[b] All thin-layer chromatograms are obtained on silica gel HF. The solvent systems used are: A, 1:9 (v/v) methanol–ethyl acetate; B, 1:4 (v/v) methanol–ethyl acetate.

[c] The completeness of reaction is followed by tlc in[b] solvent A. Compound **3** has R_f 0.30.

[d] The reaction has been repeated many times, and is generally complete in this length of time. The longer heating originally reported[1] is unnecessary.

[e] It is not necessary to use crystalline **3**, which is obtained in about 80–85% yield from the deacetonation of **2**. Using amorphous **3** (obtained in 100% yield from **2**), the overall yield from **2** → **3** → **4** is about the same.

[f] A weakly basic anion-exchange resin in the OH^- form, manufactured by Rohm and Haas Co., Philadelphia, Pa.

[g] A reaction temperature of 50° was also tried. The yields and purity of the product were comparable to those given here.

[h] Aliquots are checked at intervals (for completeness of reaction) by tlc in solvent[b] B. The R_f values for **3** and **4** are 0.27 and 0.52, respectively.

[i] All of the methanesulfonate ions are held by the resin, and purification of **4** hrough the picrate is unnecessary;[1] the product crystallizes readily.

washed with several portions of methanol (total, 800 ml.). The washings and filtrate are combined and concentrated (rotary, vacuum evaporator) to about 60 ml. The colorless crystals are collected, and dried at 56°/1 mm. Hg, to afford 5.5 g. (70%) of 9-(2,3-anhydro-β-D-lyxofuranosyl)adenine (**4**), m.p. 210–211°, $[\alpha]_D^{22}$ $-14°$ (*c* 1, water), homogeneous[j] by tlc. A second crop of 0.6 g. (8%), m.p. 208–210°, homogeneous by tlc, brings the total yield to 6.1 g. (78%).

REFERENCE

(1) E. J. Reist, A. Benitez, L. Goodman, B. R. Baker, and W. W. Lee, *J. Org. Chem.*, **27**, 3274 (1962).

[j] In addition to tlc (see footnote *b*), paper chromatography may be used to demonstrate homogeneity. The chromatograms are obtained on Whatman No. 1 paper by a descending technique in two solvent systems: (1) butanol saturated with water (**4** has $R_{Adenosine}$ 0.88), and (2) aqueous 5% disodium hydrogen phosphate, pH 8.9 (**4** has $R_{Adenosine}$ 1.32).

[46] 9-β-D-Arabinofuranosyladenine (Spongoadenosine)

Direct Synthesis of a cis *Nucleoside*

C. P. J. GLAUDEMANS and HEWITT G. FLETCHER, JR.

NATIONAL INSTITUTE OF ARTHRITIS AND METABOLIC DISEASES, NATIONAL INSTITUTES OF HEALTH, PUBLIC HEALTH SERVICE, U. S. DEPARTMENT OF HEALTH, EDUCATION, AND WELFARE, BETHESDA, MARYLAND 20014

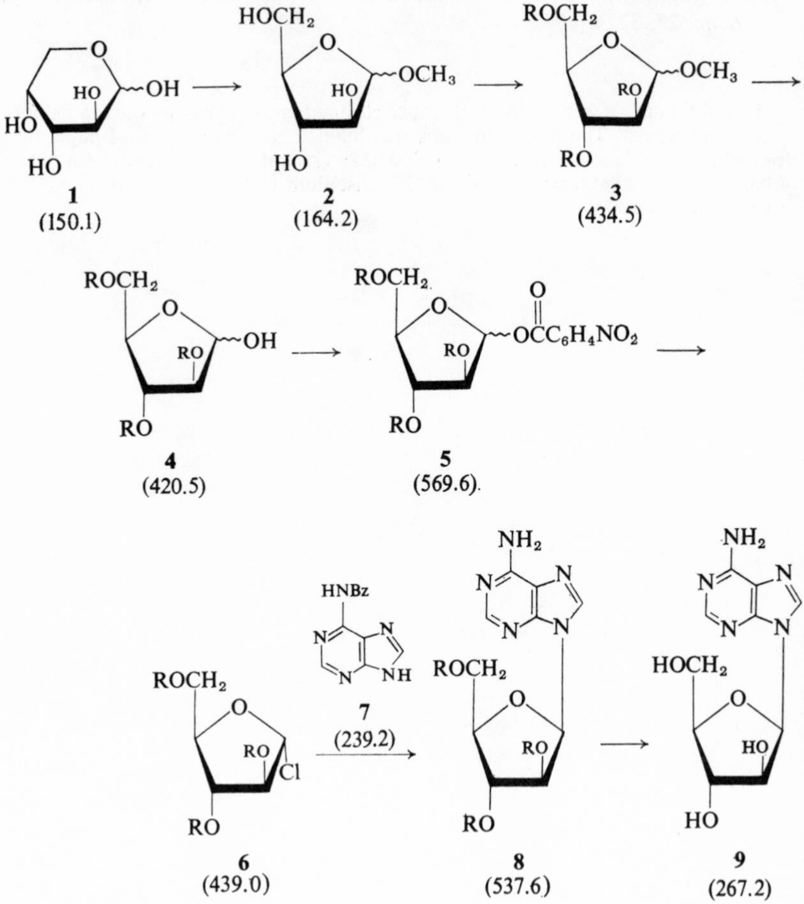

1 (150.1) → 2 (164.2) → 3 (434.5) →

4 (420.5) → 5 (569.6) →

6 (439.0) + 7 (239.2) → 8 (537.6) → 9 (267.2)

where R is $C_6H_5CH_2$.

INTRODUCTION

Owing to their cytotoxic and antiviral activities, some of the β-D-arabinofuranosyl nucleosides have attracted widespread attention.[1] As illustrated by **9**, the aglycon in these substances is *cis* to the oxygen function at C-2 in the sugar moiety, hence the designation "*cis* nucleoside." Since the condensation of fully acylated glycosyl halides with purines and pyrimidines involves participation of the acyl group at C-2, the preponderant product is normally a *trans* nucleoside. Of the various approaches to the synthesis of *cis* nucleosides, two may be mentioned. First, a synthesis may be designed involving inversion of the configuration at C-2 in the carbohydrate moiety of a pre-formed *trans* nucleoside,[2–5] and second, a nonparticipating group may be used to mask the hydroxyl group at C-2 in the glycosyl halide to be used for nucleoside synthesis; in actual practice, it is more convenient to mask all of the positions in the halide with a single type of nonparticipating group. This second strategy for the synthesis of *cis* nucleosides is here illustrated by the synthesis of 9-β-D-arabinofuranosyladenine[6] (**9**) through the use of 2,3,5-tri-*O*-benzyl-α-D-arabinofuranosyl chloride (**6**); compound **6** has also been used for the synthesis of 1-β-D-arabinofuranosylcytosine,[7] 1-β-D-arabinofuranosyl-5-(trifluoromethyl)-uracil,[7] 1-β-D-arabinofuranosyl-5-fluorouracil,[8] and 1-β-D-arabinofuranosylthymine.[9]

PROCEDURE

2,3,5-Tri-*O*-benzyl-D-arabinose[10–12] (4)

Thirty grams (200 mmoles) of powdered D-arabinose (**1**) and then 4.5 ml. of concentrated sulfuric acid are added to a mixture of 600 ml. (14.8 moles) of anhydrous methanol and 15 g. of Drierite (soluble anhydrite) and the suspension is stirred at room temperature for 5 hr. The reaction mixture then being devoid of reducing power (Fehling solution), it is filtered, and the filtrate is passed through a column containing 150 ml. of Amberlite IR-45, the column being washed with 400 ml. of methanol. The combined solution and washings are concentrated *in vacuo* to a heavy sirup (**2**) which is diluted with 50 ml. of freshly purified tetrahydrofuran[a,13] and reconcentrated (35–40° bath) to remove residual methanol. Freshly purified tetrahydrofuran (400 ml.) is added, and the mixture treated with 30 g. of Drierite, 156 g. (2.78 moles) of powdered potassium hydroxide,[b] and 200 ml. (1.74 moles) of benzyl chloride. The mixture is stirred and

[a] Fractional distillation from lithium aluminum hydride is an effective method for the purification of tetrahydrofuran.

[b] Hooker Chemical Corp., Niagara Falls, N. Y.

heated[c] under gentle reflux overnight, cooled, filtered through a thin layer of Filter-Cel, and concentrated *in vacuo*, finally at *ca.* 1 mm. Hg and 100° (bath). The crude, sirupy methyl 2,3,5-tri-*O*-benzyl-D-arabinoside (**3**) is dissolved in 400 ml. of glacial acetic acid, and the solution is diluted with 60 ml. of 6 *N* hydrochloric acid. The hydrolysis mixture is heated at 65° for 1.25 hr., concentrated *in vacuo* to one-third its volume, and poured into 1.5 l. of a mixture of ice and water. After being seeded,[d] the mixture is kept at 5° overnight. The aqueous layer is decanted from the partially crystalline mass, and the latter is dissolved in 200 ml. of dichloromethane. The solution is washed with cold, aqueous sodium hydrogen carbonate, dried (magnesium sulfate), filtered through a thin bed of decolorizing carbon, and concentrated *in vacuo* to a thin sirup which is dissolved in 200 ml. of cyclohexane. After being seeded, the solution is kept at room temperature for 1 hr. and at 5° overnight, to yield 40.1 g. (48%) of 2,3,5-tri-*O*-benzyl-D-arabinose (**4**)[e]; $[\alpha]^{20}_{D}$ −27.1° [*c* 2.0, 9:1 (v/v) *p*-dioxane–water, equilibrium value after 20 hr.].

2,3,5-Tri-*O*-benzyl-1-*O*-(*p*-nitrobenzoyl)-D-arabinose[10] (5)

Compound **4** (4 g., 9.51 mmoles) is dissolved in 15 ml. of dichloromethane, and to this solution is added a solution of 1.9 g. (10.2 mmoles) of *p*-nitrobenzoyl chloride in a mixture of 10 ml. of dichloromethane and 3 ml. of dry pyridine. The reaction mixture is kept at room temperature overnight, and is then washed successively with *N* hydrochloric acid, aqueous sodium hydrogen carbonate, and water. Moisture is removed with magnesium sulfate, and the solution is concentrated *in vacuo* to give 4.64 g. (86%) of a mixture of the anomers of **5**; m.p. 75–92° (cor.) and $[\alpha]^{20}_{D}$ −11.0° (*c* 6.8, dichloromethane) have been reported[10] for such a mixture.[f]

9-(2,3,5-Tri-*O*-benzyl-β-D-arabinosyl)adenine[6] (8)

Thoroughly dried **5** (10 g., 17.6 mmoles) is added to 165 ml. of dichloromethane presaturated with anhydrous hydrogen chloride at 0°. After 2 hr. at 0°, the precipitated *p*-nitrobenzoic acid (2.8 g., 97%) is removed by

[c] The slightly exothermic character of the reaction may be ignored when working on this scale. With much larger batches, care must be exercised, and it is well to add the benzyl chloride to the boiling reaction-mixture at such a rate as to ensure a modest rate of reflux.

[d] Seed crystals were originally obtained by chromatographing a sample of crude **4** on Woelm alumina (grade II), unchanged **3** being eluted with benzene, and **4** (which crystallized spontaneously from the sirupy form) with methanol.

[e] As thus obtained, **4** is a mixture of the anomers, but is fully suitable for use in the succeeding step. The α-D anomer has been reported[11] to have m.p. 78–80° (cor.), and the β-D anomer, m.p. 88–89° (cor.).

[f] The anomers may be separated by fractional recrystallization,[10] but the process is wasteful and unnecessary, as the mixture is suitable for the preparation of **6**.

filtration, and the filtrate is concentrated *in vacuo* to an almost colorless sirup which is then kept at *ca.* 0.08 mm. Hg/room temperature for 2 hr. The 2,3,5-tri-*O*-benzyl-α-D-arabinosyl chloride[g] (**6**) thus prepared is dissolved in 100 ml. of dry dichloromethane, and the solution is added to a mixture of 9 g. (37.6 mmoles) of dried *N*-benzoyladenine[15] (**7**) (Sec. III [44] and 60]) and 29 g. of a molecular sieve.[h] The reaction mixture is stirred in a glass-stoppered flask at room temperature for 1 week, filtered through a layer of Filter-Cel, and concentrated *in vacuo* to a sirup (10.2 g.) which is dissolved in 100 ml. of methanol. Methanolic barium methoxide (0.45 *M*, 85 ml.) is added, and the solution is boiled under reflux for 5 hr., becoming dark in the process. The solution is neutralized with carbon dioxide, filtered, and concentrated to a semisolid mass which is extracted with dichloromethane, the insoluble material being removed by centrifugation and thoroughly washed with fresh dichloromethane. The extracts and washings (350 ml.) are combined, diluted with cyclohexane (450 ml.), filtered through Filter-Cel, and boiled in an open flask until the temperature of the vapor has risen to 70°, when crystallization begins spontaneously. On being cooled slowly, the solution deposits a mass of fine needles (5.20 g.); one recrystallization from 5.2 parts of warm isopropyl alcohol gives almost pure **8**; 4.38 g. (46%), m.p. 125–128° (cor.), $[\alpha]_D^{20}$ +21.8° (*c* 2.0, dichloromethane). Two further recrystallizations from isopropyl alcohol raise the m.p. to 128–129° (cor.), but do not change the specific rotation.

9-β-D-Arabinofuranosyladenine (9) by Catalytic Hydrogenation[6] of 8

Palladium chloride (300 mg., 0.558 mmole) is suspended in 150 ml. of methanol, and reduced by shaking with hydrogen at room temperature. To the acidic suspension of palladium black is then added a solution of 300 mg. of **8** in 50 ml. of methanol. The reaction mixture is shaken with hydrogen at room temperature until absorption of the gas ceases; after the catalyst has been removed, the solution is passed through a column of Dowex-2 X8 (HCO_3^-) and concentrated *in vacuo* to a sirup (159 mg.) which crystallizes when rubbed with water. The pure nucleoside (**9**) is obtained in hydrous form by recrystallization from 12 ml. of hot water; 148 mg. (93%, on the anhydrous basis), m.p. 258–260° (cor.), $[\alpha]_D^{20}$ −1.7° (*c* 0.54, pyridine), λ_{max}^{EtOH} 258 nm. On being dried *in vacuo* at 100°, **9** is obtained in anhydrous form.[i]

[g] A preparation typically shows $[\alpha]_D^{20}$ +91.1° (*c* 2.27, dichloromethane); the α-D anomeric configuration has been shown by n.m.r. studies.[14]

[h] Type 4A, 1/16-in. pellets, Fisher Scientific Co., Pittsburgh, Pennsylvania.

[i] Typically, the loss of weight on drying is 5.6%; the theoretical value for the monohydrate is 6.3%.

9-β-D-Arabinofuranosyladenine (9) by Treatment of Compound 8 with Sodium–Liquid Ammonia[16]

Owing to the modest solubility of **8** in methanol, increase in the scale of the catalytic debenzylation described above soon involves awkwardly large volumes. Debenzylation through the use of sodium in liquid ammonia is fully as effective and better suited for larger-scale operations.

To a stirred suspension of 3.75 g. (6.98 mmoles) of **8** in 160 ml. of liquid ammonia is added a total of 600 mg. (26 mg.-atoms) of sodium in portions during 10–12 min., by which time the characteristic, deep-blue color of the sodium persists. At this point, the blue color is discharged by careful addition of ammonium chloride, and the reaction mixture is evaporated to dryness under a stream of nitrogen. The solid residue is triturated with 50 ml. of benzene (to remove bibenzyl) and then dissolved in 40 ml. of water. The aqueous solution is treated with decolorizing carbon, and then acidified with acetic acid to precipitate the product, which is dried *in vacuo* at 100°; yield 1.53 g. (82%).

REFERENCES

(1) S. S. Cohen, *Progr. Nucleic Acid Res.*, **5**, 1 (1966).

(2) D. M. Brown, A. R. Todd, and S. Varadarajan, *J. Chem. Soc.*, **1956**, 2388.

(3) J. J. Fox, N. Yung, and A. Bendich, *J. Am. Chem. Soc.*, **79**, 2775 (1957).

(4) E. R. Walwick, W. K. Roberts, and C. A. Dekker, *Proc. Chem. Soc.*, **1959**, 84.

(5) E. J. Reist, A. Benitez, L. Goodman, B. R. Baker, and W. W. Lee, *J. Org. Chem.*, **27**, 3274 (1962).

(6) C. P. J. Glaudemans and H. G. Fletcher, Jr., *J. Org. Chem.*, **28**, 3004 (1963).

(7) T. Y. Shen, H. M. Lewis, and W. V. Ruyle, *J. Org. Chem.*, **30**, 835 (1965).

(8) F. Keller, N. Sugisaka, A. R. Tyrrill, L. H. Brown, J. E. Bunker, and I. J. Botvinick, *J. Org. Chem.*, **31**, 3842 (1966).

(9) F. Keller and A. R. Tyrrill, *J. Org. Chem.*, **31**, 1289 (1966).

(10) R. Barker and H. G. Fletcher, Jr., *J. Org. Chem.*, **26**, 4605 (1961).

(11) S. Tejima and H. G. Fletcher, Jr., *J. Org. Chem.*, **28**, 2999 (1963).

(12) Compound **4** is available from Pfanstiehl Laboratories, Inc., Waukegan, Ill.

(13) Here, and in the benzylation step that follows, tetrahydrofuran could probably be replaced advantageously with *p*-dioxane, since methyl α-D-glucopyranoside is completely benzylated in 2 hr. through the use of a boiling mixture of *p*-dioxane, benzyl chloride, and potassium hydroxide: T. D. Perrine, C. P. J. Glaudemans, R. K. Ness, J. Kyle, and

H. G. Fletcher, Jr., *J. Org. Chem.*, **32**, 1664 (1967). Attention is also drawn to the improved technique (involving steam-distillation) that these authors devised for processing their benzylation mixture; doubtless, the same process could be applied here.

(14) C. P. J. Glaudemans and H. G. Fletcher, Jr., *J. Am. Chem. Soc.*, **87**, 4636 (1965).

(15) M. W. Bullock, J. J. Hand, and E. L. R. Stokstad, *J. Org. Chem.*, **22**, 568 (1957).

(16) E. J. Reist, V. J. Bartuska, and L. Goodman, *J. Org. Chem.*, **29**, 3725 (1964).

[47] The Anomeric 2′-Deoxyadenosines [9-(2-Deoxy-α-D-*erythro*-pentofuranosyl)adenine and Its Anomer

Synthesis of Nucleosides by Condensation of a Partially Protected Aldose Dithioacetal with a Heavy-Metal Salt of a Purine

CHRISTIAN PEDERSEN

POLYTEKNISK LAEREANSTALT, LYNGBY, DENMARK

and

HEWITT G. FLETCHER, JR.

NATIONAL INSTITUTE OF ARTHRITIS AND METABOLIC DISEASES, NATIONAL INSTITUTES OF HEALTH, PUBLIC HEALTH SERVICE, U. S. DEPARTMENT OF HEALTH, EDUCATION, AND WELFARE, BETHESDA, MARYLAND 20014

$Me_2CHSC(H)SCHMe_2$–CH_2–HCOH–HCOH–H_2COBz **1** (372.52) + **2** (479.29) [6-benzamido-9-(chloromercuri)purine, HNBz, HgCl] → **3** (251.25) + **4** (251.25)

INTRODUCTION

Aldose dithioacetals and alkyl 1-thioaldosides react with mercury and silver salts of carboxylic acids to give C-1 esters of the aldoses (when these are stable);[1] the reaction is not stereospecific, but gives a mixture of the anomers, even from anomerically pure 1-thioaldosides. In one case[2] (described here), this type of reaction has been applied to the synthesis

›f an anomeric pair of nucleosides, 5-*O*-benzoyl-2-deoxy-D-*erythro*-pentose liisopropyl dithioacetal (**1**) being condensed with chloromercuri-*N*-›enzoyladenine (**2**) to give (after removal of the benzoyl groups) the ınomeric 9-(2-deoxy-D-*erythro*-pentofuranosyl)adenines (**3** and **4**) (Sec. II [60] and [61]) under comparatively mild conditions. The benzoyl group on the hydroxyl group at C-5 in **1** was inserted to ensure the formaion of furanosides; whether its presence is essential for this purpose is not known. In general, it is probable that suitably masked alkyl 1-thioaldosides vould be more effective than dithioacetals for this type of synthesis.

PROCEDURE

9-(2-Deoxy-D-*erythro*-pentofuranosyl)adenines[2] (3 and 4)

One gram (2.68 mmoles) of 5-*O*-benzoyl-2-deoxy-D-*erythro*-pentose liisopropyl dithioacetal[3] (**1**) is dissolved in 50 ml. of dry acetonitrile,[a] and 5.07 g. (10.6 mmoles) of chloromercuri-*N*-benzoyladenine[4] (**2**) (Sec. III 44]) is added. The mixture is stirred and boiled under reflux for 4 hr., and hen concentrated *in vacuo*, the residue being extracted several times with lichloromethane. The extracts are combined, and concentrated *in vacuo*, ınd the resulting solid is dissolved in 20 ml. of methanol; the solution is reated with 6 ml. of *M* methanolic barium methoxide, and boiled under eflux for 4 hr. After neutralization with carbon dioxide, the solution is evaporated to dryness, and the residue is treated with 20 ml. of water. The solution is filtered, and concentrated to dryness, and the residue is extracted with methanol (3 × 25 ml.); removal of solvent from the com›ined methanolic extracts affords a residue (700 mg.) which is dissolved in a mixture of 20 ml. of methanol and 2 ml. of water. The solution is put on a column of powdered cellulose (700 g.), and the column is eluted with 32:9:2 (v/v) isopropyl ether–ethanol–water, 20-ml. fractions of eluate ›eing collected. Fractions 190 to 272 contain adenine; fractions 360 to 475 are pooled and concentrated to dryness. The residue is extracted with hot methanol (3 × 100 ml.), and the combined extracts are freed of solvent *in vacuo*. The product (150 mg.) is dissolved in *ca.* 5 ml. of water, and the solution is treated with a small amount of decolorizing carbon, filtered, and concentrated to a volume of 2 ml. Nucleating crystals of **3** are added, and the solution is stored at 5° overnight to yield 42.4 mg. (6.3%) of **3**, m.p. 165° (partial), 190–192° (cor.),[b] $[\alpha]_D^{20}$ −27.0° (*c* 0.40, water), λ_{min} 225 nm., λ_{max} 260 nm. (AM = 15,220).

[a] The acetonitrile may be dried by distillation from phosphorus pentaoxide. In a single experiment,[2] methyl sulfoxide has been used instead of acetonitrile, the reaction mixture being heated at 90° for 5 hr. Chromatography showed that the yield of products was comparable to that obtained with acetonitrile as described here.

[b] This behavior on melting is highly characteristic of **3**.

Fractions 500 to 630 are pooled, and concentrated to a dry residue which is dissolved in 40 ml. of hot, absolute ethanol. After filtration, the solution is concentrated to *ca.* 10 ml. and stored at $-5°$ overnight, to give 55.0 mg. (8.1%) of crude **4**, m.p. 201–205° (cor.). Two recrystallizations from absolute ethanol yield 35.5 mg. of pure **4**, m.p. 208–210° (cor.), $[\alpha]_D^{20}$ $+69°$ (*c* 0.32, water), λ_{min} 228 nm., λ_{max} 260 nm. (AM = 16,290).

REFERENCES

(1) See H. B. Wood, Jr., B. Coxon, H. W. Diehl, and H. G. Fletcher, Jr. *J. Org. Chem.*, **29**, 461 (1964), and references therein.
(2) C. Pedersen and H. G. Fletcher, Jr., *J. Am. Chem. Soc.*, **82**, 5210 (1960)
(3) H. Zinner and H. Nimz, *Chem. Ber.*, **91**, 1657 (1958).
(4) This substance is best prepared by a modification of the method of J. Davoll and B. A. Lowy [*J. Am. Chem. Soc.*, **73**, 1650 (1951)] that was devised by Dr. J. J. Fox and his co-workers; *cf.* B. R. Baker, K Hewson, H. J. Thomas, and J. A. Johnson, Jr., *J. Org. Chem.*, **22**, 95 (1957).

[48] Anomeric 9-D-Glucopyranosyladenines and 9-D-Glucopyranosylhypoxanthines

The Synthesis of Anomeric D-Glucosylpurines by the Trimethylsilyl Method

ISSEI IWAI, TAKUZO NISHIMURA, and BUNJI SHIMIZU

RESEARCH DEPARTMENT, SANKYO CO., TOKYO, JAPAN

AcOCH$_2$ O OAc AcO X OAc

5, X = Cl (366.7)
5a, X = Br (411.2)

$Me_3SiNHSiMe_3$ or Me_3SiCl

1, R = BzNH (239.2)
2, R = OH (136.1)

Si(Me)$_3$

3, R′ = N(Bz)(SiMe$_3$) (383.5)
4, R′ = OSiMe$_3$ (280.4)

AcOCH$_2$ OAc AcO OAc

6, R = BzNH (569.5)
7, R = OH (466.4)

HOCH$_2$ OH HO HO

8a, R″ = NH$_2$ (297.3)
9a, R″ = OH (298.3)

HOCH$_2$ OH HO OH

8b, R″ = NH$_2$ (297.3)
9b, R″ = OH (298.3)

INTRODUCTION

A D-glucosyladenine phosphate has been isolated from Coka-100 wilt resistant cotton leaves,[1] but its structure and biological role are not yet clarified. A method for the synthesis of 9-D-glucopyranosyladenine using a tetra-*O*-acetyl-α-D-glucosyl halide has been shown to give the β-D anomer;[2] however, coupling of the trimethylsilyl derivative (**3**) of *N*-benzoyladenine (**1**), followed by deacylation with methanolic sodium methoxide gives 9-α-D-glucopyranosyladenine (**8a**), together with the β-D anomer[3] (**8b**). Preparations of the anomeric 9-D-glucopyranosylhypoxanthines (**9a** and **9b**) are also given.[3]

PROCEDURE

N-Benzoyl-*N*,9-bis(trimethylsilyl)adenine[4] (3)

To a suspension of 24.0 g. (0.1 mole) of *N*-benzoyladenine (**1**) (Sec. II [44] and [60]) and 26.0 g. of chlorotrimethylsilane in 300 ml. of dry benzene is added dropwise, with mechanical stirring, a solution of 24.2 g of triethylamine in 100 ml. of benzene. Stirring is continued for 10 hr. the mixture is filtered, and the filter cake is washed with three 30-ml portions of dry benzene. The filtrate and washings are combined, the solvent is evaporated, and the resulting, viscous oil is distilled under diminished pressure to give 25.3 g. (75%, based on **1**) of **3**, b.p. 177–184°/ 8×10^{-4} mm. Hg, $\lambda_{max}^{p\text{-dioxane}}$ 237 nm. On treatment of the still residue with aqueous ethanol, 3.17 g. of unreacted **1** is recovered.

N,9-Bis(trimethylsilyl)hypoxanthine[4] (4)

A suspension of 4.2 g. (0.031 mole) of dry, powdered hypoxanthine (**2**) in 10.0 g. of hexamethyldisilazane is placed in a 100-ml., round-bottomed flask and refluxed for 12 hr. The crystals of hypoxanthine gradually dissolve, and, at the end of the reaction period, the mixture becomes homogeneous. After the solution has been cooled, the solvent is removed by evaporation, and the thick, oily residue is distilled *in vacuo* to yield 7.5 g (87%) of **4**, b.p. 113–117°/0.15 mm. Hg. The product slowly crystallizes at room temperature, m.p. 71–74°, $\lambda_{max}^{p\text{-dioxane}}$ 249 and 252 nm.

The Anomeric 9-D-Glucopyranosyladenines[3] (8a and 8b)

N-Benzoyl-*N*,9-bis(trimethylsilyl)adenine (**3**) (3.83 g., 0.01 mole) and 4.1 g. of 2,3,4,6-tetra-*O*-acetyl-α-D-glucosyl chloride[5] (**5**) are heated at 150–160° for 4 hr. The mixture is then dissolved in aqueous ethanol, the solvent is evaporated off, and the residue (**6**) is dissolved in 150 ml. of methanol containing 3 g. of sodium methoxide. The solution is refluxed for 45 min. and placed on a column of 50 ml. of Dowex-50 (H^+) ion exchange resin which is thoroughly washed with aqueous methanol. Elution with *N* ammonium hydroxide, and evaporation of the eluate, gives 1.9 g. (64%, based on **3**) of a mixture of **8a** and **8b**, which is dissolved in 50 ml. of 0.01 *M* ammonium formate buffer (pH 11). The solution is applied to the top of a column (3 × 21 cm.) of Dowex-1 (OH^-) prepacked with water. After the column is washed with 700 ml. of the same buffer, the β-D anomer is eluted with ammonium formate buffer (pH 10.2). The eluate is evaporated to dryness, and the residue is crystallized from water to give 0.8 g. (47%, based on the anomeric mixture) of **8b**, m.p. 204–207°, $[\alpha]_D^{30}$ −8.5° (*c* 1.5, water), $\lambda_{max}^{H_2O}$ 260 nm. The α-D anomer is eluted with ammonium formate buffer (pH 9.1), and evaporation of the solvent and sublimation of the inorganic salt gives 0.4 g. (21%, based on the anomeric mixture) of amorphous **8a**, $[\alpha]_D^{28}$ +91.4° (*c* 0.9, water), $\lambda_{max}^{H_2O}$ 260 nm.

Anomeric 9-D-Glucopyranosylhypoxanthines (9a and 9b)

A solution of the anomeric mixture (0.5 g.) of **8a** and **8b** in 40 ml. of water is treated with 1.0 g. of sodium nitrite and 1.3 ml. of acetic acid, and the mixture is kept at room temperature overnight. After the addition of more sodium nitrite (0.3 g.) and acetic acid (0.8 ml.), the solution is kept for an additional 7 hr. and then evaporated to dryness under diminished pressure. An aqueous solution of the residue is applied to a column (3 × 10 cm.) of IR-120 (H^+) ion-exchange resin, and the column is washed with water. 9-β-D-Glucopyranosylhypoxanthine (**9b**) is eluted from the column before the α-D anomer (**9a**), and the eluates are separately evaporated to dryness, and recrystallized from aqueous ethanol to give 260 mg. (52%) of **9b**, m.p. 232–235°, $[\alpha]_D^{26}$ +117.5° (*c* 1.3, water) and 160 mg. (32%) of **9a**, m.p. 277–280°, $[\alpha]_D^{27}$ −38.5° (*c* 2.7, *N* sodium hydroxide).

9-(2,3,4,6-Tetra-*O*-acetyl-β-D-glucosyl)hypoxanthine[3] (7)

N,9-Bis(trimethylsilyl)hypoxanthine (**4**) (1.4 g., 5 mmoles) is heated at 160–165°, and 2.05 g. of 2,3,4,6-tetra-*O*-acetyl-α-D-glucosyl bromide[6] (**5a**) is added, in portions, to the fused sirup during 10 min. After the mixture is heated for an additional 10 min., the bromotrimethylsilane produced is removed under diminished pressure. After treatment with hot, aqueous ethanol, the cooled mixture is dissolved in chloroform, and insoluble hypoxanthine (0.2 g.) is recovered. The chloroform solution is evaporated to dryness under diminished pressure, and the residue is washed several times with ether. The ether-insoluble material (*ca.* 2.1 g.) is dissolved in chloroform, and the solution is applied to a column (4 × 6 cm.) of silica gel, which is eluted with chloroform, followed by 9:1 chloroform–ethyl acetate. After removal of unreacted sugar, the solvent is changed to

19:1 chloroform–methanol, and the nucleoside fractions are collected and evaporated (at this stage, the product is not very pure). Rechromatography of the material on silica gel, by elution with chloroform–ethyl acetate and subsequent treatment with ethanol, gives crude **7**. After several recrystallizations from methanol, 322 mg. (19%, based on **4**) of the pure product is obtained, m.p. 265–267.5°, $[\alpha]_D^{28}$ $-19.2°$ (*c* 3.06, chloroform), λ_{max}^{EtOH} 244 nm.

9-β-D-Glucopyranosylhypoxanthine[3] (9b)

Compound **7** (150 mg., 0.32 mmole) is dissolved in 20 ml. of methanol, and the solution is saturated at 0° with dry ammonia. The reaction flask is stoppered, and kept in a refrigerator for 2 days; the solvent is evaporated, and the acetamide is then removed by sublimation. The residue is recrystallized twice from aqueous ethanol, giving 76 mg. (80%) of pure **9b**.

REFERENCES

(1) P. H. Plaisted and R. B. Reggio, *Nature*, **193**, 685 (1962).
(2) J. Davoll, B. Lythgoe, and A. R. Todd, *J. Chem. Soc.*, **1946**, 833.
(3) T. Nishimura and B. Shimizu, *Agr. Biol. Chem.* (Tokyo), **28**, 224 (1964).
(4) T. Nishimura and I. Iwai, *Chem. Pharm. Bull.* (Tokyo), **12**, 352 (1964).
(5) R. U. Lemieux, *Methods Carbohyd. Chem.*, **2**, 223 (1963).
(6) R. U. Lemieux, *Methods Carbohyd. Chem.*, **2**, 221 (1963).

[49] 9-(3,5-*O*-Isopropylidene-β-D-xylofuranosyl)adenine

Preparation of an Isopropylidene Acetal of a Nucleoside

OSBORNE P. CREWS, JR., and LEON GOODMAN

LIFE SCIENCES RESEARCH, STANFORD RESEARCH INSTITUTE, MENLO PARK, CALIFORNIA 94025

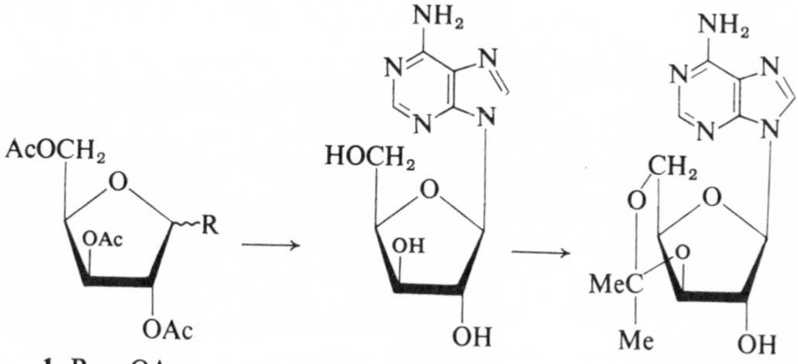

1, R = OAc (318.3)
2, R = Br (339.2)

3 (267.2)

4 (307.3)

INTRODUCTION

The antitumor activity[1] and the utility of 9-β-D-xylofuranosyladenine (**3**) as an intermediate in the preparation of other nucleosides[2] emphasize the need for a convenient synthesis of **3**. The procedure described herein utilizes the readily accessible 1,2,3,5-tetra-*O*-acetyl-D-xylose (**1**) instead of the difficultly obtainable 1,2,3,5-tetra-*O*-benzoyl-D-xylose used in the original preparation[3] of **3**. The 3,5-*O*-isopropylidene acetal[2a,3,4] (**4**) is a highly crystalline compound and provides a convenient means of purification of **3** as it can be readily reconverted into **3**. Furthermore, **4** is usually the first intermediate in the transformation of **3** into other adenine nucleosides.

PROCEDURE

2,3,5-Tri-*O*-acetyl-D-xylosyl Bromide (2)

To a solution of 180 g. of hydrogen bromide in 700 ml. of *p*-dioxane (reagent grade) is added 102.5 g. (0.322 mole) of the tetraacetate[5] (**1**), the temperature being kept below 20°. The flask is stoppered and the bright-yellow solution is kept at room temperature for 1 hr., and then evaporated below 55°, using an efficient aspirator, to about 250 ml. The solution darkens considerably during this treatment. Toluene (400 ml.) is added to the solution, and the mixture is evaporated to about 150 ml. A second portion of toluene (200 ml.) is added, and the mixture is evaporated to dryness *in vacuo*. To remove the last traces of hydrogen bromide, the residue is kept for 45 min. at 55°/1 mm. Hg. The residual dark sirup weighs 105.8 g. (97%) and is used immediately in the next step.

9-β-D-Xylofuranosyladenine (3)

In a 5-l., three-necked, round-bottomed flask is placed 223 g. of a mixture of 80 g. of Celite and 143 g. (0.30 mole) of chloromercuri-*N*-benzoyladenine (Sec. III [44]) and 4 l. of xylene (technical). The mixture is dried by azeotropic distillation of about 500 ml. of xylene, and then 105.5 g. (0.297 mole)[a] of the bromide **2**, dissolved in 250 ml. of dry xylene, is added to the hot mixture. The mixture, protected from atmospheric moisture, is boiled under reflux for 2 hr. and filtered hot through Celite, and the filtrate is evaporated *in vacuo* at 75°. The filter cake is extracted with three 1,200-ml. portions of boiling chloroform, and the extracts are combined and used for dissolving the residue from the evaporation. The solution is successively washed with two 750-ml. portions of 30% aqueous potassium iodide and 900 ml. of saturated, aqueous sodium hydrogen carbonate solution, dried (magnesium sulfate), and evaporated[b] *in vacuo* to give 136 g. of crude, acetylated nucleoside.

A solution of the acetylated nucleoside (135 g.) in 850 ml. of methanol is treated with a solution of 25 g. of sodium methoxide in 250 ml. of methanol. The flask is stoppered, and the solution is kept at room temperature for 18 hr., and then brought to pH 7 by adding glacial acetic acid (pH paper). The mixture is filtered through Celite, and the filtrate is evaporated to dryness at 55° at 25 mm. and then 0.1 mm. Hg. The residual glass is dissolved in 600 ml. of methanol, the solution is cooled to 10° and, with stirring, a solution of 118 g. of picric acid in 600 ml. of methanol is slowly added. The mixture is kept cold for 30 min., and the precipitate is collected, and successively washed with two 125-ml. portions of cold methanol and then ether. The damp picrate is added to a mixture of 206 g. of Dowex-2 (CO_3^{2-}) with 1,800 ml. of water, the mixture is stirred for 18 hr. and filtered, and the resin is extracted with two 1-l. portions of hot water. The

[a] The use of a 10% mole excess of the bromide (**2**) would probably compensate for some decomposition occurring during the preparation of **2**.

[b] A small amount of *N*-benzoyladenine is precipitated during evaporation of the chloroform, and is removed by filtration.

filtrate and extracts are combined and evaporated to dryness under diminished pressure. Absolute ethanol (500 ml.) is added to the residue, and the mixture is evaporated to dryness, giving 42.2 g. of crude nucleoside (**3**) which contains an appreciable amount of adenine.

9-(3,5-*O*-Isopropylidene-β-D-xylofuranosyl)adenine (4)

Crude **3** (42.2 g.) is added to 1.75 l. of acetone containing 45 g. of ethanesulfonic acid, in a 2-l. flask. The flask is stoppered, the mixture is stirred magnetically for 18 hr., and the solution is decanted (from some insoluble, gummy material, A), poured, with stirring, into 750 ml. of saturated aqueous sodium hydrogen carbonate solution, and stirred for 30 min. The mixture is filtered, and the filtrate is concentrated under diminished pressure to about 200 ml., and extracted with four 500-ml. portions of chloroform. The chloroform extracts are combined, and evaporated to dryness under diminished pressure. Absolute ethanol (100 ml.) is added to the residue, and the mixture is evaporated *in vacuo*, giving 23 g. of residue. The residue is dissolved in 500 ml. of hot ethanol, the hot solution is rapidly filtered, and the filtrate is concentrated to about 100 ml. and cooled, yielding 13.1 g. of crystalline **4**, m.p. 204–206°.

The acetone-insoluble material A is re-treated with 600 ml. of acetone and 10 g. of ethanesulfonic acid to give a further 3.6 g. of crystalline **4**, m.p. 204–206°. Re-treatment of the insoluble material from the second crop gives a further 1.9 g. of **4**, m.p. 203–207°; total yield, 18.6 g. [17.5 %, from the tetraacetate (**1**)]. All of this material is suitable for further transformations. The analysis of the product agrees with that calculated for $C_{13}H_{17}N_5O_4 \cdot 0.5\ C_2H_5OH$.

Recrystallization from methanol gives material that, after being dried at 130°/1 mm. Hg, has m.p. 206–208° and an analysis that agrees with that calculated for $C_{13}H_{17}N_5O_4 \cdot 0.5\ CH_3OH$.

From the mother liquors, 1.9 g. of a mixture of **4** and its α-D anomer[4] is isolated. Recrystallization from isopropyl alcohol gives 0.8 g. of the α-D anomer, m.p. 260–262°.

Material having m.p. 204–206° can readily be reconverted into **3**. Thus, 10 g. of **4** in 200 ml. of 80% aqueous acetic acid is heated for 30 min. on a steam bath, and the solution is evaporated at 55° under diminished pressure. Two 75-ml. portions of toluene are added and evaporated (to remove water and acetic acid), and the residue is dried at about 2 mm. Hg, giving a semi-crystalline mass (9.8 g., m.p. 130–134°) containing one molecule of acetic acid per molecule; yield, 92%. 9-β-D-Xylofuranosyladenine (**3**) is obtained from this solvate by passing an aqueous solution through Amberlite IR-45 (OH^-) and evaporating the eluate to dryness, giving **3** as an amorphous solid.

REFERENCES

(1) (a) D. B. Ellis and G. A. LePage, *Mol. Pharmacol.*, **1**, 231 (1965); (b) *Can. J. Biochem.*, **43**, 617 (1965).

(2) (a) A. Benitez, O. P. Crews, Jr., L. Goodman, and B. R. Baker, *J. Org. Chem.*, **25**, 1946 (1960); (b) E. J. Reist, A. Benitez, L. Goodman, B. R. Baker, and W. W. Lee, *ibid.*, **27**, 3274 (1962); (c) E. J. Reist, D. F. Calkins, and L. Goodman, *Chem. Ind.* (London), **1965**, 1561.

(3) B. R. Baker and K. Hewson, *J. Org. Chem.*, **22**, 966 (1957).

(4) W. W. Lee, A. P. Martinez, G. L. Tong, and L. Goodman, *Chem. Ind.* (London), **1963**, 2007.

(5) E. J. Reist and L. Goodman, *Biochemistry*, **3**, 15 (1964).

(6) B. R. Baker and K. Hewson, *J. Org. Chem.*, **22**, 959 (1957).

[50] 9-β-D-Lyxofuranosyladenine

*Synthesis of a 1,2-*cis *Nucleoside*

ELMER J. REIST AND DIANNE F. CALKINS

LIFE SCIENCES RESEARCH, STANFORD RESEARCH INSTITUTE, MENLO PARK, CALIFORNIA 94025

1 (313.3)

2, R = Bz or H (720.7)

3 (720.7)

trans opening

cis opening

4 (267.2)

6 (267.2) + 5 (267.2)

INTRODUCTION

The synthesis of β-D-lyxonucleosides must be accomplished by indirect means, because direct coupling of a suitably protected derivative of D-lyxose would be expected to give an α-D-lyxonucleoside. Neighboring-group participation by a vicinal *O*-benzoyl group has proved useful for inverting the configuration of a carbon atom bearing a secondary hydroxyl group in a cyclic sugar derivative[1] and is employed, in this instance, for preparing 9-β-D-lyxofuranosyladenine (**4**) from the 2-methanesulfonate (**1**) of 9-β-D-xylofuranosyladenine.[2]

PROCEDURE

N-Benzoyl-9-[3,5-di-*O*-benzoyl-2-*O*-(methylsulfonyl)-β-D-xylosyl]adenine (2)

A solution of 1.96 g. (5.7 mmoles) of 9-[2-*O*-(methylsulfonyl)-β-D-xylofuranosyl]adenine (**1**) (Sec. III [45]) in 22.3 ml. of dry pyridine is cooled to 0° in an ice bath, and to the solution is added dropwise, with stirring and continued cooling, 4.05 ml. (34.9 mmoles) of benzoyl chloride. After the addition is complete, the reaction mixture is kept at 0° for 18 hr.; it is then poured, with stirring, into 50 ml. of ice–water, and the mixture is extracted with three 50-ml. portions of chloroform. The chloroform extracts are united, washed successively with saturated aqueous sodium hydrogen carbonate and water, dried, and evaporated to dryness under diminished pressure, and the product is freed of pyridine by evaporation of the residue with two 25-ml. portions of toluene. The resulting brown oil is placed on a column (2.2 × 26 cm.) of silica gel, and eluted with chloroform. After elution of some benzoic anhydride, compound **2** is eluted. Crystallization from absolute ethanol gives 3.0 g. of **2**, m.p. 97–107°, which (from the analytical data and n.m.r. spectrum) probably contains *ca.* 40% of the *N*-benzoyl-di-*O*-benzoyl derivative, and which is of satisfactory purity for the subsequent conversion.

9-β-D-Lyxofuranosyladenine (4)

A mixture of 3.0 g. of crude **2** and 3.0 g. of sodium fluoride is dried at room temperature/0.2 mm. Hg for 1 hr.; it is then dissolved in 70 ml. of freshly dried (over alumina) *N*,*N*-dimethylformamide, and the solution is heated under an atmosphere of nitrogen at 140° for 24 hr.[a] The reaction mixture is cooled, 5 ml. of water is added, and the mixture is stirred at 55–60° for 1 hr. The dark suspension is poured into 200 ml. of water, and the mixture is extracted with 200 ml. of chloroform. The chloroform layer is washed with water, dried, and evaporated to dryness, giving a dark oil. The oil is dissolved in 50 ml. of methanol presaturated with ammonia at 0°, and the solution is kept at room temperature for 18 hr. The solution is then evaporated to dryness under diminished pressure, and the residue is partitioned between 50 ml. each of chloroform and water. The aqueous phase is evaporated to dryness under diminished pressure, to give 1.26 g. of a mixture of **4**, **5**, and **6** as a brown oil.

A solution of the latter mixture in 25 ml. of water is applied to a column

[a] On a larger scale, reaction times of up to 120 hr. are needed in order to effect complete reaction.

(2.2 × 22.5 cm.) of Dowex-1 X2 (OH^-),[3] and the column is successively eluted with 750 ml. of carbon dioxide-free water, 100 ml. of 10% aqueous methanol, and 500 ml. of 60% aqueous methanol. Evaporation of the 60% aqueous methanol eluate gives 578 mg. (55%) of the D-lyxofuranosyl-adenine (**4**) as an amorphous solid having $[\alpha]_D^{21.5}$ −21° (*c* 0.3, water); $\lambda_{max}^{pH\,1}$ 256 nm. (ϵ 14,200); $\lambda_{max}^{pH\,7}$ 259 nm. (ϵ 14,900); $\lambda_{max}^{pH\,13}$ 260 nm. (ϵ 14,300).

The material is homogeneous, as shown by chromatography on paper with 4:1:5 butyl alcohol–acetic acid–water, and has R_{Ade} 0.70. It consumes 0.975 mole of periodate per mole within 15 min., indicating the presence of a *cis*-glycol grouping.

Further elution of the ion-exchange column with 60% aqueous methanol gives an incompletely resolved mixture of 9-β-D-xylofuranosyladenine (**5**) and spongoadenosine (**6**).

REFERENCES

(1) E. M. Acton, K. J. Ryan, and L. Goodman, *J. Am. Chem. Soc.*, **86**, 5352 (1964); B. R. Baker and A. H. Haines, *J. Org. Chem.*, **28**, 438 (1963).
(2) E. J. Reist, D. F. Calkins, and L. Goodman, *J. Org. Chem.*, **31**, 169 (1967).
(3) C. A. Dekker, *J. Am. Chem. Soc.*, **87**, 4027 (1965).

[51] 9-α-D-Mannofuranosyladenine

Use of Acetal Protecting Groups in the Synthesis of Hexofuranosyl Nucleosides

LEON M. LERNER and PAUL KOHN

DEPARTMENT OF BIOLOGICAL CHEMISTRY, UNIVERSITY OF ILLINOIS AT THE MEDICAL CENTER, CHICAGO, ILLINOIS 60612

1 (180.2) $\xrightarrow{Me_2CO}$ 2 (260.3) $\xrightarrow{SOCl_2}$ 3 (278.7) →

4 (481.3) $\xrightarrow{\text{1. 70\% HOAc, 50°; 2. NaOMe}}$ 5 (337.3) $\xrightarrow{\text{25\% HOAc, 100°}}$ 6 (297.2)

where Bz is benzoyl.

INTRODUCTION

The synthesis of most nucleosides has thus far been accomplished through the condensation of a suitable derivative of a purine or pyrimidine with an *O*-acylglycosyl halide. The preparation of 9-α-D-mannofuranosyladenine[a,1] (**6**) is an example of the synthesis of a nucleoside utilizing acetal protecting groups on the carbohydrate reactant in order to maintain a furanoid structure in the product. The procedure appears to be applicable to other carbohydrates, such as (D or L)-gulose, which afford isopropylidene acetals analogous in structure to that formed by D-mannose.[2]

[a] The anomeric configuration of **6** has been demonstrated[6] to be α.

PROCEDURE

2,3:5,6-Di-*O*-isopropylidene-D-mannose[3] (2)

To 600 ml. of acetone containing 14 ml. of concentrated sulfuric acid is added 20 g. of D-mannose (**1**), and the mixture is shaken mechanically. After 4 hr., all of the carbohydrate has dissolved, and the solution is rendered neutral with sodium carbonate (warming may be necessary to initiate the neutralization reaction), and refluxed with several grams of sodium carbonate for 1 hr. The suspension is treated with decolorizing carbon (Norit A), and filtered, and the filtrate is evaporated under diminished pressure to a colorless solid. The residue is dissolved in ether, and to the solution petroleum ether is cautiously added to faint opalescence to afford 26 g. (90%) of compound **2**, m.p. 115–120°. Recrystallization in the same way yields pure product, m.p. 122°, $[\alpha]_D^{19}$ +16.6° (*c* 2.5, absolute ethanol).

2,3:5,6-Di-*O*-isopropylidene-D-mannosyl chloride[4] (3)

To a mixture of 60 ml. of dry pyridine and 190 ml. of dry chloroform cooled to 0° is added 24 ml. of thionyl chloride. Compound **2** (24 g., 0.092 mole) is added, and the mixture is kept in an ice bath for 6 hr., and then poured over 200 g. of ice. The aqueous solution is extracted with chloroform, and the chloroform layer is washed successively with two 100-ml. portions of *N* sodium hydroxide and two 100-ml. portions of water, dried (anhydrous sodium sulfate), and evaporated under diminished pressure to an oil, which is distilled; b.p. 112–114°/4 mm. Hg; yield 18.5 g. (72%). Analytically pure **3** may be obtained by two redistillations.

N-Benzoyl-9-(2,3:5,6-di-*O*-isopropylidene-α-D-mannosyl)adenine[1] (4)

Chloromercuri-*N*-benzoyladenine[5] (Sec. III [44]) (9.4 g., 19.8 mmoles), 9.4 g. of Celite 535, and 400 ml. of dry xylene are mixed, and 80 ml. of the xylene is distilled off (to remove traces of moisture). To the mixture is added a solution of 5.5 g. (19.8 mmoles) of **3** in 100 ml. of dry xylene. The mixture is refluxed, protected from moisture, and thoroughly stirred for 4 hr. The mixture is filtered while hot, and the filter cake is washed with 200 ml. of chloroform. The filtrate and washings are combined, and evaporated to dryness under diminished pressure (45°), and the residue is dissolved in 200 ml. of chloroform. A small amount of insoluble material is removed by filtration, and the filtrate is washed successively with three 100-ml. portions of 30% potassium iodide (aqueous) and three 100-ml. portions of water, dried (anhydrous sodium sulfate), and evaporated to a yellow sirup; yield 13.9 g. The sirup is dissolved in a small volume of warm benzene, and, after several hours, a small amount of solid material is removed by filtration. The filtrate is concentrated to a glass (**4**).

9-(2,3-*O*-Isopropylidene-α-D-mannofuranosyl)adenine[1] (5)

A solution of the glass (**4**) in 150 ml. of 7:3 (v/v) acetic acid–water at 50° is kept at 50° for 2.5 hr., and evaporated to a sirup, and three portions of absolute ethanol followed by one of toluene are added and evaporated off, to give a glass. This is dissolved in 100 ml. of hot, absolute methanol, and to the hot solution is added 20 ml. of *M* methanolic sodium methoxide. The solution is refluxed for 50 min., and is then rendered neutral with glacial acetic acid, and filtered. The filtrate is evaporated to dryness under diminished pressure at 45°, and the residue is partitioned between 130 ml. each of water and chloroform. The chloroform layer is extracted with five 50-ml. portions of water, and the aqueous extracts are combined, and evaporated under diminished pressure at 45°, during which process crystallization of **5** occurs. After recrystallization from water, the total yield, in three crops, is 2.6 g. (38% based on **3**), m.p. 249–250° (softens at 240–246°), $[\alpha]_D^{21}$ +32.5° (*c* 1.26, 0.1 *N* hydrochloric acid).

9-α-D-Mannofuranosyladenine[1] (6)

A mixture of 118 ml. of 1:3 (v/v) acetic acid–water and 2.36 g. (7 mmoles) of **5** is stirred at 100° for 3.5 hr., cooled quickly to room temperature, and kept at room temperature for 30 min. The solvent is removed under diminished pressure at 45°, leaving a white residue. Recrystallization from ethanol–water gives 1.09 g. (52%) of tiny rods of compound **6**, m.p. 237–237.5°, $[\alpha]_D^{21}$ +74.8° (*c* 3.05, *N* hydrochloric acid).

REFERENCES

(1) L. M. Lerner and P. Kohn, *J. Org. Chem.*, **31**, 339 (1966).
(2) K. Iwadare, *Bull. Chem. Soc. Japan*, **18**, 226 (1943).
(3) K. Freudenberg and A. Wolf, *Ber.*, **60**, 232 (1927).
(4) K. Freudenberg, A. Wolf, E. Knopf, and S. H. Zaheer, *Ber.*, **61**, 1743 (1928).
(5) J. Davoll and B. A. Lowy, *J. Am. Chem. Soc.*, **73**, 1650 (1951).
(6) P. Kohn, L. M. Lerner, and B. D. Kohn, *J. Org. Chem.*, **32**, 4076 (1967).

[52] 9-(2-Methyl-β-D-ribofuranosyl)adenine (2′-Methyladenosine)

The Chloromercuri Synthesis Applied to Purine Nucleosides of Branched-chain Sugars

SUSAN R. JENKINS and EDWARD WALTON

MERCK, SHARP AND DOHME RESEARCH LABORATORIES, DIVISION OF MERCK AND CO., INC., RAHWAY, NEW JERSEY 07065

1 (162.1) → 2 (474.5) → 3, R = Bz (476.5); 4, R = H (372.4) →

5 (580.6) → 6 (494.9) —7 (474.3)→ 8, R = Bz (697.7); 9, R = H (282.3)

where Bz is benzoyl.

INTRODUCTION

2′-Methyladenosine[1] (9) inhibits the growth of KB cells in culture,[1] and the incorporation of hypoxanthine-*8*-^{14}C into ribonucleic acid by Ehrlich ascites cells.[2] Its preparation is an example of the use of a C-2-branched poly-*O*-acyl-D-pentofuranosyl halide in the synthesis of purine C-2′-branched nucleosides *via* the chloromercuri method.[3] Deacylation of the intermediate product **8** is accomplished with methanolic sodium methoxide.

PROCEDURE

2,3,5-Tri-*O*-benzoyl-2-methyl-D-ribono-1,4-lactone (2)

A solution of 5 g. (30.8 mmoles) of 2-methyl-D-ribono-1,4-lactone[4] (**1**) in 100 ml. of dry (barium oxide) pyridine is cooled in an ice bath, stirred, and treated with 17 ml. of benzoyl chloride. During the addition, a solid is precipitated. The mixture is heated at 65–70° for 4 hr., cooled, and stirred while 20 ml. of water is added. After 25 min., the mixture is concentrated[a] to a thick semi-solid which is dissolved in 100 ml. of chloroform and successively washed with three 50-ml. portions of 10% hydrochloric acid, two 50-ml. portions of *N* sodium hydrogen carbonate, and two 50-ml. portions of water. The chloroform layer is dried (magnesium sulfate) and evaporated to dryness, and the solid residue is recrystallized from ether. The yield of **2** is 10.8 g. (74%), m.p. 141–142°, $[\alpha]_D$ −79°, $[\alpha]_{578}$ −73° (*c* 1, chloroform); λ_{max}^{Nujol} 5.57 μm. (lactone), 5.70 and 5.78 μm. (ester).

1,2,3,5-Tetra-*O*-benzoyl-2-methyl-β-D-ribose (5)

To a stirred solution of 30 g. (63 mmoles) of 2,3,5-tri-*O*-benzoyl-2-methyl-D-ribono-1,4-lactone (**2**) in 125 ml. of dry tetrahydrofuran at 0° is added dropwise, under nitrogen, 175 ml. of *M* bis(3-methyl-2-butyl)-borane (disiamylborane, Alfa Inorganics). After 16 hr. at 25°, the reaction solution is cooled to 0°, and 26 ml. of water is carefully added. After the evolution of gas has subsided, the mixture is refluxed for 30 min., the mixture is concentrated, and the residual oil is dissolved in 250 ml. of acetone and 75 ml. of water. The solution is cooled (0–5°) and stirred during the dropwise addition of 33 ml. of 30% hydrogen peroxide, while the pH is kept between 7 and 8 by the addition of 3 *N* sodium hydroxide. The excess hydrogen peroxide is decomposed at 25° by the cautious, portionwise addition[b] of 500 mg. of 5% platinum on carbon. Stirring is continued, the pH being kept at 7 to 8, until the evolution of gas subsides and a negative test for peroxide is obtained (potassium iodide test-paper). The catalyst is removed, and the filtrate is extracted with four 200-ml. portions of chloroform.[c] The chloroform solution is dried (magnesium

[a] All concentrations are conducted in a rotary evaporator under diminished pressure unless otherwise noted.

[b] If the decomposition of the hydrogen peroxide becomes too vigorous, the temperature is moderated by addition of ice to the reaction mixture.

[c] If the chloroform solution still gives a test for undecomposed peroxide, the solution is washed with 30% ferrous sulfate solution until a negative test is obtained with potassium iodide–starch solution.

sulfate), and concentrated to give an oil (42 g.). Thin-layer chromatography (tlc) on silica with 19:1 (v/v) chloroform–ethyl acetate shows zones (developed with iodine vapor) at R_f 0.7 (by-product), 0.5 (product, **3**), 0.4 (product, **4**), and 0.2 and 0.1 (by-products). The residue is chromatographed on a short column[d] of 650 g. of silica gel (J. T. Baker, 100–200 mesh) with 99:1 (v/v) chloroform–ethyl acetate. Fractions containing materials of R_f (tlc) 0.7, 0.5, and 0.4 are combined, and concentrated to an oil (32 g.). Rechromatography of the oil on 650 g. of silica with 19:1 (v/v) benzene–ethyl acetate gives, after elution of the impurity having R_f 0.7, a total of 15 g. of a mixture of 2,3,5-tri-*O*-benzoyl-2-methyl-α(and β)-D-ribofuranose (**3**) and 3,5-di-*O*-benzoyl-2-methyl-α(and β)-D-ribofuranose (**4**).

The mixture of **3** and **4** (15 g.) is dissolved in 250 ml. of dry pyridine, and the solution is cooled to 10° and treated dropwise, with stirring, with 15.2 ml. of benzoyl chloride. The mixture is heated at 80° for 5 hr., cooled in an ice bath, treated with 10 ml. of water, and stirred for 30 min. at 25° (to decompose unreacted benzoyl chloride). Most of the pyridine is removed under diminished pressure, and the residue is dissolved in 350 ml. of chloroform and washed successively with three 80-ml. portions of 10% hydrochloric acid, three 80-ml. portions of saturated, aqueous sodium hydrogen carbonate, and three 80-ml. portions of water. The chloroform layer is dried (magnesium sulfate) and evaporated to dryness, and the residue (18 g.) is dissolved in 55 ml. of ether and kept at 5° for several hours. The precipitated solid (8.7 g.) is removed, and washed by warming it with 28 ml. of ether. After being cooled, the mixture is filtered, and 8.25 g. (23%) of 1,2,3,5-tetra-*O*-benzoyl-2-methyl-β-D-ribose (**5**) is obtained; m.p. 156.5–157.5°, $[\alpha]_D$ +68°, $[\alpha]_{578}$ +72° (*c* 1, chloroform). The ether filtrates are combined and concentrated, and a residue (10 g.) consisting mainly of 1,2,3,5-tetra-*O*-benzoyl-2-methyl-α-D-ribose and some of the β-D anomer (**5**) is obtained.

2,3,5-Tri-*O*-benzoyl-2-methyl-β-D-ribosyl Chloride (6)

To 300 ml. of dry (NaPb) ether saturated at 0° with hydrogen chloride in a round-bottomed flask are added 12 ml. of acetyl chloride and 6 g. (10.3 mmoles) of 1,2,3,5-tetra-*O*-benzoyl-2-methyl-β-D-ribose (**5**). The flask is tightly stoppered and kept at 25° for 2.5 hr. [the tetrabenzoate (**5**) dissolves during the first hour]. The solvent is removed under diminished pressure, and 75 ml. of dry toluene is added and distilled from the residue. The residue is dissolved in 300 ml. of dry ether, and the solution is rapidly washed with three 120-ml. portions of cold, saturated sodium hydrogen carbonate solution and two 120-ml. portions of cold water. The ether

[d] The ratio of height to diameter of the column is about 1:1. A fritted-glass Büchner funnel of medium porosity may be used.

layer is dried (magnesium sulfate) and evaporated to dryness under diminished pressure. The product (**6**) is an oil; yield 5.5 g.; tlc on alumina with 1:1 (v/v) benzene–chloroform, R_f 0.3; $\tau(CDCl_3)$ 3.13 (singlet, C-1 proton), 3.92 (doublet, C-3 proton, $J_{3,4}$ 7.5 Hz), 5.27 (multiplet, C-4 and C-5 protons), 8.02 p.p.m. (singlet, C-2 methyl protons).[e]

9-(2-Methyl-β-D-ribofuranosyl)adenine (9)

About 400 ml. of xylene is distilled at atmospheric pressure from a suspension of 4.86 g. (10.3 mmoles) of finely powdered chloromercuri-*N*-benzoyladenine[3] (**7**) (Sec. III [44]) in 750 ml. of xylene. The last 90 ml. of xylene distilled is used to dissolve 5.5 g. of chloride **6**. The solution of **6** is added to the stirred xylene suspension of **7** at 60–80° and the mixture is heated until refluxing and refluxed for 1.25 hr. The mixture is concentrated to 150 ml., cooled, and diluted with 500 ml. of petroleum ether. After being kept at 5° for several hours, the precipitate is removed and added to 200 ml. of chloroform. A small amount of chloroform-insoluble material is removed by filtration, and the filtrate is washed with three 150-ml. portions of 30% aqueous potassium iodide and two 150-ml. portions of water. Evaporation of the dried (magnesium sulfate) chloroform layer gives a glass (5.7 g.). Tlc on silica gel with 9:1 (v/v) chloroform–ethyl acetate shows a large zone (iodine vapor) for *N*-benzoyl-9-(2,3,5-tri-*O*-benzoyl-2-methyl-β-D-ribosyl)adenine (**8**) at R_f 0.3, and faint zones due to impurities at R_f 0.0, 0.1, 0.5, and 0.9. The crude product is chromatographed on 200 g. of silica gel with 4:1 (v/v) chloroform–ethyl acetate, and 4.4 g. of purified **8** is obtained.

To a suspension of 4.4 g. (6.3 mmoles) of **8** in dry methanol is added a solution prepared from 240 mg. (10.5 mg.-atoms) of sodium and 50 ml. of dry methanol. The solution is refluxed for 30 min. and evaporated to dryness, and the residue is dissolved in 66 ml. of water. The pH is adjusted to 7 with acetic acid, and the solution is washed with four 100-ml. portions of ether. The aqueous layer is concentrated to *ca.* 20 ml., during which process the product is precipitated. After the mixture has been kept at 5° for several hours, the product (1.35 g.) is removed and recrystallized from 30 ml. of hot water. After being kept at 5° for several hours, the suspension is filtered, and 1.3 g. (44–46%, from **5**) of 9-(2-methyl-β-D-ribofuranosyl)adenine (**9**) is obtained; m.p. 256–258°, $[\alpha]_D$ −21°, $[\alpha]_{578}$ −22° (*c* 0.5, water); $\lambda_{max}^{H_2O}$ nm. ($\epsilon \times 10^{-3}$): at pH 1, 258 (15.1), at pH 7, 260 (15.1), and at pH 13, 260 (14.9); $\tau(C_5D_5N)$: 3.10 (singlet, C-1′ proton), 4.93 p.p.m. (doublet, C-3′ proton, $J_{3',4'}$ 8.8 Hz).

[e] Nuclear magnetic resonance data are obtained on a Varian Associates Model A-60 spectrometer.

REFERENCES

(1) E. Walton, S. R. Jenkins, R. F. Nutt, M. Zimmerman, and F. W. Holly, *J. Am. Chem. Soc.*, **88**, 4524 (1966).
(2) Personal communication from Dr. H. T. Shigeura of the Merck, Sharp and Dohme Research Laboratories, Division of Merck & Co., Inc., Rahway, N. J. 07065.
(3) J. Davoll and B. A. Lowy, *J. Am. Chem. Soc.*, **73**, 1650 (1951).
(4) R. L. Whistler and J. N. BeMiller, *Methods Carbohydrate Chem.*, **2**, 484 (1963).

[53] 9-(3-Methyl-β-D-ribofuranosyl)adenine (3′-Methyladenosine)

The Chloromercuri Synthesis Applied to Purine Nucleosides Containing Branched-chain Sugars

RUTH F. NUTT, MARY J. DICKINSON, FREDERICK W. HOLLY, and EDWARD WALTON

MERCK, SHARP AND DOHME RESEARCH LABORATORIES, DIVISION OF MERCK AND CO., INC., RAHWAY, NEW JERSEY 07065

INTRODUCTION

3′-Methyladenosine[1] (**9**) inhibits the growth of KB cells in culture,[1] as well as the incorporation of hypoxanthine-*8*-^{14}C into ribonucleic acid by Ehrlich ascites cells.[2] It is resistant to deamination by adenosine deaminase.[1] Its preparation is an example of the use of a 3-branched poly-*O*-acyl-D-pentofuranosyl halide in the synthesis of purine 3′-branched nucleosides *via* the chloromercuri method.[3]

PROCEDURE

5-*O*-Benzoyl-1,2-*O*-isopropylidene-α-D-*erythro*-pentofuranos-3-ulose[4,5] (2)

A solution of 125 g. of sodium metaperiodate in 1.5 l. of water is cooled in an ice-bath and added portionwise to a vigorously stirred suspension of 15 g. of ruthenium dioxide (Engelhard Industries[a]) in 1.5 l. of carbon tetrachloride cooled in an ice bath. About 20 to 30 min. after the addition has been completed, most of the black, insoluble ruthenium dioxide has been converted into soluble, yellow ruthenium tetraoxide. The carbon tetrachloride solution of ruthenium tetraoxide is separated from the water layer, and added during 15 min. to a stirred solution of 18 g. (0.06 mole) of 5-*O*-benzoyl-1,2-*O*-isopropylidene-α-D-xylofuranose[6] (**1**) in 1.5 l. of carbon tetrachloride covered by 100 ml. of water. After 1 hr., the reaction mixture, which now contains a black precipitate of ruthenium dioxide, is warmed to room temperature and stirred for a further 2 hr. Thin-layer chromatography (tlc) on silica gel in 4:1 chloroform–ethyl acetate shows zones[b] at R_f 0.35 (blue; starting material **1**), R_f 0.40 (brown;

[a] Ruthenium dioxide from other commercial sources is not oxidized by sodium metaperiodate under these conditions.

[b] Zones are made visible by spraying the plates with a solution of 100 mg. of 1,3-dihydroxynaphthalene in 100 ml. of ethanol containing 2.5 ml. of phosphoric acid, and then warming the plates on a steam cone until the colors develop.

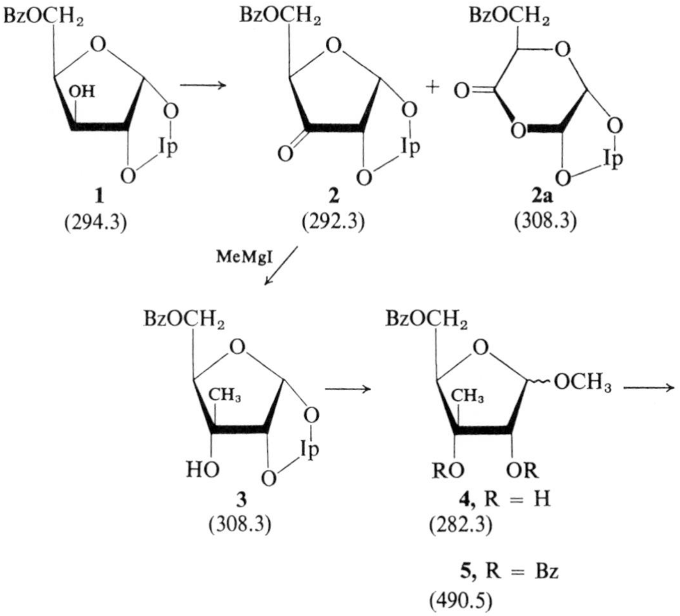

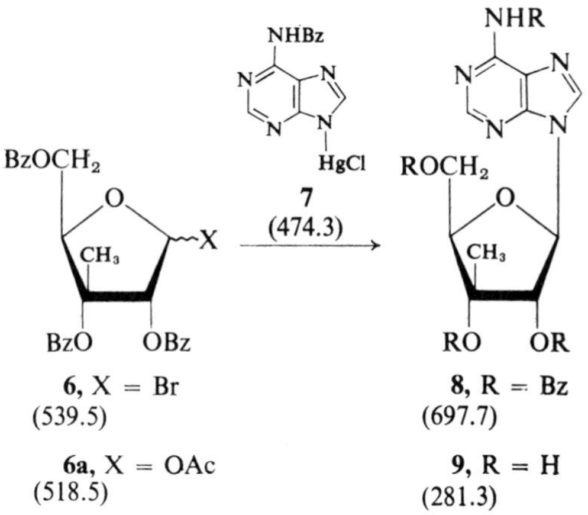

$BzOCH_2$ ... OH ... O–Ip–O → $BzOCH_2$ ... O ... O–Ip–O + $BzOCH_2$ ... O= ... O–Ip–O

1 (294.3) **2** (292.3) **2a** (308.3)

MeMgI

$BzOCH_2$ ... CH_3 ... HO ... O–Ip–O → $BzOCH_2$ ... CH_3 ... OCH_3 ... RO OR →

3 (308.3)

4, R = H (282.3)

5, R = Bz (490.5)

$BzOCH_2$ ... CH_3 ... X ... BzO OBz

NHBz ... N–HgCl

7 (474.3)

→ NHR ... $ROCH_2$ ... CH_3 ... RO OR

6, X = Br (539.5)

6a, X = OAc (518.5)

8, R = Bz (697.7)

9, R = H (281.3)

where Bz is benzoyl.

product **2**),[c] and R_f 0.80 (purple; by-product **2a**[d]). The reaction mixture is treated with 10 ml. of isopropyl alcohol in 50 ml. of carbon tetrachloride (to decompose unreacted ruthenium tetraoxide) and the black ruthenium dioxide is filtered off, and successively washed with 50 ml. of water and 100 ml. of carbon tetrachloride. The filtrates are combined, and the carbon tetrachloride layer is washed with 100 ml. of saturated sodium hydrogen carbonate solution. The carbon tetrachloride layer is evaporated[e] to dryness, and the residue is crystallized from ether. A total of 9.9 g. (55%) of **2** is obtained; m.p. 98–99°; $[\alpha]_D$ +136°, $[\alpha]_{578}$ +144° (*c* 1, chloroform); λ_{max}^{Nujol} 5.62 (cyclic ketone), 5.77 μm. (ester); $\tau(CDCl_3)$ 3.87 p.p.m. (doublet, C-1 proton, $J_{1,2}$ 4.5 Hz).[f]

5-*O*-Benzoyl-1,2-*O*-isopropylidene-3-methyl-α-D-ribofuranose (3)

A solution of methylmagnesium iodide is prepared by adding 29.6 g. (0.208 mole) of methyl iodide in 100 ml. of dry (NaPb) ether to a stirred suspension of 6.4 g. (0.264 mole) of magnesium shavings in 80 ml. of dry ether. The solution of methylmagnesium iodide is added to a stirred solution of 7.7 g. (26.4 mmoles) of compound **2** in 600 ml. of dry ether at 5°; a heavy, white precipitate is formed immediately. The reaction mixture is poured into a cold, stirred mixture of 800 ml. of ether and a solution of 250 g. of ammonium chloride in 1.2 l. of water. The aqueous layer is separated, and extracted with three 400-ml. portions of ether. The ether layers are combined, washed with 200 ml. of saturated aqueous sodium chloride, and dried with anhydrous magnesium sulfate. Evaporation of the ether solution gives a residue which, when crystallized from ether–petroleum ether, affords a total of 5.4 g. (67%) of **3**; m.p. 109–111°; $[\alpha]_D$ +12.6°, $[\alpha]_{578}$ +12.6° (*c* 2.4, chloroform).

Methyl 5-*O*-benzoyl-3-methyl-α(and β)-D-ribofuranoside (4)

To a solution of 320 ml. of 20% (w/w) hydrogen chloride–methanol in 80 ml. of water is added 7.8 g. (0.025 mole) of compound **3**, and the solution is stirred at 25° for 80 min. The acid is neutralized by the addition of 120 g. of sodium hydrogen carbonate, the mixture is evaporated to dryness, and the residue is leached with two 100-ml. portions of warm dichloromethane. The extract is evaporated to dryness, and the residual

[c] The zones for **1** and **2** are resolved only if the silica-gel plates are dried at 100° for 30 min. prior to use.

[d] M.p. 111–112°; $[\alpha]_D$ +81°, $[\alpha]_{578}$ +86° (*c* 1.03 chloroform); λ_{max}^{Nujol} 5.71 (lactone), 5.80 μm. (ester).[5]

[e] Solvent evaporations are conducted under diminished pressure in a rotary evaporator, unless otherwise noted.

[f] Nuclear magnetic resonance (nmr) spectra are obtained with a Varian Associates Model A-60 spectrometer.

oil (6.6 g.) is chromatographed on a column[g] of 170 g. of silica gel (J. T. Baker, 100–200 mesh) in 4:1 chloroform–ethyl acetate. Elution of the column is followed by tlc on silica gel in 4:1 chloroform–ethyl acetate: R_f 0.7 for **3**; R_f 0.2 for **4**. After the elution of a small amount of fast-running impurities, fractions containing compound **4** are obtained. These fractions are combined, and evaporated to a sirup; yield of **4**, 5.75 g. (80%); $\tau(CDCl_3)$ (α-**4**) 5.02 p.p.m. (doublet, C-1 proton, $J_{1,2}$ 4.5 Hz), 6.48 p.p.m. (singlet, OCH_3 protons), 8.65 p.p.m. (singlet, C-3 methyl protons); (β-**4**) 5.12 p.p.m. (doublet, C-1 proton, $J_{1,2}$ 2.5 Hz), 6.60 p.p.m. (singlet, OCH_3 protons), 8.59 p.p.m. (singlet, C-3 methyl protons).

Methyl 2,3,5-Tri-*O*-benzoyl-3-methyl-α(and β)-D-riboside (5)

To a solution of 8.65 g. (30.7 mmoles) of mixture **4** in 91 ml. of dry (barium oxide) pyridine is added 15.4 g. (0.106 mole) of benzoyl chloride, and the mixture is heated at 95 ± 5° for 16 hr. The mixture is cooled to 25°, stirred with 5 ml. of water for 30 min., and added to 450 ml. of 10% hydrochloric acid and 600 ml. of chloroform. The aqueous layer is extracted with two 500-ml. portions of chloroform, and the chloroform layers are combined, and successively washed with two 350-ml. portions of saturated aqueous sodium hydrogen carbonate and 300 ml. of saturated aqueous sodium chloride. After being dried over anhydrous magnesium sulfate, the chloroform solution is evaporated, and the residual oil (14.9 g.) is chromatographed on a short column of 375 g. of silica gel in chloroform. Elution of the column is followed by tlc on silica gel in 19:1 chloroform–ethyl acetate: R_f 0.7 for **5**; R_f 0.1 for **4**. The product (**5**) is eluted in the first several fractions and, after removal of the solvent, amounts to 14.4 g. (97%); $\tau(CDCl_3)$ (α-**5**) 4.55 p.p.m. (doublet, C-2 proton), 4.67 p.p.m. (doublet, C-1 proton, $J_{1,2}$ 4.5 Hz), 6.60 p.p.m. (singlet, OCH_3 protons), 8.17 p.p.m. (singlet, C-3 methyl protons); (β-**5**) 4.32 p.p.m. (doublet, C-2 proton), 4.93 p.p.m. (doublet, C-1 proton, $J_{1,2}$ 1.0 Hz), 6.53 p.p.m. (singlet, OCH_3 protons), 8.11 p.p.m. (singlet, C-3 methyl protons).

2,3,5-Tri-*O*-benzoyl-3-methyl-α(and β)-D-ribosyl bromide (6)

To a solution of 2.0 g. (4.08 mmoles) of mixture **5** in 10 ml. of acetic acid is added 0.5 ml. of acetyl bromide, and the solution is cooled to 10° and treated with 10 ml. of 30% (w/w) hydrogen bromide in acetic acid. The solution is warmed to 25° and, after 40 min. at 25°, is evaporated to dryness below 40°. The residue is a mixture of anomers of 2,3,5-tri-*O*-benzoyl-3-methyl-D-ribosyl bromide (**6**) and 1-*O*-acetyl-2,3,5-tri-*O*-benzoyl-

[g] The column has a ratio of height to diameter of about 1:1. A fritted-glass Büchner funnel of medium porosity may be used.

3-methyl-α(and β)-D-ribose (**6a**).[h] For the conversion of **6a** into **6**, the mixture is dissolved in 100 ml. of ether presaturated with hydrogen bromide at 0°. After being kept at 25° for 80 min., the solution is concentrated below 40° (bath temperature). Four 30-ml. portions of dry toluene are successively added to and evaporated from the residue, giving **6**. $\tau(CDCl_3)$: (α-**6**) 3.07 p.p.m. (doublet, C-2 proton), 4.63 p.p.m. (doublet, C-1 proton, $J_{1,2}$ 4.5 Hz), 8.08 p.p.m. (singlet, C-3 methyl protons); (β-**6**) 3.55 p.p.m. (doublet, C-2 proton), 3.78 p.p.m. (doublet, C-1 proton, $J_{1,2}$ 1.0 Hz), 8.05 p.p.m. (singlet, C-3 methyl protons); the ratio of α-**6** to β-**6** is about 2:1.

N-Benzoyl-9-(2,3,5-tri-*O*-benzoyl-3-methyl-β-D-ribosyl)adenine (8)

A suspension of 1.92 g. (4.04 mmoles) of finely ground chloromercuri-*N*-benzoyladenine[3] (**7**) (Sec. III [44]) in 170 ml. of xylene is dried by distilling off 90 ml. of xylene. The stirred mixture is cooled to about 60°, and a solution of compound **6** [prepared from 2.0 g. (4.04 mmoles) of compound **5**] in 20 ml. of dry xylene is added. The mixture is stirred and refluxed for 40 min., and most of the suspended solid dissolves. The hot mixture is filtered, and the filtrate is diluted with 400 ml. of petroleum ether, kept at 5° for several hours, and filtered. The solid is dissolved in 150 ml. of chloroform, and the solution is successively washed with two 30-ml. portions of 30% potassium iodide solution and two 30-ml. portions of water, and evaporated to dryness. The residue (1.93 g.) is chromatographed on a short column of 40 g. of alumina (Merck; acid-washed) in 9:1 benzene–chloroform. Fractions containing the product **8** (R_f 0.28, tlc on alumina in 9:1 benzene–chloroform; zones made visible with iodine vapor) are combined and evaporated to dryness. A total of 1.36 g. (48%) of compound **8** is obtained as a glass; $[\alpha]_D$ −146°, $[\alpha]_{578}$ −155° (*c* 1.5, chloroform); λ_{max}^{EtOH} nm. ($\epsilon \times 10^{-3}$): 232 (51), 279 (23.4); $\tau(CDCl_3)$ 3.44 p.p.m. (singlet, unresolved C-1′ and C-2′ protons), 7.92 p.p.m. (singlet, C-3 methyl protons).

9-(3-Methyl-β-D-ribofuranosyl)adenine (9)

A mixture of 8.14 g. (11.6 mmoles) of compound **8**, 200 ml. of dry (Molecular Sieves) methanol, and methanolic sodium methoxide solution [prepared from 400 mg. (17 mg.-atoms) of sodium and 10 ml. of dry methanol] is refluxed for 45 min. The solution is evaporated to dryness, the residue is dissolved in 200 ml. of water, and the solution is rendered neutral

[h] Tlc on alumina in 1:1 benzene–chloroform shows zones (made visible with iodine vapor) at R_f 0.1 for **6**; R_f 0.6 for α-**6a**; and R_f 0.85 for β-**6a**. In this system, **5** shows zones at R_f 0.8 for α, and R_f 0.9 for β. Nmr indicates that the ratio of **6** to **6a** is about 1:1. $\tau(CDCl_3)$:(α-**6a**) 3.31 p.p.m. (doublet, C-2 proton), 4.28 p.p.m. (doublet, C-1 proton, $J_{1,2}$ 5.0 Hz); (β-**6a**) 3.62 p.p.m. (doublet, C-2 proton), 4.14 p.p.m. (doublet, C-1 proton, $J_{1,2}$ 1.5 Hz).

(pH 7 to 8) with acetic acid, washed with three 100-ml. portions of chloroform, and filtered. The filtrate is concentrated to about 30 ml., nucleated, and kept at 25° for 45 hr. The crystalline **9** (2.0 g.) is removed by filtration, and washed with two 3-ml. portions of cold water. The filtrate and washings are evaporated to dryness, and the residue is leached with several portions of hot ethanol (total, 100 ml.). The ethanol extracts are evaporated to dryness, and the residue is crystallized from the minimal volume of hot water. A total of 2.7 g. (83%) of compound **9** is obtained; m.p. 213–215° (transition at 165°); $\lambda_{max}^{H_2O}$ nm. ($\epsilon \times 10^{-3}$): pH 1, 257.5 (14.8); pH 7, 260 (14.9); pH 13, 260 (14.3); $[\alpha]_D$ $-58°$, $[\alpha]_{578}$ $-61°$ (*c* 1, water); $\tau(D_2O)$ 4.08 p.p.m. (doublet, C-1′ proton), 5.43 p.p.m. (doublet, C-2′ proton, $J_{1',2'}$ 8.0 Hz), 5.83 p.p.m. (multiplet, C-4′ proton), 6.18 p.p.m. (multiplet, C-5′ protons), 8.57 p.p.m. (singlet, C-3′ methyl protons).

REFERENCES

(1) E. Walton, S. R. Jenkins, R. F. Nutt, M. Zimmerman, and F. W. Holly, *J. Am. Chem. Soc.*, **88**, 4524 (1966).
(2) Personal communication from Dr. H. T. Shigeura of the Merck, Sharp and Dohme Research Laboratories, Division of Merck & Co., Inc., Rahway, N. J. 07065.
(3) J. Davoll and B. A. Lowy, *J. Am. Chem. Soc.*, **73**, 1650 (1951).
(4) K. Oka and H. Wada, *Yakugaku Zasshi*, **83**, 890 (1963).
(5) R. F. Nutt, B. Arison, F. W. Holly, and E. Walton, *J. Am. Chem. Soc.*, **87**, 3273 (1965).
(6) P. A. Levene and A. L. Raymond, *J. Biol. Chem.*, **102**, 317 (1933).

[54] 3-β-D-Ribofuranosyladenine (3-Isoadenosine)

Direct N-Alkylation of Adenine with a Sugar Derivative

RICHARD A. LAURSEN, WOLFGANG GRIMM, and NELSON J. LEONARD

DEPARTMENT OF CHEMISTRY AND CHEMICAL ENGINEERING, UNIVERSITY OF ILLINOIS, URBANA, ILLINOIS 61801

1 (135.1) + 2 (525.4) → 3 (579.6) $\xrightarrow{NH_3,\ CH_3OH}$ 4 (267.2)

where R is benzoyl.

INTRODUCTION

The isomer of adenosine in which a β-D-ribofuranosyl residue is attached to N-3 of adenine has shown remarkable biological activity: as the nucleoside, as the related nucleotides, and as the analog of nicotinamide-adenine dinucleotide.[1–6] The synthesis of 3-β-D-ribofuranosyladenine (3-isoadenosine) (**4**) by direct alkylation of adenine with protected D-ribofuranosyl halides[1–3] is illustrative of glycosylation of adenine at the 3-position.[7]

PROCEDURE

3-(2,3,5-Tri-*O*-benzoyl-β-D-ribosyl)adenine (3)

To a solution of 50.4 g. (0.10 mole) of 1-*O*-acetyl-2,3,5-tri-*O*-benzoyl-D-ribose[a] in 100 ml. of dry dichloromethane is added 300 ml. of 30–32% hydrogen bromide in acetic acid. The solution is kept at room temperature for 75 min., and acetic acid and hydrogen bromide are removed by evaporation under diminished pressure (bath temperature, below 35°), followed by addition and evaporation of five 50-ml. portions of dry toluene. The reddish oil, consisting of 2,3,5-tri-*O*-benzoyl-D-ribosyl bromide (**2**) (Sec. V, [163]), is dissolved in 650 ml. of dry acetonitrile, and 13.5 g. (0.10 mole) of dry adenine (**1**) is added to the solution. The mixture is stirred at 50° for 36 hr., 15 ml. of concentrated ammonium hydroxide is then added, the solvent is removed under diminished pressure, and the viscous residue is triturated with 500 ml. of ether. The ether is decanted, a further 500 ml. of ether is added, and the solid is removed by filtration and washed with ether. (The ether extracts, containing 2′,3′,5′-tri-*O*-benzoyladenosine and other derivatives, may be saved for recovery of this ester.[2]) The solid is dissolved, by heating on a steam bath, in a mixture of 1.5 l. of ethyl acetate and sufficient water to effect the disappearance of all solid material. The layers are separated while hot, and the ethyl acetate layer is decolorized with charcoal and concentrated to about 300 ml. The solution is cooled, and the resulting crystals are removed by filtration, washed successively with ethyl acetate, methanol, and ether, and recrystallized from ethanol or *N,N*-dimethylformamide–methanol; yield 14.9 g. (26%), m.p. 246–247°, $[\alpha]_D^{26}$ −69° (*c* 0.89, *N,N*-dimethylformamide).

3-β-D-Ribofuranosyladenine (3-Isoadenosine) (4)

To a solution of 3.15 g. (5.4 mmoles) of 3-(2,3,5-tri-*O*-benzoyl-β-D-ribosyl)adenine (**3**) in 500 ml. of tetrahydrofuran is added, with stirring, 400 ml. of methanol presaturated with ammonia, and the solution is kept in a refrigerator for 5 days. The solvents are evaporated under diminished pressure until crystallization begins. The mixture is then kept overnight in a refrigerator, and the crystalline product **4** is collected by filtration; yield 1.39 g. (96%), m.p. 210–211° (dec.), $[\alpha]_D^{26}$ −35° (*c* 0.606, 0.05 *N* hydrochloric acid).

[a] Calbiochem, Box 54282, Los Angeles, California 90054.

REFERENCES

(1) N. J. Leonard and R. A. Laursen, *J. Am. Chem. Soc.*, **85**, 2026 (1963).
(2) N. J. Leonard and R. A. Laursen, *Biochemistry*, **4**, 354 (1965).
(3) N. J. Leonard and R. A. Laursen, *Biochemistry*, **4**, 365 (1965).
(4) R. Wolfenden, T. K. Sharpless, I. S. Ragade, and N. J. Leonard, *J. Am. Chem. Soc.*, **88**, 185 (1966).
(5) A. M. Michelson, C. Monny, R. A. Laursen, and N. J. Leonard, *Biochim. Biophys. Acta*, **119**, 258 (1966).
(6) K. Gerzon, I. S. Johnson, G. B. Boder, J. C. Cline, P. J. Simpson, C. Speth, N. J. Leonard, and R. A. Laursen, *Biochim. Biophys. Acta*, **119**, 445 (1966).
(7) L. B. Townsend, R. K. Robins, R. N. Loeppky, and N. J. Leonard, *J. Am. Chem. Soc.*, **86**, 5320 (1964).

[55] 9-β-L-Ribofuranosyladenine ("L-Adenosine")

Configurational Inversion within a Furanoid Ring

KENNETH J. RYAN and EDWARD M. ACTON

LIFE SCIENCES RESEARCH, STANFORD RESEARCH INSTITUTE, MENLO PARK, CALIFORNIA 94025

INTRODUCTION

L-Ribofuranose derivatives are obtained[1] by configurational inversion at C-3 in the furanoid ring of L-xylofuranose derivatives. This is accomplished by the displacement of a sulfonic ester group, which is, generally, a difficult and impractical process for secondary sulfonates attached to a furanoid ring, but is made feasible by participation of a neighboring *trans*-benzoate in the displacement. The method has general applicability[2] for the synthesis of rare sugars and their derivatives.

PROCEDURE

1,2-Di-*O*-acetyl-3,5-di-*O*-*p*-tolylsulfonyl-α,β-L-xylose (2)

A solution of 50 g. (0.10 mole) of 1,2-*O*-isopropylidene-3,5-di-*O*-*p*-tolylsulfonyl-L-xylose[3] (**1**) in 700 ml. of glacial acetic acid and 80 ml. of acetic anhydride in a round-bottomed flask is stirred, and treated with 48 ml. of concentrated sulfuric acid, added dropwise, while the flask is cooled intermittently to keep the temperature at 10–20°. The flask is stoppered, the solution is kept overnight at 25° and then poured into 3 l. of ice-water, and the mixture is extracted with two 600-ml. portions of chloroform. The chloroform extracts are combined, and washed successively with three 600-ml. portions of saturated aqueous sodium hydrogen carbonate and 1 l. of water, dried with magnesium sulfate, and filtered. Removal of the chloroform *in vacuo* on a rotary evaporator affords 60 g. (110%) of crude (**2**) as a sirup.

Methyl 3,5-Di-*O*-*p*-tolylsulfonyl-α,β-L-xylofuranoside (3)

A solution of 60 g. (0.10 mole) of crude **2** in 1.2 l. of 1% anhydrous methanolic hydrogen chloride is kept for 17 hr. at 25°, and then evaporated *in vacuo* to a sirup which is partitioned between 100 ml. of water and 150 ml. of chloroform. The chloroform layer is separated, dried with magnesium

Me Me C O O O $TsOH_2C$ OTs **1** (498.6) → O OR′ R″O $TsOH_2C$ OTs **2**, R′ = R″ = Ac (542.6) **3**, R′ = Me; R″ = H (472.5) →

O OMe BzO $TsOH_2C$ OTs **4** (576.6) → O OMe OR R′O $BzOH_2C$ **5**, 1:1 R = H, R′ = Bz; R = Bz, R′ = H (372.4) **6**, R = R′ = Bz (476.5) →

O OBz BzO $BzOH_2C$ OAc **7** (504.5) → [O OBz BzO Cl $BzOH_2C$] → O OR RO ROH_2C N N N N NHR **8**, R = Bz (683.7) **9**, R = H (267.2)

where Bz is benzoyl and Ts is *p*-tolylsulfonyl.

sulfate, filtered, and evaporated *in vacuo*, giving crude **3**, a sirup, weighing 39 g. (82%, based on **1**).

Methyl 2-*O*-Benzoyl-3,5-di-*O*-*p*-tolylsulfonyl-α,β-L-xyloside (4)

A stirred solution of 39 g. (0.082 mole) of crude **3** in 100 ml. of anhydrous pyridine (dried with KOH pellets) is cooled in ice, and treated dropwise with 30 ml. of benzoyl chloride. The mixture is stirred overnight at 25°, treated with 1 ml. of water, and stirred for 1 hr. (to minimize formation of benzoic anhydride). The mixture is diluted with 200 ml. of benzene, and shaken with 250 ml. of cold *N* hydrochloric acid. The benzene layer is separated, washed successively with 250-ml. portions of saturated aqueous sodium hydrogen carbonate and water, dried with magnesium sulfate, and filtered. Removal of the benzene *in vacuo* with a rotary evaporator yields 47 g. (100%) of **4**, which partially crystallizes.[a]

Methyl 2,5- and 3,5-Di-*O*-benzoyl-α,β-L-ribofuranoside (5)

A solution of 42 g. (0.073 mole) of **4** in 1 l. of *N*,*N*-dimethylformamide containing 84 g. of suspended sodium benzoate is refluxed for 6 hr. The mixture is cooled, and treated with 2 l. of ice-water (with chilling, to dissipate the heat of dilution), and the resultant, cloudy solution is extracted with four 400-ml. portions of ethyl ether.[b] The ether extracts are combined, washed successively with three 1-l. portions of saturated aqueous sodium hydrogen carbonate and 1 l. of water, dried with magnesium sulfate, filtered, and evaporated to dryness *in vacuo*. The crude dibenzoate (**5**) is a sirup, weighing 23 g. (85%).

Methyl 2,3,5-Tri-*O*-benzoyl-α,β-L-riboside (6)

A solution of 23 g. (0.062 mole) of **5** in 120 ml. of anhydrous pyridine is stirred and chilled, while 12 ml. of benzoyl chloride is added dropwise. The mixture is stirred at 25° overnight, treated with 0.5 ml. of water, and stirred for 1 hr. The mixture is diluted with 200 ml. of benzene, and washed successively with 400 ml. of cold *N* hydrochloric acid, 400 ml. of saturated aqueous sodium hydrogen carbonate, and 400 ml. of water, dried with magnesium sulfate, filtered, and evaporated to dryness *in vacuo*, giving sirupy tribenzoate **6**; yield 26 g. (87%).

1-*O*-Acetyl-2,3,5-tri-*O*-benzoyl-β-L-ribose (7)

A solution of 26 g. (0.055 mole) of **6** in 275 ml. of glacial acetic acid and 55 ml. of acetic anhydride is stirred, and kept at 10–20°, while 16 ml. of concentrated sulfuric acid is added dropwise. The solution is kept in a stoppered flask at 25° overnight, and then poured into 1 l. of ice water, and the mixture is extracted with three 400-ml. portions of chloroform. The chloroform extracts are combined, washed successively with three

[a] The anomers may be separated by chromatography on silica gel with 1:1 benzene–chloroform as the eluant, followed by fractional recrystallization from methanol; for the β-L anomer, m.p. 120–121°, $[\alpha]_D$ −23.6° (*c* 1.0, chloroform); for the α-L anomer, m.p. 86–88°, $[\alpha]_D$ −150.3° (*c* 1.0, chloroform).

[b] When ether is used for extraction, the product is accompanied by only minimal amounts of *N*,*N*-dimethylformamide; larger amounts of *N*,*N*-dimethylformamide are extracted, for example, by dichloromethane.

500-ml. portions of saturated aqueous sodium hydrogen carbonate (caution: foaming) and 1 l. of water, dried with magnesium sulfate, filtered, and evaporated to dryness *in vacuo*. The residue is crystallized from methanol; yield 13 g. (47%), m.p. 129–130°, $[\alpha]_D^{25}$ −43.6° (*c* 1.0, chloroform). The D enantiomorph has[4] m.p. 130–131°, $[\alpha]_D$ +44.2°.

N-Benzoyl-9-(2,3,5-tri-*O*-benzoyl-β-L-ribosyl)adenine (8)

A solution, at 0° and protected from moisture, of 11.0 g. (0.0218 mole, dried *in vacuo*) of compound[5] **7** in 500 ml. of anhydrous ether, presaturated at 0° with hydrogen chloride, is treated with 11 ml. of acetyl chloride (to maintain anhydrous conditions[6]), and the solution is kept at 0° for 3 days. The solution is evaporated *in vacuo* to a sirup, which is re-evaporated twice *in vacuo* with a little anhydrous benzene to remove traces of acid. The sirupy 2,3,5-tri-*O*-benzoyl-L-ribosyl chloride[7] (10.5 g., 100%) is used immediately.

A mixture of 10.4 g. (0.022 mole) of chloromercuri-*N*-benzoyladenine[8–10] (Sec. III [44]) and Celite (5.6 g.) is suspended in 700 ml. of dry xylene in a 3-necked, round-bottomed flask fitted with a stirrer, an addition funnel, and a Dean–Stark distilling receiver with a reflux condenser (drying tube). The suspension is dried by distilling off 50 ml. of xylene.

The freshly prepared halide (10.5 g., 0.0218 mole) is dissolved in 50 ml. of anhydrous xylene, and the solution is added through the addition funnel to the suspension at 100°. The mixture is refluxed for 2 hr. and filtered while hot. The Celite, which collects on the filter, is washed with 500 ml. of hot chloroform, and the filtrates are combined, and evaporated to dryness *in vacuo*. The residue is dissolved in 250 ml. of chloroform, and the solution is washed successively with 120 ml. of 30% aqueous potassium iodide (to remove mercury salts) and 120 ml. of water. The aqueous washing is extracted with 60 ml. of chloroform, and the combined chloroform solutions are dried with magnesium sulfate, filtered, and evaporated to dryness *in vacuo* to give 15 g. (100%) of crude **8** as a foamy glass.

9-β-L-Ribofuranosyladenine ("L-Adenosine") (9)

A solution of 15 g. of crude, protected nucleoside (**8**) in 240 ml. of anhydrous methanol is treated with 44 ml. of *M* methanolic sodium methoxide, and the solution is refluxed for 2 hr. The solution is cooled, rendered neutral by adding 44 ml. of *M* methanolic acetic acid, and concentrated *in vacuo* with a rotary evaporator. After two thirds of the methanol has been removed, 1.2 g. of crude L-adenosine (**9**) has crystallized out; it is collected on a filter, and recrystallized from water (5 ml./g.) to yield 0.70 g. of product.

Concentration of the *methanolic* filtrate affords additional **9** as a crude residue, from which methyl benzoate is removed by partitioning between 25 ml. of water and 50 ml. of chloroform. The crude **9** is recovered from the aqueous layer by evaporation, and dissolved in 80 ml. of methanol; the solution is treated with 100 ml. of methanolic picric acid (10%), and kept overnight at 5°. The picrate is collected on a filter, washed with methanol, and, without further drying, is slurried in 350 ml. of water. The suspension is stirred with 100 g. of Dowex 2 (CO_3^{2-}) ion-exchange resin for 5 hr. to regenerate **9**, which dissolves. The resin is removed by filtration, and the filtrate is evaporated to give a residue, which is recrystallized from water to afford 1.8 g. of **9**.

The combined yield is 2.5 g. (43%, based on **7**), m.p. 232–235°, $[\alpha]_D^{25}$ +61.4° (*c* 0.5, water); $\lambda_{max}^{pH\,1}$ 257 nm. (ϵ 14,600), $\lambda_{max}^{pH\,13}$ 260 nm. (ϵ 15,200); R_{Ade} 0.86,[c] homogeneous on Whatman No. 1 paper in 4:1:5 butyl alcohol–acetic acid–water (detection by ultraviolet light).[d]

REFERENCES

(1) E. M. Acton, K. J. Ryan, and L. Goodman, *J. Am. Chem. Soc.*, **86**, 5352 (1964).

(2) K. J. Ryan, H. Arzoumanian, E. M. Acton, and L. Goodman, *J. Am. Chem. Soc.*, **86**, 2497, 2503 (1964).

(3) L. Vargha, *Chem. Ber.*, **87**, 1351 (1954). The yield from commercially available L-xylose (Pfanstiehl Laboratories, Inc., Waukegan, Illinois) *via* 1,2:3,5-di-*O*-isopropylidene- and 1,2-*O*-isopropylidene-α-L-xylofuranose is 66%.

(4) R. K. Ness, H. W. Diehl, and H. G. Fletcher, Jr., *J. Am. Chem. Soc.*, **76**, 763 (1954).

(5) As described for the D enantiomorph by H. M. Kissman, C. Pidacks, and B. R. Baker, *J. Am. Chem. Soc.*, **77**, 18 (1955).

(6) E. J. Reist, R. R. Spencer, and B. R. Baker, *J. Org. Chem.*, **23**, 1753, 1958 (1958).

(7) G. L. Tong, K. J. Ryan, W. W. Lee, E. M. Acton, and L. Goodman, *J. Org. Chem.*, **32**, 859 (1967).

(8) J. Davoll and B. A. Lowy, *J. Am. Chem. Soc.*, **73**, 1650 (1951).

(9) J. A. Montgomery and H. J. Thomas, *J. Org. Chem.*, **28**, 2304 (1963).

(10) The present method of preparation is the same as that for chloromercuri-2,6-diacetamidopurine, given by B. R. Baker and K. Hewson, *J. Org. Chem.*, **22**, 959 (1957).

[c] Adenine (Ade) is the standard for the chromatographic comparison.

[d] Values for natural adenosine, lit.[8] m.p. 233–235°, $[\alpha]_D^{25}$ −60.2° (*c* 0.5, water); $\lambda_{max}^{pH\,1}$ 257 nm. (ϵ 14,600), $\lambda_{max}^{pH\,13}$ 260 nm. (ϵ 15,100); R_{Ade} 0.86.

[56] Anomeric Adenine Nucleosides of 2-Amino-2-deoxy-D-ribofuranose

Synthesis of the Anomeric Forms of a Purine Nucleoside of a 2-Amino Sugar, Utilizing the N-*(2,4-Dinitrophenyl) Group in the Chloromercuri Procedure*

M. L. WOLFROM, M. W. WINKLEY, and P. McWAIN

DEPARTMENT OF CHEMISTRY, THE OHIO STATE UNIVERSITY, COLUMBUS, OHIO 43210

ROCH$_2$ ... SEt, RO, NHDNP → AcOCH$_2$ ... Cl, AcO, NHDNP → ROCH$_2$ ... (adenine, NHR), RO, NHR′

1, R = H (287.3)
2, R = Ac (371.4)

3 (345.7)

4, R = Ac, R′ = DNP, α, β (486.4)
5, R = R′ = H, α, β, 1:1 (266.3)
6, R = R′ = H, α-D (266.3)
7, R = R′ = H, β-D (266.3)

where Ac is CH_3CO, and DNP is 2,4-dinitrophenyl.

INTRODUCTION

The use of the *N*-(2,4-dinitrophenyl) protecting group for the synthesis of the anomeric forms of a purine nucleoside of a 2-amino sugar was established with 2-amino-2-deoxy-D-glucopyranose and adenine.[1] It is herein extended to the anomeric furanose forms of 2-amino-2-deoxy-D-ribose in combination with adenine.[2] The nucleosides were obtained in the form of a 1:1 molecular compound, as demonstrated by x-ray powder diffraction data, and the anomers were separated by elution from a column of ion-exchange resin by the method of Dekker.[3] The anomeric nature of each of the final crystalline products was established by optical rotatory values and by nuclear magnetic resonance spectroscopy. The compounds are currently undergoing biological testing against carcinomas.

PROCEDURE

9-[3,5-Di-*O*-acetyl-2-deoxy-2-(2,4-dinitroanilino)-α,β-D-ribosyl]adenine Picrate (Picrate of 4)[2]

Ethyl 2-deoxy-2-(2,4-dinitroanilino)-1-thio-α-D-ribofuranoside[4] (**1**) (6.94 g., 24.2 mmoles) is dissolved in pyridine (75 ml.) and acetic anhydride (75 ml.), and the mixture is kept at room temperature overnight and then poured into ice and water, and extracted with dichloromethane. The extract is washed successively with water, aqueous sodium hydrogen carbonate solution (cold, saturated), and water. The dried (magnesium sulfate) extract is evaporated to a sirup under diminished pressure, the residual pyridine is removed by repeated evaporation with toluene, and the sirup is dried under vacuum (oil pump) to give the acetylated thiofuranoside (**2**) in a yield of 9.01 g.

Dry chlorine[5] is passed into a solution of **2** (9.01 g., 24.3 mmoles) in dry (Drierite) dichloromethane (100 ml.) at 0° for 10 min. The residual sirup (**3**) obtained on removal of solvent is dissolved in dichloromethane (50 ml.) and the solution is added to a hot, vigorously stirred, azeotropically dried suspension of chloromercuri-*N*-acetyladenine (20 g., 48.5 mmoles) and Celite (8 g.) in toluene (300 ml.). The dichloromethane is removed by distillation, and the mixture is heated for 6 hr. under reflux, with vigorous stirring. The cooled suspension is filtered, and the filter cake is thoroughly washed with hot chloroform and then with dichloromethane. The filtrate is evaporated to dryness, and the residue is extracted with dichloromethane. The extract is washed successively with 30% aqueous potassium iodide solution, aqueous sodium hydrogen carbonate solution (cold, saturated), and water. The dried (magnesium sulfate) solution is evaporated to yield 7.9 g. of a sirup. Thin-layer chromatography on silica gel G, with ethyl acetate as developer, reveals a major component having R_f 0.30. Preparative, thin-layer chromatography on plates (200 × 200 × 1 mm.) of silica gel (100 mg. of material per plate), with ethyl acetate as developer, is employed to purify the nucleosidic material. The zones having R_f 0.3 are excised, and eluted with acetone. The acetone is removed, the residue is extracted with dichloromethane, and removal of solvent gives crude **4** in a yield of 5.9 g. (55% from **1**).

The above sirup is dissolved in ethyl acetate (60 ml.) and methanol (440 ml.). To this solution is added picric acid (6 g., 26.2 mmoles), and the mixture is refluxed for 15 min. The crystalline picrate that separates on cooling is removed by filtration, and washed with cold methanol to yield

6.6 g. (84%) of the picrate of **4**, m.p. 190–192°, $[\alpha]_D^{25}$ −70 ± 1° (*c* 1.20, acetone).

9-(2-Amino-2-deoxy-α,β-D-ribofuranosyl)adenine [1:1 molecular compound (5)][2]

To a stirred solution of the picrate of **4** (5.0 g., 7.0 mmoles) in acetone (650 ml.) and water (150 ml.) is added portionwise, at 45–50°, Bio Rad AG-1 X2 (OH^-) resin (50–100 mesh), until the solution becomes colorless. The resin is removed by filtration and well washed with hot methanol. The filtrate and washings are combined, and evaporated under diminished pressure, and the residue is triturated with methanol–dichloromethane until a filterable solid is obtained; yield, 1.54 g. This material is dissolved in aqueous methanol, and the solution is treated with decolorizing carbon. The residue obtained on removal of solvent is crystallized from methanol–ethanol to yield 1.28 g. (72%) of **5**, m.p. 195–197°, $[\alpha]_D^{22}$ +17 ± 3° (*c* 0.62, methanol).

Separation of the Anomeric Nucleosides[2] **(6 and 7)**

Following the general technique of Dekker,[3] the above anomeric mixture (700 mg., 2.6 mmoles) in 1:9 methanol–water is siphoned onto a column (50 × 3.2 cm.) of Bio Rad AG-1 X2 (OH^-, 200–400 mesh). Elution is effected with the same solvent, and is monitored by an ultraviolet fraction-analyzer, 10-ml. fractions being collected. At tube 115, an ultraviolet-absorbing component issues from the column. At tube 189, this component is almost completely eluted and a second component appears. The second component is completely eluted at tube 310. Tubes 115–179 (fraction 1) are combined, and tubes 195–310 (fraction 2) are combined; each is separately evaporated to dryness. The residue from the first fraction is crystallized from water–ethanol to give **9-(2-amino-2-deoxy-α-D-ribofuranosyl)adenine (6)** in a yield of 230 mg. (33%), m.p. 149–151°, $[\alpha]_D^{23}$ +90 ± 2° (*c* 0.653, methanol).

The residue from fraction 2 is crystallized from methanol–ethanol to give **9-(2-amino-2-deoxy-β-D-ribofuranosyl)adenine (7)** in a yield of 215 mg. (31%), m.p. 194–196°, $[\alpha]_D^{22}$ −66 ± 2° (*c* 0.98, methanol).

Thin-layer chromatography on silica gel G, with 1:1 (v/v) ethyl acetate–methanol as developer, of the two separate anomers shows two distinct spots having R_f values of 0.28 (α-D) and 0.31 (β-D). The infrared absorption spectra of the anomers are very similar at all wavelengths except in the region of 10.5 to 12.5 μm. The α-D anomer is less soluble than the β-D form in water.

REFERENCES

(1) M. L. Wolfrom, H. G. Garg, and D. Horton, *J. Org. Chem.*, **30**, 1556 (1965).
(2) M. L. Wolfrom and M. W. Winkley, *Chem. Commun.*, **1966**, 533; *J. Org. Chem.*, **32**, 1823 (1967).
(3) C. A. Dekker, *J. Am. Chem. Soc.*, **87**, 4027 (1965).
(4) M. L. Wolfrom and M. W. Winkley, *J. Org. Chem.*, **31**, 1169 (1966).
(5) M. L. Wolfrom and W. Groebke, *J. Org. Chem.*, **28**, 2986 (1963).

[57] 3′-*O*-Acetyl-2′-deoxyadenosine

Selective 5′-O-Etherification of Basic Nucleosides, and Use of the Resulting Derivatives in the Synthesis of Specifically Substituted Nucleoside Derivatives

A. HOLÝ

INSTITUTE OF ORGANIC CHEMISTRY AND BIOCHEMISTRY,
CZECHOSLOVAK ACADEMY OF SCIENCES, PRAGUE, CZECHOSLOVAKIA

INTRODUCTION

Selective etherification of the 5′-hydroxyl group of basic ribonucleosides and 2′-deoxyribonucleosides by reaction with *p*-methoxy substituted chlorotriphenylmethane is accompanied by reaction of the amino groups attached to the heterocyclic moiety. This complication may be circumvented by selective protection of the amino groups by reaction with *N*,*N*-dimethylformamide acetals to give *N*-(dimethylamino)methylene derivatives. The latter compounds are then treated with a *p*-methoxy substituted chlorotriphenylmethane, and the resulting, 5′-*O*-protected *N*-(dimethylamino)methylene derivatives can be used as intermediates in the preparation of nucleoside derivatives[1] that are further specifically substituted.

PROCEDURE

2′-Deoxy-*N*-[(dimethylamino)methylene]adenosine (2)

A mixture of 5.0 g. (0.019 mole) of 2′-deoxyadenosine monohydrate (**1**), 25 ml. of anhydrous *N*,*N*-dimethylformamide,[a] and 10 ml. of *N*,*N*-dimethylformamide dimethyl acetal[2] in a 250-ml., round-bottomed flask, is shaken, with exclusion of atmospheric moisture, until a solution is obtained. The solution is kept at room temperature overnight, and evaporated to dryness at 40°/0.2 mm. Hg. The residue is dissolved in 25 ml. of chloroform, and the solution is added dropwise, with stirring, to 500 ml. of light petroleum (b.p. 40–60°).[b] The precipitate is collected, with suction,

[a] Commerical, pure *N*,*N*-dimethylformamide is distilled from phosphorus pentaoxide at 15–20 mm. Hg.

[b] The precipitation is almost quantitative if final traces of *N*,*N*-dimethylformamide are removed. Agitation during the dropwise addition should be vigorous, as otherwise, part of the product separates as a semisolid oil; if this happens, the precipitation procedure should be repeated.

NH_2

N N N N

$HOCH_2$

O

HO

1

(251.3)

$\xrightarrow{Me_2NCH(OMe)_2}$

N=CH—NMe_2

N N N N

$HOCH_2$

O

HO

2

(306.3)

$(p\text{-}MeOC_6H_4)_2PhCCl$ | pyridine

N=CH—NMe_2

N N N N

$(p\text{-}MeOC_6H_4)_2PhCOCH_2$

O

HO

3

(608.7)

$\xleftarrow[\text{pyridine}]{Ac_2O,}$

N=CH—NMe_2

N N N N

$(p\text{-}MeOC_6H_4)_2PhCOCH_2$

O

AcO

4

(650.7)

Acetic acid | in aq. EtOH

NH_2

N N N N

$HOCH_2$

O

AcO

5

(293.3)

on a sintered-glass funnel, washed with 50 ml. of light petroleum, and dried over phosphorus pentaoxide at 0.1 mm. Hg. The chromatographically homogeneous[c] product (**2**) weighs 5.45 g. (96%); m.p. 195–200°.

2′-Deoxy-*N*-[(dimethylamino)methylene]-5′-*O*-[bis(*p*-methoxyphenyl)phenylmethyl]adenosine (3)

A mixture of 5.45 g. (0.018 mole) of compound **2**, 6.80 g. (0.020 mole) of chlorobis(*p*-methoxyphenyl)phenylmethane,[d] and 60 ml. of anhydrous pyridine[e] is shaken in a 250-ml., round-bottomed flask until complete dissolution has occurred (about 1 hr.). The solution is kept at room temperature overnight, with exclusion of atmospheric moisture, and poured, with stirring, into 500 ml. of iced water in a 1-l. separatory funnel. The mixture is extracted with four 100-ml. portions of chloroform, and the extracts are combined, washed with 100 ml. of water, and dried with anhydrous magnesium sulfate; the suspension is filtered through a sintered-glass funnel, and the desiccant is washed with 50 ml. of chloroform. The filtrate and washings are combined, and evaporated to dryness on a rotary evaporator at 35°/15 mm. Hg. The residue is dissolved in 50 ml. of chloroform, and the solution is added dropwise, with stirring, to 500 ml. of light petroleum (b.p. 40–60°). The precipitate is collected with suction, washed with 100 ml. of light petroleum, and dried over phosphorus pentaoxide at 0.1 mm. Hg. The chromatographically homogeneous[f] product (**3**) weighs 7.40 g. (70%); m.p. 110–112°.

3′-*O*-Acetyl-2′-deoxy-*N*-[(dimethylamino)methylene]-5′-*O*-[bis(*p*-methoxyphenyl)phenylmethyl]adenosine (4)

A solution of 7.0 g. (0.018 mole) of compound **3** in 20 ml. of anhydrous pyridine[e] and 10 ml. of acetic anhydride is kept at room temperature overnight, with exclusion of atmospheric moisture, and poured onto a mixture of 200 g. of ice and 200 ml. of water. The precipitate is collected with suction, washed with 200 ml. of water, dried over phosphorus pentaoxide at 0.1 mm. Hg, and dissolved in 50 ml. of benzene; the solution is mixed with 0.1 g. of active carbon, and the suspension is filtered through a thin layer of Hyflo SuperCel (prewashed with benzene).

[c] The purity of the product is tested by (*a*) thin-layer chromatography on silica gel containing 10% of water, with 4:1 chloroform–methanol, with which the R_f values are 0.35 (2′-deoxyadenosine) and 0.55 (compound **2**), or (*b*) chromatography on Whatman No. 1 paper in butyl alcohol saturated with water, with which the R_f values are 0.30 and 0.41, respectively.

[d] For the preparation of this reagent, see Sec. V [160].

[e] Commercial pyridine is dried over molecular sieves.

[f] The chromatography is performed on a thin layer of silica gel containing 10% of water, with 9:1 chloroform–methanol; the R_f values are 0.30 (**2**), 0.60 (**3**), and 0.75 (**4**).

The material on the filter is washed with 10 ml. of benzene, and the filtrates are combined and added dropwise to 500 ml. of light petroleum (b.p. 40–60°). The precipitate is collected with suction, washed with 50 ml. of light petroleum, and dried over phosphorus pentaoxide at 0.1 mm. Hg. The chromatographically homogeneous[f] product (**4**) weighs 6.90 g. (91%); m.p. 105–107°; λ_{max} 233 and 312 nm., λ_{min} 260 nm.[g] in 96% ethanol.

3′-*O*-Acetyl-2′-deoxyadenosine (5)

In a 1-l., round-bottomed flask is placed a solution of 6.5 g. (0.01 mole) of compound **4** in 150 ml. of 96% ethanol, 150 ml. of water, and 150 ml. of acetic acid. The solution is kept at room temperature for 5 hr., and evaporated to dryness on a rotary evaporator at 30°/15 mm. Hg. Two 100-ml. portions of water are added to and evaporated from the residue, which is then shaken with a mixture of 200 ml. of water and 100 ml. of ether for 10 min. The aqueous layer is separated, washed with two 50-ml. portions of ether, and evaporated to dryness at 30°/15 mm. Hg. The residue is dissolved in 50 ml. of boiling methanol, and the solution is added dropwise, with stirring, to 300 ml. of 1:1 cyclohexane–light petroleum (b.p. 40–60°). The suspension is kept in a refrigerator at 2° for 48 hr., and the resulting precipitate is collected with suction, washed with 20 ml. of the same solvent mixture, and dried over phosphorus pentaoxide at 0.1 mm. Hg. The chromatographically homogeneous[h] product (**5**) weighs 2.10 g. (74%); m.p. 203–207°; λ_{max} 262 nm., λ_{min} 230 nm.[i] in 96% ethanol.

REFERENCES

(1) J. Žemlička and J. Holý, *Collection Czech. Chem. Commun.*, **32**, 3159 (1967).
(2) Z. Arnold and M. Kornilov, *Collection Czech. Chem. Commun.*, **29**, 645 (1964).

[g] The spectral maximum at 312 nm. indicates the presence of an *N*-[(dimethylamino)methylene] group.

[h] The chromatography is performed on Whatman No. 1 paper in 7:1:2 isopropyl alcohol–concentrated ammonium hydroxide–water; the R_f values are 0.61 (**1**) and 0.75 (**5**).

[i] The absence of a spectral maximum at 312 nm. indicates the absence of an *N*-[(dimethylamino)methylene] group.

[58] Sulfonium Salt of 5′-*S*-(3-Aminopropyl)-5′-*S*-methyl-5′-thioadenosine ("Decarboxylated *S*-Adenosyl-L-methionine")

Preparation of Substituted 5′-Thioadenosine Nucleosides

G. A. JAMIESON

BLOOD PROGRAM RESEARCH LABORATORY, THE AMERICAN NATIONAL RED CROSS, WASHINGTON, D.C. 20006

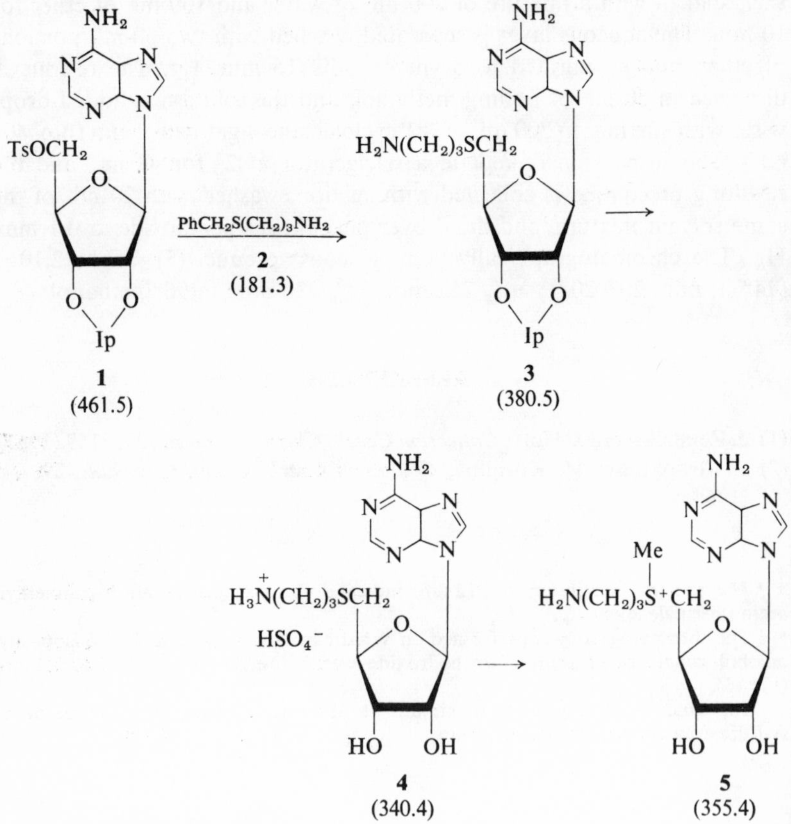

NH_2

$TsOCH_2$

1 (461.5)

$PhCH_2S(CH_2)_3NH_2$ **2** (181.3)

$H_2N(CH_2)_3SCH_2$

3 (380.5)

$H_3\overset{+}{N}(CH_2)_3SCH_2$ HSO_4^-

4 (340.4)

$H_2N(CH_2)_3S^+CH_2$ (Me)

5 (355.4)

where Ts is *p*-tolylsulfonyl.

INTRODUCTION

5′-Methylsulfonium derivatives of adenosine are central to the processes of enzymic transmethylation[1] and of polyamine synthesis.[2] Their chemical synthesis[3,4] proceeds most effectively *via* the methylation of the corresponding thio ethers prepared by condensation of the appropriate sodium thioxide derivatives with 2′,3′-*O*-isopropylidene-5′-*O*-*p*-tolylsulfonyladenosine in liquid ammonia. An illustrative example is the synthesis of the sulfonium salt of 5′-*S*-(3-aminopropyl)-5′-*S*-methyl-5′-thioadenosine[4] (**5**).

PROCEDURE

3-(Benzylthio)propylamine (2)

A solution of 3-(benzylthio)propionitrile[5] (25.0 g., 0.14 mole) in dry ether (150 ml.) is added during 1 hr. to a vigorously stirred suspension of lithium aluminum hydride (7.3 g.) in dry ether (200 ml.). A stream of helium is slowly passed through the flask during the addition, sufficient ice-cooling being applied to maintain gentle refluxing. After addition of the nitrile, the reaction mixture is refluxed for 1 hr., and then cooled in ice. Ethyl acetate (200 ml.) is cautiously added, followed by the addition of water (100 ml.). The mixture is filtered, and the precipitate is washed with a little acetone and then extensively with ether. The filtrate is extracted with ether, and the ether extracts are combined, dried with sodium sulfate, and evaporated to dryness under diminished pressure. The resulting sirup is distilled at 124°/3 mm. Hg, through a Claisen distillation head, giving **2** (15.3 g., 60%) as a colorless, mobile, almost odorless oil. The hydrochloride, prepared by passing gaseous hydrogen chloride through an ethereal solution of **2**, has m.p. 76° (after two recrystallizations from ether–ethanol).

5′-*S*-(3-Aminopropyl)-2′,3′-*O*-isopropylidene-5′-thioadenosine (3)

Dry ammonia (*ca.* 300 ml.) is condensed into a 1-l., two-necked flask provided with a stirrer and a sodium hydroxide tube to maintain anhydrous conditions. 3-(Benzylthio)propylamine (**2**) (3.2 g., 17 mmoles) is carefully added, and the mixture is stirred while sodium is added in small pieces until the resulting blue color persists for 5–10 min. 2′,3′-*O*-Isopropylidene-5′-*O*-*p*-tolylsulfonyladenosine[4] (**1**) (9.2 g., 20 mmoles) is then carefully added to the solution, stirring is continued for 10 min., and then the liquid ammonia is allowed to evaporate during several hr., final traces being removed under diminished pressure, giving a solid residue which is extracted with chloroform (100 ml.) plus cold, *N* sulfuric acid (100 ml.). The aqueous layer is separated, and the chloroform layer is re-extracted with *N* sulfuric acid. The acid extracts are combined, and rendered alkaline, and the oil that is precipitated is extracted by shaking with three 50-ml. portions of chloroform. The chloroform extracts are combined, washed once with water (50 ml.), dried with sodium sulfate, and filtered. The filtrate is evaporated to dryness under diminished pressure, giving an oil which is dissolved in hot water. The solution is treated with a little decolorizing charcoal, filtered, and cooled. Compound **3** sesquihydrate is obtained as glistening plates; yield 4.1 g. (53%), m.p. 66–68°.

5′-*S*-(3-Aminopropyl)-5′-thioadenosine Hydrogen Sulfate (4)

Compound **3** (3.0 g., 7.4 mmoles) is dissolved in *N* sulfuric acid (40 ml.), and the solution is kept at room temperature for 48 hr. Ethanol is added to incipient turbidity, and the solution is cooled in ice. The resulting crystals (2.2 g.) are filtered off, and successively washed with a little 50% aqueous ethanol, and absolute ethanol. A second crop of crystals (0.5 g.) is obtained by adding a large excess of alcohol to the mother liquors. Both crops melt at 173–175°; after recrystallizing from aqueous alcohol, m.p. 180–183° (dec.).

5′-*S*-(3-Aminopropyl)-5′-*S*-methyl-5′-thioadenosine, Sulfonium Salt ("Decarboxylated *S*-Adenosyl-L-methionine," 5)

Methyl iodide (0.5 ml.) is added to a solution of compound **4** (100 mg.) in 2.5 ml. of formic acid plus 2.5 ml. of acetic acid, and the mixture is kept in the dark at room temperature for 6 days. It is then diluted with an equal volume of water, and washed with three 10-ml. portions of ether. To the aqueous solution is added a solution (10 ml.) of flavianic acid (2%, in *N* hydrochloric acid), and the precipitated sulfonium flavianate is centrifuged off, and washed successively with two 5-ml. portions of the cold flavianic acid solution, and 5 ml. of cold water.

Without purification, the flavianate is suspended in 0.4 *N* hydrochloric acid (10 ml.) and the suspension is washed with 10-ml. portions of 2-butanone until the aqueous layer is colorless.[a] The aqueous layer is then washed with ether, treated with sufficient Dowex 1 (OAc^-) to remove the chloride ions, and lyophilized. The residue is dissolved in a little water, and the solution is adjusted to pH 5–6. The yield of sulfonium nucleoside (**5**) is about 50 μmoles[b]; the product is homogeneous, as shown (following paper electrophoresis or paper chromatography) by its ultraviolet absorption at 260 nm., and by ninhydrin, iodoplatinate, and periodate–Schiff sprays.

[a] Five to six extractions.
[b] Based on an extinction coefficient of 16,000 at 280 nm.

REFERENCES

1) G. L. Cantoni, *J. Biol. Chem.*, **204**, 403 (1953).
2) H. Tabor, S. M. Rosenthal, and C. W. Tabor, *J. Biol. Chem.*, **233**, 907 (1958).
3) J. Baddiley and G. A. Jamieson, *J. Chem. Soc.*, **1955**, 1085.
4) G. A. Jamieson, *J. Org. Chem.*, **28**, 2397 (1963).
5) C. D. Hurd and L. H. Gershbein, *J. Am. Chem. Soc.*, **69**, 2328 (1947).

[59] 2-Bromoadenosine

An Illustration of the Fusion Method for the Preparation of Purine Nucleosides

JOHN A. MONTGOMERY and KATHLEEN HEWSON

KETTERING-MEYER LABORATORY, SOUTHERN RESEARCH INSTITUTE, BIRMINGHAM, ALABAMA 35205

Br N N Br N N H + $AcOCH_2$ O OAc AcO OAc $\xrightarrow{p\text{-}CH_3C_6H_4SO_3H}$

1 (277.9) **2** (318.2)

Br N N Br N N $AcOCH_2$ O AcO OAc **3** (536.2) $\xrightarrow[\text{MeOH}]{NH_3}$ NH_2 N N Br N N $HOCH_2$ O HO OH **4** (346.2)

INTRODUCTION

The fusion method[1] for the preparation of nucleosides has been widely applied. It affords an improved synthesis[2] of an acylated derivative of 2,6-dichloropurine ribonucleoside,[3] suitable for conversion into a variety of purine ribonucleosides, including the biologically active[2] adenosine analog, 2-chloroadenosine.[4] The procedure may also be used for the preparation of 2-bromoadenosine[2] (**4**) from 2,6-dibromopurine (**1**).

PROCEDURE

2,6-Dibromo-9-(2,3,5-tri-*O*-acetyl-β-D-ribosyl)-9*H*-purine (3)

A mixture of 1,2,3,5-tetra-*O*-acetyl-β-D-ribose[5] (**2**) (1.9 g., 6.0 mmoles) and 2,6-dibromopurine (**1**) (1.6 g., 5.8 mmoles) in a round-bottomed flask, fitted with a vacuum adapter and a stirrer, is heated with continuous stirring, at 130° *in vacuo*, until an opaque solution is obtained and vigorous gas evolution ceases (*ca.* 5 min.).[a] After the reaction flask has been cooled at room temperature for 5 min., atmospheric pressure is restored, *p*-toluenesulfonic acid (30 mg.) is added, and the reaction mixture is heated, with continuous stirring, at 130–135° for 25 min. *in vacuo*. The clear, amber solution is cooled to room temperature and dissolved in chloroform (5 ml.), the solution is transferred to a separatory funnel with additional chloroform, and the solution (approximately 10 ml.) is successively washed with 5 ml. of saturated aqueous sodium hydrogen carbonate and 5 ml. of water. After being dried for several hours with anhydrous magnesium sulfate, the chloroform solution is filtered through a pad of dry Celite, and the filtrate is evaporated to dryness under diminished pressure at 35°. The sirupy residue is triturated with ether (5–10 ml.), and the crystals that form are filtered off, washed with fresh ether, and dried *in vacuo* over phosphorus pentaoxide at room temperature/0.05 mm. Hg; yield 1.5 g. (48%), m.p. 147°, λ_{max} in nm. (log ϵ): pH 1 and 7, 254 (3.70) and 277 (3.97); pH 13, 260 (4.02).

2-Bromoadenosine (4)

Under anhydrous conditions, a suspension of 2,6-dibromo-9-(2,3,5-tri-*O*-acetyl-β-D-ribosyl)-9*H*-purine (**3**) (1.4 g., 2.6 mmoles) in absolute methanol (40 ml.) is saturated at 5° with dry ammonia, and the resulting solution is kept at room temperature for 1 week. The amber solution is evaporated to dryness under diminished pressure at 35°, and the residue is triturated with chloroform, affording a suspension of a crude solid which is collected by filtration, and recrystallized from ethanol (20 ml.); yield 540 mg. (60%) of purified product **4**. An ethyl acetate solution of the purified product is filtered through dry Celite, the filtrate is evaporated to dryness under diminished pressure, and the resulting residue is recrystallized from acetone (10 ml.). Crystallization is completed by cooling, and the crystals are filtered off, washed with cold acetone, and dried at room temperature/0.05 mm. Hg; yield 260 mg. (29%), m.p. indefinite, $[\alpha]_D^{22}$ $-48.5 \pm 0.8°$ (*c* 1.03, methanol), λ_{max} in nm. (log ϵ): pH 1, 266 (4.15); pH 7 and 13, 265 (4.17).

[a] An efficient water aspirator may be used as the vacuum source (10–20 mm. Hg pressure is ideal). Use of an oil bath, of sufficient depth to permit complete submersion of the reaction flask, prevents the accumulation of unfused reactants on the walls of the flask.

REFERENCES

(1) T. Sato, T. Simadate, and Y. Ishido, *Nippon Kagaku Zasshi*, **81**, 1440 (1960).
(2) J. A. Montgomery and K. Hewson, *J. Heterocyclic Chem.*, **1**, 213 (1964); Y. Ishido, Y. Kikuchi, and T. Sato, *Nippon Kagaku Zasshi*, **86**, 240 (1965).
(3) H. J. Schaeffer and H. J. Thomas, *J. Am. Chem. Soc.*, **80**, 3738 (1958).
(4) J. Davoll, B. Lythgoe, and A. R. Todd, *J. Chem. Soc.*, **1948**, 1685.
(5) H. Zinner, *Chem. Ber.*, **83**, 517 (1950).

[60] 2′-Deoxyadenosine and its α-D Anomer

A Direct Synthesis of Anomeric Purine 2′-Deoxyribonucleosides

ROBERT K. NESS

NATIONAL INSTITUTE OF ARTHRITIS AND METABOLIC DISEASES, NATIONAL INSTITUTES OF HEALTH, PUBLIC HEALTH SERVICE, U. S. DEPARTMENT OF HEALTH, EDUCATION, AND WELFARE, BETHESDA, MARYLAND 20014

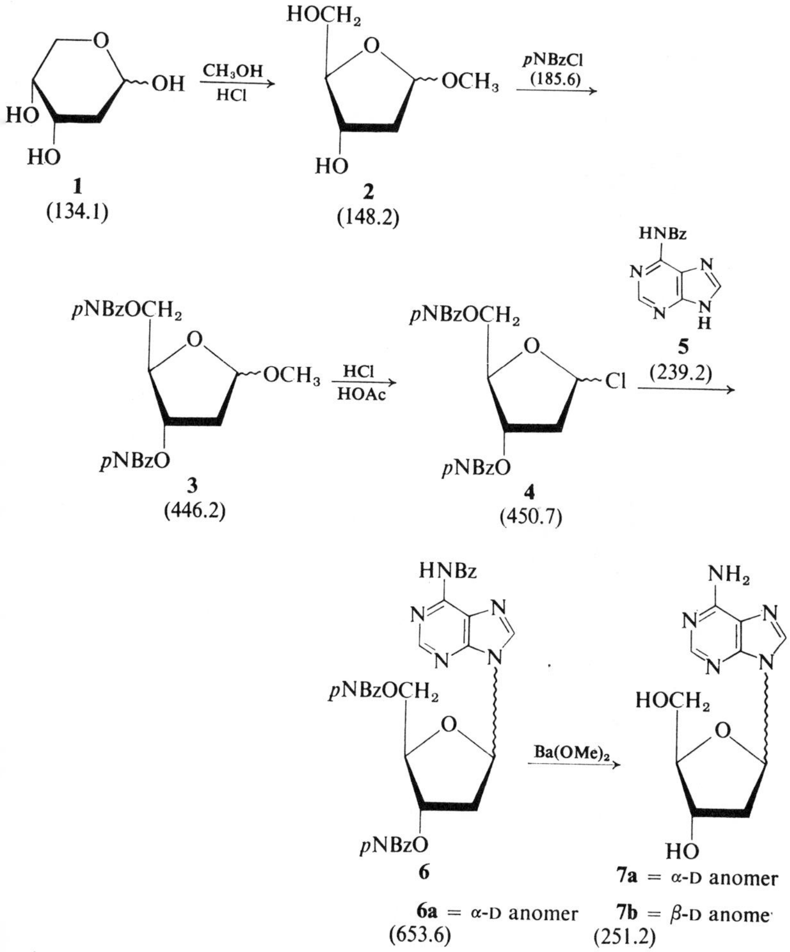

1 (134.1) $\xrightarrow[HCl]{CH_3OH}$ **2** (148.2) $\xrightarrow[(185.6)]{pNBzCl}$

3 (446.2) $\xrightarrow[HOAc]{HCl}$ **4** (450.7) $\xrightarrow[(239.2)]{\textbf{5}}$

6 (653.6) $\xrightarrow{Ba(OMe)_2}$ **7** (251.2)

6a = α-D anomer

7a = α-D anomer

7b = β-D anome

where *p*NBz is *p*-nitrobenzoyl.

INTRODUCTION

Unlike the direct synthesis of ribonucleosides from suitable derivatives of a nitrogenous base and D-ribose, that of 2′-deoxyribonucleosides cannot utilize neighboring-group participation at C-2 of the sugar derivative; thus, both anomers of the nucleoside will be formed. 2′-Deoxyadenosine (**7b**) (Sec. III [47] and [61]) is one of the nucleosides obtainable from 2′-deoxyribonucleic acid. Its α-D anomer does not occur naturally. The two compounds have been synthesized by the direct condensation[1] in methyl sulfoxide (with subsequent hydrolysis of the product) of amorphous 2-deoxy-3,5-di-*O*-(*p*-nitrobenzoyl)-D-*erythro*-pentosyl chloride (**4**) (Sec. III [111]) with chloromercuri-*N*-benzoyladenine in yields of 10% and 19%, respectively. This unpublished modification of that condensation by Ness and Fletcher[1] includes a change in the reaction solvent and the use of crystalline chloride[2] (**4**) and of *N*-benzoyladenine (**5**) (Sec. III [44]). The yields have been increased, and the isolation simplified. Other direct condensations of *O*-substituted glycosyl halides with nitrogenous bases (or their non-metal derivatives) in the presence of an acid acceptor have been reported.[3]

PROCEDURE

Methyl 2-Deoxy-3,5-di-*O*-(*p*-nitrobenzoyl)-α,β-D-*erythro*-pentoside[2] (3)

A solution of 20.00 g. (0.149 mole) of 2-deoxy-D-*erythro*-pentose[4] (2-deoxy-D-ribose, **1**) in 360 ml. of absolute methanol is treated with 3.0 ml. of 0.649 *M* hydrogen chloride in methanol. After *ca.* 40 min., when the observed rotation of the solution reaches a maximum at *ca.* α_D^{20} +8.4° (4-dm. tube), 100 ml. of dry pyridine is added (to stop the reaction). The reaction mixture is then concentrated (bath at 50°) to a small volume. An additional 100 ml. of dry pyridine is added, and the solution is evaporated under diminished pressure to a stiff sirup which is then dissolved in 120 ml. of dry pyridine. With mechanical stirring, and cooling in an ice bath, 70.2 g. (0.379 mole) of *p*-nitrobenzoyl chloride is added during 10 min. at such a rate that the temperature does not exceed 40°. The reaction mixture is kept at room temperature for 18 hr., cooled, and treated with 10 ml. of water. Ten minutes later, 1 l. of ice-water is added, and the aqueous portion is decanted from the gummy product, extracted with dichloromethane, and discarded; the extract is then used for dissolving the gum. This solution is washed successively with a saturated solution of sodium hydrogen carbonate (to remove *p*-nitrobenzoic acid, which is insoluble in dichloromethane), cold 3 *N* sulfuric acid, and saturated sodium hydrogen carbonate, and dried (magnesium sulfate). The solution is concentrated *in vacuo* to a sirup, which is dissolved in ethyl acetate, and the solution is evaporated to dryness under diminished pressure, affording a crystalline residue. This is dissolved in 60 ml. of hot ethyl acetate, hot absolute alcohol (120 ml.) is added to the solution, with stirring, and nucleating crystals[a] of the anomeric glycosides (**3**) are added. The resulting crop of **3** is removed by filtration; yield 51.8 g. (77.9%). The specific rotation of this product is variable, but approximates[b] +45°; the product may be employed directly for the preparation of the chloride (**4**).

2-Deoxy-3,5-di-*O*-(*p*-nitrobenzoyl)-D-*erythro*-pentosyl Chloride[2] (4)

A rapid stream of hydrogen chloride is passed into a vigorously stirred suspension of 4.98 g. (11.2 mmoles) of glycoside **3** in 30 ml. of glacial acetic acid while the reaction flask is cooled in an ice-water bath. After 3 min., the addition of the gas is halted, and after 7 min., the glycosyl chloride begins to crystallize. After a total of 20 min., 30 ml. of a freshly saturated solution of hydrogen chloride in ether is added, and the addition of gaseous hydrogen chloride is resumed for 1 min. Ten minutes later, 30 ml. of pentane is added, and the cooling bath is removed. After a further 10 min., the suspension is filtered, and the crystals are washed with ether and then with pentane, yielding 3.8–4.3 g. (75–85%) of **4**; m.p. (dec.[c]) 105–107° (placed in bath at 95°, and heated at 6°/min.), $[\alpha]_D^{20}$ +107° (*c* 1.13, dichloromethane dried with Drierite).

N-Benzoyladenine[5] (5)

A mixture of 25.0 g. (0.185 mole) of adenine and 100 g. of benzoic anhydride in an open flask submerged to its neck in an oil bath at 140° is stirred for 2 hr. The mixture is then cooled to 80°, 150 ml. of absolute ethanol is added, the resulting suspension is poured into 450 ml. of absolute ethanol, and the mixture is heated (to effect dissolution), and cooled to 10°, affording 36.5 g. (69%) of **5**, m.p. 242–244°; this is used directly in the preparation of **6**.

9-(2-Deoxy-α-D-*erythro*-pentofuranosyl)adenine (7a)

To a well stirred suspension of 4.35 g. (0.182 mole) of powdered **5** and 8.7 g. of Molecular Sieve[d] in 43 ml. of dichloromethane (dried with

[a] Nucleating crystals of the α-D anomer of **3** are obtained by dissolving a portion of the reaction product in 3:3:1 dichloromethane–ether–pentane (2.3 ml. per g.). After the crystalline α-D anomer has been removed by filtration, the filtrate is evaporated to a sirup which is dissolved in 1:2 ethyl acetate–absolute alcohol (2.25 ml. per g. of sirup) to afford the crystalline β-D anomer.

[b] The rotation of the pure β-D anomer is $[\alpha]_D^{20}$ −5.4° (*c* 1.16, chloroform), and that of the pure α-D anomer is $[\alpha]_D^{20}$ +115° (*c* 1.84, chloroform).

[c] When the chloride is pure, it may be stored for several weeks in a freezer, but, in a vacuum desiccator at room temperature, it decomposes fairly rapidly. Consequently, it is best to prepare **4** from **3** as needed.

[d] Type 4A, 1/16-in. pellets, Fisher Scientific Co. Later work has indicated that type 5A is more effective in removing hydrogen chloride, and may be used more advantageously.

Drierite) is added 4.35 g. (0.0965 mole) of **4**, and the suspension is stirred at room temperature for three days. The mixture is now filtered through diatomaceous earth, and the solids are well washed with dichloromethane. The filtrate and washings are combined, and evaporated to dryness, giving 6.76 g. of product that is dissolved in 100 ml. of warm amyl acetate. The addition of 100 ml. of warm absolute ethanol creates an opalescence that is removed by adding decolorizing carbon and filtering. Warm absolute ethanol (100 ml.) is added, and the solution is allowed to cool slowly, and is then cooled to 0°. The crystalline *N*-benzoyl-9-(2-deoxy-3,5-di-*O*-*p*-nitrobenzoyl-α-D-*erythro*-pentosyl)adenine[e] (**6a**) is removed by filtration and the filtrate (*A*) is saved: yield, air-dried, 3.40 g.; dried at 100°/1 mm. Hg, 3.23 g. (51.3%), m.p. 115–117°. The material in the filtrate (*A*) is converted into **7b**.

A suspension of 473 mg. (0.687 mmole) of air-dried **6a** in 1.5 ml. of 0.1 *M* methanolic barium methoxide is heated at 50° for 5 min. The resulting solution is kept at room temperature for 17 hr.; crystallization, mainly of methyl *p*-nitrobenzoate, occurs. When the reaction is complete, carbon dioxide (Dry Ice) is added, the mixture is evaporated to dryness under diminished pressure, and the residue is partitioned between dichloromethane and water made slightly basic with ammonia. The aqueous solution is filtered through diatomaceous earth, and the filtrate is evaporated to dryness *in vacuo* (at 50°). The residue is dissolved in about 10 ml. of warm, absolute ethanol, and the solution is cooled, affording crystalline **7a**; yield 109 mg. (0.434 mmole), m.p. 203–207°. On concentrating the filtrate to about 2 ml., 55 mg. of impure **7a** is obtained; this is dissolved in 2 ml. of ethanol plus 0.3 ml. of water, the suspension is filtered, and the filtrate is cooled, giving 26 mg. (0.104 mmole) of **7a** (m.p. 200–203°); yield of **7a**, 78% for the deacylation and 40% for the condensation. Recrystallization of **7a** is effected from ethanol and then from methanol; m.p. 209–211°, $[\alpha]_D^{20}$ +71° (*c* 0.54, water), $\lambda_{max}^{H_2O}$ 260 nm.

2′-Deoxyadenosine (7b)

Filtrate *A* from the isolation of **6a** is evaporated to dryness under diminished pressure, giving 2.97 g. of **6**. To this material (0.0454 mole) is added 6.6 ml. of 0.1 *M* methanolic barium methoxide, and the mixture is warmed at 50° for 1 hr., cooled, and evaporated to dryness. The residue is partitioned between dichloromethane and water made slightly basic

[e] Compound **6a** is new; it may be purified by recrystallizations from ethyl acetate–ethanol; m.p. 120–122°, $[\alpha]_D^{20}$ +4.2° (*c* 2.0, chloroform).

Anal. Calcd. for $C_{31}H_{23}N_7O_{10}$: C, 56.97; H, 3.55; N, 15.00. Found: C, 57.08; H, 3.78; N, 14.70.

with ammonia. The aqueous layer is treated with Amberlite MB-3 mixed-bed, ion-exchange resin,[f] and evaporated to dryness under diminished pressure at 50°. The residue is dissolved in 100 ml. of absolute ethanol, the solution is filtered through Whatman No. 50 filter paper (to remove a flocculent precipitate), and the filtrate is evaporated to dryness under diminished pressure. The residue is dissolved in 1.5 ml. of water slightly basic with ammonia, and **7b** crystallizes on nucleation; yield 354 mg., m.p. 125° (resolidifies, and re-melts at about 190°). When dried at 110°/2 mm. Hg for 2 hr., the product weighs 332 mg. (13.7%), m.p. 180–189° (sinters at 158–161°); it may be recrystallized from a small volume of water; pure **7b** has[2] m.p. 187–189° (usually with prior sintering at *ca.* 160°), $[\alpha]_D^{20}$ $-25°$ (*c* 0.47, water), $\lambda_{max}^{H_2O}$ 260 nm.

REFERENCES

(1) R. K. Ness and H. G. Fletcher, Jr., *J. Am. Chem. Soc.*, **82**, 3434 (1960).

(2) R. K. Ness, D. L. MacDonald, and H. G. Fletcher, Jr., *J. Org. Chem.*, **26**, 2895 (1961).

(3) A. J. Cleaver, A. B. Foster, and W. G. Overend, *J. Chem. Soc.*, **1959**, 409; C. P. J. Glaudemans and H. G. Fletcher, Jr., *J. Org. Chem.*, **28**, 3004 (1963) (see Sec. III [46]); N. Yamaoka, K. Aso, and K. Matsuda, *ibid.*, **30**, 149 (1965); T. Y. Shen, H. M. Lewis, and W. V. Ruyle, *ibid.*, **30**, 835 (1965).

(4) H. W. Diehl and H. G. Fletcher, Jr., *Biochem. Prepn.*, **8**, 49 (1961).

(5) A. Kossel, *Z. Physiol. Chem.*, **12**, 241 (1888).

[f] Product of Rohm and Haas Co., Resinous Products Division, Washington Square, Philadelphia, Pa.

[61] 2′-Deoxyadenosine and 3′-Deoxyadenosine (Cordycepin)

Synthesis of a 2′ (or 3′)-Deoxyadenosine from Adenosine via *a 2′,8- or a 3′,8-Anhydronucleoside*

MORIO IKEHARA and HIROSHI TADA

FACULTY OF PHARMACEUTICAL SCIENCES, HOKKAIDO UNIVERSITY,
SAPPORO, JAPAN

INTRODUCTION

2′-Deoxyadenosine (**8**) (Sec. III [47] and [60]) is synthesized, alternatively, by (*1*) ring opening of a 2′,3′-epithiuronium derivative of adenosine,[1] (*2*) the condensation of a chloromercuri salt of adenine with an *O*-acylglycosyl halide,[2,3] (*3*) coupling of adenine with an acylated sugar by fusion,[4] or (*4*) the reduction of an adenine anhydronucleoside.[5] Cordycepin (**9**) is an antibiotic substance obtained from *Cordyceps militaris*, and has been shown to be active against *Bacillus subtilis*. Several alternative syntheses of **9** have been reported.[6–8]

PROCEDURE

8-Bromo-2′,3′-*O*-isopropylideneadenosine[9] (2)

2′,3′-*O*-Isopropylideneadenosine[10,11] (**1**) (3.05 g., 10 mmoles) is dissolved in a mixture of *p*-dioxane (150 ml.) and 10% aqueous disodium hydrogen phosphate (150 ml.), and bromine (1.9 g., 12 mmoles) is added. The reaction mixture is shaken for 5 hr. at room temperature and kept overnight. The solution is thoroughly extracted with chloroform, and the extract is successively washed with 2 *N* sodium hydrogen sulfite and water. After being dried with sodium sulfate, the solution is evaporated to afford a solid, which is recrystallized from ethanol to give needles; yield 2.85 g. (74.2%), m.p. 221–222°, λ_{max}^{EtOH} 265 nm., R_f 0.80 (4:1:5 butyl alcohol–acetic acid–water) and R_f 0.73 (43:7 butyl alcohol–water).

5′-*O*-Acetyl-8-bromo-2′,3′-*O*-isopropylideneadenosine[12] (3)

Compound **2** (1.38 g., 3.58 mmoles) is dissolved in dry pyridine (35 ml.), and acetic anhydride (6 ml.) is added. After being kept at room temperature overnight, ethanol (20 ml.) is added, and the mixture is kept at room temperature for an additional 2 hr. The solution is evaporated under

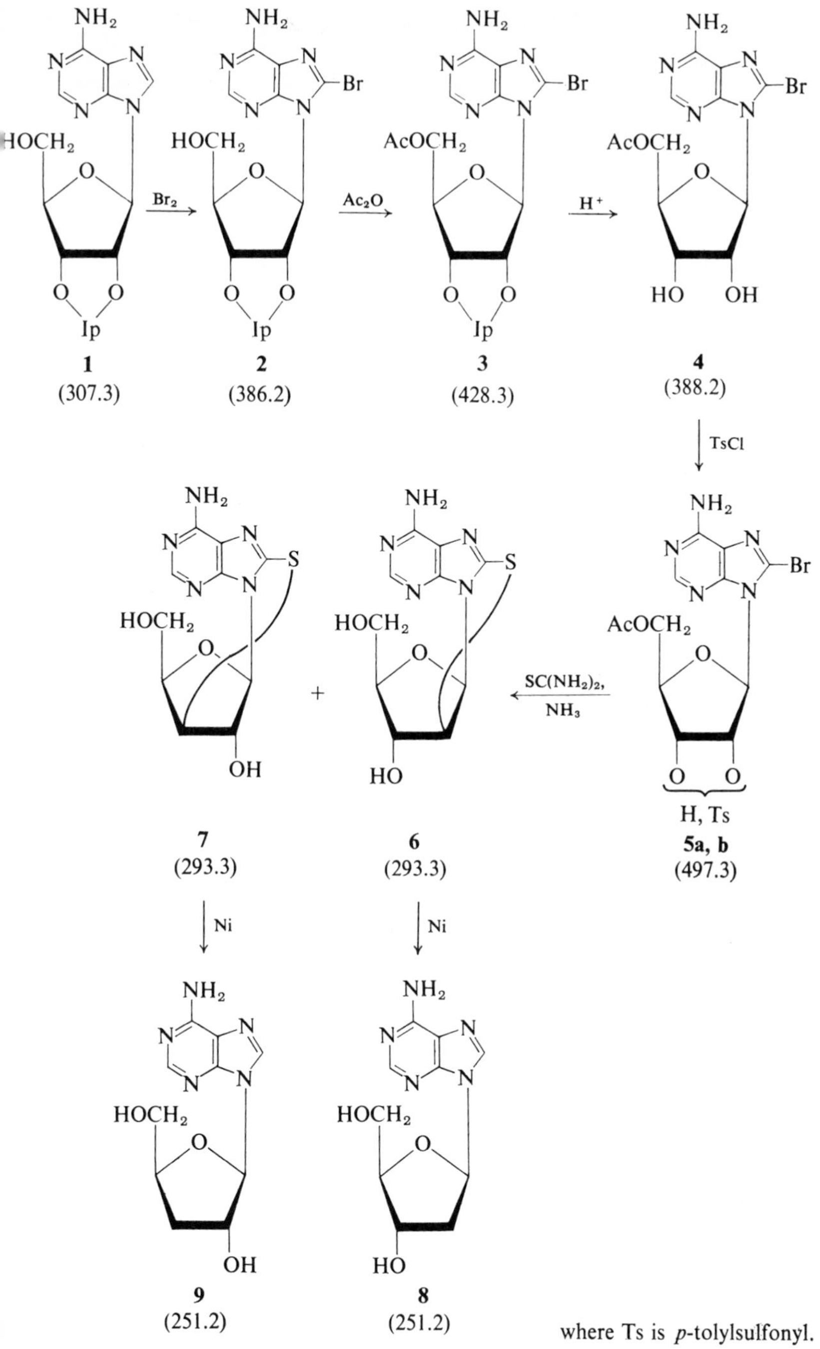

NH_2 / HOCH_2 / O / Ip — **1** (307.3) $\xrightarrow{Br_2}$ NH_2, Br / HOCH_2 / O / Ip — **2** (386.2) $\xrightarrow{Ac_2O}$ NH_2, Br / AcOCH_2 / O / Ip — **3** (428.3) $\xrightarrow{H^+}$ NH_2, Br / AcOCH_2 / HO OH — **4** (388.2)

4 $\xrightarrow{TsCl}$ NH_2, Br / AcOCH_2 / O O / H, Ts — **5a, b** (497.3)

5a, b $\xrightarrow{SC(NH_2)_2,\ NH_3}$ NH_2, S / HOCH_2 / HO — **6** (293.3) + NH_2, S / HOCH_2 / OH — **7** (293.3)

7 $\xrightarrow{Ni}$ NH_2 / HOCH_2 / OH — **9** (251.2)

6 $\xrightarrow{Ni}$ NH_2 / HOCH_2 / HO — **8** (251.2)

where Ts is *p*-tolylsulfonyl.

diminished pressure, the residual sirup is dissolved in chloroform (50 ml.), and the solution is washed with saturated aqueous sodium hydrogen carbonate, and dried with sodium sulfate. The solution is evaporated to dryness under diminished pressure, and the red residue is triturated with ethanol to induce crystallization. Compound **3** is obtained as tiny plates; yield 1.01 g. (65.6%), m.p. 158–160°, λ_{max}^{EtOH} 265 nm., R_f 0.82 (43:7 butyl alcohol–water).

5′-*O*-Acetyl-8-bromoadenosine[12] (4)

Compound **3** (1.0 g., 2.34 mmoles) is dissolved in 98% formic acid (30 ml.) and the mixture is kept at room temperature for 20 hr. Ethanol (20 ml.) is added, and the solvent is distilled off under diminished pressure. The procedure is repeated until the odor of formic acid is no longer present. The residual sirup is dissolved in a small volume of chloroform, the solution is washed twice with water, and the water layer is extracted with chloroform. The chloform solutions are combined and evaporated to dryness, and the residue is re-evaporated several times with small volumes of ethanol, to afford a colorless glass; yield 600 mg. (67%), λ_{max}^{EtOH} 266 nm., R_f 0.54 (water, adjusted to pH 10 with ammonium hydroxide) and R_f 0.57 (7:1:2 isopropyl alcohol–ammonia–water).

5′-*O*-Acetyl-8-bromo-2′(and 3′)-*O*-*p*-tolylsulfonyladenosine[12] (5a and 5b)

Compound **4** (998 mg., 2.57 mmoles) is dried by azeotropic distillation with dry pyridine (60 ml.), and dissolved in dry pyridine (60 ml.). *p*-Toluenesulfonyl chloride (499 mg., 1.02 equiv.) is added with cooling in an ice–salt bath. The reaction mixture is kept in a refrigerator for 60 hr. After the addition of 10 ml. of water, the mixture is immediately poured into saturated, aqueous sodium hydrogen carbonate. The solution is thoroughly extracted with chloroform, and the chloroform layer is washed successively with aqueous sodium hydrogen carbonate and water, and dried with sodium sulfate. The solvent is removed under diminished pressure, and the residue is dissolved in chloroform; the solution is re-extracted with water to remove starting material. Evaporation of the chloroform solution gives a glass, which is dried for 4 hr. at 0.1 mm. Hg in a vacuum desiccator containing phosphorus pentaoxide. The solid is dissolved in anhydrous methanol (20 ml.) and to the solution is added methanol (20 ml.) presaturated at 0° with dry ammonia. After the solution has been kept in a refrigerator for 21 hr., the solvent is removed under diminished pressure to afford a crystalline solid, which is extracted with hot benzene (to remove the acetamide), yielding crude, crystalline product; yield 537 mg. (42%). Repeated recrystallization from 50% aqueous isopropyl alcohol gives pure 2′-*p*-toluenesulfonate (**5a**), m.p. 220–223°

(dec.), ν_{max}^{Nujol} 1186, 1172 cm.$^{-1}$ (covalent *p*-toluenesulfonate). The mother liquors contain the 3′-*p*-toluenesulfonate (**5b**), m.p. 213° (dec.) and a small amount of 8-bromo-2′,3′-di-*O*-*p*-tolylsulfonyladenosine, m.p. 176–177°.

2′,8-Anhydro-(6-amino-9-β-D-arabinofuranosyl-9*H*-purine-8-thiol)[12] **(6)**

Compound **5a** (510 mg., 1.03 mmoles) is refluxed in butyl alcohol (80 ml.) with thiourea (81.5 mg., 1.1 equiv.) for 2 hr. Evaporation of the solvent under diminished pressure gives a powder (561 mg.), which is dissolved in ethanol (10 ml.) and the solution is placed on a column (28 × 380 mm.) of cellulose powder (70 g.). Elution is carried out with 100:1 butyl alcohol saturated with water–concentrated ammonium hydroxide, and 10-ml. fractions are collected. Fractions 11–18 are combined and evaporated, and the residual, colorless solid (328 mg.) is crystallized from 0.2 ml. of water at 0°. Recrystallization from 1 ml. of water gives 167 mg. (57.5%) of **6**, m.p. 191–194°, $[\alpha]_D^{23.5}$ −187.2° (*c* 1.0, water), λ_{max}^{EtOH} 276 nm.

3′,8-Anhydro-(6-amino-9-β-D-xylofuranosyl-9*H*-purine-8-thiol)[12] **(7)**

The mother liquors containing **5b** (1.67 g., 3.36 mmoles) are refluxed with thiourea (277 mg.) in butyl alcohol (100 ml.) for 2 hr. After evaporation of the solvent, the reddish semi-solid (1.76 g.) is dissolved in ethanol (10 ml.) and placed on a column (28 × 560 mm.) of cellulose powder (120 g.), which is eluted with 100:1 butyl alcohol saturated with water–concentrated ammonium hydroxide. Fractions 12–23 (10 ml. each) give 240 mg. of the 3′-*p*-toluenesulfonate (**5b**), which is discarded. Fractions 28–30 are combined and evaporated to dryness. The residual white powder (250 mg.) is recrystallized from 80% aqueous ethanol, giving compound **7**, which discolors at 231–232°, and decomposes at 250°, $[\alpha]_D^{20}$ −32.0° (*c* 0.75, pyridine), λ_{max}^{EtOH} 276 nm. (shoulder), 283 nm., 291 nm. (shoulder); $\lambda_{max}^{H^+}$ 284 nm.; $\lambda_{max}^{OH^-}$ 285 nm., ν_{max}^{Nujol} 1175–1185 cm.$^{-1}$ (covalent *p*-toluenesulfonate does not appear), R_f 0.25 (4:1:5 butyl alcohol–acetic acid–water). From fractions 31–34 is obtained compound **6** (110 mg.).

2′-Deoxyadenosine[12] **(8)**

Compound **6** (210 mg., 0.72 mmoles) is refluxed in water (20 ml.) with Raney nickel (1.5 g.) for 6 hr. After the catalyst has been filtered off, the solution is evaporated to a small volume and set aside. Compound **8** is obtained as tiny needles; m.p. 184–187°, yield 65 mg. (33%), $\lambda_{max}^{H_2O}$ 260 nm.

3′-Deoxyadenosine (Cordycepin)[12] **(9)**

Compound **7** (100 mg., 0.34 mmoles) is refluxed in water (10 ml.) with Raney nickel (0.75 g.) for 4 hr. The catalyst is removed by filtration, and the filtrate and washings are combined and evaporated under diminished pressure. The residue is recrystallized from isopropyl alcohol to give compound **9**; yield 30 mg. (31%), m.p. 210–212°, λ_{max}^{EtOH} 259 nm., R_f 0.37 (43:7 butyl alcohol–water).

REFERENCES

(1) C. D. Anderson, L. Goodman, and B. R. Baker, *J. Am. Chem. Soc.*, **81**, 3967 (1959).
(2) R. K. Ness and H. G. Fletcher, Jr., *J. Am. Chem. Soc.*, **81**, 4752 (1959); **82**, 3434 (1960).
(3) H. Venner, *Chem. Ber.*, **93**, 140 (1960).
(4) M. J. Robins and R. K. Robins, *J. Am. Chem. Soc.*, **87**, 4934 (1964).
(5) M. Ikehara and H. Tada, *J. Am. Chem. Soc.*, **85**, 2344 (1963); **87**, 606 (1965).
(6) A. R. Todd and T. L. V. Ulbricht, *J. Chem. Soc.*, **1960**, 3275.
(7) W. W. Lee, A. Benitez, C. D. Anderson, L. Goodman, and B. R. Baker, *J. Am. Chem. Soc.*, **83**, 1906 (1961).
(8) E. Walton, R. F. Nutt, S. R. Jenkins, and F. W. Holley, *J. Am. Chem. Soc.*, **86**, 2952 (1964).
(9) M. Ikehara, S. Uesugi, and M. Kaneko, *Chem. Commun.*, **1967**, 17.
(10) P. A. Levene and R. S. Tipson, *J. Biol. Chem.*, **121**, 131 (1937); A. Hampton, *J. Am. Chem. Soc.*, **83**, 3143 (1961).
(11) R. Kuhn and W. Jahn, *Chem. Ber.*, **98**, 1699 (1965).
(12) M. Ikehara and H. Tada, *Chem. Pharm. Bull.* (Tokyo), **15**, 94 (1967).

[62] 3′-Deoxyadenosine (Cordycepin)

The Titanium Tetrachloride Procedure for Preparation of Nucleosides

D. H. MURRAY and J. PROKOP

DEPARTMENT OF MEDICINAL CHEMISTRY, SCHOOL OF PHARMACY, STATE UNIVERSITY OF NEW YORK AT BUFFALO, BUFFALO, NEW YORK 14214

INTRODUCTION

Cordycepin (**9**), a nucleoside antibiotic having antitumor properties, may be isolated from the mold *Cordyceps militaris* Linn[1]; it has the structure 9-(3-deoxy-β-D-*erythro*-pentofuranosyl)adenine[2] (Sec. III [61]). Its preparation is an example of the application of titanium tetrachloride to the synthesis of purine nucleosides,[3] particularly those of adenine. The use of titanium tetrachloride eliminates the need for the separate preparation of an *O*-acylglycosyl halide, and frequently results in higher yields of the nucleoside. As predicted by the *trans* rule,[4] the β-D anomer is formed. Cordycepin (**9**) has also been synthesized by other procedures.[5,6]

PROCEDURE

3-Deoxy-1,2:5,6-di-*O*-isopropylidene-α-D-*ribo*-hexose[7] (2)

A mixture of 8.00 g. (4.88 mmoles) of finely powdered 3-deoxy-D-*ribo*-hexose[8] (**1**), 20 g. of anhydrous cupric sulfate, and 200 ml. of acetone (reagent grade) containing 160 mg. of concentrated sulfuric acid, is stirred continuously in a stoppered flask at room temperature for 24 hr. The mixture is filtered, and the solid is washed with three 25-ml. portions of acetone. The filtrate and washings are combined, and stirred for 1 hr. with 8 g. of powdered barium carbonate, and the suspension is filtered through Celite. The filter cake is washed with 50 ml. of acetone, and the filtrate and washings are combined and evaporated to dryness under diminished pressure at 40°, giving a pale-yellow, moderately viscous liquid; yield, 11.5 g. (96.7%). Vacuum distillation gives a clear, colorless liquid, b.p. 55–57°/0.5 mm. Hg; yield, 10.5 g. (88.4%); $[\alpha]_D^{20}$ −4° (*c* 4.7, acetone).

1 (164.2) $\xrightarrow{Me_2CO\ (H_2SO_4)}$ 2 (244.3) $\xrightarrow{H^+}$ 3 (204.2) $\xrightarrow{IO_4^-}$

[4] $\xrightarrow{NaBH_4}$ 5 (174.2) $\xrightarrow{BzCl}$ 6 (278.3) $\xrightarrow{Ac_2O,\ H_2SO_4}$

7 (322.3) $\xrightarrow{\text{HNBz-purine HgCl (474), } TiCl_4}$ 8 $\xrightarrow{NaOMe}$ 9 (251.3)

where Ac is acetyl and Bz is benzoyl.

3-Deoxy-1,2-*O*-isopropylidene-α-D-*ribo*-hexofuranose[7] **(3)**

To 495 ml. of 50% aqueous methanolic hydrochloric acid (0.01 *M*) at 40° is added 9.92 g. (40.6 mmoles) of compound **2**. The temperature of the solution is kept at 40° for 90 min., with continuous stirring, and the solution is neutralized with *N* sodium hydroxide (phenolphthalein end-point), and evaporated under diminished pressure at 40° to a sirup. The sirup is dissolved in 200 ml. of water, and the solution is extracted with 50 ml. of chloroform. [The chloroform extract is dried with magnesium sulfate and filtered, and the filtrate is evaporated to dryness under diminished pressure at 40° to give 2.10 g. (21.2%) of a clear, colorless liquid, the infrared absorption spectrum of which is identical with that of the starting material (**2**).]

The aqueous phase is evaporated to dryness under diminished pressure at 40°, and the partially crystalline residue is freed of water by the addition and distillation of 100 ml. of absolute ethanol. A mixture of the residue and 10 g. of anhydrous magnesium sulfate is extracted with 100 ml. of hot chloroform. The extract is filtered, and the filtrate is evaporated to dryness under diminished pressure at 50°, giving a viscous sirup which solidifies on cooling. Further drying of the sirup in a vacuum desiccator containing phosphorus pentaoxide yields 5.67 g. [86.8%, based on unrecovered starting material (**2**)] of crude product, m.p. 83–84°, sufficiently pure for the following reaction. Recrystallization from ethyl acetate–hexane gives pure **3**, m.p. 84.5–85.5°; $[\alpha]_D^{20}$ $-15°$ (*c* 1.3, 1,2-dichloroethane).

3-Deoxy-1,2-*O*-isopropylidene-α-D-*erythro*-pentofuranose[7] (5)

To a well stirred solution of 1.77 g. (8.69 mmoles) of compound **3** (m.p. 83–84°) in 45 ml. of water is added 1.86 g. (8.69 mmoles) of sodium metaperiodate, and the pH of the solution is adjusted to 6–6.5 (pH paper) with 0.2 *N* sodium hydroxide. The pH is maintained in this range for 1 hr. by the periodic addition of aqueous sodium hydroxide, followed by evaporation of the solution to dryness under diminished pressure at 40°. Further drying is accomplished by the addition of 50 ml. of absolute ethanol, which is evaporated under diminished pressure at 40°. To the partially crystalline residue is added 10 g. of anhydrous magnesium sulfate, and the mixture is extracted with three 25-ml. portions of chloroform. The extracts are combined, and evaporated to dryness under diminished pressure at 50°, giving a clear, colorless, moderately viscous sirup; yield, 1.74 g. of the crude pentodialdose derivative (**4**) which, without purification, is immediately reduced.

To a cold, well-stirred solution of 1.61 g. of this crude, sirupy **4** in 25 ml. of 50% aqueous methanol is added a cold solution of 0.61 g. (16 mmoles) of sodium borohydride in 25 ml. of water. The solution is stirred at room temperature for 4 hr., neutralized with 10% aqueous acetic acid, and then evaporated to dryness under diminished pressure at 40°. The partially crystalline residue is further dried by the addition of 50 ml. of absolute ethanol, followed by its removal under diminished pressure at 40°. To the residue is added 6 g. of anhydrous magnesium sulfate, the mixture is refluxed with 25 ml. of chloroform for 30 min. and filtered, and the filter cake is washed with chloroform (3 × 5 ml.). The filtrate and washings are combined, and evaporated to dryness under diminished pressure at 50°, giving crude product as a colorless, crystalline solid which is sufficiently pure for its conversion into its benzoic ester (**6**); yield, 1.29 g. (92.2%), m.p. 76–77.5°. The pure product is obtained by recrystallization from cyclohexane, m.p. 79–80°; $[\alpha]_D^{20}$ −10° (*c* 0.80, 1,2-dichloroethane).

5-*O*-Benzoyl-3-deoxy-1,2-*O*-isopropylidene-α-D-*erythro*-pentose[7] (6)

To a well stirred solution of 0.087 g. (0.5 mmole) of compound **5** (m.p. 76–77.5°) in 1 ml. of pyridine (reagent grade) is added, dropwise, 0.09 ml. (0.78 mmole) of benzoyl chloride, and the mixture is kept at room temperature for 5 hr. To the mixture is then added 1 drop of water, and the solution is poured into 10 ml. of vigorously stirred, saturated aqueous sodium hydrogen carbonate. After 30 min., an oil separates; the mixture is extracted with three 5-ml. portions of chloroform, and the combined extracts are washed with water (10 ml.), dried (magnesium sulfate), and evaporated to dryness under diminished pressure at 50°. The last traces of pyridine are removed by the addition of toluene (2 × 5 ml.), followed by its removal under diminished pressure at 50°, to give the crude product as a slightly turbid, pale-yellow sirup; yield 0.123 g. (88.5%). The product is sufficiently pure for use in the next step. Further purification may be achieved by short-path distillation, giving a clear, colorless sirup, $[\alpha]_D^{20}$ −7.1° (*c* 0.72, 1,2-dichloroethane).

1,2-Di-*O*-acetyl-5-*O*-benzoyl-3-deoxy-D-*erythro*-pentose[7] (7)

To a well stirred solution of 1.84 g. (6.64 mmoles) of crude **6** in 32 ml. of glacial acetic acid and 3.64 ml. (38.5 mmoles) of acetic anhydride, is added dropwise 1.49 ml. of concentrated sulfuric acid, the temperature being kept at 15–20°. The solution is kept overnight at room temperature, and is then poured into 250 ml. of vigorously stirred, 10% aqueous sodium acetate at 0°. After 30 min., the mixture is extracted with three 30-ml. portions of chloroform, and the extracts are combined, washed successively with saturated, aqueous sodium hydrogen carbonate (2 × 90 ml.) and water (90 ml.), dried (magnesium sulfate), and evaporated to dryness under diminished pressure at 60°. The last traces of acetic acid are removed by the addition of benzene (2 × 10 ml.), followed by its removal under diminished pressure at 60°, giving a colorless, slightly turbid sirup which is sufficiently pure for the next step; yield, 1.64 g. (77.0%). The product shows characteristic infrared absorption bands at 1750 cm^{-1} (acetate carbonyl) and 1220 cm^{-1} (acetate C-O-C).

3′-Deoxyadenosine[7] (9)

A mixture of 1.47 g. (4.71 mmoles) of crude **7**, 2.79 g. (5.89 mmoles) of chloromercuri-*N*-benzoyladenine,[9] 2.8 g. of Celite, and 200 ml. of 1,2-dichloroethane is distilled until 25 ml. of distillate has been collected. To the somewhat cooled mixture is added, dropwise, a solution of 0.65 ml. (5.9 mmoles) of titanium tetrachloride in 10 ml. of 1,2-dichloroethane, and the mixture is refluxed for 22 hr. with exclusion of moisture. While the mixture is still warm, 85 ml. of saturated, aqueous sodium hydrogen carbonate is added, with vigorous stirring. After being stirred for 2 hr., the mixture is filtered through Celite, the cake is washed with chloroform (3 × 20 ml.), and the organic layer of the filtrate is separated, and evaporated almost to dryness under diminished pressure at 40°. A solution of the residue in 40 ml. of chloroform is washed successively with 40 ml. of 30% aqueous potassium iodide and 40 ml. of water, dried (magnesium sulfate), and evaporated to dryness under diminished pressure at 60° to give the crude, protected nucleoside (**8**) as a yellow glass; yield, 1.71 g. (74.7%).

A solution of 1.66 g. of crude **8** in 35 ml. of 0.1 *M* methanolic sodium methoxide is refluxed for 2.5 hr. The cooled solution is neutralized with glacial acetic acid, diluted with an equal volume of water, and chromatographed on a column (38 × 500 mm.) of Bio-Rad AG-1 X8 (OH^-) with 60% aqueous methanol. The fractions containing the major ultraviolet-absorbing peak are combined, and evaporated to dryness under diminished pressure at 50°, giving a colorless, crystalline solid; yield, 437 mg. (39%, from **7**), m.p. 227–228°. Recrystallization from water, and thorough drying under diminished pressure at 100° (Drierite), gives pure **9**, m.p. 228–229°, $[\alpha]_D^{20}$ −47° (*c* 0.64, water), λ_{max} 259 nm. (ϵ 14,900).

REFERENCES

(1) K. G. Cunningham, S. A. Hutchinson, W. Manson, and J. H. Williams, *J. Chem. Soc.*, **1951**, 2299.

(2) E. A. Kaczka, N. R. Trenner, B. Arison, R. W. Walker, and K. Folkers, *Biochem. Biophys. Res. Commun.*, **14**, 456 (1964).

(3) B. R. Baker, R. E. Schaub, J. P. Joseph, and J. H. Williams, *J. Am. Chem. Soc.*, **77**, 12 (1955).

(4) R. S. Tipson, *J. Biol. Chem.*, **130**, 55 (1939); B. R. Baker, *Ciba Found. Symp., Chem. Biol. Purines, 1957*, 120.

(5) A. R. Todd and T. L. V. Ulbricht, *J. Chem. Soc.*, **1960**, 3275; W. W. Lee, A. Benitez, C. D. Anderson, L. Goodman, and B. R. Baker, *J. Am. Chem. Soc.*, **83**, 1906 (1961).

(6) E. Walton, F. W. Holly, G. E. Boxer, R. F. Nutt, and S. R. Jenkins, *J. Med. Chem.*, **8**, 659 (1965).

(7) D. H. Murray and J. Prokop, *J. Pharm. Sci.*, **54**, 1468 (1965).

(8) D. H. Murray and J. Prokop, *J. Pharm. Sci.*, **54**, 1637 (1965).

(9) J. Prokop and D. H. Murray, *J. Pharm. Sci.*, **54**, 359 (1965).

[63] 3′,5′-Di-*O*-acetyladenosine

Synthesis of a Partially Protected Purine Nucleoside by Transacetylation

L. SZABÓ

INSTITUT DE BIOCHIMIE, FACULTE DES SCIENCES, ORSAY, FRANCE

1 (393.4) + **2** (267.2) ⟶ **3** (351.3)

INTRODUCTION

3′,5′-Di-*O*-acetyladenosine (**3**) is an intermediate in the synthesis of 2′-substituted derivatives of adenosine[1] (**2**). It has been obtained from 5′-*O*-acetyladenosine by partial acetylation,[1,2] from a mixture of 5′-*O*-acetyladenosine and 2′,3′,5′-tri-*O*-acetyladenosine (**1**) by transacetylation,[2] and from a mixture of adenosine and 2′,3′,5′-tri-*O*-acetyladenosine (**1**), also by transacetylation.[3] Because the preparation of 5′-*O*-acetyladenosine by the methods described is not entirely satisfactory, the advantage of the last procedure over those previously used is that the two starting materials are readily accessible.

PROCEDURE

3′,5′-Di-*O*-acetyladenosine (3)

An intimate mixture of 7.9 g. (0.02 mole) of 2′,3′,5′-tri-*O*-acetyladenosine (**1**) and 5.3 g. (0.014 mole) of commercial adenosine (**2**) monohydrate is placed in a 25-ml., round-bottomed flask. The flask is evacuated, and immersed in an oil bath preheated to 200°. The temperature of the bath is slowly raised to 220–225° (not higher); the mixture changes to a homogeneous melt within 5–10 min. The flask is removed from the bath, kept at 140° for 48 hr., and cooled, and the transparent, brown glass is dissolved in about 50 ml. of hot water. The solution is treated with charcoal, filtered, and concentrated under diminished pressure to about 30 ml. Equal portions of the latter solution are placed in each of the first two tubes of a countercurrent extraction apparatus (bottom and top phase, 20 ml. each), each is diluted with water to 20 ml., and the mixture is submitted to 100 transfers in a water (stationary phase)–ethyl acetate system at 20°. The contents (both phases) of tubes 26–57 are combined and evaporated to dryness under diminished pressure, and the crystalline residue is recrystallized from a small volume of water; yield 3.9 g. of 3′,5′-di-*O*-acetyladenosine (**3**), m.p. 172–173°. Tubes 76–95 contain 2′,3′,5′-tri-*O*-acetyladenosine (**1**) (0.9 g., m.p. 173–174°) and tubes 58–75 contain 2′,3′-di-*O*-acetyladenosine (0.2 g., m.p. 178–179°). The efficiency of the separation can be checked by paper chromatography in a butyl alcohol–acetic acid–water system [4:1:5 (v/v), freshly prepared upper phase] with which the acetylated derivatives of adenosine are satisfactorily separated.

If desired, the tri-*O*-acetyladenosine (**1**) need not be crystallized for the above preparation. Instead, 5.34 g. of anhydrous adenosine (**2**) is acetylated with a mixture of 50 ml. of acetic anhydride and 70 ml. of anhydrous pyridine[4] for a few hours (with vigorous shaking). After removal of the solvents under diminished pressure, the residue is dissolved in ethanol, and the solvent is evaporated. The residue, obtained after two successive evaporations with ethanol, is thoroughly mixed with 5.3 g. of adenosine (**1**) monohydrate, and the mixture is treated as described in the foregoing.

REFERENCES

(1) D. M. Brown, G. D. Fasman, D. I. Magrath, and A. R. Todd, *J. Chem. Soc.*, **1954**, 1448.
(2) A. M. Michelson, L. Szabó, and A. R. Todd, *J. Chem. Soc.*, **1956**, 1546.
(3) L. Szabó, *Bull. Soc. Chim. France*, **1966**, 3159.
(4) H. Bredereck, *Chem. Ber.*, **80**, 401 (1947).

[64] *N,N*-Dimethyladenosine

A Facile Synthesis of N^6,N^6-*Dimethyl Purine Nucleosides*

CLARITA C. BHAT

DEPARTMENT OF CHEMISTRY, GEORGETOWN UNIVERSITY, WASHINGTON, D.C. 20007

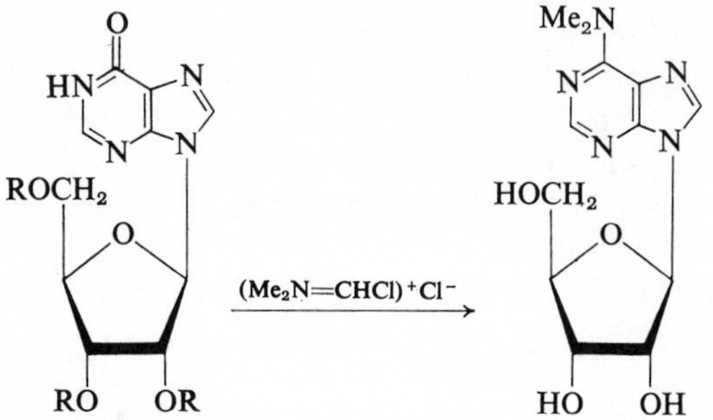

$(Me_2N{=}CHCl)^+Cl^-$

1, R = H (268.2)

2, R = Ac (394.3)

3 (295.3)

INTRODUCTION

The title compound (**3**) is an intermediate in the synthesis of 6-(dimethylamino)-9-β-D-hexopyranosylpurines (Sec. III [82]). Its preparation[1] constitutes an improved procedure for the introduction of the *N,N*-dimethyl group at N-6 of a purine nucleoside.

PROCEDURE

2′,3′,5′-Tri-*O*-acetylinosine[2] (2)

To a solution of 5 g. (19 mmoles) of inosine[a] (**1**) in 60 ml. of dry pyridine is added 50 ml. of acetic anhydride. The mixture is stirred at 37° until the solid has dissolved, and the solution is evaporated to dryness at 50°/2 mm. Hg. The solid residue is recrystallized from absolute ethanol to give 6.6 g. (90%) of **2**, m.p. 241°, $[\alpha]_D^{20}$ −38.3° (*c* 1.07, chloroform).

***N,N*-Dimethyladenosine[1] (3)**

A solution of 7.89 g. (0.02 mole) of compound **2** in 100 ml. of chloroform is treated with 20 ml. (0.04 mole) of a 2 *M* solution of (chloromethylene)dimethylammonium chloride in chloroform. The mixture is refluxed

[a] Cyclo Chemical Corporation, P.O. Box 71557, Los Angeles, California 90001.

for 1 hr., cooled, and added dropwise, with stirring and cooling with ice, to 50 ml. (0.092 mole) of a 1.84 *M* solution of ammonia in chloroform. The precipitate is filtered off, and the filtrate is evaporated to dryness under diminished pressure (bath temperature $<80°$). The residual sirup is dissolved in 50 ml. of chloroform, and the solution is washed with three 25-ml. portions of water, dried (anhydrous magnesium sulfate), and evaporated to dryness under diminished pressure. The residue is dissolved in 10 ml. of methanol presaturated with ammonia, and the solution is kept for 16 hr. at 0°. The mixture is then evaporated to dryness under diminished pressure, and the residue is crystallized from hot methanol; yield 4.25 g. (72%), m.p. 182–184°.

REFERENCES

(1) J. Žemlička and F. Šorm, *Collection Czech. Chem. Commun.*, **30**, 1880 (1965).
(2) P. A. Levene and R. S. Tipson, *J. Biol. Chem.*, **111**, 313 (1935); H. Bredereck and A. Martini, *Chem. Ber.*, **80**, 401 (1947).

[65] *N*-[(Dimethylamino)methylene]-2′,3′-*O*-(ethoxymethylene)adenosine

2′,3′-O-(Ethoxymethylene) Ribonucleosides Having the Amino Group Protected by a Dimethylaminomethylene Group; Specifically Substituted Intermediates, Bearing a Free 5′-Hydroxyl Group, in the Preparation of Ribo-oligonucleotides Possessing the Naturally Occurring 3′ → 5′ Internucleotide Linkage

J. ŽEMLIČKA

INSTITUTE OF ORGANIC CHEMISTRY AND BIOCHEMISTRY, CZECHOSLOVAK ACADEMY OF SCIENCES, PRAGUE, CZECHOSLOVAKIA

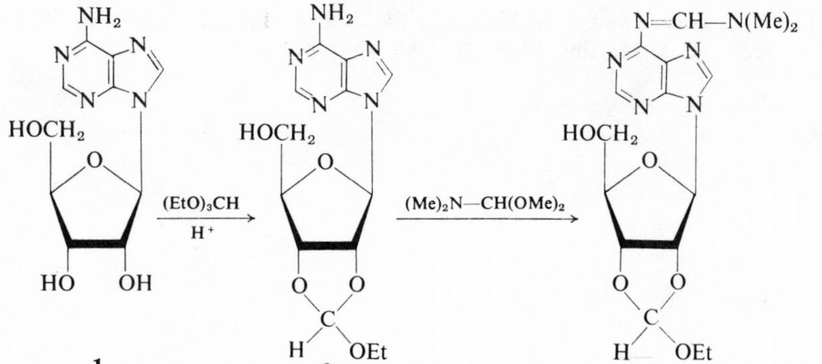

$\xrightarrow[H^+]{(EtO)_3CH}$ $\xrightarrow{(Me)_2N{-}CH(OMe)_2}$

1 (267.2) **2** (323.3) **3** (378.4)

INTRODUCTION

The reactive amino group of nucleosides may successfully be protected by the (dimethylamino)methylene group.[1,2] Introduction of this group is effected by reaction of the corresponding nucleoside with *N*,*N*-dimethylformamide dialkyl acetals[3] in *N*,*N*-dimethylformamide or chloroform as the solvent.[1,2,4] Thus, treatment of 2′,3′-*O*-(ethoxymethylene) ribonucleosides[4–6] that bear a reactive amino group with *N*,*N*-dimethylformamide dimethyl acetal[3] affords *N*-[(dimethylamino)methylene]-2′,3′-*O*-(ethoxymethylene) ribonucleosides that may be used in, for example, the synthesis of ribo-oligonucleotides, because all of their reactive groups except the 5′-hydroxyl group are protected.

The present procedure has been used in the preparation of the *N*-(dimethylamino)methylene derivatives of 2′,3′-*O*-(ethoxymethylene)adenosine, -guanosine, -cytidine,[4] and -6-azacytidine.[7] The presence of an *N*-(dimethylamino)methylene group in the molecule is shown by a characteristic ultraviolet spectrum, and especially by an absorption maximum at 300–310 nm. (ϵ_{max} 20,000–30,000).

PROCEDURE

2′,3′-*O*-(Ethoxymethylene)adenosine (2)

A mixture of 5.34 g. (0.02 mole) of adenosine (**1**), 100 ml. of *N,N*-dimethylformamide,[a] 6.7 ml. (0.04 mole) of ethyl orthoformate, and 4.4 ml. (0.025 mole) of a 5.65 *M* solution[b] of anhydrous hydrogen chloride in *N,N*-dimethylformamide is placed in a 250-ml., round-bottomed flask, and is shaken at room temperature until the adenosine has dissolved. The resulting solution is kept at room temperature for 3 days,[c] two 1.67-ml. (0.025 mole) portions of ethyl orthoformate being added after 24 and 48 hr. The mixture is then treated with 5 ml. (0.036 mole) of triethylamine, and cooled to 0°. The precipitate of triethylamine hydrochloride is filtered off with suction, and washed with 5 ml. of *N,N*-dimethylformamide. The filtrates are combined, and evaporated[d] to dryness at 40° (bath temperature)/ $<$ 1 mm. Hg. The resulting sirup is treated with 20 ml. of water[e] and two drops of triethylamine, and the mixture is kept at 0° overnight. The product (**2**) that separates is collected with suction, and dried at 20°/0.1 mm. Hg; yield, 5.88 g. (91%) of chromatographically homogeneous **2**, R_f 0.69 in 7:1:2 isopropyl alcohol–concentrated aqueous ammonia–water, and 0.67 in butyl alcohol saturated with water (Whatman No. 1 paper, descending technique). This product is pure enough to be used in the next step. Analytically pure **2**, recrystallized from ethanol, melts at 228–230°.

[a] Commercial *N,N*-dimethylformamide is dried by distillation from 10% (by weight) of phosphorus pentaoxide, and is stored over molecular sieves.

[b] The solution is prepared by passing dry hydrogen chloride into *N,N*-dimethylformamide cooled in ice. If a precipitate separates (probably *N,N*-dimethylformamide hydrochloride), it is dissolved by adding a further portion of *N,N*-dimethylformamide.

[c] The course of the reaction is checked by descending chromatography on Whatman No. 1 paper in 7:1:2 isopropyl alcohol–concentrated ammonium hydroxide–water. The reaction is complete when no adenosine (R_f 0.48) is present in the sample of the reaction mixture. Prior to chromatography, the sample of the reaction mixture is made alkaline with triethylamine.

[d] During the evaporation, the receiver is cooled in Dry Ice–ethanol.

[e] The addition of water is necessary, in order to convert the by-product (R_f 0.85 in 7:1:2 isopropyl alcohol–concentrated aqueous ammonia–water) into **2**. The by-product is probably formed by reaction of the orthoformate and the free 5′-hydroxyl group of **2**.

N-[(Dimethylamino)methylene]-2′,3′-O-(ethoxymethylene)adenosine (3)

A solution of 1.62 g. (5 mmoles) of compound **2** in 10 ml. of *N*,*N*-dimethylformamide[a] is placed in a 50-ml., round-bottomed flask, and treated with 2 ml. (*ca.* 0.02 mole) of *N*,*N*-dimethylformamide dimethyl acetal. The mixture is kept at room temperature for 24 hr., and evaporated[d] to dryness at 40° (bath temperature)/ < 1 mm. Hg. The residual sirup is dissolved in 20 ml. of chloroform, and the solution is added dropwise, with stirring, to 300 ml. of light petroleum (b.p. 40–60°) in a 1-l., three-necked flask fitted with a stirrer, dropping funnel, and a calcium chloride drying-tube. The resulting amorphous, hygroscopic precipitate is collected by centrifugation (2000 r.p.m. for 10 min.), dissolved in 100 ml. of *p*-dioxane,[f] and freeze-dried; yield, 1.95 g. (100%) of **3**, analytically (as well as chromatographically) pure; R_f 0.44 in 9:1 chloroform–methanol on a thin layer of loose silica gel containing 10% of water (the developed layer is sprayed, while moist, with a 1% solution of quinine sulfate in methanol, and the spot is detected under ultraviolet light); λ_{max} 232 nm. (ϵ_{max} 11,500) and 312 nm. (ϵ_{max} 29,500) in 96% ethanol.

REFERENCES

(1) J. Žemlička, *Collection Czech. Chem. Commun.*, **28**, 1060 (1963).
(2) J. Žemlička and A. Holý, *Collection Czech. Chem. Commun.*, **32**, 3159 (1967).
(3) Z. Arnold and M. Kornilov, *Collection Czech. Chem. Commun.*, **29**, 645 (1964).
(4) J. Žemlička, S. Chládek, A. Holý, and J. Smrt, *Collection Czech. Chem. Commun.*, **31**, 3198 (1966).
(5) J. Žemlička, *Chem. Ind.*(London), **1964**, 581.
(6) S. Chládek, J. Žemlička, and F. Šorm, *Collection Czech. Chem. Commun.*, **31**, 1785 (1966).
(7) A. Holý, J. Smrt, and F. Šorm, *Collection Czech. Chem. Commun.*, **32**, 2980 (1967).

[f] Commercial *p*-dioxane is dried by distillation from sodium.

[66] 2-Fluoroadenosine

Application of the Schiemann Reaction to 2-Aminoadenosine

JOHN A. MONTGOMERY and KATHLEEN HEWSON

KETTERING-MEYER LABORATORY, SOUTHERN RESEARCH INSTITUTE, BIRMINGHAM, ALABAMA 35205

$NaNO_2$, HBF_4

1 (282.3) → **2** (285.2)

INTRODUCTION

A number of 2-fluoropurines have been prepared by a modification of the Schiemann reaction.[1] One of the purines prepared in this way, 2-fluoroadenosine,[2] has shown a high degree of activity in many diverse biological systems. This activity may result from the ability of 2-fluoroadenosine, after conversion into 2-fluoroadenylic acid, to serve as a feedback inhibitor of the biosynthesis of purine ribonucleotides *de novo*.[3]

PROCEDURE

2-Fluoroadenosine (2)

A solution of sodium nitrite (4.3 g., 62 mmoles) in water (8 ml.) is added dropwise, with stirring, to a solution of 2-aminoadenosine (**1**) (Sec. III [81]) (10.4 g., 37 mmoles) in 48–50% fluoroboric acid (110 ml.) contained in a 600-ml. beaker and kept below $-10°$ by means of a Dry Ice–acetone bath. Fifteen minutes after addition of nitrite is complete, 210 ml. of water-saturated butanol is added to the reaction mixture, which is then neutralized (pH 6.5) with 20–30% sodium hydroxide at a temperature below $-5°$. The neutral mixture is extracted with five 600-ml. portions of water-saturated butanol, and the combined extracts are washed with four 250-ml. portions of butanol-saturated water before concentration *in vacuo* to *ca.* 75 ml. The gelatinous solid that forms on keeping the solution overnight in a refrigerator is collected by filtration, washed successively with cold butanol and ether, and dried *in vacuo* over phosphorus pentaoxide; yield of crude **2**, 3.2 g. (56% pure).

A slurry of the crude product mixed with an equal weight of silica gel in 9:1 chloroform–methanol is added to the top of a silica gel (50 g.) column[a] (2.6 × 35 cm.), which has been wet-packed with 9:1 chloroform–methanol. The column is eluted with 9:1 chloroform–methanol until (250–500 ml.) all of the 2-fluoroadenine is removed. The eluant is changed to 4:1 chloroform–methanol, and elution is continued until all of the 2-fluoroadenosine has been eluted.[b] Evaporation of the eluate *in vacuo* gives 97–99% pure 2-fluoroadenosine (**2**); yield 950 mg. (9%), m.p. 200° (dec.); $[\alpha]_D^{26}$ $-60.3 \pm 11.1°$ (*c* 0.127, ethanol); λ_{max} in nm. (log ϵ): pH 1, 260.5 (4.14); pH 7, 260.5 (4.16); pH 13, 260.5 (4.17).

REFERENCES

(1) J. A. Montgomery and K. Hewson, *J. Am. Chem. Soc.*, **82**, 463 (1960).
(2) J. A. Montgomery and K. Hewson, *J. Am. Chem. Soc.*, **79**, 4559 (1957).
(3) L. L. Bennett, Jr., and D. Smithers, *Biochem. Pharmacol.*, **13**, 1331 (1964).

[a] SiliCAR 7 (Mallinckrodt) or Silica gel H (Merck) makes a suitable column.

[b] The composition of the column fractions (100 ml.) is followed either by paper chromatography with 43:7 butyl alcohol–water or by thin-layer chromatography on silica gel with 3:1 chloroform–methanol. If 3:1 chloroform–methanol is used to elute the column, significant quantities of sodium fluoroborate are eluted.

[67] 2′-*O*-Methyladenosine, 3′-*O*-Methyladenosine, and 2′,3′-Di-*O*-methyladenosine

Preferential Methylation of Sugar Hydroxyl Groups by Diazomethane

JERRY B. GIN and CHARLES A. DEKKER

DEPARTMENT OF BIOCHEMISTRY, UNIVERSITY OF CALIFORNIA, BERKELEY, CALIFORNIA 94720

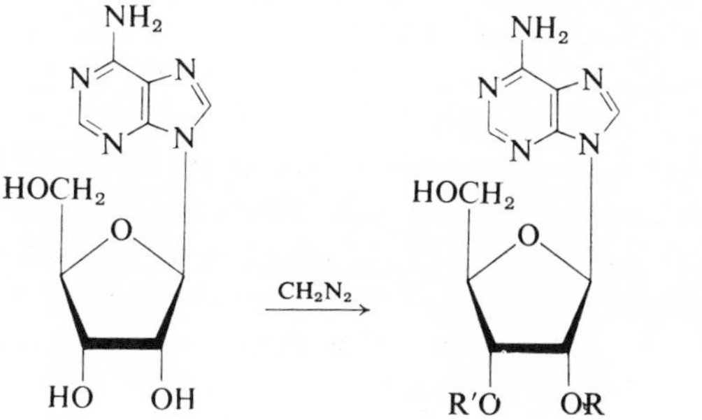

NH_2 N N N N $HOCH_2$ O HO OH

1
(267.2)

$\xrightarrow{CH_2N_2}$

NH_2 N N N N $HOCH_2$ O R′O OR

2, R = CH_3, R′ = H
(281.3)
3, R = H, R′ = CH_3
(281.3)
4, R = CH_3, R′ = CH_3
(295.3)

INTRODUCTION

2′-*O*-Methyladenosine (**2**), 3′-*O*-methyladenosine (**3**), and 2′,3′-di-*O*-methyladenosine (**4**) may be prepared from adenosine by reaction with diazomethane in partially aqueous media. The 2′-hydroxyl group is preferentially methylated, as shown by Robins *et al.*,[1] but substantial proportions of **3** and **4** are also formed. The compounds are readily separated by column chromatography on Dowex-1 2X (OH^-).

PROCEDURE

Diazomethane

The diazomethane is prepared by carefully adding 20 g. (192 mmoles) of 1-methyl-1-nitrosourea to 105 ml. of 1,2-dimethoxyethane and 45 ml. of 50% (w/w) potassium hydroxide[a] in a 250-ml., Erlenmeyer flask in an

[a] Wear gloves and goggles. To prevent explosions, avoid both excessive light and scratched glassware. The nitrosomethylurea is added during 5 min., with continuous mixing (magnetic stirrer).

ice–salt–water bath. After 15 min., the aqueous layer is frozen in a Dry Ice–acetone bath, and the ethereal diazomethane solution is decanted into a 250-ml., Erlenmeyer flask containing pellets of potassium hydroxide. After being kept at 0° for 20 min., the diazomethane solution is ready for use.[b]

O-Methylation

To a solution of 1 g. (3.74 mmoles) of adenosine (**1**) in 40 ml. of water at 80° is added 5.5 g. (131 mmoles) of diazomethane in 100 ml. of 1,2-dimethoxyethane. The mixture is allowed to cool to room temperature, and is then stirred overnight. Evaporation of the solution under diminished pressure gives a thick sirup, which is dissolved in 20 ml. of absolute ethanol, and the solution evaporated to dryness; this drying process is repeated once.

The products are separated into unmethylated, monomethylated, and dimethylated adenosine by cellulose chromatography with 86:14:5 butyl alcohol–water–ammonia.[c] The R_f values observed are: adenosine, 0.27; mono-*O*-methyladenosine, 0.47; di-*O*-methyladenosine, 0.65. After elution with water, the eluate containing monomethylated material is flash-evaporated, and the residue is dissolved in 3 ml. of 30% methanol and applied to a column (1.5 × 20 cm.[d]) of Bio-Rad AG-1 2X (OH^-) ion-exchange resin (200–400 mesh) pre-equilibrated with the eluant, *viz.*, 30% aqueous methanol.[2] Fractions (10 ml. each) are collected, at 1 ml./min., and the elution pattern is followed by the optical absorbance at 260 nm. After the initial removal of a small amount of as-yet unidentified material from the column, compound **2** (yield, 38%) is completely eluted with 170 ml. of eluant. The next zone, eluted by 350 ml. of 30% methanol, is compound **3** (yield, 11%). The pure products are obtained by evaporating the respective eluates to dryness under diminished pressure, and crystallizing from methanol; for **2**, m.p. 201.5°, $[\alpha]_{589}$ −57.5°; and for **3**, m.p. 178–180°, $[\alpha]_{589}$ −59°.

Compound **4** is similarly isolated from the dimethylated material on a column of Bio-Rad AG-1 2X (OH^-) ion-exchange resin equilibrated with water. The initial zone, eluted completely with 280 ml. of water, is evaporated to dryness, and the residue is dissolved in 0.5 ml. of 0.2 *N* sodium hydroxide and heated in a boiling-water bath for 40 min. After

[b] The concentration of diazomethane may be determined by adding a 5-ml. aliquot to 1.3 g. of benzoic acid dissolved in a small volume of 1,2-dimethoxyethane, and titrating the excess benzoic acid with 0.2 *N* sodium hydroxide.

[c] Approximately 75 mg. of mixed products may conveniently be separated on a sheet (46 × 57 cm.) of Whatman 3 MM paper.

[d] Up to 700 mg. of reaction products has been successfully separated on a column of this size.

neutralization with 0.5 ml. of 0.2 *N* hydrochloric acid, the solution is applied to a column (1.1 × 100 cm.) of Bio-Rad AG-1 2X (OH^-); fractions eluted by water and having A_{280}/A_{260} of ~0.15 are combined, concentrated, and recycled through the column. The middle portion of the emerging zone (A_{280}/A_{250} ~0.15) is evaporated to dryness, and the residue is crystallized from methanol; yield,[e] 3.5%. The resulting compound **4** is chromatographically homogenous and gives, after hydrolysis, the expected amount of formaldehyde on oxidation with periodate; m.p. 177°, $[\alpha]_{589}$ −49°.

REFERENCES

(1) A. D. Broom and R. K. Robins, *J. Am. Chem. Soc.*, **87**, 1145 (1965); T. A. Khwaja and R. K. Robins, *ibid.*, **88**, 3640 (1966).
(2) C. A. Dekker, *J. Am. Chem. Soc.*, **87**, 4027 (1965).

[e] A 10% yield may be obtained by evaporating the initial, unfractionated, reaction products to dryness, redissolving in water, and repeating the treatment with diazomethane.

[68] *N*-(3-Methyl-2-butenyl)adenosine

N-*Alkylation of a Purine Nucleoside*

ROSS H. HALL and MORRIS J. ROBINS

DEPARTMENT OF EXPERIMENTAL THERAPEUTICS, ROSWELL PARK MEMORIAL INSTITUTE, BUFFALO, NEW YORK 14203

Cl; N; N; N; N; $HOCH_2$; O; HO; OH
1 (286.7)

+ $Me_2C{=}CHCH_2NH_2$ ⟶
2 (85.2)

$HNCH_2CH{=}CMe_2$; N; N; N; N; $HOCH_2$; O; HO; OH
3 (335.4)

INTRODUCTION

N-(3-Methyl-2-butenyl)adenosine[a] (**3**) occurs in the soluble ribonucleic acid of yeast,[1,2] animal tissue,[3] and plant tissue[4]; it exhibits a selective toxicity towards various mammalian cell-lines grown in culture,[5] and possesses exceptionally high cytokinin activity,[1,6] that is, it promotes cell division, growth, and organ formation in plant tissue-culture systems. The preparation of compound **3** described herein is an alternative procedure (see Sec. III [69]), in which 3-methyl-2-butenylamine (**2**) is condensed with 6-chloro-9-β-D-ribofuranosyl-9*H*-purine (**1**).

PROCEDURE

N-(3-Methyl-2-butenyl)adenosine (3)

A solution of 3.0 g. (10.5 mmoles) of 6-chloro-9-β-D-ribofuranosyl-9*H*-purine (**1**) (Cyclo Chemical Corporation, P.O. Box 71557, Los Angeles, Calif. 90001) and 2.0 g. (23.6 mmoles) of 3-methyl-2-butenylamine (**2**) (Sec. I [4]) in 150 ml. of absolute ethanol is refluxed for 3 hr., and the solution is kept at 4° overnight, giving 3.0 g. of crude product. Recrystallization of the material, once from absolute ethanol, and once from 3:1

[a] This compound has also been referred to as *N*-(2-isopentenyl)adenosine.

acetonitrile–ethanol, yields 2.0 g. (57%) of colorless needles, m.p. 145–146°, $[\alpha]_{546}^{25} -97°$ (*c* 0.07, ethanol).

REFERENCES

1) R. H. Hall, L. Stasiuk, M. J. Robins, and R. Thedford, *J. Am. Chem. Soc.*, **88**, 2614 (1966).
2) H. G. Zachau, D. Dütting, and H. Feldmann, *Angew. Chem.*, **78**, 392 (1966).
3) M. J. Robins, R. H. Hall, and R. Thedford, *Biochemistry*, **6**, 1837 (1967).
4) R. H. Hall, L. Csonka, H. David, and B. D. McLennan, *Science*, **156**, 69 (1967).
5) J. T. Grace, Jr., M. T. Hakala, R. H. Hall, and J. Blakeslee, *Proc. Am. Assoc. Cancer Res.*, **8**, 23 (1967).
6) H. Q. Hamzi and F. Skoog, *Proc. Natl. Acad. Sci. U. S.*, **51**, 76 (1964).

[69] *N*-(3-Methyl-2-butenyl)adenosine and 1-(3-Methyl-2-butenyl)adenosine [*N*-(Isopentenyl)adenosine and 1-(Isopentenyl)adenosine]

Preparation of N^6*-Substituted Adenosines by Alkylation at the 1-Position of Adenosine Followed by Rearrangement*

WOLFGANG GRIMM, TOZO FUJII, and NELSON J. LEONARD

DEPARTMENT OF CHEMISTRY AND CHEMICAL ENGINEERING, UNIVERSITY OF ILLINOIS, URBANA, ILLINOIS 61801

1 (267.2) + $(CH_3)_2C{=}CHCH_2Br$ **2** (149.0) ⟶ **3a**, X = Br (416.3); **3b**, X = $SO_4/2$ (384.4)

3 —OH^-⟶ **4** (335.4)

NH_2, HOCH$_2$, HO, OH, $(Me)_2C{=}CHCH_2$, X^-, $NHCH_2CH{=}C(Me)_2$

INTRODUCTION

N-(3-Methyl-2-butenyl)adenosine (**4**) (Sec. III [68]) has assumed importance with its isolation from yeast soluble ribonucleic acid (sRNA) and calf-liver sRNA[1] and with the finding that this "minor base" is attached directly to the anticodon sequence of serine transfer ribonucleic acid.[2] The synthesis of **4** from adenosine[3,4] is representative of alkylation of adenosine, 2′-deoxyadenosine, and 9-benzyladenine at the 1-position,[5-7] followed by rearrangement to the corresponding N^6-substituted derivative.[5,6,8-11] It is also possible to isolate, by ion-exchange chromatography and by avoidance of basic conditions, the intermediate 1-(3-methyl-2-butenyl)adenosine as a salt (*e.g.*, **3b**).[3,12]

PROCEDURE

N-(3-Methyl-2-butenyl)adenosine (4)

To a solution of 5.34 g. (20 mmoles) of lyophilized adenosine (**1**) in 75 ml. of anhydrous *N,N*-dimethylformamide is added 4.32 g. (29 mmoles) of freshly distilled 1-bromo-3-methyl-2-butene (Sec. V [156]) (**2**). The reaction mixture is kept in the dark at room temperature for 24 hr., and is then evaporated to dryness under diminished pressure. If the material is not solid at this stage, it may be advantageous to treat it with dry acetone and partially re-evaporate. The solid (**1** plus **3a**) is then collected by filtration, and is dissolved in 100 ml. of water. The aqueous solution is adjusted to pH 7.5, and is heated on a steam bath for 2.5 hr. while 0.1 *N* sodium hydroxide is added periodically to keep the pH at 7.5. The solution is then extracted with four 100-ml. portions of ethyl acetate, the extracts are combined, dried with anhydrous sodium sulfate, and filtered, and the filtrate is evaporated until crystallization starts. Recrystallization from 1:1 ethanol–acetonitrile gives colorless prisms of **4**; over-all yield 2.53 g. (38%), m.p. 146–147°, $[\alpha]_D^{28}$ $-103°$ (*c* 0.14, ethanol).

1-(3-Methyl-2-butenyl)adenosine Acid Sulfate (3b)

At the stage in the above procedure at which the initial reaction-mixture is evaporated to dryness, the mixture of adenosine (**1**) and 1-(3-methyl-2-butenyl)adenosine hydrobromide (**3a**), which is hygroscopic, is dissolved in water and treated with Dowex-1 X8 (HCO_3^-) anion-exchange resin to remove most of the hydrogen bromide. The solution is concentrated to about 50 ml., and added to the top of a 2.5 × 60 cm. column of Dowex 50-W X8 (Na^+), and elution is conducted with *N* sodium sulfate. The (uncharged) adenosine is eluted first, followed by the desired product (**3b**). The fractions containing **3b** are combined, and concentrated to about 60 ml. Ethanol is added to cause precipitation of sodium sulfate, and this is removed by filtration. The filtrate is evaporated and the residue is re-treated with ethanol. The mixture is filtered, the filtrat is evaporated to dryness, the product is dissolved in water, and th solution is lyophilized. The material is dissolved in boiling methanol, th solution is kept overnight, and the resulting precipitate is collected b filtration, dissolved in water, and the solution relyophilized. These step (which are employed to remove sodium sulfate and then water) produce colorless product (**3b**), which is pulverized, under exclusion of moisture i a dry-box, and then dried at 65° overnight *in vacuo* over phosphoru pentaoxide; yield 3.92 g. (51%), m.p. 150° (dec.).

REFERENCES

(1) R. H. Hall, M. J. Robins, L. Stasiuk, and R. Thedford, *J. Am. Chem Soc.*, **88**, 2614 (1966).

(2) H. G. Zachau, D. Dütting, and H. Feldmann, *Angew. Chem.*, **78**, 39 (1966).

(3) N. J. Leonard, *Trans. Morris County Res. Council.*, **1**, 11 (1965).

(4) N. J. Leonard, A. Achmatowicz, R. N. Loeppky, K. L. Carraway W. A. H. Grimm, A. Szweykowska, H. Q. Hamzi, and F. Skoog, *Proc Natl. Acad. Sci. U. S.*, **56**, 709 (1966).

(5) J. W. Jones and R. K. Robins, *J. Am. Chem. Soc.*, **85**, 193 (1963).

(6) P. Brookes and P. D. Lawley, *J. Chem. Soc.*, **1960**, 539.

(7) A. Coddington, *Biochim. Biophys. Acta*, **59**, 472 (1962).

(8) G. B. Elion, *J. Org. Chem.*, **27**, 2478 (1962).

(9) E. C. Taylor and P. K. Loeffler, *J. Am. Chem. Soc.*, **82**, 3147 (1960).

(10) N. J. Leonard and T. Fujii, *Proc. Natl. Acad. Sci. U. S.*, **51**, 73 (1964)

(11) N. J. Leonard, K. L. Carraway, and J. P. Helgeson, *J. Heterocylic Chem.* **2**, 291 (1965).

(12) W. Grimm and N. J. Leonard, unpublished results.

[70] (+)-*S*-Adenosyl-L-Methionine

Enzymic Resolution of a Sulfonium Coenzyme

G. A. JAMIESON

BLOOD PROGRAM RESEARCH LABORATORY, THE AMERICAN NATIONAL RED CROSS, WASHINGTON, D.C. 20006

Adenosyl-S(R) **1** (384.4) ⟶ (±)-Adenosyl-S^+(Me)(R) **2** (399.4)

$\xrightarrow[\text{Guanidinoacetate methyltransferase (EC 2.1.1.2)}]{HN{=}C(H_2N)-NH-CH_2-COOH}$

(+)-Adenosyl-S^+(Me)(R) **3** (399.4) + Adenosyl-S(R) + HN=C(H_2N)–N(Me)–CH_2–COOH

where Adenosyl is [5'-deoxyadenosin-5'-yl structure: adenine (6-NH₂ purine) N9-linked to ribofuranose with –CH₂ at C5', HO and OH at C2', C3'] and R is $-CH_2-CH_2-C(NH_2)(H)-COOH$.

INTRODUCTION

Sulfonium nucleosides are intermediates in enzymic transmethylation[1] and in the biosynthesis of spermidine and spermine.[2] When isolated from natural sources, the nucleosides are fully active biologically, but chemical synthesis of the intermediates leads to the racemic sulfonium diastereoisomers, which are only partially active in transfer reactions.[3,4] In the following procedure is described the isolation of the enzymically inactive sulfonium diastereoisomer, (+)-*S*-adenosyl-L-methionine, prepared by enzymic treatment of (±)-*S*-adenosyl-L-methionine.[5] No chemical resolution of this substance has so far been achieved.

PROCEDURE

(±)-*S*-Adenosyl-L-methionine (2)

S-Adenosyl-L-homocysteine (**1**) (75.6 mg., 0.2 mmole), prepared by chemical[6,7] or by enzymic[8] synthesis, is dissolved in a mixture of formic acid (2 ml.) and acetic acid (2 ml.). Methyl iodide (0.5 ml., 8 mmoles) is added to the solution, which is shaken vigorously and then kept in the dark at room temperature for about three days.[a] Water (*ca.* 5 ml.) is then added to the reaction mixture, and the unreacted methyl iodide is extracted with ether (2 × 5 ml.) by use of a small separatory funnel. The aqueous layer is freeze-dried, and the resulting powder is dissolved *immediately* in a few ml. of water.[b]

The aqueous solution is then applied to a column (1 × 6 cm.) of Amberlite CG-50 ion-exchange resin in 0.01 *M* potassium phosphate buffer at pH 7. The unreacted thioether is eluted with 20 ml. of the same buffer, and an intermediate fraction is eluted with 20 ml. of 0.25 *N* acetic acid. The sulfonium nucleoside is then eluted with 20 ml. of 4 *N* acetic acid; this eluate is freeze-dried, and the product is dissolved *immediately* in water. Because acetate ions appear to have an inhibitory effect in the subsequent enzymic resolution, the nucleoside is converted into the chloride form by applying it to a column (1 × 6 cm.) of Amberlite CG-50 (H^+), which is washed successively with water and 0.1 *N* hydrochloric acid (10 ml.). The sulfonium nucleoside (**2**) is eluted with 20 ml. of 0.3 *N* hydrochloric acid, and the eluate is freeze-dried; the product is dissolved in a few ml. of water, and the pH of the solution is carefully adjusted to 6 with dilute aqueous sodium hydroxide.[c]

Enzymic Resolution

A reaction mixture is prepared that contains potassium phosphate buffer (pH 7.4, 1,000 μmoles), the racemic sulfonium nucleoside (**2**) as the chloride salt (98 μmoles), guanidinoacetic acid (64 μmoles, adjusted to pH 7), freshly neutralized, reduced glutathione (163 μmoles), and 11 units of guanidinoacetate methyltransferase[9] (specific activity, 0.7) in a final volume of 22 ml., and the mixture is kept at 37° for 2 hr. The reaction is stopped by addition of 1.2 ml. of 100% trichloroacetic acid, and the mixture is cooled in an ice-bath and centrifuged. The supernatant liquor is transferred to an Erlenmeyer flask, and a saturated solution of ammonium reineckate in 5% aqueous trichloroacetic acid (20 ml.) is added.

[a] After three days, approximately 50% of the nucleoside (**1**) has been methylated and, after an additional four days, the proportion has risen to only 60%.

[b] The sulfonium nucleoside (**2**) is unstable in the dry state; therefore, it is advantageous to conduct the freeze-drying in a cold-room.

[c] Sulfonium nucleosides are very unstable under alkaline conditions; hence, care must be taken to avoid local excesses of this base.

The reineckate salt is immediately precipitated; the suspension is kept at ice-bath temperature for 1 hr., and the precipitate is filtered off (sintered-glass funnel), and washed with a small volume of dilute reineckate solution. The wet reineckate is transferred to a separatory funnel, and decomposed by suspending it in 0.5 *N* hydrochloric acid (25 ml.) and washing the aqueous layer with 2-butanone (*ca.* 4 × 10 ml.) until the organic layer is no longer pink. The aqueous layer is washed once with ether, and is then freeze-dried. The residue is dissolved in water, and the pH of the solution is adjusted to 7.4. The resolution procedure and precipitation of reineckate are repeated, and the residue then remaining after freeze-drying of the decomposed reineckate is dissolved in the minimal volume of water.

A trace of colored impurity is removed by descending paper-chromatography with 6:4:3 butyl alcohol–acetic acid–water (top phase) for 8–10 hr. The (+)-*S*-adenosyl-L-methionine remains near the origin as an ultra-violet-absorbing strip, which is cut out, and eluted by descending chromatography with water. The aqueous eluate (*ca.* 20 ml.) is treated in the usual way on a buffered column of Amberlite CG-50 ion-exchange resin,[d] and the acetic acid solution is freeze-dried. The residue is dissolved in the minimal volume (*ca.* 0.5 ml.) of 5 *N* acetic acid,[e] and the pH of the solution is carefully adjusted to 3.6 by addition of dilute, aqueous sodium hydroxide. The clear solution contains about 25 μmoles (50%) of (+)-*S*-adenosyl-L-methionine (**3**). It has[f] $[\alpha]_{589}$ $+48 \pm 1°$ and $[\alpha]_{436}$ $+101 \pm 1°$. It is homogeneous, and indistinguishable from (−)-*S*-adenosyl-L-methionine by paper electrophoresis at pH 5.2, and by paper chromatography in 1:100:300 concentrated aqueous hydrochloric acid–water–ethanol and in 1:12:112 acetic acid–water–methyl Cellosolve. The resolved product is inactive as a substrate for the *S*-adenosyl-L-methionine-cleaving enzyme of yeast[5] and for catechol-*O*-methyl transferase,[10] as well as for guanidino-acetate methyltransferase.[9]

REFERENCES

(1) G. L. Cantoni, *J. Biol. Chem.*, **204**, 403 (1953).
(2) H. Tabor, S. M. Rosenthal, and C. W. Tabor, *J. Biol. Chem.*, **233**, 907 (1958).

[d] In order to remove any impurities eluted from the chromatography paper.

[e] To ensure reproducible optical rotational data, it has been found essential to recrystallize the acetic acid by filtering off the crystalline portion of the partially frozen solvent, and discarding the filtrate.

[f] The specific rotation of the cation is calculated on the basis of a molecular weight of 387, and the concentration of the solution is determined from its molar absorbance of 16,000 at 260 nm. A 5 *M* solution of equal amounts of acetic acid and sodium acetate is used as the blank.

(3) J. Baddiley, G. L. Cantoni, and G. A. Jamieson, *J. Chem. Soc.*, **1954**, 4280.
(4) G. A. Jamieson, *J. Org. Chem.*, **28**, 2397 (1963).
(5) G. de la Haba, G. A. Jamieson, S. H. Mudd, and H. H. Richards, *J. Am. Chem. Soc.*, **81**, 3975 (1959).
(6) J. Baddiley and G. A. Jamieson, *J. Chem. Soc.*, **1955**, 1085.
(7) W. Sakami, *Biochem. Prepn.*, **8**, 8 (1961).
(8) G. de la Haba and G. L. Cantoni, *J. Biol. Chem.*, **234**, 603 (1959).
(9) G. L. Cantoni and P. J. Vignos, Jr., *Methods Enzymol.*, **2**, 260 (1955).
(10) J. Axelrod and R. Tomchick, *J. Biol. Chem.*, **233**, 702 (1958).

[71] 1-(Adenin-9-yl)-1-deoxy-1-*S*-ethyl-1-thio-*aldehydo*-D-galactose Aldehydrol

An Acyclic Sugar–Purine Nucleoside Analog Prepared by the Chloromercuri Procedure

M. L. WOLFROM, P. McWAIN, and A. THOMPSON

DEPARTMENT OF CHEMISTRY, THE OHIO STATE UNIVERSITY, COLUMBUS, OHIO 43210

INTRODUCTION

Several types of nucleoside analogs containing an acyclic sugar residue have been prepared[1,2] in which C-1 is substituted by various groups in addition to the base. Herein is described such a derivative (**9**) in which C-1 is so substituted by an ethylthio group.[1] The acyclic bromide (**4**) was first reported by Wolfrom and associates,[3] but it is more easily prepared from penta-*O*-acetyl-D-galactose diethyl dithioacetal[4] (**3**) by the method of Gauthier[5] as adapted by Weygand and co-workers.[6] The bromide (**4**) readily reacts with chloromercuri-*N*-acetyladenine[7] (**5**), in a Davoll–Lowy[7] type of procedure, to give the condensation product (**6**) in high yield. Compound **6** can then be *N*-deacetylated by the picric acid method,[8] and the product *O*-deacetylated by butylamine[9] to yield the acyclic, sulfur-containing nucleoside analog (**9**), in which the sugar entity is a D-galactitol residue. The high yield in the Davoll–Lowy procedure is probably due to the higher reactivity and steric availability of the acyclic bromide in comparison with the cyclic halides (where the molecular conformations and the group interactions of the ring substituents can produce some hindrances to reaction). Unfortunately, efforts to close the ring, preferably to a furanoid type, in these acyclic nucleoside analogs, have thus far been unsuccessful. The compounds are very labile, and the groups attached to C-1 are removed without occurrence of ring closure. Nevertheless, the substances are an addition to the many acyclic sugar derivatives now known. The configuration at C-1 is not yet established, but the two isomers predictable have been isolated in many other cases. The compounds are perhaps best named as derivatives of the hydrated aldehyde or aldehydrol.

Compound **9** is nontoxic and is inactive against adenocarcinoma 755 and L-1210 lymphoid leukemia.

Penta-*O*-acetyl-D-galactose diethyl dithioacetal (**3**) is trimorphous,[10]

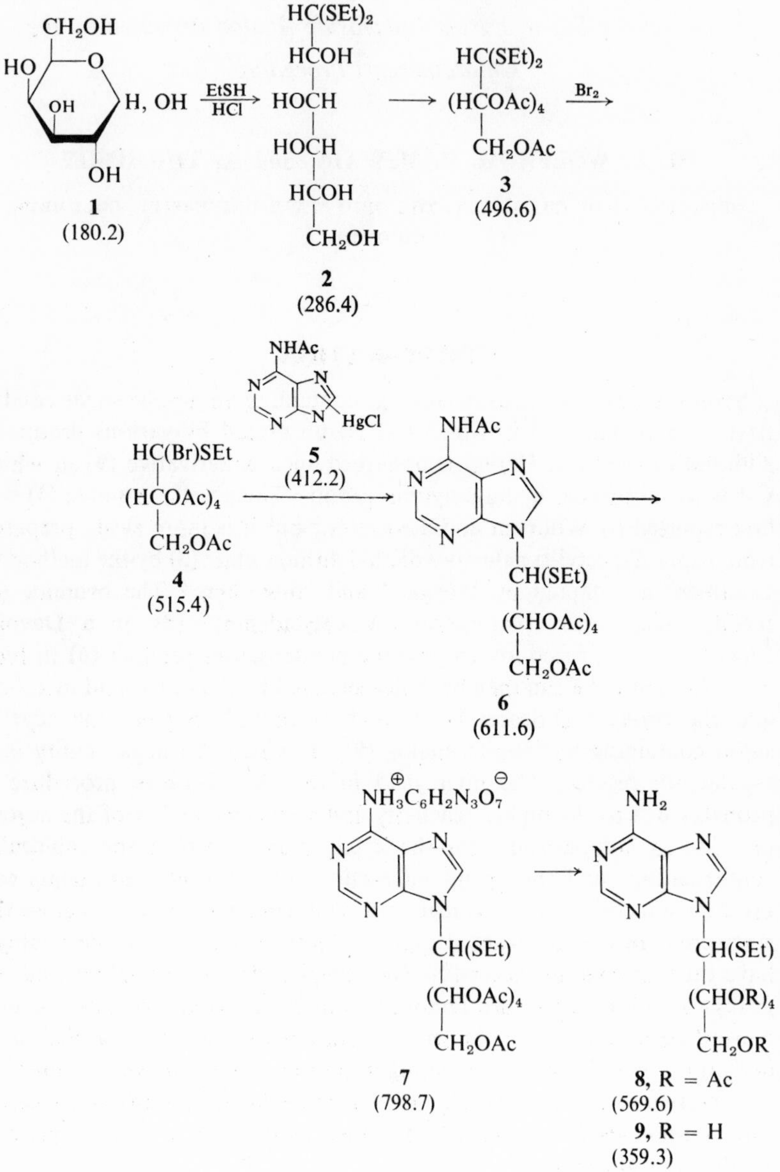

CH_2OH, HO, O, OH, H, OH, OH

1
(180.2)

$\xrightarrow[HCl]{EtSH}$

$HC(SEt)_2$
HCOH
HOCH
HOCH
HCOH
CH_2OH

2
(286.4)

$\longrightarrow$

$HC(SEt)_2$
$(HCOAc)_4$
CH_2OAc

3
(496.6)

$\xrightarrow{Br_2}$

HC(Br)SEt
$(HCOAc)_4$
CH_2OAc

4
(515.4)

NHAc, N, N, N, N, HgCl

5
(412.2)

$\longrightarrow$

NHAc, N, N, N, N
CH(SEt)
$(CHOAc)_4$
CH_2OAc

6
(611.6)

$\longrightarrow$

$\overset{\oplus}{N}H_3C_6H_2N_3\overset{\ominus}{O_7}$, N, N, N, N
CH(SEt)
$(CHOAc)_4$
CH_2OAc

7
(798.7)

$\longrightarrow$

NH_2, N, N, N, N
CH(SEt)
$(CHOR)_4$
CH_2OR

8, R = Ac
(569.6)
9, R = H
(359.3)

where Ac is acetyl.

and the polymorphous forms exhibit remarkable stability. Consequently, the melting point of **3** varies from one preparation to another.

PROCEDURE

D-Galactose Diethyl Dithioacetal[4,11] (2)

D-Galactose (**1**) (50 g., 0.28 mole) is placed in a 500-ml., glass-stoppered, wide-mouthed bottle and dissolved at room temperature in 75 ml. of concentrated hydrochloric acid (sp. gr. 1.19). Ethanethiol (ethyl mercaptan, 50 ml.) is then added, and the mixture is shaken vigorously, with occasional release of the pressure. After 3–5 min., a definite increase in temperature should be noticeable, and a little ice and ice–water is then added. The contents of the bottle solidify almost immediately to a mass of white crystals. More ice–water is added, and the mass is immediately filtered and washed with a small amount of ice–water. The material is recrystallized from absolute ethanol, and then from hot water; yield 37 g. (47%), m.p. 140–142°.

Penta-*O*-acetyl-D-galactose Diethyl Dithioacetal[4] (3)

D-Galactose diethyl dithioacetal (**2**) (50 g., 0.18 mole) is treated with 175 ml. of dry pyridine, and the mixture is cooled in ice. After most of the material has dissolved, 250 ml. of acetic anhydride is gradually added; this addition causes the separation of a considerable quantity of solid. The mixture is kept in ice for about 1 hr. and then at room temperature for 18 hr.; during this period, the undissolved material dissolves on being shaken occasionally. The solution is then poured into about 10 l. of ice and water in a precipitating jar. The sirup that separates crystallizes readily; yield 85 g. (98%). The product is recrystallized by dissolving it in methanol, adding water to incipient opalescence, placing the solution in a refrigerator, and, from time to time, adding more water until no further crystallization occurs; yield 81 g. (95%), $[\alpha]_D^{25}$ +9.8° (*c* 4.5, chloroform). This substance is trimorphous,[10] and the melting point is variable but is generally in the range of 77–91°. The pure forms, and their (capillary) melting points, are reported[10] to be: elongated prisms (most stable), m.p. 80.5–81°; hexagonal prisms, m.p. 90.5–91°, after sudden shrinkage at 76.5–77°; rectangular plates, m.p. 90.5–91°. These melting-point behaviors do not necessarily indicate the true melting points of the crystalline phases, as transitions may occur on heating.

Penta-*O*-acetyl-1-bromo-1-deoxy-1-*S*-ethyl-1-thio-*aldehydo*-D-galactose Aldehydrol[6] (4)

Penta-*O*-acetyl-D-galactose diethyl dithioacetal (**3**) (36.4 g., 73 mmoles) is dissolved in 300 ml. of anhydrous ether, and to the stirred solution is added dropwise, during 40 min., 4 ml. of bromine in 120 ml. of anhydrous ether. This is followed by the dropwise addition of cyclohexene until the solution is colorless. Hexane (250 ml.) is added, and the solution is kept at 0° for 1 hr., giving a crystalline precipitate. Filtration and washing with hexane affords 31.0 g. of **4** (82%), m.p. 104–105°, $[\alpha]_D^{22}$ $-17.5 \rightarrow +28.0°$ (*c* 3, chloroform, 3 hr.).

1-(6-Acetamidopurin-9-yl)penta-*O*-acetyl-1-deoxy-1-*S*-ethyl-1-thio-*aldehydo*-D-galactose Aldehydrol[1] (6)

A mixture of 24.0 g. (58 mmoles) of chloromercuri-*N*-acetyladenine[7] (**5**), 20 g. of cadmium carbonate, 5 g. of Celite, and 300 ml. of toluene is dried by the codistillation of 100 ml. of toluene. To this suspension, 31 g. (60 mmoles) of freshly prepared **4** is added with stirring, and the mixture is refluxed for 4.5 hr. The hot suspension is filtered, and the filtrate is concentrated under diminished pressure to a sirup. Both the filter cake and the residue from the filtrate are extracted with hot chloroform, and the extracts are combined and washed twice with 30% aqueous potassium iodide, and thrice with water. The chloroform solution is dried, and evaporated under diminished pressure to a sirup which crystallizes; yield, 34.7 g. (97.5%, based on **5**). The light-brown solid is recrystallized from ether. Pure material is obtained by recrystallizing from toluene, after treatment with carbon, to form light cream-colored platelets of **6**; m.p. 151–152°, $[\alpha]_D^{23}$ $-62°$ (*c* 0.82, chloroform).

Picrate of Penta-*O*-acetyl-1-(adenin-9-yl)-1-deoxy-1-*S*-ethyl-1-thio-*aldehydo*-D-galactose Aldehydrol[1] (7)

A solution of compound **6** (10.4 g., 17 mmoles) in 60 ml. of warm ethanol is treated with 41 ml. (1 molar proportion) of 10% picric acid (ethanolic), and the mixture is heated to boiling for 1 min. and cooled to 0°. A bright-yellow, crystalline precipitate forms, and is removed by filtration and washed with cold ethanol; yield 6.34 g. (47%) of crude **7**. Recrystallization from chloroform–ethanol by slow evaporation gives pure material, m.p. 196–197°.

Penta-*O*-acetyl-1-(adenin-9-yl)-1-deoxy-1-*S*-ethyl-1-thio-*aldehydo*-D-galactose Aldehydrol[1] (8)

The crystalline picrate (**7**) (6.2 g., 7.8 mmoles) is suspended in 350 ml. of warm, 50% aqueous acetone and stirred with an excess (100 ml.) of moist Dowex-1 (CO_3^{2-}) anion-exchange resin. The resulting, faintly yellow solution is passed through a column (20 × 100 mm.) of Dowex-1 (CO_3^{2-}) and concentrated to 200 ml. under diminished pressure. The suspension is extracted with two 100-ml. portions of chloroform, and the extracts are combined, dried with anhydrous sodium sulfate, and evaporated under diminished pressure to a crystalline solid (**8**); yield 2.91 g. (66%). Recrystallization from ethanol (carbon) affords colorless needles; m.p. 186–187°, $[\alpha]_D^{21}$ $-60°$ (*c* 0.51, chloroform).

1-(Adenin-9-yl)-1-deoxy-1-*S*-ethyl-1-thio-*aldehydo*-D-galactose Aldehydrol[1] (9)

Crystalline **8** (5.0 g., 8.8 mmoles) is dissolved in 75 ml. of hot, dry methanol, 1.5 ml. of butylamine is added, and the solution is refluxed for 6 hr. A crystalline solid (**9**) separates on cooling the solution; yield 2.66 g. (91%). The material is redissolved in water, treated with carbon, and evaporated to a sirup. Addition of ethanol and storage at 0° induces crystallization; yield 2.59 g. (88%), m.p. 217–218°, $[\alpha]_D^{22}$ $-114°$ (*c* 0.53, water).

REFERENCES

(1) M. L. Wolfrom, P. McWain, and A. Thompson, *J. Org. Chem.*, **27**, 3549 (1962).
(2) M. L. Wolfrom, A. B. Foster, P. McWain, W. von Bebenburg, and A. Thompson, *J. Org. Chem.*, **26**, 3095 (1961); M. L. Wolfrom, W. von Bebenburg, R. Pagnucco, and P. McWain, *ibid.*, **30**, 2732 (1965).
(3) M. L. Wolfrom, D. I. Weisblat, and A. R. Hanze, *J. Am. Chem. Soc.*, **62**, 3246 (1940).
(4) M. L. Wolfrom, *J. Am. Chem. Soc.*, **52**, 2464 (1930).
(5) B. Gauthier, *Ann. Pharm. France*, **12**, 281 (1954).
(6) F. Weygand, H. Ziemann, and H. J. Bestmann, *Chem. Ber.*, **91**, 2534 (1958).
(7) J. Davoll and B. A. Lowy, *J. Am. Chem. Soc.*, **73**, 1650 (1951); B. R. Baker, K. Hewson, H. J. Thomas, and J. H. Johnson, Jr., *J. Org. Chem.*, **22**, 954 (1957).
(8) J. R. Parikh, M. E. Wolff, and A. Burger, *J. Am. Chem. Soc.*, **79**, 2778 (1957).
(9) E. J. Reist and B. R. Baker, *J. Org. Chem.*, **23**, 1083 (1958).
(10) L. H. Welsh and G. L. Keenan, *J. Am. Chem. Soc.*, **64**, 183 (1942).
(11) E. Fischer, *Ber.*, **27**, 673 (1894).

[72] 9-(3-Deoxy-β-D-*erythro*-pentofuranosyl)guanine

A New Method of Synthesis of Guanine Nucleosides

WILLIAM W. LEE and GEORGE L. TONG

LIFE SCIENCES RESEARCH, STANFORD RESEARCH INSTITUTE, MENLO PARK, CALIFORNIA 94025

$HOCH_2$ OMe O O
1 (146.1)

$ROCH_2$ OMe O OR
2, R = H (148.2)
3, R = Bz (356.4)

$BzOCH_2$ Cl O OBz
4 (360.8)

Cl N N AcHN N N $BzOCH_2$ O OBz
5 (535.9)

[SCH_2CH_2OH N N AcHN N N $BzOCH_2$ O OBz]
6 (577.6)

OH N N H_2N N N $HOCH_2$ O OH
7 (267.2)

INTRODUCTION

9-Substituted-guanine nucleosides have been synthesized by a number of indirect routes utilizing 2,6-dichloropurine[1] and 2,8-dichloroadenine[2] as precursors. The direct condensation of a protected guanine derivative with a sugar derivative has been used[3] to prepare **7**; this required a careful separation of the 7-substituted from the 9-substituted-guanine nucleosides that were formed. The present preparation of **7** features a useful combination of directness, convenience, and yield. The procedure has been used in these laboratories for the preparation of several other 9-substituted-guanine nucleosides, and seems to have general applicability.

PROCEDURE

Methyl 3-Deoxy-β-D-*erythro*-pentofuranoside (2)

A literature procedure,[4] used for the synthesis of other deoxy sugars, was modified for this synthesis (Sec. III [98]). A suspension of 3.90 g. (103 mmoles) of lithium aluminum hydride in 100 ml. of dry ether is stirred under a nitrogen atmosphere and cooled in an ice bath. To this suspension is added dropwise a solution of 5.00 g. (34.2 mmoles) of methyl 2,3-anhydro-β-D-ribofuranoside[5] (**1**) in 100 ml. of dry[a] tetrahydrofuran at such a rate that the temperature remains below 25° (addition time, about 1 hr.). Another 0.58 g. (15.3 mmoles) of lithium aluminum hydride is added, and the mixture is stirred at room temperature for 16 hr.[b] The stirred mixture is cooled in an ice bath, and 37 ml. of isopropyl alcohol is added dropwise, followed by 100 ml. of water. The suspension is filtered through Celite,[c] and the filter pad is well washed with water and then methanol. The filtrate is stirred with 75 g. (wet weight) of Amberlite IRC-50[d] until neutral, filtered, and evaporated.[e] The residue is dissolved in 75 ml. of chloroform, filtered, and re-evaporated, to afford 4.34 g. (85%) of[6] **2** as a yellow sirup that is homogeneous by thin-layer chromatography (tlc)[f] in solvent A, R_f 0.13, and whose nmr spectrum[g] is compatible with the structure indicated.

Several preparations, using 5 to 24 g. of starting material **1**, have given yields of 85–92%.

Methyl 2,5-Di-*O*-benzoyl-3-deoxy-β-D-*erythro*-pentoside (3)

By the literature procedure,[6] a solution of 4.26 g. (28.7 mmoles) of the methyl deoxyglycoside (**2**) in 40 ml. of dry pyridine is treated with 10 ml. (86 mmoles) of benzoyl chloride at room temperature for 16 hr. The solution is then cooled in an ice bath, diluted with 40 ml. of water, and evaporated to a sirup. This is dissolved in 100 ml. of toluene, and the solution is washed successively with six 25-ml. portions of saturated sodium hydrogen carbonate solution and two 50-ml. portions of water,

[a] Dried with calcium hydride.

[b] Reaction is not complete after 1 hr.

[c] A diatomaceous earth from Johns–Manville Products Corp.

[d] A cationic (H^+) type of ion-exchange resin from Rohm and Haas Co., Philadelphia, Pa.

[e] Evaporations are performed under diminished pressure (water aspirator or vacuum pump) in a rotary evaporator, at a bath temperature of 40–50°.

[f] All thin-layer chromatograms are run on silica gel HF in the following solvent systems: A, ethyl ether; B, 1:1 (v/v) benzene–ethyl ether. The spots are detected with sulfuric acid spray or under ultraviolet light.

[g] The use of lithium aluminum hydride instead of Raney nickel affords **2** that, by nmr, contains no detectable proportion of isomeric product. See footnote 18 of Ref. 6.

dried,[h] filtered, and evaporated. The residue is crystallized from 100 ml. of hexane, to afford 8.55 g. (84%) of[6] **3** (first crop, 8.07 g., m.p. 80.5–81.5°; second crop, m.p. 78.5–80.5°) that is homogeneous by tlc; R_f 0.90 in solvent B.[f] Several preparations, starting with 4 to 23 g. of **2**, have given comparable yields.

2,5-Di-*O*-benzoyl-3-deoxy-β-D-*erythro*-pentosyl Chloride (4)

A solution of 3.25 g. (9.1 mmoles) of the methyl deoxyglycoside (**3**) in 1 ml. of acetyl chloride and 10 ml. of acetic acid is protected from moisture, cooled in an ice bath, and stirred while 10 ml. of cold (10–15°) acetic acid presaturated with hydrogen chloride is added. After being stirred for 1 hr. at 10–15°, the solution is evaporated. The residue is treated with 10 ml. of dry toluene and evaporated; this treatment is repeated three more times, to remove all traces of acetic acid. The residue is dissolved in 100 ml. of dry ether, and the solution is saturated with dry hydrogen chloride at 0°, stored at 0° for 2 days in a closed flask, and then evaporated. The residue is treated with 5 ml. of dry benzene, re-evaporated, and then crystallized from 40 ml. of 1:3 (v/v) benzene–Skellysolve C[i] to afford 2.06 g. (62%) of[7] **4**, and a second crop of 0.29 g., m.p. 88.5–92°; total, 2.35 g. (81%). The product **4** is homogeneous by tlc; R_f 0.70 in solvent B.[f] Experiments starting with 3 g. to 27 g. of **3** have afforded comparable yields.

2-Acetamido-6-chloro-9-(2,5-di-*O*-benzoyl-3-deoxy-β-D-*erythro*-pentosyl)-9*H*-purine (5)

A suspension of 3.23 g. of a mixture of 1.02 g. of Celite and 2.21 g. (5.0 mmoles) of the mercury derivative of 2-acetamido-6-chloropurine (Sec. I [9]) in 200 ml. of dry xylene is dried by distilling off 50 ml. of xylene. The suspension is cooled to room temperature, and to it are added a solution of 1.81 g. (5.0 mmoles) of the glycosyl chloride (**4**) in 20 ml. of dry xylene, and 5.0 g. of molecular sieves (Linde type 4A).[j] The mixture is protected from moisture, heated with stirring for 6 hr. at reflux temperature, and filtered hot. The filter cake is washed with two 25-ml. portions of chloroform, and the filtrates are combined and evaporated. The residual gum is dissolved in 75 ml. of dichloromethane, washed successively with two 20-ml. portions of 30% aqueous potassium iodide and two 25-ml. portions of water, and dried. The solution is then treated with charcoal, filtered, and evaporated to a foam which is dissolved in 10 ml. of warm benzene, filtered, and allowed to cool, to afford 0.876 g. (33%) of[7] **5**, m.p. 197.5–203.5° and a second crop, weighing 0.042 g., m.p. 191.5–

[h] Anhydrous magnesium sulfate is used as the drying agent, unless otherwise specified.

[i] A petroleum fraction, essentially heptane, b.p. 90–100°.

[j] A product of Union Carbide Corp., 270 Park Ave., New York, N. Y. 10017.

198.5°; total yield 0.918 g. (34%). This experiment has been performed several times, starting with 1 g. to 13 g. of the glycosyl chloride (**4**), to afford **5** in 28–34% yield.

9-(3-Deoxy-β-D-*erythro*-pentofuranosyl)guanine (7)

To a suspension of 0.54 g. (1.0 mmole) of 2-acetamido-9-(2,5-di-*O*-benzoyl-3-deoxy-β-D-*erythro*-pentofuranosyl)-6-chloro-9*H*-purine (**5**) in 50 ml. of methanol is added a solution of 0.30 ml. (4.3 mmoles) of 2-mercaptoethanol in 3.0 ml. of *M* methanolic sodium methoxide. The suspension is stirred and heated at reflux temperature under a nitrogen atmosphere for 3 hr., during which time **5** dissolves and product **7** is precipitated.[k] The mixture is evaporated to dryness, the residue is dissolved in 15 ml. of water, and the solution is washed with three 5-ml. portions of ether. The aqueous phase is treated with charcoal, adjusted to pH 6.5–7 with acetic acid, and cooled, to afford 0.16 g. of **7**, m.p. >300° (browns from 240°) and a second crop, 0.03 g., m.p. >300° (browns from 240°); total yield 0.19 g. (70%). Recrystallization of the combined crops from 10 ml. of water gives 0.17 g. (63%) of[7] **7** as white, fibrous needles, m.p. >300° (browns from 250°); $\lambda_{max}^{pH\,1}$ 255 nm. (ϵ, 11,800), 275 (shoulder) (~7,800); $\lambda_{max}^{pH\,7}$ 252 nm. (12,900), 270 (shoulder) (~9,000); $\lambda_{max}^{pH\,13}$ 260 nm. (shoulder) (~11,600), 267 (11,700); $[\alpha]_{589}^{23}$ −41° (*c* 0.3, water); homogeneous by paper chromatography on Whatman No. 1 paper in water; with R_{Ado}[l] 1.56.

REFERENCES

(1) See E. J. Reist and L. Goodman, *Biochemistry*, **3**, 15 (1964), and references therein.
(2) J. Davoll, B. Lythgoe, and A. R. Todd, *J. Chem. Soc.*, **1948**, 1685.
(3) S. R. Jenkins, F. W. Holly, and E. Walton, *J. Org. Chem.*, **30**, 2851 (1965).
(4) M. Dahlgard, B. H. Chastain, and R.-J. L. Han, *J. Org. Chem.*, **27**, 929 (1962).
(5) C. D. Anderson, L. Goodman, and B. R. Baker, *J. Am. Chem. Soc.*, **80**, 5247 (1958).
(6) E. Walton, F. W. Holly, G. E. Boxer, R. F. Nutt, and S. R. Jenkins, *J. Med. Chem.*, **8**, 659 (1965).
(7) G. L. Tong, K. J. Ryan, W. W. Lee, E. M. Acton, and L. Goodman, *J. Org. Chem.*, **32**, 859 (1967).
(8) T. P. Johnston, L. B. Holum, and J. A. Montgomery, *J. Am. Chem. Soc.*, **80**, 6265 (1958).

[k] Presumably, the initial product is **6**. The unstable 6-(2-hydroxyethylthio) grouping[8] is rapidly hydrolyzed off under the basic conditions, with concomitant deacylation to afford the final product **7**.

[l] Adenosine (Ado) is the standard for the chromatographic comparison.

[73] 8-Bromoguanosine

Bromination of the Purine Ring of a Purine Nucleoside

ROBERT A. LONG, ROLAND K. ROBINS, and LEROY B. TOWNSEND

DEPARTMENT OF CHEMISTRY, UNIVERSITY OF UTAH, SALT LAKE CITY, UTAH 84112

Br_2—H_2O

1 (283.2) → **2** (362.3)

INTRODUCTION

8-Bromoguanosine (**2**), first reported by Holmes and Robins,[1] is an important intermediate for the synthesis of various 8-substituted guanosine derivatives by nucleophilic displacement of bromine therefrom.[1,2] The procedure described here is the most direct and the simplest of those reported,[1,3] and affords 8-bromoguanosine in a yield of 88% (based on guanosine).

PROCEDURE

8-Bromoguanosine[1,3] (2)

To a suspension of guanosine (**1**) (5.0 g., 0.018 mole) in 30 ml. of water is added 140 ml. of saturated bromine–water[a] in aliquots of *ca.* 15 ml. at such a rate that the yellow color of the reaction mixture disappears between each addition, the total time for addition being 4–6 min. The colorless solid is then quickly filtered off, successively washed with 30 ml.

[a] To 100 ml. of water at room temperature is added 1 ml. of bromine, the flask is stoppered, and the mixture is stirred very vigorously (magnetically) for 15 min., or until all the bromine has dissolved.

of cold water and 30 ml. of cold acetone, and dried; yield 6.65 g. (88.3%) of colorless crystals; it is recrystallized from water, and dried at 110°/0.1 mm. Hg, to give 5.30 g. (82.8%) of chromatographically[b] pure **2**, m.p. 193° (softens), 201–203° (dec.); $\lambda_{max}^{pH\,1}$ 261 nm., ϵ 18,470; $\lambda_{max}^{pH\,11}$ 270 nm., ϵ 15,900; R_{Ade} in solvent systems A, B, C, and D, 0.99, 1.49, 0.75, and 1.17, respectively.

REFERENCES

(1) R. E. Holmes and R. K. Robins, *J. Am. Chem. Soc.*, **86**, 1242 (1964).

(2) R. A. Long, R. K. Robins, and L. B. Townsend, unpublished data; *J. Org. Chem.*, **32**, 2751 (1967); R. E. Holmes and R. K. Robins, *J. Am. Chem. Soc.*, **87**, 1772 (1965); M. Ikehara, H. Tada, and K. Muneyama, *Chem. Pharm. Bull.* (Tokyo), **13**, 1140 (1965).

(3) R. Shapiro and S. C. Agarwal, *Biochim. Biophys. Acta*, **24**, 401 (1966); M. Ikehara, H. Tada, and K. Muneyama, *Chem. Pharm. Bull.* (Tokyo), **13**, 639 (1965).

[b] Chromatograms on Whatman No. 1 chromatography paper (descending technique) are developed with: [A] 7:3 (v/v) ethanol–water; [B] 5% aqueous ammonium hydrogen carbonate; [C] 14:1:5 (v/v) isopropyl alcohol–ammonia–water; [D] 1:1 (v/v) acetone–water; the chromatographic mobilities are referred to that of adenine (Ade) as unity.

[74] 2′,3′-*O*-Isopropylideneguanosine

*2′, 3′-*O*-Alkylidene Acetals as Specifically Substituted Intermediates in the Synthesis of Ribonucleoside 5′-Phosphates*

S. CHLÁDEK

INSTITUTE OF ORGANIC CHEMISTRY AND BIOCHEMISTRY, CZECHOSLOVAK ACADEMY OF SCIENCES, PRAGUE, CZECHOSLOVAKIA

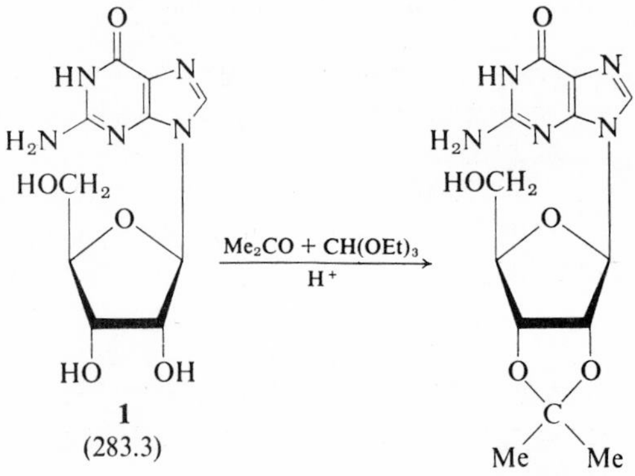

HN, H_2N, N, N, N, $HOCH_2$, O, HO, OH

1
(283.3)

$\xrightarrow[H^+]{Me_2CO + CH(OEt)_3}$

HN, H_2N, N, N, N, $HOCH_2$, O, O, O, C, Me, Me

2
(323.3)

INTRODUCTION

2′,3′-*O*-Isopropylidene acetals of nucleosides are important intermediates in the synthesis of 5′-substituted D-ribonucleosides, especially D-ribonucleoside 5′-phosphates.[1] The first synthetic nucleotide was prepared in this way.[1a] The acetals are usually prepared by the condensation of D-ribonucleosides with acetone in the presence of an acid catalyst,[1a] or by reaction with acetone and 2,2-diethoxypropane in the presence of bis(*p*-nitrophenyl) hydrogen phosphate.[2] An advantageous procedure for the preparation of D-ribonucleoside 2′,3′-*O*-isopropylidene acetals consists in the reaction, in *N,N*-dimethylformamide,[3] with acetone and ethyl orthoformate in the presence of anhydrous hydrogen chloride. The latter procedure has been used in the preparation of the 2′,3′-*O*-isopropylidene acetals of adenosine, 6-azacytidine, 6-azauridine, cytidine, guanosine, inosine, and uridine.[4] The reaction of acetone and ethyl orthoformate with D-ribonucleosides having an amino group on the heterocyclic moiety requires more than one equivalent of hydrogen chloride.

PROCEDURE

2′,3′-*O*-Isopropylideneguanosine (2)

Into a 100-ml., round-bottomed flask is introduced 1.42 g. (5 mmoles) of guanosine (**1**), 30 ml. of *N,N*-dimethylformamide,[a] 0.5 ml. (5 mmoles) of acetone, 1.5 ml. (0.01 mole) of ethyl orthoformate, and 2.5 ml. (0.01 mole) of a 4 *M* solution of hydrogen chloride in *p*-dioxane,[b] and the mixture is kept at room temperature for 2 days.[c] The mixture is then poured into 10 ml. of 1:1 concentrated ammonium hydroxide–water, and the alkaline solution is placed on a column (1 × 10 cm.) of 30 ml. of Dowex-1 (OH^-) ion-exchange resin.[d] The column is eluted with 2:1:1 water–concentrated ammonium hydroxide–methanol,[e] and the eluate is evaporated to dryness at 35° (bath temperature)/15 mm. Hg. The residual sirup is dried by repeated evaporation with ethanol and acetone. The crystalline product **2** weighs 1.4 g. (90%) and is chromatographically homogeneous; R_f 0.42 (butyl alcohol saturated with water) and 0.71 (7:1:2 isopropyl alcohol–concentrated ammonium hydroxide–water).

The crude product may be purified by dissolving it in concentrated ammonium hydroxide and concentrating the solution under diminished pressure to about 5 ml. The crystals are collected with suction, and washed with anhydrous acetone. Analytically pure material is obtained by recrystallization from water. Compound **2** does not show a definite melting point, becoming brown at 240°.

REFERENCES

(1) (a) P. A. Levene and R. S. Tipson, *J. Biol. Chem.*, **106**, 113 (1934);
(b) A. M. Michelson, *The Chemistry of Nucleosides and Nucleotides*, Academic Press Inc. (London) Ltd., 1963, pp. 111–116.

(2) A. Hampton, *J. Am. Chem. Soc.*, **83**, 3640 (1961).

(3) S. Chládek and J. Smrt, *Collection Czech. Chem. Commun.*, **28**, 1301 (1963).

(4) A. Holý, unpublished results.

[a] *N,N*-Dimethylformamide is dried by vacuum distillation from phosphorus pentaoxide (10% by weight), and the distillate is stored over molecular sieves.

[b] Dry hydrogen chloride is introduced (with ice-cooling) into *p*-dioxane predried by distillation from sodium.

[c] The course of the reaction is followed by descending chromatography in 7:1:2 isopropyl alcohol–concentrated ammonium hydroxide–water. A sample of the reaction mixture is made alkaline with ammonium hydroxide and spotted on Whatman No. 1 paper.

[d] The column is packed with Dowex-1 X-2 (Cl^-) ion-exchange resin (200–400 mesh) and washed with 5% aqueous sodium hydroxide, and then with water until the effluent no longer gives an alkaline reaction.

[e] The elution is discontinued when the eluate no longer absorbs ultraviolet light.

[75] 9-(3-Acetamido-3-deoxy-β-D-talopyranosyl)-hypoxanthine

Inversion at C-2′ in 3-Amino-3-deoxyaldopyranosyl Nucleosides

FRIEDER W. LICHTENTHALER and PETER EMIG

INSTITUT FÜR ORGANISCHE CHEMIE, TECHNISCHE HOCHSCHULE,
61 DARMSTADT, GERMANY

1 (375.3) → 2 (357.3) → 3 (445.4) → 4 (505.4) → 5 (436.4) → 6 (339.3)

where Ac is acetyl, and Ms is methylsulfonyl.

INTRODUCTION

Inversion of attachment of a hydroxyl group adjacent and *trans* to an amino group, *via* the *N*-acetyl-*O*-methylsulfonyl derivative and de-*O*-methylsulfonylation through an oxazoline intermediate has found extensive use in the amino sugar[1] and amino cyclitol[2] field. The reaction was first applied to amino sugar nucleosides, to convert 9-(3-amino-3-deoxy-α-D-arabinofuranosyl)-*N*,*N*-dimethyladenine into the corresponding α-D-ribofuranosyl derivative.[3] That the reaction has considerable utility has been shown by the conversion of 9-(2-amino 2-deoxy-β-D-glucopyranosyl)-*N*,*N*-dimethyladenine, by inversion[4] at C-3, into the D-*allo* derivative; by inversion at C-2 of (3-amino-3-deoxy-β-D-glucopyranosyl)-uracil,[5] -hypoxanthine,[6] and -theophylline[7] to form the corresponding D-*manno* compounds; and by conversion of 1-(3-amino-3-deoxy-β-D-glucopyranosyl)-uracil into the *N*-acetate having the D-*galacto* configuration, wherein the attachment of the 4-hydroxyl group is inverted. By this type of inversion, nucleosides derived from 3-amino-3-deoxy-β-D-galactopyranose may be converted into the corresponding D-*talo* compounds (as in **1** → **6**).

PROCEDURE

9-(3-Acetamido-3-deoxy-β-D-galactopyranosyl)hypoxanthine (2)

To an ice-cooled suspension of 9.5 g. (25.3 mmoles) of 9-(3-amino-3-deoxy-β-D-galactopyranosyl)hypoxanthine acetate monohydrate[6] (**1**) (Sec. III [76]) in 150 ml. of absolute methanol is added 5.8 ml. of acetic anhydride through a dropping funnel, with stirring. The mixture is allowed to warm to room temperature, giving an almost clear solution after 30 min. On continued stirring, a precipitate separates; the suspension is kept at room temperature overnight, cooled to 0°, and filtered with suction. One recrystallization from 14:3 methanol–water gives 8.2 g. (91%) of pure **2** monohydrate as needles, m.p. 225–227°, $[\alpha]_D^{25}$ +6.9° (*c* 0.4, water); λ_{max}^{MeOH} 244 nm.

9-(3-Acetamido-4,6-*O*-benzylidene-3-deoxy-β-D-galactopyranosyl)-hypoxanthine (3)

A mixture of 8.0 g. (22.4 mmoles) of compound **2**, 40 ml. of freshly distilled benzaldehyde, and 7.5 g. of freshly fused and pulverized zinc chloride is stirred vigorously at room temperature for 24 hr. The clear solution is then diluted with 700 ml. of ether, resulting in the separation of a cloudy precipitate; the suspension is stirred for 30 min., and the precipitate is collected by suction filtration. For removal of zinc chloride, the solid is treated with 250 ml. of cold 0.1 *N* disodium (ethylenedinitrilo)-tetraacetate (EDTA) for 30 min. The insoluble material is filtered off, and dissolved in 150 ml. of water, and the solution is stirred with a small amount of mixed-bed, ion-exchange resin. After removal of the resin, the solution is evaporated to dryness *in vacuo*, and the residue is recrystallized twice from 2:1 ethanol–water. It is dried at 150°/0.1 mm. Hg for 6 hr., giving 7.0 g. (70%) of **3** monohydrate, m.p. 218–222°, $[\alpha]_D^{25}$ +10.5° (*c* 0.4, *N*,*N*-dimethylformamide).

9-[3-Acetamido-4,6-*O*-benzylidene-3-deoxy-2-*O*-(methylsulfonyl)-β-D-galactosyl]hypoxanthine (4)

A suspension of compound **3** (3.95 g., 8.86 mmoles) in 70 ml. of dry pyridine containing 3 drops of water[a] is cooled to −5°, and 1 ml. of methanesulfonyl chloride is added gradually, with stirring. After 1 hr., the mixture is allowed to warm to room temperature, and stirring is continued for 24 hr. A small amount of insoluble material is then filtered off, 20 ml. of water is gradually added (to decompose the excess of sulfonyl chloride), and the brown solution is kept for 3 hr., and evaporated to dryness *in vacuo*. The residue is treated with 10 ml. of water, and the insoluble material is filtered off, dissolved in ethanol, and decolorized by refluxing the solution with charcoal and silica gel for 30 min. The suspension is filtered and the filtrate is evaporated to dryness, giving a yellow residue that is triturated with water; the suspension is filtered, and the insoluble material is thoroughly washed successively with water and cold methanol. One recrystallization from absolute methanol gives 2.05 g. (46%) of **4** as colorless needles, m.p. 241–245°, $[\alpha]_D^{25}$ −18.0° (*c* 0.4, *N,N*-dimethylformamide).

9-(3-Acetamido-4,6-*O*-benzylidene-3-deoxy-β-D-talopyranosyl)hypoxanthine (5)

A suspension of 1.72 g. (3.4 mmoles) of compound **4** and 2.5 g. of sodium acetate in a mixture of 140 ml. of 2-methoxyethanol and 12.5 ml. of water is refluxed for 3 days. A small turbidity is removed by filtration, and the filtrate is evaporated to dryness *in vacuo* (1 mm. Hg), giving a brown residue that is extracted with five 35-ml. portions of acetone at 30°. The extracts are combined, and evaporated to dryness *in vacuo*. Ethanol (75 ml.) is added, the resulting solution is decolorized by heating with charcoal and silica gel, and the suspension is filtered. A mixed-bed ion-exchange resin is added, with stirring, to the filtrate (to remove the last traces of sodium acetate and sodium methanesulfonate), the resin is filtered off, and washed thoroughly with ethanol, and the filtrates are combined, and evaporated to dryness *in vacuo*. The residue is suspended in a small volume of acetone and, with shaking, a small volume of ethanol is added, such that only the yellowish impurity is dissolved. The insoluble material is recrystallized from ethanol, yielding 0.95 g. (66%) of **5** hemihydrate, m.p. 218–221°, $[\alpha]_D^{25}$ +5.5° (*c* 0.4, *N,N*-dimethylformamide).

9-(3-Acetamido-3-deoxy-β-D-talopyranosyl)hypoxanthine (6)

A solution of 0.90 g. (2.06 mmoles) of compound **5** in 70 ml. of glacial acetic acid and 35 ml. of water is heated on a steam bath for 1 hr. The solu-

[a] Use of absolute pyridine leads to incomplete methanesulfonylation.

tion is cooled, and washed with three 150-ml. portions of ligroin, and the aqueous layer is evaporated to dryness *in vacuo* at 30°. The brownish residue is dissolved in the minimal volume of water, and reprecipitated by addition of ethanol. One recrystallization from 14:3 methanol–water gives 0.65 g. (93%) of pure **6** as needles, m.p. 289–290°, $[\alpha]_D^{25}$ −3.3° (*c* 0.4, water).

REFERENCES

(1) B. R. Baker, *Methods Carbohyd. Chem.*, **2**, 444 (1963); A. C. Richardson and H. O. L. Fischer, *J. Am. Chem. Soc.*, **83**, 1182 (1961).
(2) F. W. Lichtenthaler, *Chem. Ber.*, **94**, 3071 (1961); **96**, 945 (1963).
(3) B. R. Baker and R. E. Schaub, *J. Am. Chem. Soc.*, **77**, 2396, 5900 (1955).
(4) F. J. McEvoy, M. J. Weiss, and B. R. Baker, *J. Am. Chem. Soc.*, **82**, 205 (1960).
(5) K. A. Watanabe and J. J. Fox, *J. Org. Chem.*, **31**, 211 (1966).
(6) F. W. Lichtenthaler, P. Emig, and D. Bommer, *Chem. Ber.*, **101**, in press (1968).
(7) F. W. Lichtenthaler, T. Nakagawa, and J. Yoshimura, *Chem. Ber.*, **100**, 1833 (1967).

[76] 9-(3-Amino-3-deoxy-β-D-galactopyranosyl)-hypoxanthine

The Dialdehyde–Nitromethane Cyclization Applied to "Inosine Dialdehyde"

FRIEDER W. LICHTENTHALER and PETER EMIG

INSTITUT FÜR ORGANISCHE CHEMIE, TECHNISCHE HOCHSCHULE DARMSTADT, 61 DARMSTADT, GERMANY

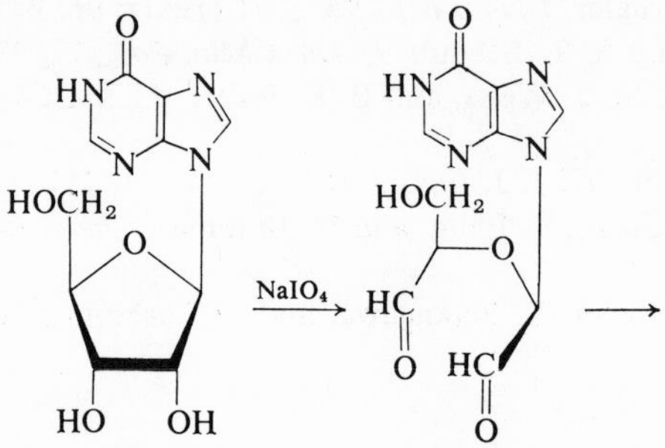

$NaIO_4$

1 (268.2)

2 (266.2)

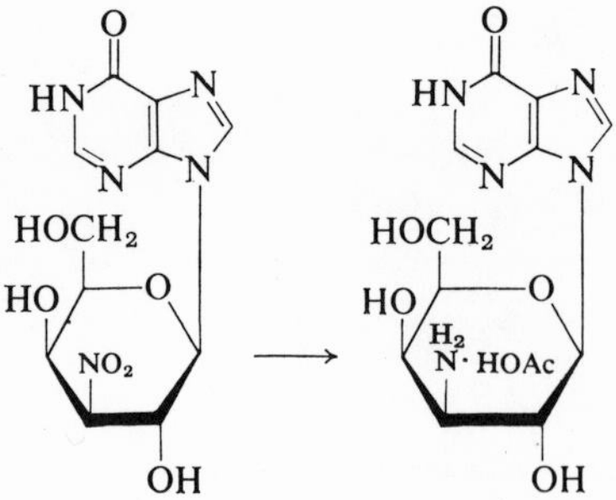

3 (343.3)

4 (375.3)

INTRODUCTION

Nitromethane cyclization of the dialdehyde **2**, obtained by periodate oxidation of inosine[1] (**1**), leads to a mixture of two (*C*-nitrohexosyl)hypoxanthines, from which the *galacto* isomer (**3**) may be separated by fractional recrystallization.[2] Subsequent hydrogenation gives **4** (see Sec. III [75]) in an overall yield of 17%.

PROCEDURE

2-*O*-[(*R*)-Formyl-(hypoxanthin-9-yl)methyl]-(*R*)-glyceraldehyde ("Inosine Dialdehyde," 2)

To an ice-cooled, magnetically stirred solution of 39.5 g. (185 mmoles) of sodium metaperiodate in 400 ml. of water is added, in small portions during 10 min., 48.75 g. (183 mmoles) of inosine (**1**). Stirring is continued for 30 min. at 0° and 8 hr. at room temperature. The mixture is poured into 1.1 liters of ethanol, the resulting precipitate of sodium iodate is filtered off and washed with cold ethanol, and the filtrate and washings are combined, and concentrated under diminished pressure at 30–35° to *ca.* 100 ml. Ethanol (600 ml.) is added, and traces of inorganic material are removed by filtration. The solution is evaporated to dryness under diminished pressure at 30–35° and dried at 1 mm. Hg, giving 46 g. (96%) of **2** as a solid mass, m.p. 110–125°.

9-(3-Deoxy-3-nitro-β-D-galactopyranosyl)hypoxanthine (3)

"Inosine dialdehyde" (**2**) (46 g.) is suspended in a magnetically stirred mixture of 250 ml. of methanol, 80 ml. of water, and 19.4 ml. (0. 35 mole), of nitromethane cooled to −10°, and 194 ml. of *M* methanolic sodium methoxide is added dropwise. Stirring is continued for 8 hr. at 0–5°, 25 ml. of nitromethane is added, and the mixture is stirred for an additional hr. The solution is deionized with a strongly acidic, ion-exchange resin (Merck I), which, after removal, is thoroughly washed with 1:1 methanol–water. The filtrate and washings are combined and evaporated under diminished pressure at 35–40°. The resulting crystalline mass is filtered off, giving 40.95 g. (66%, based on **1**) of a mixture of the isomers of (3-*C*-nitrohexosyl)hypoxanthine[a] that is suspended in 300 ml. of boiling methanol and dissolved by the addition of a few ml. of water; the solution is decolorized by treatment with charcoal, and cooled. The resulting, colorless needles are filtered off after 12 hr.; two similar recrystallizations from methanol–water give 11.8 g. (19%, based on **1**) of chromatographically[b] pure **3**, crystallizing with 0.5 molecule of methanol, m.p. 195–197° (dec.), $[\alpha]_D^{25}$ +28.5° (*c* 0.4, water), λ_{max}^{MeOH} 244 nm.

9-(3-Amino-3-deoxy-β-D-galactopyranosyl)hypoxanthine Acetate (4)

To a prehydrogenated suspension of 8 g. of 10% palladium-on-charcoal in a mixture of 215 ml. of water, 70 ml. of methanol, and 1.8 ml. of glacial acetic acid is added 10.0 g. (29.1 mmole) of **3**, and the hydrogenation is

[a] This is a *ca.* 1:1 mixture consisting of **3** and the corresponding *gluco* isomer. The latter may be isolated, by repeated fractional recrystallization,[1] in a yield of 15%.

[b] Good results are obtained with 0.2-mm layers of "Kieselgel PF_{254}" (E. Merck AG, Darmstadt) and 15:2:1 ethyl acetate–ethanol–water.

continued. After the uptake of 2 l. of hydrogen (4–5 hr.), the catalyst is filtered off, and thoroughly washed with 50% aqueous methanol. The filtrates are combined, and evaporated to dryness under diminished pressure. The residue is dissolved in a small volume of water, methanol is slowly added, and the resulting precipitate is filtered off. Addition of ethyl acetate to the mother liquor affords a second crop; total yield 9.5 g. (86%) of **4** as the monohydrate, m.p. 251–253°, $[\alpha]_D^{25}$ +34° (*c* 0.4, water), $\lambda_{max}^{H_2O}$ 248 nm.

REFERENCES

(1) F. W. Lichtenthaler and H. P. Albrecht, *Chem. Ber.*, **99**, 575 (1966).
(2) F. W. Lichtenthaler, P. Emig, and D. Bommer, *Chem. Ber.*, **101**, in press (1968).

[77] 9-(2-Acetamido-3,5,6-tri-*O*-acetyl-2-deoxy-D-glucosyl)-2,6-dichloro-9*H*-purine

Synthesis of a Nucleoside Derivative through an Oxazoline by Acid-catalyzed Fusion

M. L. WOLFROM, M. W. WINKLEY, and P. McWAIN

DEPARTMENT OF CHEMISTRY, THE OHIO STATE UNIVERSITY, COLUMBUS, OHIO 43210

AcOCH$_2$ / AcOCH / OAc / SEt / HNAc — 1. Cl_2, 2. Ag_2CO_3 → AcOCH$_2$ / AcOCH / OAc / O–C(CH$_3$)=N — **3** (189.0), H$^+$ → **4**

Cl, N, N, Cl, N, N, H

1 (391.4) **2** (329.3) **4** (518.3)

where Ac is CH_3CO.

INTRODUCTION

The utilization of the 2-phenyl-2-oxazoline derivative of 2-amino-2-deoxy-D-glucose in the synthesis of glycosides[1,2] suggested the potential of oxazolines for nucleoside synthesis. A 2-methyl-2-oxazoline derivative of the acetylated furanose form of 2-amino-2-deoxy-D-glucose (**2**) was obtained[3] by the successive action of chlorine and silver carbonate on ethyl 2-acetamido-3,5,6-tri-*O*-acetyl-2-deoxy-1-thio-α-D-glucoside (**1**).[4] This oxazoline derivative was the first representative of such an oxazoline containing an aliphatic methyl group attached to the ring. Reaction of **2** with 2,6-dichloropurine (**3**), in a fusion catalyzed by *p*-toluenesulfonic acid, led to the formation of 9-(2-acetamido-3,5,6-tri-*O*-acetyl-2-deoxy-D-glucosyl)-2,6-dichloro-9*H*-purine (**4**).[3] (Such fusion techniques had been used in the synthesis of phenolic glycosides[5] of sugars, including 2-amino-2-deoxy sugars.[6]) The anomeric nature of the product (**4**) was not definitively established, although its mode of formation would indicate that it is probably β-D. Attempts to remove the *O*-acetyl groups and to replace the chlorine atom in this nucleoside derivative did not lead to crystalline products, and a purine nucleoside of D-glucofuranose was synthesized by the same workers[7] by another route. Nevertheless, the reaction indicates a potential for the future employment of oxazoline derivatives of amino sugars in the synthesis of nucleosides.

PROCEDURE

3′,5′,6′-Tri-*O*-acetyl-2-methyl-α-D-gluco[2′,1′:4,5]-2-oxazoline[3] (2)

Dry chlorine is passed for 10 min. at 0° into a solution of **1** (10 g., 25.6 mmoles)[4] in dichloromethane (100 ml., predried over Drierite), protected from moisture. The solution is evaporated to dryness at 30°, and the residue is dissolved in dichloromethane (100 ml.). To this solution is added excess silver carbonate (15 g.), portionwise, with stirring and cooling in ice and water. After effervescence has largely ceased, the mixture is stirred for a further 30 min. at room temperature. The silver salts are removed by centrifugation and by repeated filtration through sintered glass. The resulting clear, yellow solution is evaporated at 30° to a sirup which is crystallized from ether; yield 4.73 g. (56%), m.p. 84–89°. Several recrystallizations produce pure material; m.p. 96–97°, $[\alpha]_D^{20}$ +25 ± 1° (*c* 6.96, chloroform). The compound is unstable in the air at room temperature, but, if kept dry, can be stored for several weeks.

9-(2-Acetamido-3,5,6-tri-*O*-acetyl-2-deoxy-D-glucosyl)-2,6-dichloro-9*H*-purine[3] (4)

Finely ground **2** (4.73 g., 14.3 mmoles) and 2,6-dichloropurine (**3**) (2.30 g.) are mixed, and fused at 110–120°. *p*-Toluenesulfonic acid (60 mg.) is added, and the melt is well stirred and heated for 10 min. at 110–120°. The cooled melt is extracted with chloroform, and the extract is filtered. The filtrate is washed successively with cold, saturated, aqueous sodium hydrogen carbonate solution and water, and dried (magnesium sulfate). The residue obtained on removal of solvent at 40° is dissolved in methanol, and the solution is decolorized with activated carbon. The solution is evaporated to a sirup, which is crystallized from methanol–ether; yield 0.75 g. (10%), m.p. 148–150°. Further recrystallizations produce pure material; m.p. 152–153°, $[\alpha]_D^{25}$ +7 ± 1° (*c* 1.56, methanol).

REFERENCES

(1) F. Micheel and H. Köchling, *Chem. Ber.*, **90**, 1597 (1957); **91**, 673 (1958); **93**, 2372 (1960).

(2) S. Konstas, I. Photaki, and L. Zervas, *Chem. Ber.*, **92**, 1288 (1959).
(3) M. L. Wolfrom and M. W. Winkley, *J. Org. Chem.*, **31**, 3711 (1966).
(4) M. L. Wolfrom, S. M. Olin, and W. J. Polglase, *J. Am. Chem. Soc.*, **72**, 1724 (1950); M. L. Wolfrom and M. W. Winkley, *J. Org. Chem.*, **31**, 1169 (1966).
(5) B. Helferich and E. Schmitz-Hillebrecht, *Ber.*, **66**, 378 (1933).
(6) S. Fujisa and K. Yokoyama, *Nippon Kagaku Zasshi*, **72**, 728 (1951); *Chem. Abstr.*, **46**, 11116 (1952).
(7) M. L. Wolfrom and M. W. Winkley, *Chem. Commun.*, **1966**, 533.

[78] 2-Amino-6-chloro-9-(2,3,5-tri-*O*-acetyl-β-D-ribosyl)-9*H*-purine

Replacement of the 6-Keto Group of Guanosine

JOHN F. GERSTER, ARTHUR F. LEWIS, and ROLAND K. ROBINS

DEPARTMENT OF CHEMISTRY, UNIVERSITY OF UTAH, SALT LAKE CITY, UTAH 84112

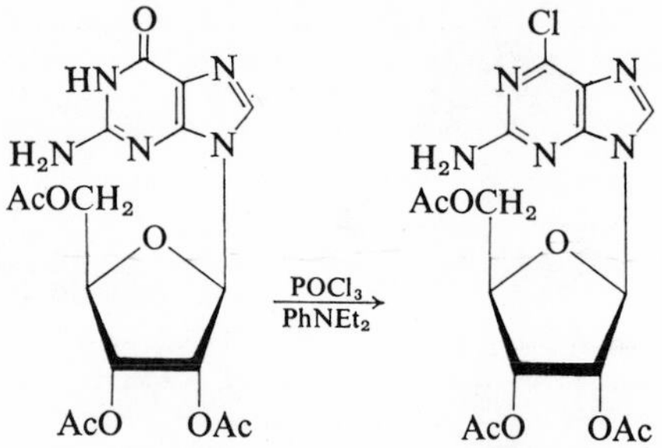

1 (409.4) $\xrightarrow[PhNEt_2]{POCl_3}$ 2 (428.9)

INTRODUCTION

The preparation of 2-amino-6-chloro-9-β-D-ribofuranosyl-9*H*-purine has been accomplished by treating 2-amino-6-(methylthio)-9-β-D-ribofuranosyl-9*H*-purine or 2-amino-9-β-D-ribofuranosyl-9*H*-purine-6(1*H*)-thione with chlorine in methanol.[1,2] A more convenient procedure involves the treatment of 2′,3′,5′-tri-*O*-acetylguanosine (**1**) with phosphoryl chloride in the presence of *N*,*N*-diethylaniline,[2] to give 2-amino-6-chloro-9-(2,3,5-tri-*O*-acetyl-β-D-ribosyl)-9*H*-purine (**2**) (Sec. III [79]); deacetylation of **2** with methanolic ammonia affords[2] over 90% of 2-amino-6-chloro-9-β-D-ribofuranosyl-9*H*-purine. Compound **2** is readily available, and serves as a convenient intermediate for various reactions involving nucleophilic substitution and as the starting material for the preparation of other important purine nucleosides.[2,3]

PROCEDURE

2-Amino-6-chloro-9-(2,3,5-tri-*O*-acetyl-β-D-ribosyl)-9*H*-purine (2)

2′,3′,5′-Tri-*O*-acetylguanosine[a] (**1**) (100 g., 0.244 mole) is added to a mixture of 750 ml. of phosphoryl chloride and 38 ml. of *N*,*N*-diethylaniline, and the flask is swirled to mix the material thoroughly. The mixture is then refluxed for 3 min.,[b,c] the vacuum from a water aspirator is applied immediately,[d,e] and *ca.* 500 ml. of phosphoryl chloride is distilled off. The solution is then slowly poured over crushed ice, with efficient stirring,[f] and the cold, aqueous solution is stirred for 10–15 min. (to ensure that all of the phosphoryl chloride has been hydrolyzed). The solution is extracted with five 200-ml. portions of dichloromethane, and the extracts are combined, washed with 1-l. portions of cold water until the aqueous wash is neutral, dried overnight (anhydrous sodium sulfate), and concentrated under diminished pressure below 30° to 300–500 ml.[g,h] The solution is slowly poured, with efficient stirring, into 4 l. of anhydrous ether, giving a light-yellow solid that is filtered off, washed with anhydrous ether, and air-dried; yield 51.5 g., m.p. 145–148°.

An additional 15.6 g. (m.p. 145–148°) is obtained by concentrating[h] the combined filtrates to *ca.* 100 ml.; total yield 67.1 g. (63.5%). This material is sufficiently pure for most synthetic purposes, but must be used soon after its preparation, as it slowly decomposes on long standing unless further purified.

REFERENCES

(1) R. K. Robins, *J. Am. Chem. Soc.*, **82**, 2654 (1960).
(2) J. F. Gerster, J. W. Jones, and R. K. Robins, *J. Org. Chem.*, **28**, 945 (1963).
(3) J. F. Gerster and R. K. Robins, *J. Am. Chem. Soc.*, **87**, 3752 (1965); *J. Org. Chem.*, **31**, 3258 (1966).

[a] Commercially available from C. F. Boehringer u. Soehne, Mannheim-Waldhof, Germany.

[b] The reaction is conducted in a 1-l., single-necked, round-bottomed flask, assembled for vacuum distillation through a short still-head before heating is begun. The mixture is heated with a heating mantle, and the refluxing time is counted from the time when liquid starts refluxing in the lower part of the still-head.

[c] All of the solid has usually dissolved at the end of the 3-min. period of refluxing. In those cases where it has not all dissolved, refluxing is continued (within reason) until the solid has all dissolved.

[d] The vacuum is at first applied judiciously, to avoid loss of product because of foaming. However, the phosphoryl chloride is removed as rapidly as possible, the heating mantle being used for heating. The heating mantle is not shut off, or turned down, until the excess phosphoryl chloride has been removed.

[e] A cold trap must be used between the receiving flask and the water aspirator, and both the trap and the receiver must be well cooled.

[f] An excess of ice must always be present.

[g] The amount of material obtained after the solution in dichloromethane has been poured into ether is somewhat dependent on the extent to which the dichloromethane has been evaporated off.

[h] A rotary evaporator is used for this evaporation.

[79] 2-Amino-9-β-D-ribofuranosyl-9*H*-purine

Catalytic Dehalogenation of a Nucleoside of a Chlorinated Purine

RICHARD S. VICKERS, JOHN F. GERSTER, and ROLAND K. ROBINS

DEPARTMENT OF CHEMISTRY, UNIVERSITY OF UTAH, SALT LAKE CITY, UTAH 84112

Cl
H_2N AcOCH$_2$ O Pd/C → H_2N AcOCH$_2$ O NH_3/MeOH → H_2N HOCH$_2$ O

AcO OAc — AcO OAc — HO OH

1 (427.5) — **2** (393.0) — **3** (267.0)

INTRODUCTION

2-Aminopurine is incorporated into the 2′-deoxyribonucleic acid of certain microorganisms,[1] and, as is well known, causes subsequent mutations.[2] 2-Amino-9-β-D-ribofuranosyl-9*H*-purine has been prepared by treatment of 2-amino-9-β-D-ribofuranosyl-9*H*-purine-6(1*H*)-thione with Raney nickel,[3] and by the reaction of chloromercuri-2-benzamidopurine with 2,3,5-tri-*O*-benzoyl-D-ribosyl chloride.[4] In our hands, the procedure described here gives the best results. 2-Amino-6-chloro-9-(2,3,5-tri-*O*-acetyl-β-D-ribosyl)-9*H*-purine[5] (**1**) (Sec. III [78]) is treated with palladium-on-carbon to give 2-amino-9-(2,3,5-tri-*O*-acetyl-β-D-ribosyl)-9*H*-purine (**2**), and this is readily deacetylated with methanolic ammonia to 2-amino-9-β-D-ribofuranosyl-9*H*-purine (**3**) in excellent yield.

PROCEDURE

2-Amino-9-β-D-ribofuranosyl-9*H*-purine (3)

To 100 ml. of absolute ethanol are added 250 mg. of 10% palladium-on-carbon catalyst, 3.0 g. (7 mmoles) of[5] **1**, and 1.2 g. of anhydrous sodium acetate in the pressure bottle of a Parr hydrogenation apparatus. The mixture is shaken at 40 lb./in.^{2} of hydrogen for 10 hr., and then warmed to 50°, and filtered through a Celite pad. The filtrate is evaporated to dryness under diminished pressure, the resulting solid is dissolved in 200 ml. of water, and the solution is extracted with four 50-ml. portions of ethyl acetate. The extracts are combined, washed with 25 ml. of water, dried (anhydrous sodium sulfate), and evaporated to dryness under diminished pressure, giving a pale sirup. This is dissolved in 50 ml. of hot isopropyl alcohol, and the solution is cooled; yield 2.1 g. (76.5%) of crystalline **2**, m.p. 142–143°.

Treatment of **2** with excess methanolic ammonia at room temperature, followed by evaporation to dryness, affords compound **3** as a glass; this is crystallized from absolute ethanol; yield $>80\%$. The product is identical with that prepared by the method of Fox and co-workers.[3]

REFERENCES

(1) H. Gottschling and E. Freese, *Z. Naturforsch.*, **16b**, 515 (1961).

(2) E. Freese, *Proc. Natl. Acad. Sci. U. S.*, **45**, 622 (1959); M. Demerec, *ibid.*, **46**, 1075 (1960); R. Rudner, *Biochem. Biophys. Res. Commun.*, **3**, 275 (1960).

(3) J. J. Fox, I. Wempen, A. Hampton, and I. L. Doerr, *J. Am. Chem. Soc.*, **80**, 1669 (1958).

(4) H. J. Schaeffer and H. J. Thomas, *J. Am. Chem. Soc.*, **80**, 4896 (1958).

(5) J. F. Gerster, J. W. Jones, and R. K. Robins, *J. Org. Chem.*, **28**, 945 (1963).

[80] 9-(2-Deoxy-β-D-*erythro*-pentofuranosyl)-9*H*-purine and 9-(2-Deoxy-α-D-*erythro*-pentofuranosyl)-9*H*-purine

The Fusion Synthesis Applied to the Preparation of 2-Deoxy-D-ribofuranosylpurines

MORRIS J. ROBINS and ROLAND K. ROBINS

DEPARTMENT OF CHEMISTRY, UNIVERSITY OF UTAH, SALT LAKE CITY, UTAH 84112

$AcOCH_2$ … OAc; AcO

$\xrightarrow[\Delta]{ClCH_2CO_2H}$ $\xrightarrow[MeOH]{NH_3}$

1 (120.1) + **2** (260.2)

$HOCH_2$ … HO

3 (236.2) + **4** (236.2)

INTRODUCTION

The alleged biological activity of certain purine nucleosides has been found to be due to contaminating mercuric ions in concentrations as low as 10 nanomolar.[1] The fusion synthesis of 2′-deoxyribofuranosyl-nucleosides[2] avoids the often-tedious preparation of heavy-metal salts of purines, and this potential source of contamination. The use of the stable 1,3,5-tri-*O*-acetyl-2-deoxy-D-*erythro*-pentose (**2**) (Sec. V [157]) also obviates the preparation and use of labile, acylated 2-deoxypentofuranosyl halides. 9-(2-Deoxy-β-D-*erythro*-pentofuranosyl)-9*H*-purine[2,3] (**3**) is the 2′-deoxy analog of the antibiotic Nebularine.[4] The fusion synthesis of this 2′-deoxynucleoside and its α-D anomer[2,3] (**4**) is representative of the method.

PROCEDURE

9-(2-Deoxy-α-D-*erythro*-pentofuranosyl)-9*H*-purine (4) and its β-D Anomer (3)

Finely powdered purine[5] (**1**) (2.77 g., 0.023 mole) and 12 g. (0.046 mole) of 1,3,5-tri-*O*-acetyl-2-deoxy-D-*erythro*-pentose (**2**) (Sec. V [157]) are well mixed in a 25-ml., round-bottomed flask, and the mixture is heated in an oil bath preheated to 145°. Chloroacetic acid (75 mg.) is added, and stirred into the melt. An efficient aspirator[a] is connected to the flask, and the fusion is continued for 18 min. The flask is removed from the bath, and cooled to *ca.* 100°, and the clear, brown melt is dissolved in ethyl acetate; the solution is washed successively with two 30-ml. portions of ice-cold, saturated aqueous sodium carbonate and ice-water (to pH 7), dried (sodium sulfate), and filtered through a carbon–Celite bed. The filtrate is evaporated to dryness under diminished pressure, the resulting sirup is dissolved in 20 ml. of absolute ethanol, and the solution is treated with 100 ml. of absolute ethanol presaturated with ammonia at −10°. The solution is kept at room temperature overnight, and evaporated to a tan sirup that is dissolved in a small volume of hot methanol. The solution is cooled at −18°, and the tan-colored crystals (1.7 g., 31%) that separate are filtered off, and dried. Three recrystallizations from methanol give 1 g. (18%) of compound **3**, m.p. 181–182°, $[\alpha]_D^{25.5}$ −29.3° (*c* 1.06, water), $\lambda_{max}^{H_2O}$ 262.5 nm. (ε 7,320).

The first mother liquor is evaporated to a sirup, which is dissolved in 20 ml. of methanol. The solution is applied to a column (3.5 × 8 cm.) of neutral alumina (packed in methanol), and the column is eluted with 500 ml. of methanol. The eluate is concentrated under diminished pressure to 50 ml., and the solution is allowed to evaporate slowly. Colorless needles (1.13 g., 21%) of the α-D anomer (**4**) are collected by filtration. Recrystallization from acetonitrile affords 0.85 g. (16%) of the pure product, m.p. 135–136°, $[\alpha]_D^{26}$ +73.4° (*c* 1.09, water), $\lambda_{max}^{H_2O}$ 262.5 nm. (ε 7,800).

REFERENCES

(1) J. Škoda, I. Bartošek, and F. Šorm, *Collection Czech. Chem. Commun.*, **27**, 906 (1962); J. J. K. Novak and F. Šorm, *ibid.*, **27**, 902 (1962).

[a] To remove acetic acid evolved.

(2) M. J. Robins, W. A. Bowles, and R. K. Robins, *J. Am. Chem. Soc.*, **86**, 1251 (1964); M. J. Robins and R. K. Robins, *ibid.*, **87**, 4934 (1965).
(3) R. H. Iwamoto, E. M. Acton, and L. Goodman, *J. Org. Chem.*, **27**, 3949 (1962).
(4) N. Lofgren and B. Luning, *Acta Chem. Scand.*, **7**, 225 (1953).
(5) A. G. Beaman, *J. Am. Chem. Soc.*, **76**, 5633 (1954).

[81] 2,6-Diamino-9-β-D-ribofuranosyl-9H-purine

The Preparation of a Nucleoside Intermediate

H. JEANETTE THOMAS, JAMES A. JOHNSON, JR.,
WILLIAM E. FITZGIBBON, JR., SARAH J. CLAYTON, and
B. R. BAKER

KETTERING-MEYER LABORATORY, SOUTHERN RESEARCH INSTITUTE,
BIRMINGHAM, ALABAMA 35205

1 (150.2) $\xrightarrow{(C_6H_5CO)_2O}$ 2 (358.4)

2 $\xrightarrow{HgCl_2,\ NaOH}$ 3 (593.4)

4 (504.5) $\xrightarrow{HCl}$ 5 (480.9)

3 + 5 $\rightarrow$ 6 (802.8) $\xrightarrow{NaOMe}$ 7 (282.3)

where Ac is acetyl and Bz is benzoyl.

INTRODUCTION

2,6-Diamino-9-β-D-ribofuranosyl-9*H*-purine[1] (7) is useful as an intermediate in the preparation of such other nucleosides as 2-fluoroadenosine and crotonoside. Its preparation is a good example of the preparation of nucleosides by the coupling of purine and sugar moieties. This work is on a much larger scale, and the yield is somewhat higher, than reported in the original preparation.[1]

PROCEDURE

2,6-Bis(benzamido)purine[1] (2)

A dry mixture of 60 g. (0.4 mole) of 2,6-diaminopurine (**1**)[a] and 600 g. (2.56 moles) of benzoic anhydride[b] is placed in a 12-l., 3-necked, round-bottomed flask fitted with a thermometer, an efficient stirrer, and a heating mantle. The mixture is stirred vigorously, and heated until a melt is obtained. The temperature is rapidly raised to 180° with continuous stirring, and held there for 15 min. The mixture is then cooled to 90°, and 5.4 l. of absolute ethanol is added. The mixture (containing suspended crystals) is stirred and refluxed for 10 min., kept at 0–5° for 1 hr., and filtered. The solid is collected and suspended in 3.7 l. of fresh, absolute ethanol, and the suspension is stirred and refluxed for 1 hr., refrigerated overnight, and filtered. The crystalline 2,6-bis(benzamido)purine (**2**), after being dried *in vacuo* over phosphorus pentaoxide to constant weight, weighs 114 g. (80%), m.p. 320°.

2,6-Bis(benzamido)-*x*-(chloromercuri)purine[1] (3)

To a solution of 81.4 g. (0.3 mole) of mercuric chloride in 3.3 l. of ethanol in a 12-l., 3-necked flask (fitted with a reflux condenser, mechanical stirrer, addition funnel, and heating mantle) is added 107.5 g. (0.3 mole) of 2,6-bis(benzamido)purine (**2**), 3.3 l. of water, and 178 g. of Celite (analytical grade). The suspension is vigorously stirred, and heated under reflux for 1 hr.; then a solution of 12 g. (0.3 mole) of sodium hydroxide in 100 ml. of water is slowly added at such a rate that no permanent yellow color develops. Heating and stirring are continued for 1 hr. after completion of the addition. The mixture is thoroughly chilled in an ice-bath, and filtered. The white filter-cake is thoroughly washed with water until free from chloride ion, and then with ethyl alcohol, and dried at room temperature *in vacuo* over phosphorus pentaoxide to constant weight. The gross

[a] Procurable from Nutritional Biochemicals Corporation, Cleveland, Ohio 44128.
[b] Gross contamination with benzoic acid gives poor yields.

weight of the Celite-containing solid is 346 g. and the net weight of product **3** is 168 g. (94%).

2,3,5-Tri-*O*-benzoyl-D-ribosyl Chloride[2] (5)

Anhydrous hydrogen chloride is bubbled into anhydrous ether at 4° until the ether is completely saturated.[c] To the cold solution is added 100.9 g. (0.2 mole) of 1-*O*-acetyl-2,3,5-tri-*O*-benzoyl-β-D-ribose[3] (**4**). Dissolution is complete after stirring for about 2 min.[d]; then 101 ml. of acetyl chloride is added. The flask is securely stoppered and the solution is kept at 0–4° for at least 72 hr.[e]; it is then used in the next step.

2,6-Bis(benzamido)-9-(2,3,5-tri-*O*-benzoyl-β-D-ribosyl)-9*H*-purine (6)

The ether is removed from the solution of 96 g. (0.2 mole) of 2,3,5-tri-*O*-benzoyl-D-ribosyl chloride (**5**) under diminished pressure at 20°, with exclusion of moisture. Residual acetic acid is removed from the sirupy residue by co-distillation *in vacuo* at 20° with two successive 200-ml. portions of dry toluene,[f] and the thick, almost colorless sirup is dissolved in 500 ml. of dry xylene. (This procedure must be conducted on the day that the following coupling is to be performed.) A suspension of 245 g. of 2,6-bis(benzamido)-*x*-(chloromercuri)purine (**3**)–Celite [containing 118.7 g. (0.2 mole) of **3**] in 3.5 l. of xylene is placed in a 5-l., 3-necked flask fitted with an addition funnel, a mechanical stirrer, a condenser set for distillation, and a heating mantle. The mixture is rapidly stirred, and xylene is distilled off until all traces of moisture have been removed. The condenser is then replaced by a dry condenser (drying tube) set for reflux. To the rapidly stirred, refluxing solution is added the previously prepared solution of chloride **5** in xylene, and refluxing and stirring are continued for 3 hr. The suspension is then filtered, and the filtrate is evaporated *in vacuo* to a thick, yellow-brown sirup.[g] The filter cake is extracted with six 550-ml. portions of boiling chloroform, and the extracts are combined and used for dissolving the residual sirup. The amber solution is successively washed with two 825-ml. portions of 30% aqueous potassium iodide and two 800-ml. portions of water, and dried (magnesium sulfate). The

[c] Although the ether is in an ice-bath, the heat of dissolution causes the temperature to rise. When saturation is complete, the temperature drops below 4°, even though hydrogen chloride is added rapidly.

[d] If the acylated sugar does not dissolve completely, the ether had not been saturated, and the yield will be decreased.

[e] The ether solution may be stored for several weeks at 0° without deterioration.

[f] If there is still an odor of acetic acid, the treatment is repeated until no odor is detectable.

[g] If there is insufficient time to filter the mixture and complete the extraction with chloroform, the mixture may be allowed to stand overnight at room temperature.

solution is evaporated *in vacuo* to a thick sirup, which is used, without purification, in the subsequent step.

2,6-Diamino-9-β-D-ribofuranosyl-9*H*-purine[1] **(7)**

The sirup containing **6** is dissolved in 2 l. of anhydrous methanol in a 5-l., 3-necked flask equipped with a heating mantle, a mechanical stirrer, and a reflux condenser bearing a drying tube (Drierite). A solution of 14.1 g. (0.26 mole) of sodium methoxide in 1 l. of anhydrous methanol is added, and the solution is stirred and refluxed for 18 hr., and cooled. The pH is adjusted to 7 with glacial acetic acid, the methanol is removed under diminished pressure, and the resulting sirup is dissolved in 3 l. of warm water. The solution is washed with five 500-ml. portions of chloroform, treated with decolorizing charcoal, filtered, and concentrated *in vacuo* to approximately 1.5 l., at which point crystallization occurs. The mixture is kept at 0–5° overnight, and then filtered, giving a first crop which weighs 19.7 g., m.p. 238–240°. The filtrate is concentrated to approximately 700 ml. and kept at 0–5° overnight, affording a second crop, wt. 10.7 g., m.p. 238–240°. The two crops are combined, and recrystallized from water (with charcoal treatment); yield 26.2 g. (46.5%), m.p. 242–244°.[h] After a second recrystallization, the product **7** is colorless and weighs 24.9 g. (44%), m.p. 242–244°, $\lambda_{max}^{pH\ 6.7}$ 255 nm. (log ε 3.97), 280 nm. (log ε 4.00).

REFERENCES

(1) J. Davoll and B. A. Lowy, *J. Am. Chem. Soc.*, **73**, 1650 (1951).
(2) N. Yung and J. J. Fox, *Methods Carbohyd. Chem.*, **2**, 109 (1963).
(3) E. F. Recondo and H. Rinderknecht, *Helv. Chim. Acta*, **42**, 1171 (1959).

[h] If the material is colorless and chromatographically homogeneous, the second recrystallization may be omitted.

[82] 6-(Dimethylamino)-9-β-D-hexopyranosyl-9*H*-purines

The Dialdehyde–Nitromethane Cyclization Applied to Dialdehydes from Purine Nucleosides

FRIEDER W. LICHTENTHALER and
HANS PETER ALBRECHT

INSTITUT FÜR ORGANISCHE CHEMIE, TECHNISCHE HOCHSCHULE
DARMSTADT, 61 DARMSTADT, GERMANY

INTRODUCTION

Purine nucleosides, derived from amino hexoses, may be prepared by adaptation of the Hilbert–Johnson or the mercury procedure to poly-*O*-acyl-aminoglycosyl halides, properly protected at the amino group.[1] An alternative procedure leading to (3-amino-3-deoxyhexopyranosyl)purines is the application of the three-step sequence of periodate oxidation → nitromethane cyclization[2] → hydrogenation, either to pentofuranosyl- or hexopyranosyl-purines. Stereochemical preference in the cyclization step gives rise to three isomers only: the all-equatorial, D-*gluco* compound (major product) and, as minor components, the D-*galacto* and D-*manno* compounds, each possessing one axial hydroxyl group. The separation of the individual isomers may be accomplished at the nitro stage by fractional recrystallization,[3–6] by thin-layer chromatography,[7] or, after hydrogenation of the nitrohexoside mixture, by column chromatography of the aminonucleosides on an ion-exchange resin.[8]

That the reaction has considerable utility in the purine nucleoside field has been demonstrated by the synthesis of the 9-β-D-gluco-, -galacto-, and -manno-pyranosyl nucleosides of adenine,[8] hypoxanthine,[4] and 6-(dimethylamino)purine,[6] and the 7-substituted derivatives of theophylline.[5]

PROCEDURE

2-*O*-[(R)-Formyl-(*N*,*N*-dimethyladenin-9-yl)methyl]-(R)-glyceraldehyde (2)

To a magnetically stirred, ice-cooled solution of 7.90 g. (37 mmoles) of sodium metaperiodate in 200 ml. of water is added 11.0 g. (37 mmoles) of *N*,*N*-dimethyladenosine[9] (**1**) (Sec. III [64]) in small portions during 10 min. Stirring is continued at room temperature for 3 hr., and the

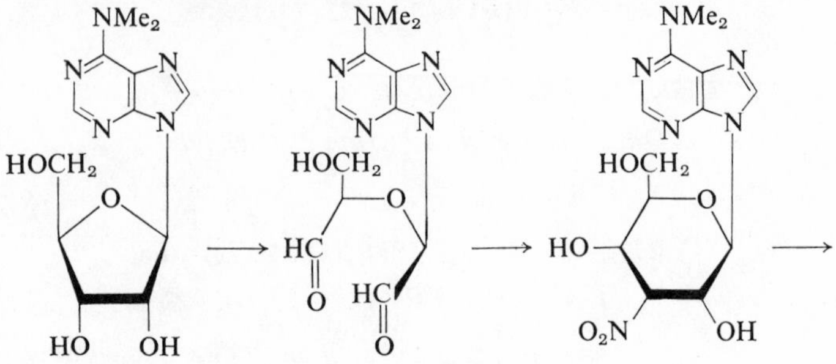

1 (295.3) → **2** → **3** →

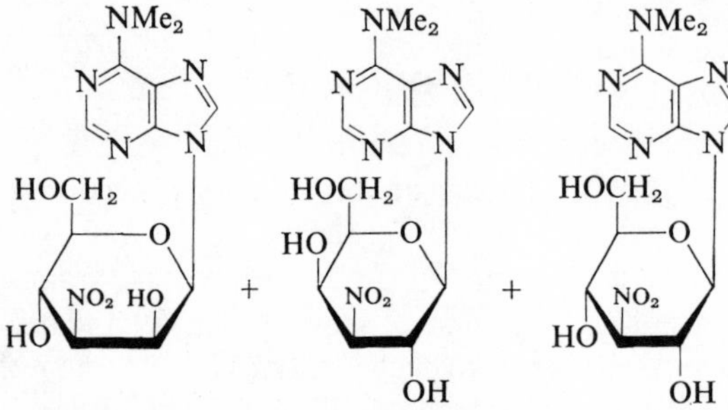

4 (354.3) + **5** (354.3) + **6** (354.3)

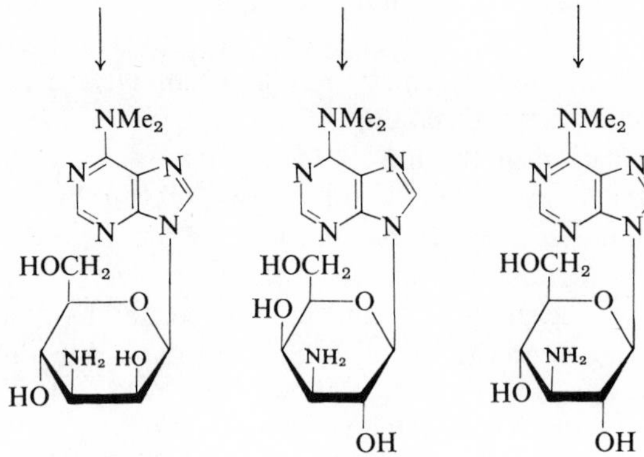

↓ ↓ ↓

7 (324.3) **8** (324.3) **9** (324.3)

solution is concentrated *in vacuo* at 40° to about 100 ml. Addition of methanol (300 ml.) precipitates most of the sodium iodate formed; this is removed by filtration. After concentration to about 50 ml., the filtrate is again treated with 300 ml. of methanol, the suspension is filtered, and the filtrate is evaporated to dryness *in vacuo* at 35°, giving 11.6 g. of a solid residue.

Nitromethane Cyclization of Dialdehyde 2

The solid **2** (11.6 g.) from the previous experiment is dissolved in 400 ml. of methanol, nitromethane (2.1 ml., 39 mmoles) is added, and the solution is magnetically stirred, with ice-cooling. *M* Sodium methoxide in methanol (37 ml.) is added dropwise during 30 min., and stirring is continued at 0° for 3 hr. and at room temperature for 1 hr. The solution is deionized with a weakly acidic, ion-exchange resin (Merck IV); after removal, this is thoroughly washed with 1:1 methanol–water. The filtrate and washings are combined, and evaporated to dryness *in vacuo* at about 40°. The residue crystallizes upon treatment with water; the mixture is kept overnight in a refrigerator, and the crystals are collected by suction filtration, and dried *in vacuo* at room temperature, giving 9.8 g. (75%, based on **1**) of an isomeric mixture (**3**), consisting of **4**, **5**, and **6**.

9-(3-Deoxy-3-nitro-β-D-mannopyranosyl)-*N*,*N*-dimethyladenine (4)

The isomeric mixture **3** is recrystallized from 3:2 methanol–water, resulting in a methanol-insoluble precipitate that is kept for 4 hr. and filtered off (giving filtrate A). The crude product obtained (2 g.) is slightly contaminated with **5** and[a] **6**; it is chromatographically pure after two further recrystallizations from methanol; yield 1.5 g. (15% based on **3**), m.p. 152–154° (dec.), $[\alpha]_D^{20}$ +84° (*c* 0.25, methanol).

9-(3-Deoxy-3-nitro-β-D-galactopyranosyl)-*N*,*N*-dimethyladenine (5)

Filtrate A, remaining after isolation of the *manno* isomer **4**, is concentrated to 60 ml. and kept overnight in a refrigerator. A crystalline product [consisting mainly[a] of **5**, with some of the *manno* (**4**) and *gluco* (**6**) isomers] separates and is filtered off (giving filtrate B). The crystals (3.1 g.) are dissolved in the minimal volume of hot methanol, and the solution is allowed to cool to room temperature. After 1 hr., the precipitate (1.1 g.)[b] is removed, and the filtrate is kept in a refrigerator for 12 hr. Suction filtration gives 1.8 g. of crude **5**, which is chromatographically pure after two recrystallizations from methanol; yield 800 mg. (8% based on the isomeric mixture **3**), m.p. 158–163° (dec.), $[\alpha]_D^{20}$ −1° (*c* 0.5, methanol).

[a] It is advisable to follow the separation of the individual isomers by thin-layer chromatography. Satisfactory results are obtained with 0.2-mm. layers of Kieselgel $HF_{254+366}$ (E. Merck A.G., Darmstadt) and 55:16:5 butyl acetate–acetic acid–water. After three consecutive developments on a 20-cm. plate, the nitronucleosides are separated by about 1 cm., and appear in the order **5**, **4**, and **6**, from the starting point.

[b] This product, consisting of an approximately 1:1 mixture of **4** and **5**, cannot be purified by further recrystallization.

9-(3-Deoxy-3-nitro-β-D-glucopyranosyl)-*N*,*N*-dimethyladenine (6)

Filtrate B, remaining after isolation of **4** and **5**, contains mainly[a] **6**. It is kept for 3 days in a refrigerator, and the resultant precipitate is collected by suction filtration and recrystallized twice from methanol, the solution being cooled each time for 24 hr. in a refrigerator; yield 2.3 g. (23%, based on **3**), m.p. 202–204° (dec.), $[\alpha]_D^{20}$ −13° (*c* 0.6, methanol).

9-(3-Amino-3-deoxy-β-D-mannopyranosyl)-*N*,*N*-dimethyladenine (7)

A suspension of 500 mg. of 10% palladium-on-charcoal in 300 ml. of methanol is prehydrogenated. Compound **4** (500 mg.) is added, and the hydrogenation is continued; after 6 hr., 110 ml. of hydrogen (calc., 106 ml.) has been absorbed. The catalyst is removed, and washed with methanol, and the filtrate and washings are combined, and evaporated to dryness *in vacuo*. The residue is re-evaporated with absolute ethanol, giving a colorless solid which is suspended in ethanol and filtered off; yield 390 mg. (85%), m.p. 135–137°, $[\alpha]_D^{20}$ +47° (*c* 0.45, methanol).

9-(3-Amino-3-deoxy-β-D-galactopyranosyl)-*N*,*N*-dimethyladenine (8)

Hydrogenation of 350 mg. of **5** under the conditions described for the preparation of **7** gives 290 mg. (90%) of **8**, m.p. 148–150°, $[\alpha]_D^{20}$ +26° (*c* 0.45, methanol).

9-(3-Amino-3-deoxy-β-D-glucopyranosyl)-*N*,*N*-dimethyladenine (9)

To a prehydrogenated suspension of 2.0 g. of 10% palladium-on-charcoal in 800 ml. of methanol is added 2.0 g. (5.7 mmoles) of **6**, and the hydrogenation is continued. After the uptake of 425 ml. of hydrogen (6–8 hr.), the catalyst is removed and washed thoroughly with methanol. The filtrate and washings are combined, and evaporated to dryness *in vacuo*, and the residue (1.6 g.) is dissolved in a small volume of hot methanol and the solution kept at 0° overnight. The resulting crystals are filtered off and dried; yield 1.20 g. (65%), m.p. 145–147° $[\alpha]_D^{20}$ −10° (*c* 0.4, methanol).

REFERENCES

(1) B. R. Baker, J. P. Joseph, R. E. Schaub, and J. H. Williams, *J. Org. Chem.*, **19**, 1786 (1954); F. J. McEvoy, M. J. Weiss, and B. R. Baker, *J. Am. Chem. Soc.*, **82**, 205 (1960); T. Sugawa, Y. Kuwada, K. Imai, M. Morinaga, K. Kajiwara, and K. Tanaka, *Takeda Kenkyusho Nempo*, **20**, 7 (1961); *Chem. Abstracts*, **58**, 3497 (1963); M. L. Wolfrom, H. G. Garg, and D. Horton, *J. Org. Chem.*, **30**, 1557 (1965); S. Fukatsu and S. Umezawa, *Bull. Chem. Soc. Japan*, **38**, 1443 (1965).

(2) F. W. Lichtenthaler, *Angew. Chem.*, **76**, 84 (1964); *Angew. Chem. Intern. Ed. Eng.*, **3**, 211 (1964); in W. Foerst (Ed.), *Neuere Methoden der Präparativen Organischen Chemie*, Vol. IV, Verlag Chemie GmbH, Weinheim/Bergstrasse, Germany, 1966, p. 140.
(3) K. A. Watanabe, J. Beránek, H. A. Friedman, and J. J. Fox, *J. Org. Chem.*, **30**, 2735 (1965).
(4) F. W. Lichtenthaler and H. P. Albrecht, *Chem. Ber.*, **99**, 575 (1966); F. W. Lichtenthaler, P. Emig, and D. Bommer, *ibid.*, **101**, in press (1968).
(5) F. W. Lichtenthaler, T. Nakagawa, and J. Yoshimura, *Chem. Ber.*, **100**, 1833 (1967).
(6) F. W. Lichtenthaler and H. P. Albrecht, *Chem. Ber.*, **101**, in press (1968).
(7) F. W. Lichtenthaler and H. P. Albrecht, *Chem. Ber.*, **100**, 1845 (1967).
(8) J. Beránek, H. A. Friedman, K. A. Watanabe, and J. J. Fox, *J. Heterocyclic Chem.*, **2**, 188 (1965).
(9) J. Žemlička and F. Šorm, *Collection Czech. Chem. Commun.*, **30**, 1880 (1965).

[83] 6-(Pentylthio)- and 6-(Cyclopentylthio)-purine and 6-(Methylthio)- and 6-(Ethylthio)-9-β-D-ribofuranosyl-9*H*-purine

S-*Alkylation of Purine-6(1*H*)-thione (6-Mercaptopurine) and its Ribonucleoside*

THOMAS P. JOHNSTON, SARAH JO CLAYTON, and JOHN A. MONTGOMERY

KETTERING-MEYER LABORATORY, SOUTHERN RESEARCH INSTITUTE, BIRMINGHAM, ALABAMA 35205

$\cdot H_2O$ $\xrightarrow[K_2CO_3,\ HCONMe_2]{RX}$

1 (170.2)

2, R = n-C_5H_{11} (222.3)

2a, R = cyclopentyl (220.3)

HOCH$_2$ ... HO OH $\xrightarrow[NaOH,\ H_2O\ (\text{or } K_2CO_3,\ HCONMe_2)]{RX}$ HOCH$_2$ ... HO OH

3 (284.3)

4, R = CH_3 (298.3)

4a, R = C_2H_5 (312.3)

INTRODUCTION

Many *S*-substituted derivatives of purine-6(1*H*)-thione (**1**) (6-mercaptopurine, 6-MP) and 9-β-D-ribofuranosyl-9*H*-purine-6(1*H*)-thione (**3**)

(6-mercaptopurine ribonucleoside, 6-MPR) have been prepared[1–8] and evaluated[9,10] in experimental, animal-tumor systems. Most of such modifications have resulted in compounds that are active against Carcinoma 755. Preparations of these compounds have involved reaction of 6-chloropurine with thiols,[4–6] or reaction of **1** and **3** with alkyl halides and related compounds.[1–3,5–8] The scope and convenience of the latter method has been extended by the use of *N,N*-dimethylformamide (DMF) as the solvent, with potassium carbonate as the base,[5,8] instead of the aqueous alkaline medium used to advantage in the preparation of *S*-methyl and *S*-benzyl derivatives.[1,5,7] Procedural variations necessitated by differences in the reactivity of the halide and in the solubility of the product are illustrated by the conversion of **1** into the *S*-pentyl and *S*-cyclopentyl derivatives (**2**) and (**2a**), and by the conversion of **3** into *S*-methyl and *S*-ethyl ribonucleosides (**4**) and (**4a**).

PROCEDURE[a]

6-(Pentylthio)purine[5] (2)

A solution of 1.51 g. (10.0 mmoles) of 1-bromopentane in 1.5 ml. of DMF is added dropwise to a stirred mixture of 1.51 g. (8.85 mmoles) of 6-MP monohydrate (**1**), 1.23 g. (8.90 mmoles) of anhydrous potassium carbonate,[b] and 5 ml. of DMF. The reaction mixture, which shows a 5° rise in temperature during the first 10 min., is stirred at 40–50° for 30 min., and then mixed with 70 ml. of water. The product is precipitated as an oil which solidifies, and, after the mixture has been cooled, is collected, dried *in vacuo* over phosphorus pentaoxide, and recrystallized from 25 ml. of carbon tetrachloride. The yield of **2** as platelets, m.p. 116°, is 1.53 g. (78%).

6-(Cyclopentylthio)purine[5] (2a)

A solution of 14.0 g. (93.8 mmoles) of freshly distilled bromocyclopentane in 13 ml. of DMF is added to a stirred mixture of 12.0 g. (70.6 mmoles) of 6-MP monohydrate (**1**), 10.9 g. (78.9 mmoles) of anhydrous potassium carbonate, and 42 ml. of DMF. The resulting mixture is heated at 80° for 3 hr., poured into 600 ml. of water, acidified to pH 6 with concentrated hydrochloric acid, and refrigerated overnight. The precipitated

[a] Melting points determined with a Kofler hot-stage.

[b] In such alkylations, an excess of potassium carbonate (up to 10%) is often used in conjunction with an excess of halide to ensure complete reaction of **1** and **3**, and, in the case of **1**, leads to a minor proportion of *N,S*-dialkylated product, which is readily removed in the recrystallization process. When an excess of potassium carbonate is used, the pH of the water-diluted reaction mixture is adjusted to 5–7 with acetic acid or hydrochloric acid before the precipitated product is collected.

product is thoroughly washed with water, dried *in vacuo* over phosphorus pentaoxide and sodium hydroxide, and recrystallized from 500 ml. (or more) of ethyl acetate. The yield of crystalline **2a**, m.p. 228°, that has been dried *in vacuo* over phosphorus pentaoxide at 78°, is 9.74 g. (63%).

6-(Methylthio)-9-β-D-ribofuranosyl-9*H*-purine[7] (4)

A stirred solution of 85 g. (0.30 mole) of 6-MPR (**3**)[7] in 285 ml. (0.31 mole) of 1.1 *N* sodium hydroxide is diluted with 450 ml. of water and treated, dropwise and steadily, with 45 g. (0.32 mole) of methyl iodide. After the initial exothermic reaction (as evidenced by a 5° rise in temperature) is over, the solution is stirred at room temperature for 3 hr., chilled, and neutralized with glacial acetic acid. The hydrated crystals, which form after the inner wall of the flask has been scratched with a glass rod, are dried *in vacuo* over phosphorus pentaoxide, and recrystallized from 1 l. of absolute ethanol; yield of anhydrous **4**, 70 g. (78%), m.p. 167°.

6-(Ethylthio)-9-β-D-ribofuranosyl-9*H*-purine[8] (4a)

Ethyl bromide (1.59 g., 14.6 mmoles) is added to a stirred mixture of 4.00 g. (14.1 mmoles) of 6-MPR (**3**),[7] 2.02 g. (14.6 mmoles) of anhydrous potassium carbonate, and 50 ml. of DMF; the resulting mixture is heated at 55° for 1 hr. The insoluble, inorganic salts are removed by filtration, and the filtrate is evaporated to dryness *in vacuo* at 60° with the aid of two additions of ethanol. The residue is extracted with 30 ml. of hot, dry acetone, and the crystals that form in the refrigerated filtrate are recrystallized from 30 ml. of acetone, and dried *in vacuo* over phosphorus pentaoxide at 78°; yield of anhydrous **4a**, as hygroscopic needles of indefinite m.p., 2.30 g. (52%); λ_{max} in nm. (log ε): 225 (4.02) and 294 (4.23) at pH 1, 224 (4.04) and 292 (4.27) at pH 7, and 225 (4.02) and 292 (4.28) at pH 13.

REFERENCES

(1) G. B. Elion, E. Burgi, and G. H. Hitchings, *J. Am. Chem. Soc.*, **74**, 411 (1952).

(2) C. G. Skinner, W. Shive, R. G. Ham, D. C. Fitzgerald, Jr., and R. E. Eakin, *J. Am. Chem. Soc.*, **78**, 5097 (1956).

(3) C. G. Skinner, R. G. Ham, D. C. Fitzgerald, Jr., R. E. Eakin, and W. Shive, *J. Org. Chem.*, **21**, 1330 (1956).

(4) M. W. Bullock, J. J. Hand, and E. L. R. Stokstad, *J. Am. Chem. Soc.*, **78**, 3693 (1956).

(5) T. P. Johnston, L. B. Holum, and J. A. Montgomery, *J. Am. Chem. Soc.*, **80**, 6265 (1958).

(6) H. C. Koppel, D. E. O'Brien, and R. K. Robins, *J. Org. Chem.*, **24**, 259 (1959).
(7) J. J. Fox, I. Wempen, A. Hampton, and I. L. Doerr, *J. Am. Chem. Soc.*, **80**, 1669 (1958).
(8) J. A. Montgomery, T. P. Johnston, A. Gallagher, C. R. Stringfellow, Jr., and F. M. Schabel, Jr., *J. Med. Pharm. Chem.*, **3**, 265 (1961).
(9) H. E. Skipper, J. A. Montgomery, J. R. Thomson, and F. M. Schabel, Jr., *Cancer Res.*, **19**, 425 (1959).
(10) F. M. Schabel, Jr., J. A. Montgomery, H. E. Skipper, W. R. Laster, Jr., and J. R. Thomson, *Cancer Res.*, **21**, 690 (1961).

[84] 6-(2-Pyridylmethylthio)-9-β-D-ribofuranosyl-9*H*-purine

*Alkylation of 9-β-D-Ribofuranosyl-9*H*-purine-6(1*H*)-thione*

ARTHUR F. LEWIS and ROLAND K. ROBINS

DEPARTMENT OF CHEMISTRY, UNIVERSITY OF UTAH, SALT LAKE CITY, UTAH 84112

1 (284.3) → **2** (375.4)

INTRODUCTION

6-(2-Pyridylmethylthio)purine has been shown to be entirely curative against adenocarcinoma 755 experimentally induced in mice,[1] for which it has a therapeutic index of 128. Alkylation of 9-β-D-ribofuranosyl-9*H*-purine-6(1*H*)-thiones has been achieved by use of the appropriate alkyl halide in a basic medium.[2] The following is a specific example of this type of reaction.

PROCEDURE

6-(2-Pyridylmethylthio)-9-β-D-ribofuranosyl-9*H*-purine (2)

A mixture of 10 g. (0.0352 mole) of 9-β-D-ribofuranosyl-9*H*-purine-6(1*H*)-thione[3] (**1**), 5.75 g. (0.0351 mole) of 2-(chloromethyl)pyridine hydrochloride, 6.5 g. (0.0793 mole) of anhydrous sodium acetate, and 250 ml. of absolute ethyl alcohol is refluxed for 18 hr., and then the solvent is evaporated off under diminished pressure (rotary evaporator, water aspirator). The solid residue is washed with water (to remove salts), collected by filtration, rewashed with water, and air-dried. After two recrystallizations from aqueous ethanol, the product is dried at 80–90° under diminished pressure; yield 7.3 g. of **2**, m.p. 212–214°, $\lambda_{max}^{pH\,1}$ 279 nm. (ϵ 19,500); λ_{max}^{EtOH} 291 nm. (sh) (ϵ 20,300) and 284 nm. (ϵ 21,000); $\lambda_{max}^{pH\,11}$ 287.5 nm. (ϵ 20,100).

Anal. Calcd. for $C_{16}H_{17}N_5O_4S$: C, 51.2; H, 4.53; N, 18.7. Found; C, 51.01; H, 4.68; N, 18.79.

REFERENCES

(1) L. R. Lewis, C. W. Noell, A. G. Beaman, and R. K. Robins, *J. Med. Pharm. Chem.*, **5**, 607 (1962).
(2) C. W. Noell and R. K. Robins, *J. Med. Pharm. Chem.*, **5**, 1074 (1962); J. A. Montgomery, T. P. Johnston, A. Gallagher, C. R. Stringfellow, and F. M. Schabel, Jr., *ibid.*, **3**, 265 (1961).
(3) J. J. Fox, I. Wempen, A. Hampton, and I. L. Doerr, *J. Am. Chem. Soc.*, **80**, 1669 (1958).

[85] D-Ribofuranosyl-9*H*-purine Nucleosides (Purine Ribonucleosides)

The Fusion Reaction Applied to Preparation of Ribofuranosylpurines

TETSUO SATO

DEPARTMENT OF CHEMISTRY, TOKYO INSTITUTE OF TECHNOLOGY, TOKYO, JAPAN

INTRODUCTION

The Fischer–Helferich procedure,[1] which involves the coupling reaction of an *O*-acylglycosyl halide with a metallic salt of a purine, has been widely applied to the preparation of various purine nucleosides. Recently, the author has developed the fusion reaction (recognized as a representative, synthetic procedure for phenyl glycosides[2]) for the preparation of purine nucleosides. Thus, the synthesis of purine nucleosides may be executed by the simple expedient of fusing an acetylated sugar with a purine in the presence of an acidic catalyst under diminished pressure.[3] Because of the ease with which such versatile intermediates as the ribonucleosides (**8** and **9**) are formed in the fusion reaction, the preparation of variously substituted ribonucleosides is now markedly simplified.

PROCEDURE

6-Chloro-2-(methylthio)-9-β-D-ribofuranosyl-9*H*-purine[4] (2a)

A mixture of 3.2 g. (0.01 mole) of 1,2,3,5-tetra-*O*-acetyl-β-D-ribose[5] (**1**) and 2.4 g. (0.01 mole) of 6-chloro-2-(methylthio)purine (**2**) is fused at 145–150° (bath temperature). One drop (3–5 mg.) of concentrated sulfuric acid is added from a capillary tube to the fused mixture, and the mixture is stirred with a glass rod. The mixture is heated *in vacuo* until the rapid evolution of acetic acid has ceased (*ca.* 15 min.), and is then cooled. The resulting sirup is dissolved in 50 ml. of chloroform, and the insoluble material is removed by filtration and discarded. The filtrate is evaporated to dryness under diminished pressure below 40°, the residue is dissolved in 50 ml. of methanol, and the solution is treated with 50 ml. of methanol presaturated with ammonia at 0°. The mixture is kept in a refrigerator overnight, and is then evaporated to dryness under diminished pressure below 40°. One crystallization of the residue from 40 ml. of hot water and one recrystallization from water (with decolorizing with active charcoal) gives colorless needles; yield 1.3 g. (39%), m.p. 180–181°, $[\alpha]_D^{24}$ +2.0°

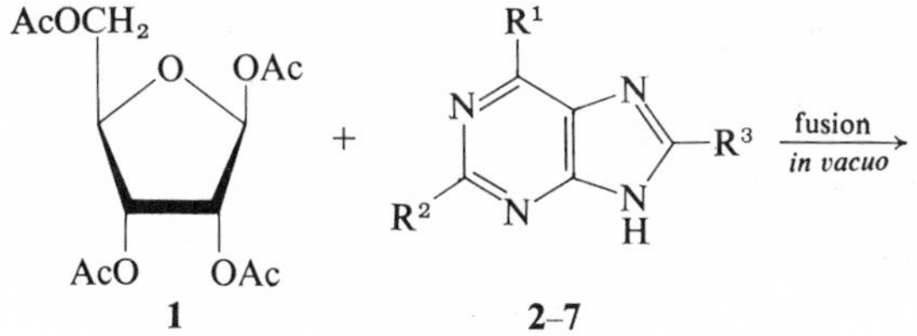

$AcOCH_2$ O OAc; AcO OAc — **1** (318.3) + R^1, N, N, R^3, R^2, N, N, H — **2–7** $\xrightarrow{\text{fusion, in vacuo}}$

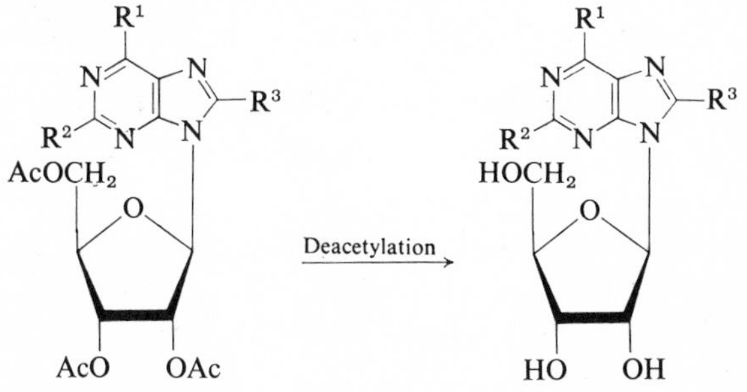

R^1, N, N, R^3, R^2, N, N, $AcOCH_2$, O, AcO, OAc — **8, 9** $\xrightarrow{\text{Deacetylation}}$ R^1, N, N, R^3, R^2, N, N, $HOCH_2$, O, HO, OH — **2a–5a, 10**

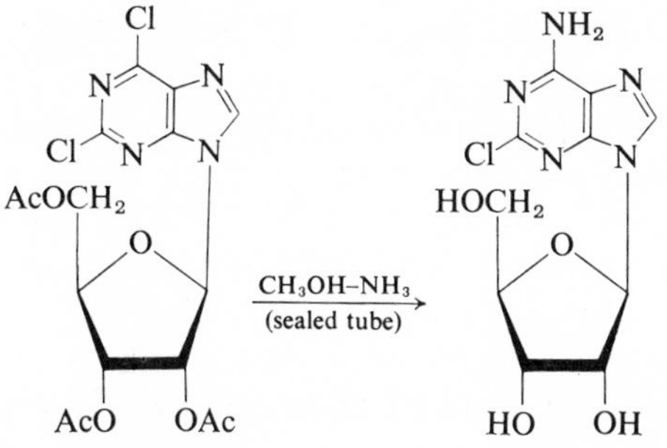

Cl, N, N, Cl, N, N, $AcOCH_2$, O, AcO, OAc — **8** (447.2) $\xrightarrow[\text{(sealed tube)}]{CH_3OH–NH_3}$ NH_2, N, N, Cl, N, N, $HOCH_2$, O, HO, OH — **10** (301.7)

where Ac is acetyl, and R^1, R^2, and R^3 are:

	2 (200.7)	**2a** (332.8)	**3** (245.1)	**3a** (377.2)	**4** (292.1)	**4a** (424.2)
R^1	Cl	Cl	Br	Br	I	I
R^2	SMe	SMe	SMe	SMe	SMe	SMe
R^3	H	H	H	H	H	H

	5 (285.3)	**5a** (313.3)	**6** (447.2)	**7** (223.5)	**8** (447.2)	**9** (481.7)	**10** (301.7)
R^1	NHBz	NH_2	Cl	Cl	Cl	Cl	NH_2
R^2	SMe	SMe	Cl	Cl	Cl	Cl	Cl
R^3	H	H	H	Cl	H	Cl	H

(*c* 0.23, 50% aqueous methanol), λ_{max}^{EtOH} 236.5 nm. (ϵ 19,400) and 263 nm. (ϵ 12,000), λ_{min}^{EtOH} 250 nm. (ϵ 8,400).

6-Bromo-2-(methylthio)-9-β-D-ribofuranosyl-9*H*-purine[4] (3a)

A mixture of 3.2 g. (0.01 mole) of 1,2,3,5-tetra-*O*-acetyl-β-D-ribose[5] (**1**) and 2.5 g. (0.01 mole) of 6-bromo-2-(methylthio)purine[3] (**3**) is treated for 10 min. as described in the preceding experiment, giving colorless needles; yield 1.2 g. (31%), $[\alpha]_D^{24}$ $-10.5°$ (*c* 0.79, 50% aqueous methanol), λ_{max}^{EtOH} 238 nm. (ϵ 19,500) and 264 nm. (ϵ 11,500), λ_{min}^{EtOH} 250 nm. (ϵ 8,900).

6-Iodo-2-(methylthio)-9-β-D-ribofuranosyl-9*H*-purine[4] (4a)

A mixture of 3.2 g. (0.01 mole) of 1,2,3,5-tetra-*O*-acetyl-β-D-ribose[5] (**1**) and 2.9 g. (0.01 mole) of 6-iodo-2-(methylthio)purine[4] (**4**) is treated for 6 min. as described in the preceding experiments, giving colorless needles; yield 1.9 g. (45%), $[\alpha]_D^{24}$ $-24.1°$ (*c* 0.21, ethanol), λ_{max}^{EtOH} 238 nm. (ϵ 17,500), 262 nm. (ϵ 14,800), and 310 nm. (ϵ 5,300), λ_{min}^{EtOH} 250 nm. (ϵ 13,300) and 287 nm. (ϵ 2,300).

2-(Methylthio)adenosine[4] (5a)

A mixture of 3.2 g. (0.01 mole) of 1,2,3,5-tetra-*O*-acetyl-β-D-ribose[5] (**1**) and 2.3 g. (0.01 mole) of *N*-benzoyl-2-(methylthio)adenine[4] (**5**) is heated at 170–175° (bath temperature), and about 20 mg. of concentrated sulfuric acid is added to the fused mixture. The mixture is stirred, and heated under diminished pressure. After repeating the operation several times (10–15 min.), the mixture begins to fuse, and heating under diminished pressure is continued for 20 min. The mixture is cooled, slurried with 30 ml. of methanol, and filtered. The filtrate is treated with 20 ml. of methanol to which had previously been added about 100 mg. of metallic sodium, and the mixture is kept at room temperature overnight. It is then refluxed for 15 min., diluted with 20 ml. of water, rendered neutral with 2 *N* hydrochloric acid, cooled in an ice bath (causing crystallization of **5a**), and filtered. Recrystallization of the deacetylated nucleoside (**5a**) from water (with decolorizing with active charcoal) gives colorless needles; yield 1.0 g. (32%), m.p. 222–223°, $[\alpha]_D^{18}$ $-2.0°$ (*c* 0.51, 0.1 *N* hydrochloric acid), $\lambda_{max}^{pH\,5.5}$ 237 nm. (ϵ 24,300) and 277 nm. (ϵ 14,500), $\lambda_{min}^{pH\,5.5}$ 252 nm. (ϵ 8,300).

2,6-Dichloro-9-(2,3,5-tri-*O*-acetyl-β-D-ribosyl)-9*H*-purine[4,5] (8)

A mixture of 3.2 g. (0.01 mole) of 1,2,3,5-tetra-*O*-acetyl-β-D-ribose[5] (**1**) and 1.9 g. (0.01 mole) of 2,6-dichloropurine[6] (**6**) is fused at 150–155° (bath temperature) to a clear melt, after 3–5 mg. of concentrated sulfuric acid has been added from a capillary tube, with stirring. The mixture is heated under diminished pressure until the vigorous evolution of acetic acid has ceased (about 4 min.), and is then cooled in an ice bath. The residue is heated in a boiling-water bath, 15 ml. of ethanol is added, and the solution is cooled in an ice bath, giving crystals of **8**. Recrystallization of the crude nucleoside from 100 ml. of ethanol[a] (with decolorizing with active charcoal) gives transparent, slender, scaly crystals; yield 2.4 g. (54%), m.p. 158–159°, $[\alpha]_D^{21.5}$ $-5.7°$ (*c* 1.05, chloroform), λ_{max}^{EtOH} 253 nm. (ϵ 6,100) and 274.5 nm. (ϵ 9,800), λ_{min}^{EtOH} 258 nm. (ϵ 5,800).

When the fusion is performed in the presence of sulfamic acid (100 mg.), *p*-toluenesulfonic acid (10–20 mg.), or sulfanilic acid (100 mg.), the product is obtained in 76, 52, and 80% yield, respectively.[5] Under non-catalytic conditions,[7] the product is obtained in only 50–55% yield.

2-Chloroadenosine[4] (10)

A suspension of 4.5 g. (0.01 mole) of compound **8** in 60 ml. of methanol in a tube (3 cm. diameter) is saturated with ammonia gas at 0°, causing dissolution of the crystalline, protected nucleoside. After being sealed, the tube is kept at room temperature overnight, warmed at 30–40° (bath temperature) for 3 hr., and then cooled in an ice–salt bath. The sealed tube is cautiously opened, and the solvent is removed *in vacuo*. The partially crystalline sirup is dissolved in 50 ml. of water, and treated with a hot, aqueous solution of picric acid (2.3 g., 0.01 mole) to give 4.4 g. of the picrate, m.p. 144–146°. The picrate (3.0 g.) is suspended in 300 ml. of distilled water, and treated with 10 g. of Dowex 1 (CO_3^{2-}) for 2 hr., with stirring. The mixture is filtered, and the filtrate is evaporated to dryness under diminished pressure at 40°, giving a colorless powder; yield 1.5 g. (84%), m.p. 142–143°, $[\alpha]_D^{12}$ $-49.8°$ (*c* 1.0, ethanol), R_f 0.31 (43:7 butyl alcohol–water), $\lambda_{max}^{pH\,5.5}$ 265 nm. (ϵ 15,000).

2,6,8-Trichloro-9-(2,3,5-tri-*O*-acetyl-β-D-ribosyl)-9*H*-purine[3,7] (9)

A finely powdered mixture of 3.18 g. (0.01 mole) of 1,2,3,5-tetra-*O*-acetyl-β-D-ribose[5] (**1**) and 2.23 g. (0.01 mole) of 2,6,8-trichloropurine[8] (**7**) is fused at 150–155° (bath temperature) under diminished pressure (in the absence of an acidic catalyst) for 10 min. The sirupy, pale-yellow residue is triturated with 10 ml. of hot ethanol, whereupon the crude product crystallizes readily. The mixture is cooled in an ice–bath and filtered; the crystalline nucleoside is recrystallized from ethanol (100 ml.)[a] (with decolorizing with active charcoal) to give fine needles; yield 2.74 g. (64%), m.p. 161–162° (slight sintering at 155°), $[\alpha]_D^{12.8}$ $-10.2°$ (*c* 1.00, chloroform) (calculated from $[\alpha]_{546}^{12.8}$ $-13.0°$ and $[\alpha]_{578}^{12.8}$ $-24.0°$).

[a] If smaller volumes of solvent are used, the product crystallizes at once on filtration through the funnel.

REFERENCES

(1) E. Fischer and B. Helferich, *Ber.*, **47**, 210 (1914).

(2) B. Helferich und R. Gootz, *Ber.*, **62**, 2788 (1929); B. Helferich and E. Schmitz–Hillebrecht, *ibid.*, **66**, 378 (1933); **67**, 1667 (1934); **68**, 790 (1934).

(3) T. Sato, T. Shimadate, and Y. Ishido, *Nippon Kagaku Zasshi*, **81**, 1440, 1442 (1961).

(4) Y. Ishido, Y. Kikuchi, and T. Sato, *Nippon Kagaku Zasshi*, **86**, 240 (1965).

(5) H. Zinner, *Chem. Ber.*, **83**, 517 (1950).

(6) G. B. Elion and G. H. Hitchings, *J. Am. Chem. Soc.*, **78**, 3508 (1956); J. A. Montgomery, *ibid.*, **78**, 1928 (1956).

(7) Y. Ishido, T. Matsuba, A. Hosono, K. Fujii, H. Tanaka, K. Iwabuchi, S. Isome, A. Maruyama, Y. Kikuchi, and T. Sato, *Bull. Chem. Soc. Japan*, **38**, 2019 (1965); Y. Ishido, T. Matsuba, A. Hosono, K. Fujii, T. Sato, S. Isome, A. Maruyama, and Y. Kikuchi, *ibid.*, **40**, 1007 (1967).

(8) J. Davoll and B. A. Lowy, *J. Am. Chem. Soc.*, **73**, 2936 (1951).

[86] 9-(Tetrahydropyran-2-yl)-9*H*-purines

Purine Deoxypentopyranose Nucleoside Models

EDWARD Y. SUTCLIFFE and ROLAND K. ROBINS

DEPARTMENT OF CHEMISTRY, UNIVERSITY OF UTAH, SALT LAKE CITY, UTAH 84112

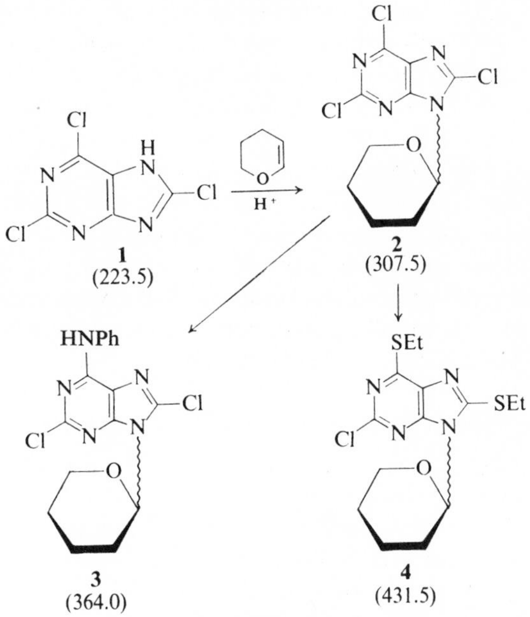

1 (223.5) → (H⁺) **2** (307.5)

2 → **3** (364.0)

2 → **4** (431.5)

INTRODUCTION

Dihydropyran has been shown to react with purines under conditions of acid catalysis, to yield the corresponding 9-(tetrahydropyran-2-yl)-9*H*-purine.[1] Especially good yields of 9-(tetrahydropyran-2-yl)-9*H*-purines are obtained from various chloropurines.[1,2] Dihydrofuran may also be used, in a similar manner, to give 9-(tetrahydro-2-furyl)-9*H*-purines.[3,4] The 9-(tetrahydropyran-2-yl) derivative of kinetin (*N*-furfuryladenine) has been prepared by the reaction of 6-chloro-9-(tetrahydropyran-2-yl)-9*H*-purine with furfurylamine. Recently, *N*-benzyl-9-(tetrahydropyran-2-yl)-adenine has proved to be a synthetic kinin that is very effective for prompting parthenocarpy in the calimyrna fig.[5] Several of the 9-(tetrahydropyran-2-yl)-9*H*-purines exhibit significant antitumor activity.[6] The tetrahydropyran-2-yl group has also proved to be a most effective protecting group for controlling the orientation of nucleophilic substitution on the purine ring.[2] The preparation of 2,6,8-trichloro-9-(tetrahydropyran-2-yl)-9*H*-purine (**2**), and certain nucleophilic substitution reactions, are described here.

PROCEDURE

2,6,8-Trichloro-9-(tetrahydropyran-2-yl)-9*H*-purine[2] (2)

Anhydrous 2,6,8-trichloropurine[7] (**1**) is recrystallized from dry benzene; 50 g. (0.224 mole) is dissolved in 500 ml. of anhydrous ethyl acetate, and the solution is heated to 35°, with stirring. *p*-Toluenesulfonic acid (100 mg.) is added, and then 29 g. (0.345 mole) of freshly distilled dihydropyran is added dropwise during 10 min., the temperature rising to 50°. Heating is stopped, stirring is continued for 20 min., and the solution is rapidly cooled to room temperature and washed successively with four 25-ml. portions of saturated, aqueous sodium carbonate, and five 25-ml. portions of water until the solution is neutral. The solution is dried (anhydrous sodium sulfate, 5 hr.), filtered, and evaporated to dryness under diminished pressure at 50°. The resulting solid is recrystallized from heptane; yield 47.0 g. (68%) of colorless crystals of **2**, m.p. 117–119°.

2,8-Dichloro-*N*-phenyl-9-(tetrahydropyran-2-yl)adenine (3)

Compound **2** (3 g., 9.75 mmoles) and 2.72 g. (29.3 mmoles) of aniline are dissolved in 45 ml. of ethanol, and the solution is stirred at room temperature for 3 hr. The resulting suspension is filtered, and the residue is washed with water, and then recrystallized from aqueous ethanol; yield 2.4 g. (69.5%) of **3**, colorless crystals, m.p. >300°; $\lambda_{max}^{pH\ 11}$ 234 nm. (ϵ 10,000) and 297.5 nm. (ϵ 21,700).

Anal. Calcd. for $C_{16}H_{15}Cl_2N_5O$: C, 52.95; H, 4.12; N, 19.25. Found: C, 52.80; H, 4.40; N, 19.1.

Reduction of **3** with hydrogen in the presence of triethylamine and palladium-on-carbon catalyst, followed by hydrolysis of the residue with hydrochloric acid gives *N*-phenyladenine, m.p. 282.5–284°, identical with an authentic sample.[7]

2-Chloro-6,8-bis(ethylthio)-9-(tetrahydropyran-2-yl)-9*H*-purine (4)

Compound **2** (3 g., 9.75 mmoles) is dissolved in 45 ml. of ethanethiol containing 2.45 g. (29.2 mmoles) of sodium ethanethioxide. The solution is stirred in a hood at room temperature for 3 hr. (in a flask fitted with a reflux condenser), the resulting mixture is filtered with suction, and the filtrate is permitted to evaporate to dryness in the air in a hood. The residue is washed with water, triturated several times with ethanol, and air-dried until the odor of ethanethiol becomes negligible. The product is recrystallized from ethanol; yield 2.3 g. (55%) of **4**, colorless crystals, m.p. 119.5–120.5°; λ_{max}^{pH11} 251 nm. (ϵ 15,000), 313 nm. (sh.) (ϵ 17,200), and 318 nm. (ϵ 17,600).

Anal. Calcd. for $C_{14}H_{19}ClN_4OS_2$: C, 46.80; H, 5.29; N, 15.6. Found: C, 46.55; H, 5.37; N, 15.7.

Treatment of **4** in methanol with hydrogen in the presence of Raney nickel, followed by hydrolysis with mineral acid, gives 2-chloropurine.

REFERENCES

(1) R. K. Robins, E. F. Godefroi, E. C. Taylor, L. R. Lewis, and A. Jackson, *J. Am. Chem. Soc.*, **83**, 2574 (1961).

(2) E. Y. Sutcliffe and R. K. Robins, *J. Org. Chem.*, **28**, 1662 (1963).

(3) L. R. Lewis, F. H. Schneider, and R. K. Robins, *J. Org. Chem.*, **26**, 3837 (1961).

(4) W. A. Bowles, F. H. Schneider, L. R. Lewis, and R. K. Robins, *J. Med. Chem.*, **6**, 471 (1963).

(5) J. C. Crane and J. Van Overbeek, *Science*, **147**, 1468 (1965).

(6) R. K. Robins, *J. Med. Chem.*, **7**, 186 (1964).

(7) R. K. Robins, *J. Org. Chem.*, **26**, 447 (1961).

[87] Anomeric 2-Amino-9-(2-deoxy-D-*erythro*-pentofuranosyl)-9*H*-purine-6-thiols [2′-Deoxy-6-thioguanosine and its α-D Anomer]

Formation and Transformation of Deoxypentosylpurines

EDWARD M. ACTON and ROBERT H. IWAMOTO

LIFE SCIENCES RESEARCH, STANFORD RESEARCH INSTITUTE, MENLO PARK, CALIFORNIA 94025

INTRODUCTION

The mercury salt (**2**) of 2-acetamido-6-chloropurine[1] is useful for the preparation of purine nucleosides having different functional groups at C-2 and C-6. Nucleosides of guanine, thioguanine, and other 2-amino-6-substituted purines have been prepared[1,2] from this salt and various glycosyl chlorides.

PROCEDURE

2-Acetamido-6-chloro-9-(2-deoxy-3,5-di-*O*-*p*-toluoyl-β-D-*erythro*-pentosyl)-9*H*-purine (3)

A mixture of 31.8 g. (0.072 mole, calcd. as $C_7H_5ClHgN_5O_3$) of the mercury salt (**2**) of 2-acetamido-6-chloropurine (Sec. I [9]) and 70 g. of Celite is suspended in 2.5 l. of benzene in a 3-necked, round-bottomed flask fitted with a stirrer, addition funnel, and Dean–Stark distilling receiver bearing a reflux condenser and a drying tube. The stirred mixture is dried by azeotropic distillation of about 500 ml. of the benzene, and to the stirred, boiling mixture is added 30.1 g. (0.0774 mole) of 2-deoxy-3,5-di-*O*-*p*-toluoyl-D-*erythro*-pentosyl chloride[3] (**1**) (Sec. V [158]). Stirring and refluxing are continued for 30 min., and the hot mixture is filtered. The filtrate is concentrated to about 900 ml. and then diluted to 4 l. with petroleum ether (b.p. 30–60°). The resultant precipitate is collected by filtration, washed with petroleum ether, dried, and dissolved in 500 ml. of chloroform. The chloroform solution is successively washed with two 400-ml. portions of 30% aqueous potassium iodide (to remove mercury salts) and 200 ml. of water, dried with magnesium sulfate, filtered, concentrated to 300–400 ml., and kept overnight at 5°. The crystalline β-D anomer separates and is collected; yield 12.7 g. (31%), m.p. 180–187°. (The mother liquor is saved; see the next paragraph.) Recrystallization

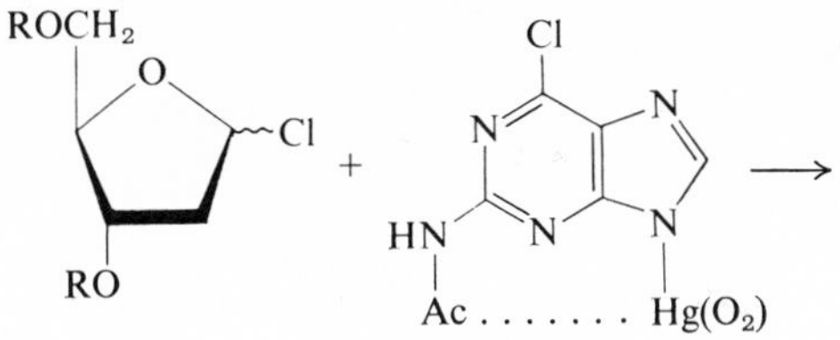

ROCH$_2$ … Cl + Cl, N, N, HN, N, N, Ac .... Hg(O$_2$) →

1 (388.8) **2** (443.2)

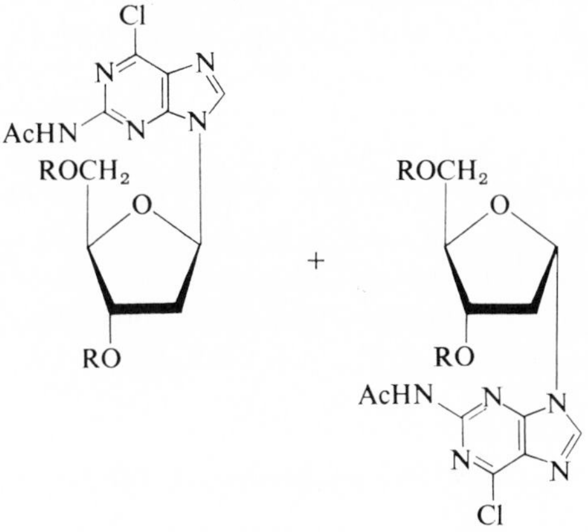

Cl, N, N, AcHN, N, N, ROCH$_2$, O, RO + ROCH$_2$, O, RO, AcHN, N, N, N, N, Cl

3 (564.0) **4** (564.0)

where R = *p*-toluoyl.

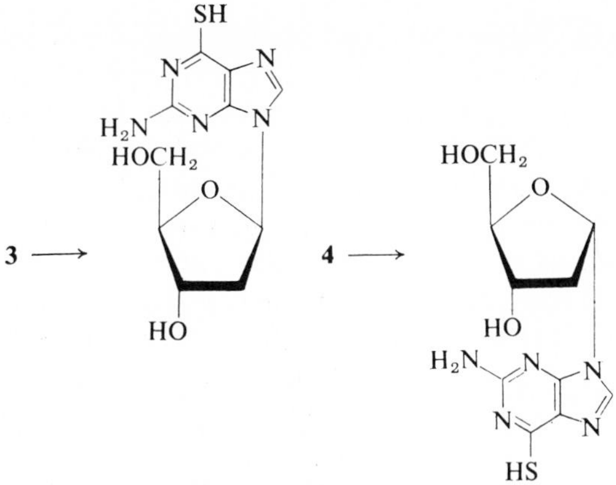

SH, N, N, H_2N, N, N, HOCH$_2$, O, HO

3 → **5** **4** → **6**

HOCH$_2$, O, HO, H_2N, N, N, N, N, HS

5 (301.3) **6** (301.3)

from chloroform gives 11.1 g. (27%),[a] m.p. 189–190°, $[\alpha]_D^{24}$ −27.2° (chloroform).

2-Acetamido-6-chloro-9-(2-deoxy-3,5-di-*O*-*p*-toluoyl-α-D-*erythro*-pentosyl)-9*H*-purine (4)

The mother liquor (350 ml. of chloroform, plus washings), remaining after the initial separation of 12.7 g. of **3** (preceding paragraph), is diluted to 1 l. with ether. After being kept at 3° overnight, a small amount of gum (containing some **3**) separates and is discarded, and the supernatant liquor is decanted, diluted with 2 l. of petroleum ether (b.p. 30–60°), and kept at 3° overnight. Amorphous **4** separates, and is collected by filtration; yield 10.5 g. (26%), $[\alpha]_D^{24}$ −51.1° (chloroform). Further purification (*ca.* 60% recovery) is achieved by reprecipitation from chloroform solution (20 ml./g., treated with decolorizing carbon) with petroleum ether (250 ml./g.); the precipitate is dissolved in benzene (30 ml./g.), and the solution is washed with aqueous potassium iodide, dried, and diluted with petroleum ether (500 ml./g.). The resultant, amorphous precipitate, $[\alpha]_D^{24}$ −54.8°, softens to a clear glass at *ca.* 92°. It is readily soluble in chloroform, in marked contrast to **3**.

2-Amino-9-(2-deoxy-β-D-*erythro*-pentofuranosyl)-9*H*-purine-6-thiol (5)

In a 3-necked, round-bottomed flask (fitted with a stirrer, a gas inlet-tube, and reflux condenser closed with a drying tube), a suspension of 5.08 g. (9 mmoles) of crystalline nucleoside **3** in 540 ml. of anhydrous methanol is saturated with hydrogen sulfide and heated to reflux. While a slow stream of hydrogen sulfide is maintained, 27 ml. of *M* methanolic sodium methoxide (presaturated with hydrogen sulfide) is added to the solution, and refluxing is continued for 2 hr. to effect displacement of the chlorine atom at C-6. The flow of hydrogen sulfide is then stopped, refluxing is continued for 15 min., 13.5 ml. of *M* methanolic sodium methoxide is added, and refluxing is continued for 1 hr. (to complete the cleavage of the *p*-toluoyl groups). The solvent is removed *in vacuo*, and the residue is partitioned between 45 ml. of water and 45 ml. of chloroform. The aqueous layer is separated, washed with two 40-ml. portions of chloroform, clarified by filtration, rendered neutral with a little acetic acid, and kept overnight at 0°. The product which separates weighs 2.24 g. (87%), $[\alpha]_D^{26}$ −32.0° (0.1 *M* sodium hydroxide).[b]

[a] Chloroform of solvation is present; it may be removed by recrystallization from methanol (250 ml./g.), m.p. unchanged, $[\alpha]_D^{24}$ −28.7°.

[b] An analytical sample is prepared by recrystallization from hot water (125 ml./g.).

2-Amino-9-(2-deoxy-α-D-*erythro*-pentofuranosyl)-9*H*-purine-6-thiol (6)

By the procedure given for **5**, the α-D nucleoside (**4**) affords the product **6** in 76% yield, $[\alpha]_D^{25}$ +58.2° (0.1 *M* sodium hydroxide). After one recrystallization from water, the yield is 45%, $[\alpha]_D^{24}$ +68.6°.

REFERENCES

(1) R. H. Iwamoto, E. M. Acton, and L. Goodman, *J. Med. Chem.*, **6**, 684 (1963).
(2) G. L. Tong, K. J. Ryan, W. W. Lee, E. M. Acton, and L. Goodman, *J. Org. Chem.*, **32**, 859 (1967).
(3) M. Hoffer, *Chem. Ber.*, **93**, 2777 (1960).

[88] Acylated Derivatives of 9-β-D-Ribofuranosyl-9*H*-purine-6-thiol and of 2-Amino-9-β-D-ribofuranosyl-9*H*-purine-6-thiol

Acylation of D-*Ribofuranosylpurine-6-thiol Nucleosides by a Modified Schotten–Baumann Reaction*

LELAND R. LEWIS, ROLAND K. ROBINS, and C. C. CHENG

MIDWEST RESEARCH INSTITUTE, KANSAS CITY, MISSOURI 64110

1, X = H
(284.3)

2, X = NH_2
(299.3)

3, X = H, R = CH_3
(410.4)

4, X = H, R = $CH_3(CH_2)_7CH \quad CH(CH_2)_7$
(1077.7)

5, X = H, R = p-CH_3O-C_6H_4
(686.7)

6, X = NH_2, R = CH_3
(425.4)

7, X = NH_2, R = $CH_3(CH_2)_{14}$
(1014.6)

8, X = NH_2, R = p-Cl-C_6H_4
(715.0)

INTRODUCTION

Acylation of a purine or a pyrimidine nucleoside usually modifies its character for transport through the cell membrane[1] as compared with the parent compound; consequently, the pattern of absorption *in vivo* is drastically altered.[2] The acylated derivatives may be prepared in high yield, under relatively mild reaction conditions, by the procedure described here.[3]

PROCEDURE

General Preparation of the Acyl Derivatives

To a stirred suspension of 0.02 mole of the nucleoside (**1** or **2**) in 250 ml. of anhydrous pyridine is added 0.08 mole of the appropriate acid chloride. The resulting solution is heated at 50–60° for 3 hr.[a] with stirring, cooled, and evaporated to dryness *in vacuo* (bath temperature < 50°), and the residue is treated with 500 ml. of distilled water.[b] The crystalline product is filtered off, recrystallized from an appropriate solvent,[c] and dried at 100°/ 0.1 mm. Hg.

9-β-D-Ribofuranosyl-9*H*-purine-6-thiol 2′,3′,5′-Triacetate (3)

This compound, recrystallized from a mixture of water and acetone, is obtained in 90% yield, m.p. 252–253° (lit.[4] m.p. 255–256°), $\lambda_{max}^{pH\,1}$ 320 nm. (ϵ 25,800); $\lambda_{max}^{pH\,11}$ 236 nm. (ϵ 13,500) and 309 nm. (ϵ 23,800).

9-β-D-Ribofuranosyl-9*H*-purine-6-thiol 2′,3′,5′-Trioleate (4)

The compound, recrystallized from methanol–dichloromethane, is obtained in 71% yield, m.p. 202–206°, $\lambda_{max}^{pH\,1}$ 321 nm. (ϵ 26,700); $\lambda_{max}^{pH\,11}$ 235 nm. (ϵ 16,200) and 310 nm. (ϵ 24,200).

9-β-D-Ribofuranosyl-9*H*-purine-6-thiol 2′,3,′5′-Tri-*p*-anisate (5)

Compound **5**, recrystallized from ethyl acetate, is obtained in 94% yield, m.p. 176–177°, $\lambda_{max}^{pH\,1}$ 261 nm. (ϵ 40,500) and 323 nm. (ϵ 22,000); $\lambda_{max}^{pH\,11}$ 236 nm. (ϵ 24,400), 258 nm. (ϵ 46,400), and 310 nm. (ϵ 23,300).

2-Amino-9-β-D-ribofuranosyl-9*H*-purine-6-thiol 2′,3′,5′-Triacetate (6)

This compound, recrystallized from ethanol, is obtained in 75% yield as a hemihydrate, m.p. 203–205° (lit.[4] m.p. 209–211°) $\lambda_{max}^{pH\,1}$, 264 nm. (ϵ 3,800) and 344 nm. (ϵ 15,100); $\lambda_{max}^{pH\,11}$ 251 nm. (ϵ 8,300) and 317 nm. (ϵ 13,600).

2-Amino-9-β-D-ribofuranosyl-9*H*-purine-6-thiol 2′,3′,5′-Tripalmitate (7)

The compound, recrystallized from methanol, is obtained in 89% yield as a hydrate, m.p. 170–172°, $\lambda_{max}^{pH\,1}$ 262 nm. (ϵ 19,000) and 343 nm. (ϵ 34,700); $\lambda_{max}^{pH\,11}$ 250 nm. (ϵ 26,200) and 318 nm. (ϵ 30,800).

[a] A shorter reaction time (1–2 hr.), or a lower reaction temperature, results in the recovery either of starting material or of mono- or di-acylated derivatives.

[b] With R = $CH_3(CH_2)_7CH{=}CH(CH_2)_7$ (**4**), the water, which is added to induce crystallization of the residue, is decanted, and the residual semi-solid crystallizes on adding 200 ml. of anhydrous methanol.

[c] Attempted recrystallization of the long-chain acyld erivatives (compounds **4** and **7**) from relatively nonpolar solvents (ether, heptane, or benzene) caused the formation of a gel. This difficulty was overcome by recrystallizing the compounds from more polar solvents.

2-Amino-9-β-D-ribofuranosyl-9*H*-purine-6-thiol 2′,3′,5′-Tris(*p*-chlorobenzoate) (8)

This compound, recrystallized from ethyl acetate–methanol, is obtained in 87% yield, m.p. 227–228°, $\lambda_{max}^{pH\,1}$ 246 nm. (ϵ 44,300) and 340 nm. (ϵ 20,000); $\lambda_{max}^{pH\,11}$ 244 nm. (ϵ 41,400) and 323 nm. (ϵ 14,300).

REFERENCES

(1) (a) R. W. Brockman, *Clin. Pharmacol. Therap.*, **2**, 237 (1961); (b) A. R. P. Peterson, *Can. J. Biochem. Physiol.*, **40**, 195 (1962); (c) A. R. P. Peterson and A. Hori, *ibid.*, **41**, 1339 (1963); (d) A. L. Bieber and A. C. Sartorelli, *Federation Proc.*, **22**, 184 (1963); (e) J. A. Montgomery, G. J. Dixon, E. A. Dulmage, H. J. Thomas, R. W. Brockman, and H. E. Skipper, *Nature*, **199**, 769 (1963).

(2) (a) H. Hoeksema, G. B. Whitfield, and L. E. Rhuland, *Biochem. Biophys. Res. Commun.*, **6**, 213 (1961); (b) E. S. Perkins, R. M. Wood, M. L. Sears, W. H. Prusoff, and A. D. Welch, *Nature*, **194**, 985 (1962); (c) K. L. Mukherjee and C. Heidelberger, *Cancer Res.*, **22**, 815 (1962); (d) W. A. Creasey, M. E. Fink, R. E. Handschumacher, and P. Calabresi, *ibid.*, **23**, 444 (1963).

(3) L. R. Lewis, R. K. Robins, and C. C. Cheng, *J. Med. Chem.*, **7**, 200 (1964).

(4) J. F. Gerster, J. W. Jones, and R. K. Robins, *J. Org. Chem.*, **28**, 945 (1963).

[89] 9-(2,3-*O*-Isopropylidene-β-D-ribofuranosyl)-9*H*-purine-6(1*H*)-thione (2′,3′-*O*-Isopropylidene-6-thioinosine)

An Intermediate in the Synthesis of Nucleotides

SARAH J. CLAYTON and WILLIAM E. FITZGIBBON, JR.

KETTERING-MEYER LABORATORY, SOUTHERN RESEARCH INSTITUTE, BIRMINGHAM, ALABAMA 35205

$$\mathbf{1}\ (284.3) \xrightarrow[\text{HClO}_4]{(\text{Me})_2\text{C(OMe)}_2,\ \text{MeCOMe}} \mathbf{2}\ (324.4)$$

INTRODUCTION

Prior to preparation of the 5′-phospate of a furanosyl nucleoside, the hydroxyl groups at C-2′ and C-3′ are frequently protected by formation of the 2′,3′-isopropylidene acetal. The use of perchloric acid[1] as a catalyst gives consistently reproducible results in the preparation of 9-(2,3-*O*-isopropylidene-β-D-ribofuranosyl)-9*H*-purine-6(1*H*)-thione (**2**)[1,2] on either a small or a large scale.

PROCEDURE

9-(2,3-*O*-Isopropylidene-β-D-ribofuranosyl)-9*H*-purine-6(1*H*)-thione[1,2] (2)

To a 5-l., round-bottomed flask containing 2.36 l. of acetone (predried with magnesium sulfate) are added 65 ml. (0.53 mole) of 2,2-dimethoxypropane and then 87 ml. (1.40 mole) of 70% perchloric acid. The flask is stoppered, and the solution is stirred for 5 min. at room temperature. Then, 47.3 g. (0.166 mole) of 9-β-D-ribofuranosyl-9*H*-purine-6(1*H*)-thione[3] (**1**) is added quickly, the resulting clear, yellow solution is stirred for 20 min., and the reaction is terminated by adding 194 ml. of pyridine (predried with calcium hydride); a white precipitate is immediately formed. Water (920 ml.) is added, followed by 65 ml. of 15 *N* ammonium hydroxide, and the suspension is concentrated under diminished pressure at room temperature until all of the acetone has been removed. Addition of another 65-ml. portion of 15 *N* ammonium hydroxide produces a clear solution, which is washed with two 325-ml. portions of dichloromethane, filtered, cooled to 0–5°, and adjusted to pH 6 with glacial acetic acid. The resulting suspension is refrigerated overnight, and the white solid is collected by filtration, washed with two 100-ml. portions of water, and dried for 16 hr. at 78° *in vacuo*. The product **2** weighs 46.5 g. (86%), m.p. 238–240° (dec.), λ_{max} pH 1, 322 nm. (log ϵ 4.38); pH 7, 318 nm. (log ϵ 4.36); pH 13, 310 nm. (log ϵ 4.37). A single spot is shown by paper chromatography (descending) in water-saturated butyl alcohol.

REFERENCES

(1) J. A. Zderic, J. G. Moffatt, D. Kau, K. Gerzon, and W. E. Fitzgibbon, *J. Med. Chem.*, **8**, 275 (1965).

(2) (a) A. Hampton and M. H. Maguire, *J. Am. Chem. Soc.*, **83**, 150 (1961); (b) A. Hampton, *ibid.*, **83**, 3640 (1961).

(3) (a) J. A. Johnson, Jr., H. J. Thomas, and H. J. Schaeffer, *J. Am. Chem. Soc.*, **80**, 699 (1958); (b) J. J. Fox, I. Wempen, A. Hampton, and I. L. Doerr, *ibid.*, **80**, 1669 (1958).

[90] *N*-(2-Amino-9-β-D-ribofuranosyl-9*H*-purin-6-yl)glycine

Preparation of a Nucleoside Derivative of an Amino Acid

RICHARD S. VICKERS, JOHN F. GERSTER, and ROLAND K. ROBINS

DEPARTMENT OF CHEMISTRY, UNIVERSITY OF UTAH, SALT LAKE CITY, UTAH 84112

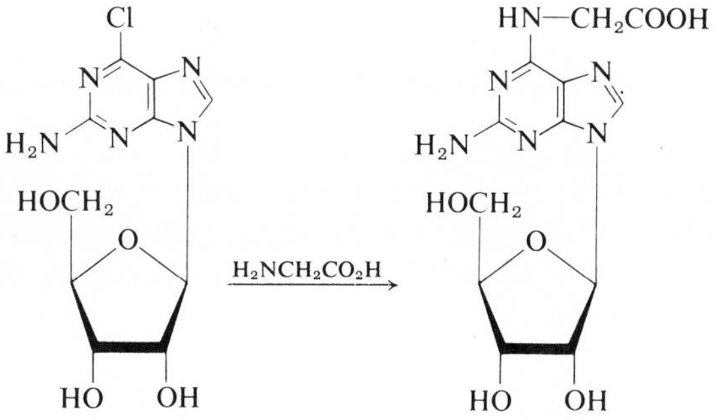

Cl
N N
H_2N N N
$HOCH_2$
O
HO OH
1
(301.5)

$H_2NCH_2CO_2H$ →

HN—CH_2COOH
N N
H_2N N N
$HOCH_2$
O
HO OH
2
(340.0)

INTRODUCTION

The isolation of *N*-(1,6-dihydro-6-oxo-9-β-D-ribofuranosyl-9*H*-purin-2-yl)alanine from acid-soluble nucleotides of[1] a *Fusarium* species, and the isolation of *N*-(aminoacyl)adenosine from yeast ribonucleic acid[2] has focused attention on nucleosides having amino acid residues attached to the purine moiety. A number of *N*-(purin-6-yl)amino acids have been prepared from the corresponding amino acid and 6-chloropurine.[3] The present procedure is an extension of that of Carter,[3] and has been adapted to purine nucleosides.

PROCEDURE

N-(2-Amino-9-β-D-ribofuranosyl-9*H*-purin-6-yl)glycine (2)

2-Amino-6-chloro-9-β-D-ribofuranosyl-9*H*-purine[4] (**1**) (3 g., 0.01 mole) is added to a solution of 1.0 g. of potassium carbonate and 1.5 g. of glycine in 30 ml. of water and 30 ml. of *N*,*N*-dimethylformamide. The mixture is heated on a steam bath for 3 hr., and completion of the reaction is ascertained by disappearance of the ultraviolet absorbance at λ_{max} 310 nm. (due to **1**). The solution is evaporated to dryness under diminished pressure, and the residue is dissolved in the minimal volume of water. The pH of the solution is adjusted to 4 with formic acid, and the precipitate is filtered off, and washed with water; yield 2.5 g. (74%) of colorless powder of **2**. The crude product is recrystallized from water; m.p. 170–175° (dec.); $\lambda_{max}^{pH\,1}$ 251 and 289 nm.; $\lambda_{max}^{pH\,11}$ 260 (sh.) and 284 nm.

Anal. Calcd. for $C_{12}H_{16}N_6O_6$: C, 42.3; H, 4.74; N, 24.7. Found: C, 42.2; H, 4.91; N, 24.6.

REFERENCES

(1) A. Ballio, W. Russi, and G. Serlupi-Crescenzi, *Gazz. Chim. Ital.*, **94**, 156 (1964).
(2) R. Hall, *Biochemistry*, **3**, 769 (1964).
(3) C. E. Carter, *J. Biol. Chem.*, **223**, 139 (1956); D. N. Ward, J. Wade, E. F. Walborg, Jr., and T. S. Osdene, *J. Org. Chem.*, **26**, 5000 (1961).
(4) J. F. Gerster, J. W. Jones, and R. K. Robins, *J. Org. Chem.*, **28**, 945 (1963).

Section III

Nucleosides

B. Pyrimidine Nucleosides and Analogs

[91] *N*-Acyl Derivatives of 2′-Deoxycytidine*

The Selective Acylation of the 4-Amino Group of a Cytosine Nucleoside

BRIAN A. OTTER and JACK J. FOX

DIVISION OF BIOLOGICAL CHEMISTRY, SLOAN-KETTERING INSTITUTE FOR CANCER RESEARCH, SLOAN-KETTERING DIVISION OF CORNELL UNIVERSITY MEDICAL COLLEGE, NEW YORK, NEW YORK 10021

NH_2 ... HOCH$_2$ ... HO

1 (227.2)

acid anhydride / ethanol, reflux →

HNR ... HOCH$_2$ ... HO

2, R = $COCH_3$ (269.3)

3, R = COC_6H_5 (331.3)

4, R = $COC_6H_4OCH_3$-*p* (361.4)

INTRODUCTION

Derivatives of 2′-deoxycytidine (**1**) having a protected 4-amino group, for example the *N*-anisoyl derivative (**4**), are important intermediates in the synthesis of phosphoric esters of 2′-deoxycytidine and of oligonucleotides containing 2′-deoxycytidine residues.[1,2] Previously, compound **4** has been prepared[1(a)] by a two-step procedure involving complete anisoylation of 2′-deoxycytidine (**1**) followed by selective removal of the *O*-anisoyl groups under carefully controlled, alkaline conditions. The *N*-benzoyl derivative (**3**) has been prepared in this way, and also by benzoylation of **1** with benzoic anhydride in pyridine.[3] The method of preparation described herein is an extension of that reported[4] for the selective *N*-acylation of cytidine. This method gives higher yields and is simpler than those previously reported.[1(a),3]

* The authors acknowledge the support of the United States Public Health Service (Grant No. CA-08748).

PROCEDURE

General Method of Acylation

To a stirred, refluxing solution of 500 mg. (2.2 mmoles) of 2′-deoxycytidine (**1**) in 50 ml. of dry ethanol is added 500 mg. of the acid anhydride.[a] During the course of refluxing, five further additions of 500-mg. portions of the anhydride are made hourly. After the final addition, the solution is refluxed for 1 hr. The solution is cooled and, depending on the acid anhydride employed, is processed as described below.

N-Acetyl-2′-deoxycytidine (2)

The reaction mixture is concentrated under diminished pressure at 30° to about 10 ml. Ether is added until the solution is faintly turbid, and the mixture is kept overnight in a refrigerator. The crystals are removed by filtration and well washed with cold ether; yield 490–530 mg. (83–89%), m.p. 154–176°. The product is chromatographically pure,[b] and the melting point remains unchanged after repeated recrystallization; $[\alpha]_D^{25}$ +89° (*c* 0.4, ethanol), $\lambda_{max}^{H_2O}$ 213, 245, 296 nm. (ϵ 19,250, 15,050, 8,700), $\lambda_{min}^{H_2O}$ 226, 269 nm. (ϵ 6,200, 3,700).

N-Benzoyl-2′-deoxycytidine (3)

The reaction mixture is concentrated almost to dryness under diminished pressure, and the sirupy residue is triturated with ether, affording crystals. The suspension is filtered, and the solid is triturated three times with 100-ml. portions of warm ether (to dissolve benzoic acid). The residue is dried, and dissolved in 50 ml. of hot water, and the solution is kept for several hours at room temperature and then in a refrigerator. The crystals are filtered off, washed with cold water, and dried; yield, 547 mg. (75%), m.p. 205–206°; solidifies at 208° and darkens, without remelting, above 230°, $[\alpha]_D^{25}$ +89° (*c* 0.2, ethanol). After concentration of the filtrate, a second crop of equal purity is obtained (110 mg., 15%). $\lambda_{max}^{H_2O}$ 257, 302 nm. (ϵ 22,100, 12,200); $\lambda_{min}^{H_2O}$ 219, 282 nm. (ϵ 8,350, 9,100).

N-Anisoyl-2′-deoxycytidine (4)

The residue remaining after evaporation of the reaction mixture is triturated with 100 ml. of warm ether. The suspension is filtered, and trituration is repeated twice. The dried residue is dissolved in 70 ml. of hot water, and the solution is kept at room temperature overnight. The

[a] Acetic anhydride is used in 0.5-ml. portions. *p*-Anisic anhydride (m.p. 99°, from ether–petroleum ether) is prepared according to the method described for benzoic anhydride.[5]

[b] On silica gel GF_{254} (Merck) in two solvent systems: 5:1 (v/v) chloroform–methanol and 4:1 (v/v) ethyl acetate–chloroform.

crystals are filtered off, washed with cold water, and dried; yield, 635 mg. (80%), m.p. 173–175°, resolidifies at 180° and darkens, without remelting, above 220°, $[\alpha]_D^{25}$ +51° (*c* 0.2, ethanol). After concentration of the filtrate, a second crop of equal purity is obtained (80 mg., 10%). $\lambda_{max}^{H_2O}$ 302 nm. (ϵ 25,100); shoulder at 250–260 nm.

REFERENCES

(1) (a) H. Schaller, G. Weimann, B. Lerch, and H. G. Khorana, *J. Am. Chem. Soc.*, **85**, 3821 (1963); (b) H. Schaller and H. G. Khorana, *ibid.*, **85**, 3841 (1963).

(2) See T. Ueda and J. J. Fox, *Advan. Carbohydrate Chem.*, **22**, 307 (1967) for a general review of the use of N^4-acylated nucleosides in nucleotide syntheses.

(3) E. Benz, N. F. Elmore, and L. Goldman, *J. Org. Chem.*, **30**, 3067 (1965).

(4) K. A. Watanabe and J. J. Fox, *Angew. Chem. Intern. Ed. Engl.*, **5**, 579 (1966).

(5) H. Adkins and Q. E. Thompson, *J. Am. Chem. Soc.*, **71**, 2242 (1949).

[92] 2′,5′-Dideoxycytidine

Conversion of a Hydroxymethyl Group into a Methyl Group

ELIZABETH BENZ, NORMAN F. ELMORE, and LEON GOLDMAN

ORGANIC CHEMICAL RESEARCH SECTION, LEDERLE LABORATORIES, A DIVISION OF AMERICAN CYANAMID COMPANY, PEARL RIVER, NEW YORK 10965

INTRODUCTION

The utility of such 2′-deoxynucleosides as 2′-deoxy-5-iodouridine, 2′-deoxy-5-fluorouridine, 2′-deoxy-5-fluorocytidine, and 3′-deoxyadenosine (cordycepin) is well established.[1] The synthesis here described for 2′,5′-dideoxycytidine (**9**) from a suitably protected 2′-deoxycytidine illustrates a synthesis of 2′-deoxynucleosides in which a hydroxymethyl group is converted, *via* its 5′-*p*-toluenesulfonic ester, into an iodo derivative that, on catalytic hydrogenolysis, gives a methyl group.

PROCEDURE[a]

2′-Deoxycytidine[2] (2)

To a suspension of 20.0 g. (0.076 mole) of 2′-deoxycytidine hydrochloride (**1**) in 1 l. of chloroform is added 10.6 ml. (7.69 g., 0.076 mole) of triethylamine. The resulting suspension is stirred at room temperature for 20 hr. and then filtered. The colorless, crystalline precipitate is removed by filtration, washed with chloroform, and air-dried. The yield of **2**, m.p. 185–195°, is 16.4 g. (95%).

N-Benzoyl-2′-deoxycytidine[3,4] (3) and *N*-Benzoyl-5′-*O*-benzoyl-2′-deoxycytidine[b] (4)

To a suspension of 13.0 g. (0.057 mole) of 2′-deoxycytidine (**2**) in 250 ml. of anhydrous pyridine is added 14.0 g. (0.062 mole) of benzoic anhydride, and the resulting suspension is stirred at room temperature for 1.75 hr., when a clear solution results. The solution is then stirred for 2.25 hr., and evaporated to dryness *in vacuo*, and the semisolid residue is added, with vigorous stirring, to 1 l. of ice–water. The resulting solution is washed with ether, and the aqueous solution is evaporated to dryness *in vacuo*, to

[a] Evaporations are conducted with a rotary evaporator, at a bath temperature below 25°.

[b] For the isolation of **4**, see ref. 4.

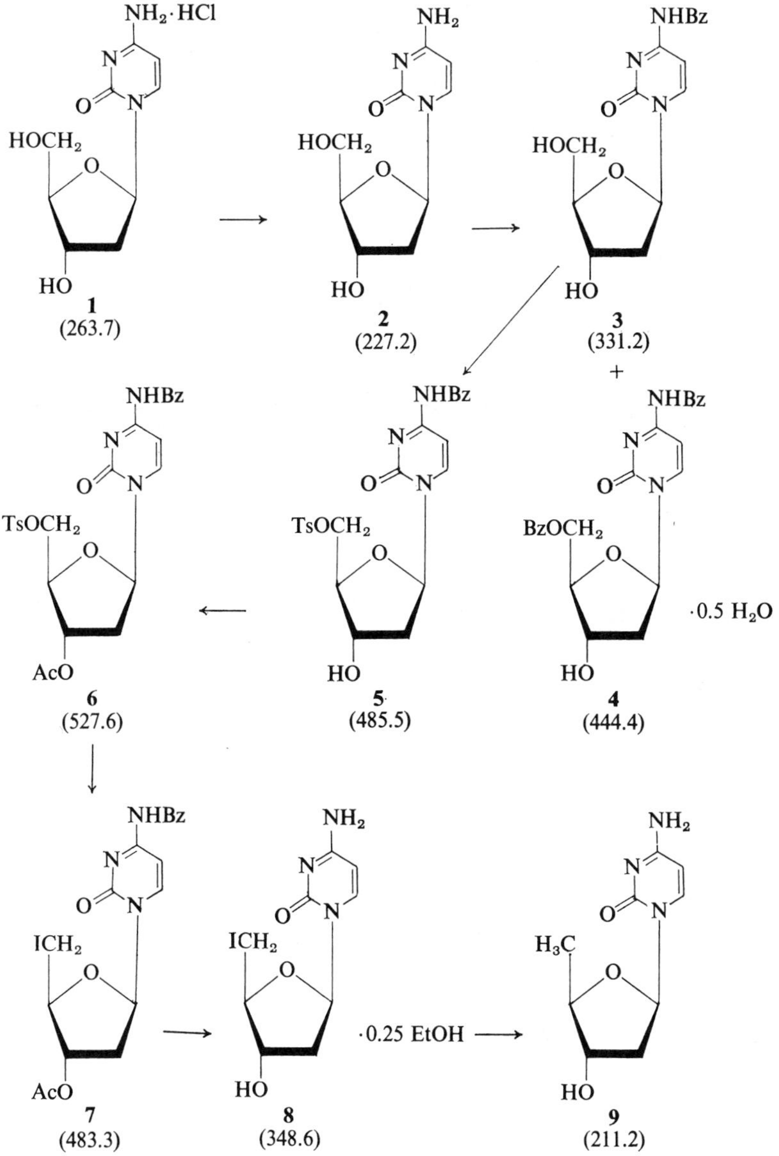

$NH_2 \cdot HCl$, N, O, N, $HOCH_2$, O, HO
1 (263.7)

→

NH_2, N, O, N, $HOCH_2$, O, HO
2 (227.2)

→

NHBz, N, O, N, $HOCH_2$, O, HO
3 (331.2)

\+

NHBz, N, O, N, $BzOCH_2$, O, HO, $\cdot 0.5\ H_2O$
4 (444.4)

NHBz, N, O, N, $TsOCH_2$, O, HO
5 (485.5)

←

NHBz, N, O, N, $TsOCH_2$, O, AcO
6 (527.6)

↓

NHBz, N, O, N, ICH_2, O, AcO
7 (483.3)

→

NH_2, N, O, N, ICH_2, O, HO, $\cdot 0.25$ EtOH
8 (348.6)

→

NH_2, N, O, N, H_3C, O, HO
9 (211.2)

where Bz is benzoyl, and Ts is *p*-tolylsulfonyl.

yield 14.9 g. of a colorless solid. Crystallization from absolute ethanol gives 11.7 g. (62%) of **3** as colorless needles, m.p. 205–207°, $[\alpha]_D^{25}$ +82° (*c* 0.6, ethanol), $\lambda_{max}^{H_2O}$ 258 and 301 nm. (ϵ 20,500 and 10,900).

N-Benzoyl-2′-deoxy-5′-*O*-*p*-tolylsulfonylcytidine[4] (5)

A suspension of 9.00 g. (0.027 mole) of *N*-benzoyl-2′-deoxycytidine (**3**) in 100 ml. of anhydrous pyridine is cooled to 0° in an ice bath, 5.18 g. (0.027 mole) of *p*-toluenesulfonyl chloride is added, and the mixture is stirred for 18 hr. while being cooled in an ice bath. The solution is then concentrated *in vacuo* to one-third of its original volume, and the resulting sirup is poured, with stirring, into 500 ml. of ice–water. The resulting gummy precipitate is extracted with ethyl acetate, and the extract is dried (magnesium sulfate), and evaporated to dryness *in vacuo*, to yield 10.8 g. of a colorless solid, m.p. 143–150° (dec.). Recrystallization from 250 ml. of absolute ethanol yields 5.87 g. (44%) of **5** as colorless needles, m.p. 146–149° (dec.). Further recrystallization from absolute ethanol gives colorless needles, m.p. 149.5–151° (dec.), $[\alpha]_D^{25}$ +105° (*c* 0.55, ethanol), λ_{max}^{EtOH} 223, 260, and 305 nm. (ϵ 21,800, 22,300, and 9,700).

3′-*O*-Acetyl-*N*-benzoyl-2′-deoxy-5′-*O*-*p*-tolylsulfonylcytidine[4] (6)

A solution of 1.00 g. (2.06 mmoles) of compound **5** and 1.00 ml. (1.08 g., 10.6 mmoles) of acetic anhydride in 10 ml. of anhydrous pyridine is kept at room temperaure for 18 hr. Methanol (5 ml.) is added and, after 30 min., the solution is concentrated *in vacuo* to a gel which is triturated with water to give a colorless solid, removed by filtration. Crystallization from 75 ml. of absolute ethanol gives 772 mg. (71%) of **6** as colorless plates, m.p. 160–163° (dec.), $[\alpha]_D^{25}$ +48° (*c* 1.0, *N*,*N*-dimethylformamide), λ_{max}^{MeOH} 224, 260, and 302 nm. (ϵ 24,200, 24,700, and 10,500).

3′-*O*-Acetyl-*N*-benzoyl-2′,5′-dideoxy-5′-iodocytidine[4] (7)

A magnetically stirred mixture of 4.00 g. (7.58 mmoles) of compound **6** and 2.28 g. (15.2 mmoles) of sodium iodide in 200 ml. of 2-butanone is refluxed under nitrogen for 2 hr. The suspension is cooled to room temperature, and filtered to remove 1.5 g. (100%) of sodium *p*-toluenesulfonate. The filtrate is evaporated to dryness *in vacuo*, giving an orange gum that solidifies on trituration with a little absolute ethanol. The solid is filtered off, and recrystallized from 250 ml. of absolute ethanol to yield 3.10 g. (85%) of **7** as colorless needles, m.p. 177–180° (dec.). Recrystallization from absolute ethanol gives colorless needles, m.p. 179–181° (dec.), $[\alpha]_D^{25}$ +42° (*c* 1.0, *N*,*N*-dimethylformamide), λ_{max}^{MeOH} 260 and 303 nm. (ϵ 23,000 and 10,400).

2′,5′-Dideoxy-5′-iodocytidine[4] (8)

A suspension of 3.10 g. (6.42 mmoles) of compound **7** in 300 ml. of half-saturated methanolic ammonia is stirred at 3° for 18 hr. The resulting solution is evaporated *in vacuo* to a semisolid, which is triturated with ether and filtered, to yield 1.99 g. (92%) of **8** as colorless needles, m.p. 179–183° (dec.). Recrystallization from absolute ethanol gives colorless needles, m.p. 178–180° (dec.), $[\alpha]_D^{25}$ +41° (*c* 1.0, *N*,*N*-dimethylformamide), λ_{max}^{MeOH} 235 and 270 nm. (ϵ 8,350 and 9,260).

2′,5′-Dideoxycytidine[4] (9)

A solution of 391 mg. (1.12 mmoles) of compound **8** in 20 ml. of water and 40 ml. of ethanol is adjusted to pH 10 with concentrated ammonium hydroxide. The solution is then hydrogenated at room temperature and atmospheric pressure in the presence of 1.60 g. of 5% palladium-on-barium sulfate. After 3.75 hr., the uptake of hydrogen ceases. The catalyst is removed by filtration through diatomaceous earth, and the filtrate is evaporated to dryness *in vacuo*. The resulting, yellow oil is dissolved in water, and the solution is washed with ether and then evaporated to dryness *in vacuo*, to yield 157 mg. (41%) of the hydriodide of **9** as a colorless solid, m.p. 162.5–163°. A suspension of this salt (157 mg.) in 10 ml. of chloroform and 0.065 ml. (47 mg., 0.46 mmole) of triethylamine is stirred at room temperature for 1 hr. The precipitate is collected by filtration, and washed with chloroform, to yield 110 mg. (46%) of **9** as colorless needles, m.p. 187.5–193°. Recrystallization from absolute ethanol–ether gives colorless needles, m.p. 189–191°, $[\alpha]_D^{25}$ +105° (*c* 1.0, ethanol), λ_{max}^{MeOH} 230 and 272 nm. (ϵ 7,380 and 8,220).

REFERENCES

(1) I. H. Leopold, *Ann. N. Y. Acad. Sci.*, **130**, 181 (1965), and references cited therein; J. H. Burchenal, E. A. D. Holmberg, J. J. Fox, S. C. Hemphill, and J. A. Reppert, *Cancer Res.*, **19**, 494 (1959); E. A. Kaczka, N. R. Trenner, R. Arison, R. W. Walker, and K. Folkers, *Biochem. Biophys. Res. Commun.*, **14**, 456 (1964).

(2) J. J. Fox, N. C. Yung, I. Wempen, and M. Hoffer, *J. Am. Chem. Soc.*, **83**, 4066 (1961).

(3) H. Schaller, G. Weimann, B. Lerch, and H. G. Khorana, *J. Am. Chem. Soc.*, **85**, 3821 (1963).

(4) E. Benz, N. F. Elmore, and L. Goldman, *J. Org. Chem.*, **30**, 3067 (1965).

[93] 2′,3′-*O*-(*p*-Methoxybenzylidene)cytidine

*2′,3′-*O*-Alkylidene Acetals of Ribonucleosides, Specifically Substituted Intermediates for the Preparation of Oligoribonucleotides Having the Naturally Occurring 3′ → 5′ Internucleotide Linkage*

S. CHLÁDEK

INSTITUTE OF ORGANIC CHEMISTRY AND BIOCHEMISTRY,
CZECHOSLOVAK ACADEMY OF SCIENCES, PRAGUE, CZECHOSLOVAKIA

$NH_2 \cdot 0.5\ H_2SO_4$; HOCH$_2$; HO ; OH

1 (292.3)

p-MeO-C$_6$H$_4$-CHO ; $HC(OEt)_3$; H^+ →

2 (361.4)

INTRODUCTION

The 2′,3′-*O*-*p*-methoxybenzylidene acetals of ribonucleosides are important, specifically substituted intermediates having a free 5′-hydroxyl group. 2′,3′-*O*-(*p*-Methoxybenzylidene)uridine is obtained by the reaction of uridine with *p*-anisaldehyde in the presence of zinc chloride.[1] The 2′,3′-*O*-*p*-methoxybenzylidene acetals of adenosine, guanosine,[2] and cytidine[3] are prepared by reaction of the corresponding nucleosides with *p*-anisaldehyde, ethyl orthoformate, and anhydrous hydrogen chloride in *N*,*N*-dimethylformamide.

PROCEDURE

2′,3′-*O*-(*p*-Methoxybenzylidene)cytidine (2)

In a 50-ml., round-bottomed flask is placed a mixture of 2.92 g. (0.01 mole) of cytidine hemisulfate (**1**), 10 ml. of anhydrous *N*,*N*-dimethylformamide,[a] 6.0 ml. (0.05 mole) of *p*-anisaldehyde,[b] 12.0 ml. (0.08 mole) of ethyl orthoformate, and 1.5 ml. of 7.1 *M* hydrogen chloride in *p*-dioxane.[c] The flask is stoppered, and the mixture is shaken until complete dissolution has occurred. The solution is kept at room temperature for 5–8 hr., and poured, with vigorous stirring, into a mixture of aqueous sodium carbonate (4 g. in 200 ml. of water) and 200 ml. of ether. The resulting precipitate is collected with suction, washed with 50 ml. of ether, and air-dried. The chromatographically homogeneous product (**2**) weighs 2.5 g. (90%), m.p. 220.5–222°, R_f 0.69 in butyl alcohol–water. Recrystallization from 50 ml. of 2:1 ethanol–water affords an analytically pure product, m.p. 231–233°.

REFERENCES

(1) M. Smith, D. H. Rammler, I. H. Goldberg, and H. G. Khorana, *J. Am. Chem. Soc.*, **84**, 430 (1962).
(2) S. Chládek and J. Smrt, *Collection Czech. Chem. Commun.*, **28**, 1301 (1963).
(3) S. Chládek, unpublished results.

[a] *N*,*N*-Dimethylformamide is dried by vacuum distillation from phosphorus pentaoxide (10% by weight), and the distillate is stored over molecular sieves.

[b] *p*-Anisaldehyde is freshly distilled under diminished pressure.

[c] Dry hydrogen chloride is introduced, with cooling in ice, into *p*-dioxane predried by distillation over sodium.

[d] The course of the reaction is followed by descending chromatography on Whatman No. 1 paper in butyl alcohol–water. A sample of the reaction mixture is made alkaline with aqueous ammonia, and then applied to the paper.

[94] *N*-Acetyl-*O*-(2,3,4,6-tetra-*O*-acetyl-β-D-glucosyl)cytosine and *N*-acetyl-1-(2,3,4,6-tetra-*O*-acetyl-β-D-glucosyl)cytosine

N- *and* O-*Glycosyl Derivatives of a Tautomeric Pyrimidine*

G. T. ROGERS, DAVID THACKER, and T. L. V. ULBRICHT

TWYFORD LABORATORIES, ELVEDEN ROAD, LONDON, N.W. 10, ENGLAND

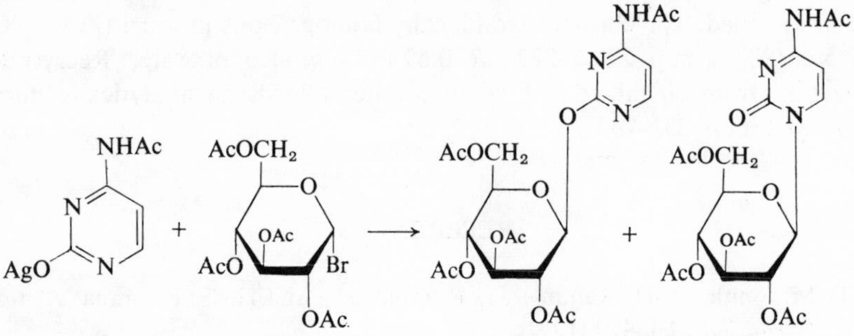

NHAc AcOCH$_2$ NHAc NHAc

N O AcOCH$_2$ AcOCH$_2$

AgO N + OAc Br → OAc + OAc

AcO OAc AcO OAc AcO OAc

1 (260.0) **2** (411.2) **3** (483.4) **4** (483.4)

INTRODUCTION

The reaction of silver salts of such tautomeric bases as cytosine with poly-*O*-acylglycosyl halides leads exclusively or preponderantly to the formation of an *O*-glycosyl derivative,[1–4] which may be accompanied by a small proportion of the *N*-glycosyl derivative. Treatment of the *O*-glycosyl derivative (or the mixture) with mercuric bromide gives the *N*-glycosyl derivative in high yield.[2]

PROCEDURE

N-Acetyl-O-(2,3,4,6-tetra-O-acetyl-β-D-glucosyl)cytosine (3) and N-Acetyl-1-(2,3,4,6-tetra-O-acetyl-β-D-glucosyl)cytosine (4)

The silver salt of *N*-acetylcytosine[2] (**1**) (5.11 g., 0.02 mole) is azeotropically dried by distillation of xylene from a suspension in xylene (120 ml.); tetra-*O*-acetyl-α-D-glucopyranosyl bromide[5] (**2**) (7.63 g., 0.019 mole) is added, and the mixture is refluxed, with stirring, for 20 min. The precipitated silver bromide is filtered off, and washed with dichloromethane (100 ml.), and the filtrate and washing are combined, and evaporated

(40°) to dryness under diminished pressure to a yellow glass. This is dissolved in dichloromethane (200 ml.), isopropyl ether is added and the solution is cooled, giving the *N*-β-D-glucosyl derivative[2] (**4**) (1.5 g., 16%), m.p. 223–224° (softens at 140–150° and resolidifies).

The rest of the product is purified by preparative, thin-layer chromatography with ethyl acetate on silica gel GF_{254} (Merck, Darmstadt). The component of R_f 0.2 is compound **4** [total yield 2.0 g. (22%)], and that having R_f 0.8 is the *O*-β-D-glucosyl derivative (**3**). The latter is further purified by preparative, thin-layer chromatography in ethyl acetate, followed by multiple elution with ether (peroxide-free, not dried) in the reverse direction, and crystallization from dichloromethane–isopropyl ether; yield 3.4 g. (37%), m.p. 133–135°, λ_{max} 231 nm. (ϵ 10,500) and 272 nm. (ϵ 12,000) in 95% EtOH.

REFERENCES

(1) T. L. V. Ulbricht, *Proc. Chem. Soc.*, **1962**, 298.
(2) T. L. V. Ulbricht and G. T. Rogers, *J. Chem. Soc.*, **1965**, 6123.
(3) H. G. Garg and T. L. V. Ulbricht, *J. Chem. Soc.* (*C*), **1967**, 51.
(4) D. Thacker and T. L. V. Ulbricht, *Chem. Commun.*, **1967**, 122.
(5) R. U. Lemieux, *Methods Carbohydrate Chem.*, **2**, 221 (1963).

[95] 1-β-D-Arabinofuranosylcytosine

Epimerization of the Glycosyl Moiety of a Nucleoside

WALDEN K. ROBERTS and CHARLES A. DEKKER

DEPARTMENT OF BIOCHEMISTRY AND VIRUS LABORATORY, UNIVERSITY OF CALIFORNIA, BERKELEY, CALIFORNIA 94720

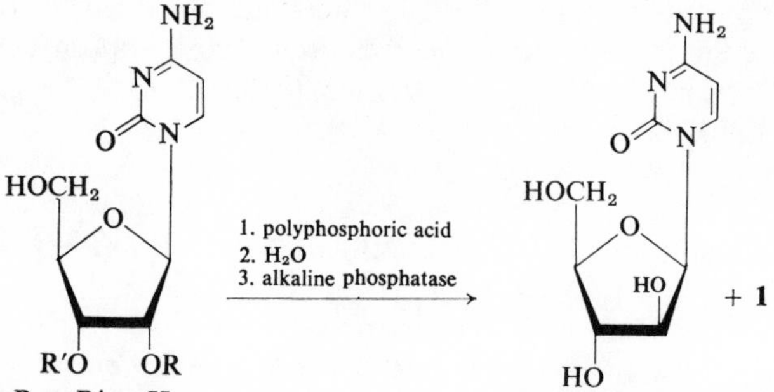

NH_2 N N O $HOCH_2$ O R'O OR

1, R = R' = H (243.2)

2, R = PO_3H_2; R' = H (323.2)

3, R = H; R' = PO_3H_2 (323.2)

1. polyphosphoric acid
2. H_2O
3. alkaline phosphatase

NH_2 N N O $HOCH_2$ O HO HO

4 (243.2) + **1**

INTRODUCTION

Treatment of cytidine (**1**) or of 2′(3′)-cytidylic acid (**2** and **3**) with polyphosphoric acid leads to the formation of phosphorylated derivatives of 2,2′-anhydro-β-D-arabinofuranosylcytosine. Subsequent hydrolysis and dephosphorylation yields a mixture of nucleosides, of which 1-β-D-arabinofuranosylcytosine (**4**) and **1** are the major components.[1] The separation of the mixture is achieved by ion-exchange chromatography.[2] Alternative procedures of differing complexities are available[3] for the synthesis of **4**.

PROCEDURE

1-β-D-Arabinofuranosylcytosine (4)

2′(3′)-Cytidylic acid (**2** and **3**) (6.65 g., 20.6 mmoles) is suspended in 100 g. of polyphosphoric acid in a stoppered flask,[a] and the suspension is

[a] Prewarming the polyphosphoric acid lowers its viscosity, and facilitates handling.

kept at 80° for 30 hr.[b] Water (200 ml.) is added, and the solution is heated at 100° for 60 min. (to hydrolyze the pyrophosphate bonds), and cooled. The pH of the solution is adjusted to 9.0 by the slow addition of 10% (w/v) aqueous lithium hydroxide, resulting in hydrolysis of the 2,2′-anhydro intermediate and giving, as the major product, the 3′,5′-diphosphate of **4**.

The solution is kept at 0° for several hours, the lithium phosphate is filtered off, and to the filtrate are added 20 g. of magnesium chloride, 20 ml. of 30% ammonium chloride solution, and then sufficient concentrated ammonium hydroxide to give a pH of 9.5, resulting in the precipitation of residual inorganic phosphate as ammonium magnesium phosphate, which is filtered off. Alkaline phosphatase[c] (20 mg.) is added to the filtrate, and the solution is incubated under toluene at 37° for 24 hr.[d] The new crop of ammonium magnesium phosphate is filtered off, and the filtrate is concentrated to 200 ml. (to remove excess of ammonia). The solution is diluted to 400 ml. with water, placed on a column (4.7 × 15 cm.) of Dowex-50 X-8 (H^+) ion-exchange resin (50–100 mesh), and the column is eluted with water until the mineral acids (derived from the anions of the salts) and the uracil nucleosides (60 mg.)[e] have emerged. Compounds **1** and **4** are then simultaneously displaced with *N* ammonium hydroxide (1 l.), and the fractions containing this mixture are evaporated to dryness (to remove ammonia and water). The residue is dissolved in 35 ml. of 30% aqueous methanol, and the solution is applied to a column (3.6 × 13 cm.) of Dowex-1 X-2 (OH^-) ion-exchange resin (200–400 mesh) pre-equilibrated with 30% aqueous methanol. Elution with the same solvent (1.5 l.) elutes **1**, which is isolated in pure form (423 mg.) by evaporation of this eluate.

The column is then eluted with 0.1 *M* ammonium hydrogen carbonate; compound **4** appears in coincidence with the bicarbonate front, and is completely eluted in a volume of *ca.* 200 ml. Repeated flash-evaporation of the aqueous solution decomposes the ammonium hydrogen carbonate, leaving crystalline **4** (2.44 g., 49%) of purity >95% as established by

[b] A clear solution is obtained by occasional swirling during the first 30 min. The progress of the reaction may be followed by periodic dilution of a small aliquot of the reaction mixture with cold, 0.1 *N* hydrochloric acid and examination of the A_{280}/A_{260} ratio. When the value has decreased to *ca.* 0.7, extensive formation of the desired 2,2′-anhydro intermediate has occurred.

[c] Commercial preparations of bovine, intestinal phosphatase are adequate.

[d] Paper chromatography in 7:1:2 isopropyl alcohol–ammonia–water may be used for ascertaining complete conversion of the nucleoside mono-, di-, and tri-phosphates into the free nucleosides.

[e] Deamination is a minor side-reaction during the incubation with polyphosphoric acid.

chromatography. The pure compound is obtained by recrystallization from alcohol; m.p. 212–213°, $[\alpha]_D^{25}$ +151° (*c* 0.5, water).

Alternatively, the hemisulfate may be prepared by addition of a slight excess of sulfuric acid to a 10% aqueous solution followed by precipitation with alcohol; for the hemisulfate, $[\alpha]_D^{25}$ +126° (*c* 0.5, water).

REFERENCES

(1) E. R. Walwick, W. K. Roberts, and C. A. Dekker, *Proc. Chem. Soc.*, **1959** 84; W. K. Roberts and C. A. Dekker, *J. Org. Chem.*, **32**, 816 (1967).

(2) C. A. Dekker, *J. Am. Chem. Soc.*, **87**, 4027 (1965).

(3) J. S. Evans, E. A. Musser, G. D. Mengel, K. R. Forsblad, and J. H. Hunter, *Proc. Soc. Exptl. Biol. Med.*, **106**, 350 (1961); T. Y. Shen, H. M. Lewis, and W. V. Ruyle, *J. Org. Chem.*, **30**, 835 (1965); J. J. Fox and I. Wempen, *Tetrahedron Letters*, **1965**, 643; H. P. M. Fromageot and C. B. Reese, *ibid.*, **1966**, 3499; J. J. Fox, N. Miller, and I. Wempen, *J. Med. Chem.*, **9**, 101 (1966).

[96] 1-β-D-Arabinofuranosyl-5-fluorocytosine*

Conversion of Uracil Nucleosides into their Cytosine Analogs via *the Thiation Process*

IRIS WEMPEN and JACK J. FOX

DIVISION OF BIOLOGICAL CHEMISTRY, SLOAN-KETTERING INSTITUTE FOR CANCER RESEARCH, SLOAN-KETTERING DIVISION OF CORNELL UNIVERSITY MEDICAL COLLEGE, NEW YORK, NEW YORK 10021

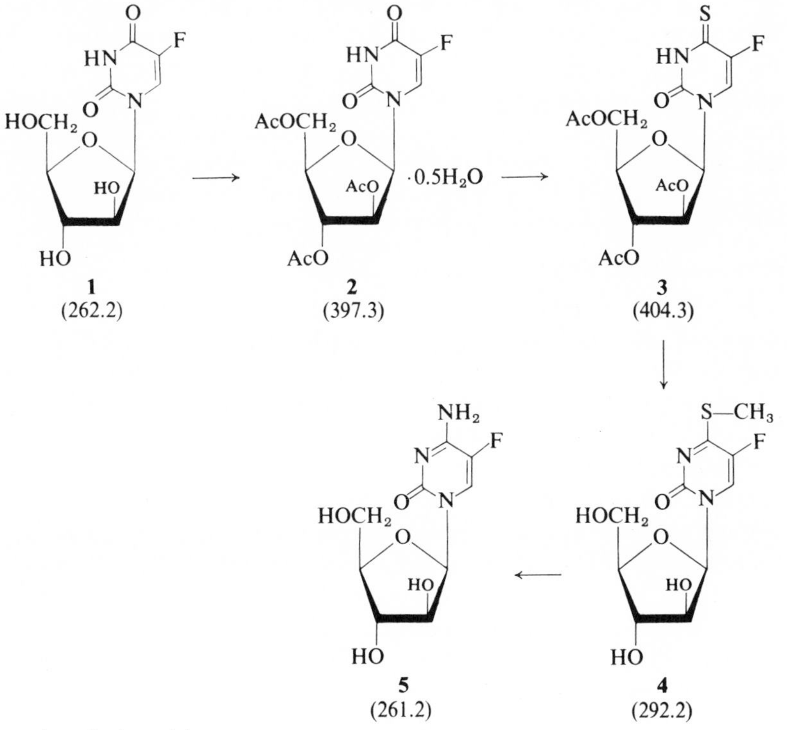

1 (262.2) → **2** $\cdot 0.5H_2O$ (397.3) → **3** (404.3) → **4** (292.2) → **5** (261.2)

where Ac is acetyl.

INTRODUCTION

1-β-D-Arabinofuranosyl-5-fluorocytosine[1] (**5**) is active against transplanted mouse leukemias, and exhibits antiviral activity against *Herpes keratitis* in rabbits.[1] The synthesis of 1-β-D-arabinofuranosyl-5-fluorocytosine illustrates a general method of conversion of various uracil

* The authors acknowledge the support of the United States Public Health Service (Grant No. CA-08748).

nucleosides into their cytosine analogs *via* the thiation procedure previously employed for nucleosides.[2] A modification, *S*-alkylation of the 4-thione (**3**), is introduced, to permit conversion into the title compound (**5**) without significant loss of the 5-fluorine atom.

PROCEDURE

5-Fluoro-1-(2,3,5-tri-*O*-acetyl-β-D-arabinosyl)uracil[1] (2)

A solution of 5.4 g. (0.02 mole) of 1-β-D-arabinofuranosyl-5-fluorouracil (**1**) (Sec. III [115]) in 60 ml. of anhydrous pyridine (distilled from barium oxide) is treated with 8.0 ml. of acetic anhydride in a tightly stoppered flask. The clear reaction mixture is kept for 16 hr. at room temperature, ethanol (*ca.* 10 ml.) is added, and the solution is evaporated under diminished pressure to a thin sirup. Pyridine is removed by repeated addition and evaporation of 50% aqueous ethanol, whereupon crystallization of **2** ensues. The mixture is cooled to complete the crystallization, and the product is filtered off by suction, well washed with water, and dried. The yield is 7.8 g. (98%), m.p. 139–140°. Recrystallization from 50% aqueous ethanol gives pure product (**2**), m.p. 142–143°.

5-Fluoro-4-thio-1-(2,3,5-tri-*O*-acetyl-β-D-arabinosyl)uracil[1] (3)

In a 100-ml., round-bottomed flask (equipped with a stirrer and a reflux condenser protected by a drying tube) is placed a solution of 3.97 g. (0.01 mole) of the acetylated nucleoside (**2**) in *ca.* 40 ml. of pyridine (reagent grade), 4.4 g. (0.02 mole) of phosphorus pentasulfide is added, and the mixture is stirred and heated for 4 hr.

The progress of the reaction is monitored as follows. A small aliquot is removed from the reaction mixture and evaporated to dryness. A small amount of 50% aqueous ethanol is added, and the evaporation is repeated; this addition of solvent and subsequent evaporation is repeated until the odor of pyridine is no longer evident. The gummy residue is then dissolved in 50% aqueous ethanol, and the ultraviolet absorption at 267 and 334 nm. is measured. The ratio of the A_{267}/A_{334} readings (starting material/thiated product) determines the progress of the reaction. When this ratio ceases to drop in value, it is necessary to add another portion (4.4 g., 0.02 mole) of phosphorus pentasulfide (with caution)[a] and continue the heating.

One or more additions of the thiating agent are usually necessary (during a total heating time of 12–16 hr.).[b] The ratio A_{267}/A_{334} of the completed

[a] The temperature of the reaction mixture should be allowed to drop to *ca.* 100° and the phosphorus pentasulfide added in portions to prevent excess foaming.

[b] It should be noted that conditions for the thiation of this compound are atypical as regards the difficulty of achieving complete reaction. It is more usual for such a thiation to be complete after one treatment with the thiating agent during 4–6 hr. of refluxing.

thiation reaction mixture should now be *ca.* 0.2. The reaction mixture is cooled, and the pyridine is decanted from a semi-solid residue. The residue is washed with a small volume of pyridine, and the decantate and washings are combined and evaporated to dryness *in vacuo*. Water (250 ml.) is added to the resulting sirupy residue, and the mixture is stirred until a yellow, granular solid separates. The solid is removed by filtration, and mixed with chloroform, and the suspension is filtered. The filtrate is then evaporated to dryness *in vacuo*, the resulting sirup is dissolved in hot methanol, and the solution is allowed to cool slowly. The yield of yellow needles varies from 3.0 to 3.6 g. (75–90%).[c] The crude product is used directly in the next step.

1-β-D-Arabinofuranosyl-5-fluoro-4-(methylthio)-2(1*H*)-pyrimidinone[1] (4)

To a solution of the thione (**3**) (13.6 g., 0.034 mole) in 250 ml. of methanol are added 50 ml. of water and 9.0 g. (0.063 mole) of methyl iodide. The solution is stirred magnetically, and 34.5 ml. of *N* sodium hydroxide is added, dropwise, during 40 min. The pH of the reaction mixture (pH 8) is then adjusted to 6 with glacial acetic acid, and part of the methanol is removed by evaporation *in vacuo*, whereupon a precipitate of light-yellow needles appears; this precipitate is removed by filtration, and the filtrate is re-evaporated. In this way, several crops are obtained; the combined yield is 8.3 g. (87%). The crude material is recrystallized from hot water, giving almost colorless pure product (**4**), m.p. 140–141°, $[\alpha]_D^{23}$ +219° (*c* 0.22, methanol). Ultraviolet absorption data: in 50% ethanol, λ_{max} at 202.5, 277.5, and 315 nm., λ_{min} at 237.5 and 292.5 nm., shoulder at 220 nm.

1-β-D-Arabinofuranosyl-5-fluorocytosine (5)

Compound **4** (5.0 g., 0.017 mole) is placed in a glass liner (of a steel bomb or autoclave) which is then cooled in a Dry Ice–acetone bath. Anhydrous ammonia is now passed into the cold container until *ca.* 25 ml. of liquid ammonia has been condensed. The cold container is now transferred as rapidly as possible into the steel casing, and this is sealed. The reaction mixture is kept at room temperature for 12–15 hr., the bleeder valve of the bomb is opened slightly, and the methanethiol liberated is allowed to escape into a hood. The bomb is now opened, and the residual ammonia is evaporated by directing a stream of dry nitrogen or dry air into the container. The resulting sirup is dissolved in 50 ml. of water, the pH is lowered to *ca.* 5 by adding dilute acetic acid, and the solution is evaporated to dryness *in vacuo*, giving a crystalline residue which is treated

[c] The crude product does not have a distinctive melting point, because some deacetylation may occur during the difficult thiating treatment.

with *ca.* 10 ml. of water. The mixture is filtered, and the insoluble material (3.4 g.), m.p. 232–233° (efferv.) is purified as follows. The crude solid is dissolved in the minimal volume of water, and the solution is placed on a column of clean[d] Dowex 50 (H^+) resin (100–200 mesh). The column is washed with water until it is free from ultraviolet-absorbing material,[e] and the product is then eluted with *N* ammonium hydroxide. The ammonia eluates that contain ultraviolet-absorbing material (280 nm.) are combined, and the ammonia is removed rapidly[f] by evaporation *in vacuo* at a bath temperature below 40°. After the ammonia has been removed, the solution is evaporated to dryness, and the residue is crystallized from hot, 90% ethyl alcohol; yield 3.2 g. (73%), m.p. 237–238° (browns at 225°), $[\alpha]_D^{23}$ $+163 \pm 2°$ (*c* 0.18, water). Ultraviolet absorption data: in *N* hydrochloric acid, λ_{max} at 221 and 290.5 nm. (ϵ_{max} 10,300 and 11,900); at pH 5–7, λ_{max} at 235 and 280 nm. (ϵ_{max} 7,860 and 5,240); pK_a 2.33 ± 0.05 (determined spectrophotometrically).

REFERENCES

(1) J. J. Fox, N. Miller, and I. Wempen, *J. Med. Chem.*, **9**, 101 (1966).
(2) J. J. Fox, D. van Praag, I. Wempen, I. L. Doerr, L. Cheong, J. E. Knoll, M. L. Eidinoff, A. Bendich, and G. B. Brown, *J. Am. Chem. Soc.*, **81**, 178 (1959).
(3) J. J. Fox, N. Miller, and R. Cushley, *Tetrahedron Letters*, **1966**, 4927.

[d] The resin used should be prewashed with 2 *N* hydrochloric acid until the ultraviolet-absorption reading becomes negligible, and then with water until the eluate gives a negative test for chloride ion.

[e] Any unconverted thione intermediate will be removed with this wash.

[f] The ammonia should be removed as rapidly as possible, and without excessive heating, to avoid side-reactions which can result in complete loss of the product.[3]

[97] 1-(2-Deoxy-β-D-*arabino*-hexofuranosyl)cytosine and 1-(2-Deoxy-β-D-*arabino*-hexofuranosyl)uracil

2-Deoxyaldohexofuranosyl Nucleosides by the Hilbert–Johnson Method

K. VENKATRAMANA BHAT

DEPARTMENT OF CHEMISTRY, GEORGETOWN UNIVERSITY, WASHINGTON, D.C. 20007

INTRODUCTION

The title compounds (**8** and **10**) are the first known nucleosides that contain a 2-deoxyaldohexose residue in the furanoid form.[1] The synthesis of these difficultly obtainable compounds here described requires, as an intermediate, a stable, crystalline *O*-acyl-2-deoxyhexofuranosyl halide,[2] for which there is no precedent. The uracil nucleoside (**10**) is the furanoid isomer of 1-(2-deoxy-β-D-*arabino*-hexopyranosyl)uracil, a competitive inhibitor of a nonspecific, pyrimidine phosphorylase.[3]

PROCEDURE

Methyl 2-Deoxy-α-D-*arabino*-hexofuranoside (2)

To a solution of 2.462 g. (0.015 mole) of 2-deoxy-β-D-*arabino*-hexose (**1**) in 100 ml. of anhydrous methanol is added 0.3 ml. of a 35% solution of hydrogen chloride in methanol. The mixture is stirred for 15 min., and the acid is neutralized with an excess (2 g.) of silver carbonate. The suspension is filtered through a bed of Darco G-60 decolorizing carbon, and the filtrate is concentrated to about 30 ml., stirred with 2 g. of Rexyn 300 (H^+—OH^-) mixed-bed, ion-exchange resin,[a] and filtered. The solvent is removed by evaporation under diminished pressure at 45°, the residual sirup is dissolved in 5 ml. of anhydrous methanol, and to this solution is added 100 ml. of dry acetone. The resulting solution is kept in a refrigerator for two days, to allow most of the unreacted **1** to crystallize out, dry ether is added to incipient turbidity, and the mixture is kept in a refrigerator for an additional two days. The solution is decanted off, the solid in the flask is rinsed with two 10-ml. portions of dry methanol, and the decanted

[a] Fisher Scientific Co.

1 (164.2) $\xrightarrow{\text{HCl, MeOH}}$ **2** (178.2) $\xrightarrow{\text{COCl}_2}$ **3** (204.2) $\xrightarrow{p\text{NBzCl}}$

4 (353.3) $\xrightarrow{\text{HBr}}$ **5** (402) $\xrightarrow{\textbf{6}\ (140.1)}$

7 (447.4) $\xrightarrow{\text{NH}_3\text{–MeOH}}$ **8** (257.2)

7 → **9** (433.3) $\xrightarrow{\text{NaOMe–MeOH}}$ **10** (258.2)

where pNBz is p-nitrobenzoyl.

liquor and rinses are combined, and evaporated to dryness. A solution of the sirupy residue in 10 ml. of acetone is placed on a column (5 × 45 cm.) of Whatman No. 1 cellulose powder, packed with acetone and prewashed with the upper layer of 10:6:1:3 (v/v) ethyl acetate–isobutyl alcohol–toluene–water. Elution is conducted with the same solvent, 3-ml. fractions being collected. From the beginning, the fractions are continuously monitored by spotting on Whatman No. 1 filter paper and testing with a boric acid spray reagent.[4] When a positive test (pink to violet coloration) is obtained, the subsequent fractions are monitored by ascending paper-chromatography with the solvent used for elution of the column. The desired furanoside (**2**) has R_f 0.511,[b] and the fractions containing it are combined and evaporated to dryness. The sirup obtained crystallizes on standing, and recrystallization from ether–absolute ethanol gives 610 mg. (32%, based on unreacted **1**) of pure **2**, m.p. 80–81°, $[\alpha]_D^{24}$ +117.1° (*c* 0.99, ethanol).

Methyl 5,6-*O*-Carbonyl-2-deoxy-α-D-*arabino*-hexofuranoside (3)

A solution of 1.782 g. (0.01 mole) of the furanoside (**2**) in 20 ml. of dry pyridine and 15 ml. of dry carbon tetrachloride is cooled to −10°, and to the cold solution is slowly added dropwise, with stirring, 5 ml. of a 20% (w/w) solution of phosgene in dry toluene. The mixture is stirred at −10° for 1 hr., and at room temperature for 1 hr., and is then poured, with stirring, into a mixture of 4 g. of freshly prepared barium carbonate and about 150 ml. of crushed ice. The mixture is stirred until all of the ice has melted, and is then filtered through a bed of Hyflo Super Cel. The filtrate is extracted with six 150-ml. portions of ethyl acetate, and the extracts are combined, dried with anhydrous sodium sulfate, and evaporated to dryness. The residue is dissolved in warm tetrahydrofuran and, on addition of pentane to incipient turbidity, the product (**3**) crystallizes out; yield 1.65 g. (80%), m.p. 91–92°, $[\alpha]_D^{24}$ +132.4° (*c* 1.0, ethanol).

Methyl 5,6-*O*-Carbonyl-2-deoxy-3-*O*-*p*-nitrobenzoyl-α-D-*arabino*-hexoside (4)

To a solution of 742 mg. (4 mmoles) of *p*-nitrobenzoyl chloride in 10 ml. of pyridine is added 613 mg. (3 mmoles) of compound **3**. The

[b] The furanoside (**2**) is attended by some unreacted starting-material (**1**) (R_f 0.12) because of incompleteness of precipitation in processing the reaction mixture in the preceding experiments, and substantial proportions of the α-D-pyranoside (R_f 0.391) and the β-D-pyranoside (R_f 0.325).

mixture is stirred at room temperature for 1 hr., and kept in a refrigerator for 24 hr. The mixture is slowly added, with stirring, to 10 ml. of saturated, aqueous sodium hydrogen carbonate, and ice (about 150 ml.) is then added, with stirring. When the ice has melted, the solid is filtered off, well washed with water, and dried for 24 hr. in a vacuum desiccator over phosphorus pentaoxide. The solid is crystallized from acetone–ether; two recrystallizations from the same solvent give 850 mg. (80%) of pure **4**, m.p. 212–213°, $[\alpha]_D^{24}$ +22.4° (*c* 1.0, dichloromethane).

5,6-*O*-Carbonyl-2-deoxy-3-*O*-*p*-nitrobenzoyl-α-D-*arabino*-hexosyl Bromide (5)

To a solution of 707 mg. (2 mmoles) of compound **4** in 15 ml. of dry dichloromethane is added 25 ml. of a saturated solution of hydrogen bromide in dichloromethane. The mixture is stirred at room temperature for 25 min. with rigorous exclusion of moisture, 50 ml. of dry ether is added, and the mixture is kept in a refrigerator for 4 hr. The separated crystals are filtered off, and washed with dry ether, to give 650 mg. (81%) of pure **5**, m.p. 125° (dec.), $[\alpha]_D^{24}$ −35.4° (*c* 0.436, acetone).

1-(5,6-*O*-Carbonyl-2-deoxy-3-*O*-*p*-nitrobenzoyl-β-D-*arabino*-hexosyl)-4-methoxy-2(1*H*)-pyrimidinone (7)

A mixture of 302 mg. (0.75 mmole) of the halide (**5**) and 2.5 g. (17.6 mmoles) of 2,4-dimethoxypyrimidine (**6**) (Sec. II [30]) is heated at 75°/20 mm. Hg for 5 min., and then kept at room temperature and atmospheric pressure overnight. To the mixture is added 20 ml. of ether, and the solid is crushed, repeatedly washed with ether, and filtered off. It is dissolved in 10 ml. of dichloromethane, and, on addition of ether, a gel-like mass separates; the crystalline mass (with solvent entrapped) is filtered off, washed with ether, and dried by suction. Recrystallization from ether–dichloromethane gives 205 mg. (60%) of pure product (**7**), m.p. 209–211°, $[\alpha]_D^{24}$ −7.35° (*c* 0.55, dichloromethane).

1-(2-Deoxy-β-D-*arabino*-hexofuranosyl)cytosine (8)

A suspension of 224 mg. (0.5 mmole) of the protected nucleoside (**7**) in 20 ml. of dry methanol presaturated with ammonia is placed in a pressure bottle and heated at 85° for 9 hr. It is then kept at room temperature overnight, and the solvent is evaporated off under diminished pressure. The residue is mixed with 10 ml. of water and 10 ml. of chloroform, the mixture is shaken, the chloroform layer is discarded, and the water layer is washed with 10 ml. of chloroform and two 10-ml. portions of ether. The aqueous solution is evaporated to dryness at 45° under diminished pressure, the residue is dissolved in 1 ml. of water, and to the solution 2 ml. of ethanol is added. It is then stirred with a little Darco G-60 decolorizing carbon, and the suspension is filtered. The filtrate is evaporated to dryness at 45° under diminished pressure, the residue is dissolved in a few drops of water, and 2 ml. of ethanol is added. On addition of ether, the nucleoside (**8**) crystallizes; two more recrystallizations give 76 mg. (59%) of pure product (**8**), m.p. 217–218°, $[\alpha]_D^{24}$ −41.0° (*c* 0.95, water).

1-(5,6-*O*-Carbonyl-2-deoxy-3-*O*-*p*-nitrobenzoyl-β-D-*arabino*-hexosyl)uracil (9)

To a solution of 448 mg. (1 mmole) of the protected nucleoside (**7**) in 10 ml. of dichloromethane and 10 ml. of methanol is added 10 ml. of methanol presaturated with hydrogen chloride (34%), and the mixture is stirred at room temperature for 6 hr. The acid is neutralized with an excess of silver carbonate (15 g.), and the suspension is filtered; the filtrate is evaporated to dryness, and the residue is crystallized from ethanol–water to yield 330 mg. (76%) of **9**, m.p. 138–140°.

1-(2-Deoxy-β-D-*arabino*-hexofuranosyl)uracil (10)

A solution of 175 mg. (0.4 mmole) of compound **9** in 20 ml. of dry methanol is stirred with 0.20 ml. of 4 *M* methanolic sodium methoxide at room temperature for 2 hr. Acetic acid (0.2 ml.) is added, and the solvent is evaporated off under diminished pressure. The residue is mixed with 15 ml. of water and 15 ml. of ether, and the mixture is shaken; the ether layer is discarded, and the aqueous layer is washed with three 15-ml. portions of ether, stirred with 2 g. of Rexyn 300 (H^+—OH^-) mixed-bed ion-exchange resin[a] for 10 min., and the suspension is filtered. The resin is stirred with 10 ml. of water containing 1 ml. of acetic acid, the suspension is filtered, and the filtrates are combined, stirred with a little Darco G60 decolorizing carbon, and filtered. The filtrate is evaporated to dryness at 45° under diminished pressure, the residue is dissolved in 5 ml. of ethanol, and ether is added to incipient turbidity. The mixture is kept at room temperature for 4 hr., pentane (5 ml.) is added, and the mixture is kept in a refrigerator overnight. The crystals are filtered off, and washed with ether; yield of **10**, 88 mg. (85%), m.p. 180–183° $[\alpha]_D^{24}$ −16.8° (*c* 1.1, water).

REFERENCES

(1) K. V. Bhat and W. W. Zorbach, *Carbohyd. Res.*, **6**, 63 (1968).
(2) K. V. Bhat and W. W. Zorbach, *Carbohyd. Res.*, **1**, 93 (1965).
(3) Personal communication from Dr. M. Zimmerman of Merck, Sharp and Dohme Research Laboratories, Rahway, N. J.
(4) M. Pöhm and R. Weiser, *Naturwissenschaften*, **43**, 582 (1956).

[98] 1-(2-Deoxy-β-D-*ribo*-hexopyranosyl)cytosine [(2-Deoxy-D-allosyl)cytosine]

The Direct Synthesis of 2-Deoxyaldopyranosyl Pyrimidines

K. VENKATRAMANA BHAT and W. WERNER ZORBACH

DEPARTMENT OF CHEMISTRY, GEORGETOWN UNIVERSITY, WASHINGTON, D.C. 20007

INTRODUCTION

The preparation of 1-(2-deoxy-β-D-*ribo*-hexopyranosyl)cytosine[1] (**8**) serves to illustrate the utility of stable, crystalline 2-deoxypoly-*O*-(*p*-nitrobenzoyl)glycosyl halides in the direct synthesis of cytosine 2′-deoxynucleosides by the Hilbert–Johnson procedure.[2] Incorporated herein are marked improvements in the conversion of methyl 2,3-anhydro-4,6-*O*-benzylidene-α-D-alloside[3] (**1**) into 2-deoxy-D-*ribo*-hexose[4] (**3**).

PROCEDURE

Methyl 4,6-*O*-benzylidene-2-deoxy-α-D-*ribo*-hexopyranoside[4(a)] (2)

Into a 1-l., 3-necked flask, fitted with a mechanical stirrer, a 500-ml. dropping funnel, and a condenser arranged for distillation, are introduced 500 ml. of *absolute* tetrahydrofuran[a] and 3.8 g. (0.1 mole) of lithium aluminum hydride. A suspension of 26.4 g. (0.1 mole) of methyl 2,3-anhydro-4,6-*O*-benzylidene-α-D-alloside[3] (**1**) in 500 ml. of absolute tetrahydrofuran[a] is put in the dropping funnel, and the solution in the flask is brought to boiling and caused to distil. With efficient stirring, the suspension in the dropping funnel is added dropwise during 2 hr., with distillation of the solvent from the reaction flask being maintained at a rate equal to that of the addition of the suspension. The condenser is re-inserted in the vertical position, the solution is refluxed for 1 hr., and allowed to cool to room temperature. To the cooled solution is added 50 ml. of ethyl acetate, followed by the cautious addition of 20 ml. of water. The aluminum hydroxide formed is removed by suction, and the filtrate is evaporated to dryness under diminished pressure. The residue is dissolved in a small volume of dichloromethane, ether is added, and the

[a] Mallinckrodt grade, dried overnight with anhydrous potassium carbonate, filtered, refluxed overnight with lithium aluminum hydride, and distilled at 66° under rigorous exclusion of moisture. The solvent should be used promptly after being dried.

1 (264.2) → 2 (266.2) → 3 (164.2) →

4 (760.7) → 5 (674.4) —6 (168.2)→

7 (733.6) → 8 (257.2)

where Ph is phenyl and *p*NBz is *p*-nitrobenzoyl.

resulting crystals are removed by suction, and washed with a little dry ether; yield 21.1 g. (80%), m.p. 128–130°, $[\alpha]_D^{16}$ +156° (*c* 0.65, chloroform).

Extraction of the aluminum hydroxide residue with two 100-ml. portions of dichloromethane gives an additional 1.5 g. of pure **2**, bringing the total yield to 85%.

2-Deoxy-D-*ribo*-hexose[4(b)] **(3)**

A solution of 8.0 g. (0.03 mole) of methyl 4,6-*O*-benzylidene-2-deoxy-α-D-*ribo*-hexopyranoside (**2**) in 820 ml. of 0.05 *N* sulfuric acid is heated for 2 hr. at 80°, cooled to room temperature, and extracted with three 100-ml.

portions of chloroform. To the aqueous solution is added 20 g. of freshly prepared barium carbonate, the suspension is stirred for 20 min. at 50°, 8 g. of Celite 535 is added, and the suspended material is filtered off by suction. The volume of the solution is diminished by one half *in vacuo* at 55°, 1 g. of barium carbonate is added, and the suspension is evaporated to approximately 50 ml. To the latter is added 2 g. of Darco G-60, and the suspension is heated at 60° for 15 min., under magnetic stirring. The suspended material is filtered off and rinsed with water, and the filtrate is evaporated under diminished pressure to a thin sirup. To the latter is added 100 ml. of absolute ethanol, from which the product slowly crystallizes; yield 3.5 g. (70%), m.p. 138–140°, $[\alpha]_D^{16}$ +58° (*c* 1.2, water). By carefully processing the mother liquors, an additional 0.3 g. is secured, bringing the total yield to 76%.

2-Deoxy-1,3,4,6-tetra-*O*-*p*-nitrobenzoyl-D-*ribo*-hexose[5] (4)

A warm solution of 5.55 g. (30 mmoles) of *p*-nitrobenzoyl chloride in 50 ml. of dry pyridine is cooled to 0° under stirring, and to the stirred suspension is added, in small portions during 15 min., 0.98 g. (6 mmoles) of finely divided 2-deoxy-D-*ribo*-hexose (**3**). Stirring is continued at 0° for 1.5 hr., and the mixture is kept in a refrigerator for 4 days. The excess *p*-nitrobenzoyl chloride is neutralized by the careful addition of 60 ml. of saturated, aqueous sodium hydrogen carbonate, and the mixture is added to 1 l. of ice–water. The precipitate is removed by filtration, washed with water, and dried in the open at room temperature, and then in a desiccator over phosphorus pentaoxide. Most of the water of crystallization is removed by heating the material at 110°/0.1 mm. Hg for 15 hr. The dry product is crystallized from nitromethane, and the solution decolorized with decolorizing carbon, giving 3.46 g. (76%) of **4**, m.p. 201–202°, of satisfactory quality for the following conversion. It has $[\alpha]_D$ +106° (*c* 1.1, chloroform).

2-Deoxy-3,4,6-tri-*O*-*p*-nitrobenzoyl-α-D-*ribo*-hexosyl Bromide[5] (5)

To 72 ml. of a solution of anhydrous hydrogen bromide in dichloromethane[b] is added 2.28 g. (3 mmoles) of the tetrakis-*p*-nitrobenzoate (**4**). The suspension is stirred magnetically for 2 hr. in a stoppered flask, and the separated *p*-nitrobenzoic acid is removed by filtration through a sintered-glass funnel. The filtrate is evaporated at room temperature under diminished pressure to a volume of about 5 ml., and 10 ml. of dry ether is added. The solution is kept overnight at room temperature, and the separated crystals are filtered by suction, washed with a little ether, and

[b] Containing about 0.25 meq. of hydrogen bromide per ml.

dried in a vacuum desiccator over phosphorus pentaoxide; yield 1.93 g. (90%), m.p. 119° (dec.), $[\alpha]_D^{20}$ +198° (*c* 0.8, dichloromethane).

1-(2-Deoxy-3,4,6-tri-*O-p*-nitrobenzoyl-β-D-*ribo*-hexosyl)-4-ethoxy-2(1*H*)-pyrimidinone[1] (7)

To 1.35 g. (8 mmoles) of 2,4-diethoxypyrimidine (**6**) (Sec. II [30]) is added 675 mg. (1 mmole) of the bromide (**5**). The mixture is stirred, with exclusion of moisture, for 2 hr., after which time it solidifies. It is kept overnight at room temperature, and the solid is crushed, and extracted with five 30-ml. portions of ether. The residue (410 mg.) is recrystallized from ether–dichloromethane; yield 360 mg. (49%), m.p. 295–296°, $[\alpha]_D^{24}$ +36° (*c* 0.8, dichloromethane).

1-(2-Deoxy-β-D-*ribo*-hexopyranosyl)cytosine[1] (8)

A suspension of 200 mg. (0.27 mmole) of the protected nucleoside (**7**) in 20 ml. of anhydrous methanol saturated with ammonia is heated in a pressure bottle at 80° for 8 hr. The solution is kept overnight at room temperature, and the solvent is evaporated under diminished pressure at 40°. The residue is dissolved in water, and the solution is thoroughly extracted with ether to remove the *p*-nitrobenzamide. The aqueous layer is evaporated under diminished pressure at 50°, and the residue is dissolved in 5 ml. of 90% aqueous ethanol. To the solution is added 10 ml. of ether, and the solution is kept in a refrigerator overnight; yield 52 mg. (69%), m.p. 236–238°, $[\alpha]_D^{24}$ +43° (*c* 0.4, water), λ_{max}^{MeOH} 271 nm. (log ε 4.0).

REFERENCES

(1) W. W. Zorbach, H. R. Munson, and K. V. Bhat, *J. Org. Chem.*, **30**, 3955 (1965).
(2) G. E. Hilbert and T. B. Johnson, *J. Am. Chem. Soc.*, **52**, 4489 (1930).
(3) N. K. Richtmyer, *Methods Carbohydrate Chem.*, **1**, 107 (1962).
(4) (a) D. A. Prins, *J. Am. Chem. Soc.*, **70**, 3955 (1948); (b) W. W. Zorbach and T. A. Payne, *ibid.*, **80**, 5564 (1958). An alternative synthesis of 2-deoxy-D-*ribo*-hexose (3) is given by W. W. Zorbach and A. P. Ollapally, *J. Org. Chem.*, **29**, 1790 (1964).
(5) W. W. Zorbach and W. Bühler, *Ann.*, **670**, 116 (1963).

[99] Isocytidine

Ammonolysis of a 2,5′-Anhydronucleoside

BARBARA F. WEST

DEPARTMENT OF CHEMISTRY, GEORGETOWN UNIVERSITY,
WASHINGTON, D.C. 20007

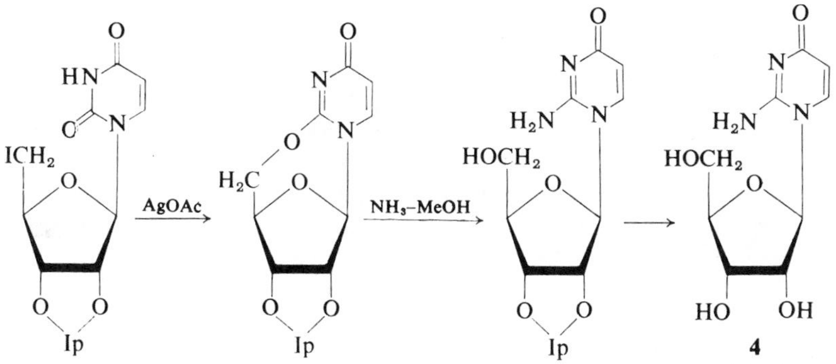

1 (394.2) —AgOAc→ **2** (266.3) —NH_3–MeOH→ **3** (283.3) ⟶ **4** (243.1)

INTRODUCTION

The preparation of the title compound[1] (**4**) is an example of how such versatile intermediates as the anhydronucleoside **2** may be converted into nucleoside derivatives. The 2′,3′-*O*-isopropylidene acetal (**3**) is an intermediate in the synthesis of isocytidine 5′-phosphate (Sec. IV [146]).

PROCEDURE

2,5′-Anhydro-2′,3′-*O*-isopropylideneuridine (2)

A solution of 2.43 g. (6.2 mmoles) of 5′-deoxy-5′-iodo-2′,3′-*O*-isopropylideneuridine[2] (**1**) (Sec. III [126]) in 600 ml. of anhydrous methanol is refluxed with 4.5 g. of silver acetate for 15 min., and the mixture is filtered through Hyflo Super Cel. Silver ions are removed by saturating the filtrate with hydrogen sulfide, and, after concentrating the solution to 20 ml., benzene is added. The crystalline product, weighing 1.3 g. (79%), is separated and recrystallized from ethanol, forming colorless plates which sinter at 190° and darken without melting; $\lambda_{max}^{95\%\ EtOH}$ 237 nm. (ϵ 14,000), $\lambda_{min}^{95\%\ EtOH}$ 212 nm. (ϵ 3,290), R_f 0.62.[a]

[a] Solvent system, 43:7 butyl alcohol–water.

2′,3′-*O*-Isopropylideneisocytidine (3)

A solution of 0.3 g. (1.1 mmoles) of compound **2** in 10 ml. of anhydrous methanol is added to 35 ml. of saturated methanolic ammonia. After five days, formation of the product is complete (R_f 0.58).[a] The solution is evaporated to dryness, and the residue is crystallized from ethanol, to form thick, colorless prisms, m.p. 206–207°; $\lambda_{max}^{H_2O}$ 254–255 and 205 nm. (ϵ 5,820 and 25,400); $\lambda_{min}^{H_2O}$ 248 nm. (ϵ 5,690); $\lambda_{max}^{0.1\,N\ HCl}$ 256 and 220 nm. (ϵ 7,110 and 8,390); $\lambda_{min}^{0.1\,N\ HCl}$ 239 nm. (ϵ 4,790); $\lambda_{max}^{0.1\,N\ NaOH}$ 223–224 nm. (ϵ 16,500); ν_{max} 1,645 cm.$^{-1}$ (carbonyl).

Isocytidine (4)

The isopropylidene acetal (**3**) (0.2 g., 7.1 mmoles) is dissolved in 20 ml. of 98% formic acid. After the solution has been kept at room temperature for 4 hr., the formic acid is removed by evaporation under diminished pressure, with repeated additions of ethanol. Compound **4** is precipitated by concentrating a 98% ethanol solution under diminished pressure, giving a glass having R_f 0.12.[a]

REFERENCES

(1) D. M. Brown, A. R. Todd, and S. V. Varadarjan, *J. Chem. Soc.*, **1957**, 868.
(2) J. P. H. Verheyden and J. G. Moffatt, *J. Am. Chem. Soc.*, **86**, 2093 (1964). The original preparation of this compound was given by P. A. Levene and R. S. Tipson, *J. Biol. Chem.*, **106**, 113 (1934).

[a] Solvent system, 43:7 butyl alcohol–water.

[100] 2-Amino-5-(*N*-methylformamido)-6-(D-ribofuranosylamino)-4-pyrimidinol and 5-(*N*-Methylformamido)-6-(D-ribofuranosylamino)uracil

Cleavage of a Purine Ring of Purine Nucleosides

LEROY B. TOWNSEND and ROLAND K. ROBINS

DEPARTMENT OF CHEMISTRY, UNIVERSITY OF UTAH, SALT LAKE CITY, UTAH 84112

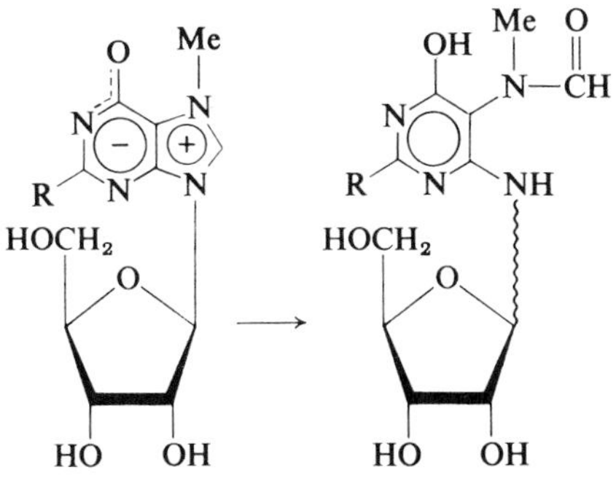

1, R = NH_2 (297.3)
3, R = OH (298.3)

2, R = NH_2 (315.3)
4, R = OH (316.3)

INTRODUCTION

Cleavage of a ring of the purine residue of purine nucleosides to furnish imidazole nucleosides occurs[1] *via* labilization of the normally stable pyrimidine ring by alkylation. Methylation at N-7 of guanosine and xanthosine conveys a degree of instability to the imidazole ring of these nucleosides, and ring cleavage at C-8 occurs[2] in the presence of a weak base to afford the title compounds. Although the β-D anomeric configuration and the furanoid ring might be inferred for compounds **2** and **4**, these details of structure have not yet been substantiated. It is of interest that the biosynthesis of pteridines from purines in a variety of biological systems has been shown to occur[3] *via* ring cleavage, namely, by enzymic labilization at C-8 with subsequent expulsion of C-8 as formate.

PROCEDURE

2-Amino-5-(*N*-methylformamido)-6-(D-ribofuranosylamino)-4-pyrimidinol (2)

A solution of 7-methylguanosine (**1**) (5 g., 16.8 mmoles) in 14% ammonium hydroxide (100 ml.) at room temperature is monitored frequently by thin-layer chromatography (ammonium hydrogen carbonate) until the blue fluorescent spot for **1** has disappeared completely (*ca.* 2 hr.) and a dark, ultraviolet-absorbing spot has appeared. The solution is evaporated to dryness under diminished pressure, and the resulting residue is dissolved in anhydrous methanol (50 ml.) at room temperature. The solution is treated with Norit decolorizing carbon, the Norit is filtered off, the filtrate is cooled to 5°, and acetone (1 l.) is slowly added, with rapid stirring, at 5–10°. The vessel is covered, the mixture is stirred at 5° for 1 hr., and the solid is filtered off, washed successively with cold (5°) acetone (100 ml.) and cold (5°) ethyl ether (100 ml.), transferred to a round-bottomed flask, and dried under diminished pressure in a rapidly rotated, rotary flash evaporator for 1 hr.; yield 3.1 g. (56.8%) of colorless crystals, decomposes at 175–180° when placed on a melting-point block at 160°; $[\alpha]_D^{25}$ +32.45° (*c* 1, water); $\lambda_{max}^{pH\,1}$ 270.5 nm. (ϵ 22,300) and $\lambda_{max}^{pH\,11}$ 265 nm. (ϵ 16,300).

5-(*N*-Methylformamido)-6-(D-ribofuranosylamino)uracil (4)

A solution of 7-methylxanthosine (**3**) (1 g., 3.36 mmoles) in 14% ammonium hydroxide (20 ml.) at room temperature is monitored frequently by thin-layer chromatography (ammonium hydrogen carbonate) until the blue fluorescent spot for **3** has disappeared completely and a dark, ultraviolet-absorbing spot has appeared. The solution is evaporated to dryness under diminished pressure at room temperature, the resulting residue is dissolved in anhydrous methanol (20 ml.), the solution is treated with Norit decolorizing carbon, the Norit is filtered off, and the filtrate is cooled to 5°. Cold (5°) acetone (200 ml.) is slowly added, with rapid stirring, at 5–10°, the vessel is covered, and the mixture is stirred at 5° for 1 hr. The solid is filtered off, successively washed with cold (5°) acetone (100 ml.) and cold (5°) ethyl ether (100 ml.), and air-dried; yield 0.6 g. (56.5%) of colorless crystals, m.p. 155–160° (dec.); $[\alpha]_D^{25}$ +20.8° (*c* 1, water); $\lambda_{max}^{pH\,1}$ 268 nm. (ϵ 20,700); $\lambda_{max}^{pH\,11}$ 269 nm. (ϵ 15,000).

REFERENCES

(1) L. B. Townsend, *Chem. Rev.*, **67**, 533 (1967).
(2) L. B. Townsend and R. K. Robins, *J. Am. Chem. Soc.*, **85**, 242 (1963); J. A. Haines, C. B. Reese, and A. R. Todd, *J. Chem. Soc.*, **1962**, 5281.
(3) A. W. Burg and G. M. Brown, *Biochim. Biophys. Acta*, **117**, 275 (1966), and references cited therein.

[101] 4-Ethoxy-1-[2,3,4-tri-*O*-acetyl-6-*O*-(3,4,6-tri-*O*-acetyl-2-benzamido-2-deoxy-β-D-glucosyl)-β-D-glucosyl]-2(1*H*)-pyrimidinone

Synthesis of a Pyrimidine Nucleoside by the Koenigs–Knorr Coupling Reaction with an Amino Sugar Derivative

CALVIN L. STEVENS and PETER BLUMBERGS

DEPARTMENT OF CHEMISTRY, WAYNE STATE UNIVERSITY, DETROIT, MICHIGAN 48202

$Cl_3CCO_2CH_2$ OAc AcO OAc X OEt N EtO N OEt N O N $ROCH_2$ OAc AcO OAc

1a, X = β-OAc (493.6)
1b, X = α-Br (514.5)

2a, R = Cl_3CCO (573.8)
2b, R = H (428.4)
2c, R = Ac (470.4)

↓ **3**

$AcOCH_2$ O OAc AcO Br PhCONH

3 (472.3)

$AcOCH_2$ O OAc AcO PhCONH O—CH_2 O OAc AcO OAc OEt N O N

4 (819.7)

INTRODUCTION

The naturally occurring amicetin, a cytosine nucleoside of an amino disaccharide, has important biological activity against bacteria and as an anticancer agent in animals. Through the use of the trichloroacetyl protecting group, the pyrimidine nucleoside (**2b**) containing a free hydroxyl group on C-6′ is readily prepared. This intermediate may then be used in the Koenigs–Knorr reaction with the appropriate 2-amino-2-deoxy-D-glucose derivative (**3**).

PROCEDURE

1,2,3,4-Tetra-*O*-acetyl-6-*O*-(trichloroacetyl)-β-D-glucose (1a)

To a solution of 7 g. (0.017 mole) of 1,2,3,4-tetra-*O*-acetyl-β-D-glucose[1] in 60 ml. of benzene is added 6 ml. of pyridine, followed by a solution of 4.3 g. (0.022 mole) of trichloroacetyl chloride in 10 ml. of benzene. The reaction mixture is kept at room temperature for 6 hr. and poured into 100 ml. of water, and the benzene layer is separated, washed successively with water, dilute acetic acid, and water, dried, and evaporated to dryness under diminished pressure. The solid residue is washed with 15 ml. of ethanol, and dried; yield 7.22 g. (86%) of **1a**, m.p. 104–105°. Concentration of the washings affords additional material; total yield 7.41 g. (88%). Recrystallization from ligroine–isopropyl ether does not raise the melting point.

2,3,4-Tri-*O*-acetyl-6-*O*-(trichloroacetyl)-α-D-glucosyl Bromide (1b)

A solution of 4 g. (8.1 mmoles) of **1a** in 20 ml. of acetic acid and 4 ml. of acetic anhydride is saturated at 0° with anhydrous hydrogen bromide. The solution is kept at 0° for 4.5 hr., and then poured onto ice, and the mixture is extracted three times with chloroform. The chloroform extracts are combined, washed successively with ice–water, ice-cold saturated sodium hydrogen carbonate solution, and ice–water, dried immediately (anhydrous sodium sulfate), and evaporated to dryness under diminished pressure. The crystalline residue is triturated with 10 ml. of isopropyl ether, the suspension is filtered, and the crystals are washed with 10 ml. of isopropyl ether, and dried; yield 3.80 g. (91%) of crude **1b**, m.p. 167–168.5°. For purification, the compound is dissolved in the minimal volume of warm chloroform, and the solution is diluted with four volumes of isopropyl ether. Two such recrystallizations give analytically pure product, m.p. 168.5–169°, $[\alpha]_D^{24}$ +161° (*c* 1.3, chloroform).

4-Ethoxy-1-[2,3,4-tri-*O*-acetyl-6-*O*-(trichloroacetyl)-β-D-glucosyl]-2(1*H*)-pyrimidinone (2a)

A mixture of 2.75 g. (5.34 mmoles) of **1b** and 5.5 ml. of 2,4-diethoxypyrimidine (Sec. II [30]) is heated at 80–85° for 20 hr., and then at 90–95°

for 24 hr. The mixture is mixed with 40 ml. of 1:3 ether–isopropyl ether, and the suspension is filtered; the solid is recrystallized from absolute ethanol; yield 1.15 g. of crude **2a**, m.p. 221–222.5°. Additional product is obtained by concentrating the mother liquors. The total yield is 1.25 g. (41%). Two recrystallizations from ethanol give analytically pure product, m.p. 221.5–223°, $[\alpha]_D^{25}$ +43.8° (*c* 1.1, chloroform), λ_{max}^{EtOH} 276 nm. (ϵ 6,410).

4-Ethoxy-1-(2,3,4-tri-*O*-acetyl-β-D-glucopyranosyl)-2(1*H*)-pyrimidinone (2b)

To a solution of 1 g. (1.74 mmole) of **2a** in 50 ml. of benzene is added 25 ml. of toluene presaturated at 0° with anhydrous ammonia. The mixture is kept in an ice-bath for 2 hr., and then evaporated to dryness under diminished pressure to a solid residue which is washed with 5 ml. of ether, and recrystallized from ethanol–isopropyl ether; yield 0.58 g. (78%) of **2b**, m.p. 218–219°, $[\alpha]_D^{26}$ +46.5° (*c* 1, chloroform), λ_{max}^{EtOH} 276 nm. (ϵ 6,430).

4-Ethoxy-1-(2,3,4,6-tetra-*O*-acetyl-β-D-glucosyl)-2(1*H*)-pyrimidinone (2c)

Compound **2b** (70 mg., 0.16 mmole) is mixed with an excess of acetic anhydride and pyridine, and the mixture is kept at room temperature for 12 hr. The resulting solution is poured onto a mixture of potassium carbonate and ice, and the mixture is extracted with chloroform. The extract is washed with water, dried (anhydrous sodium sulfate), evaporated to dryness under diminished pressure, and the solid product recrystallized from 3 ml. of ethanol; yield 65 mg. (86%) of **2c**, m.p. 203.5–204.5°, $[\alpha]_D^{26}$ +46.5° (*c* 1, chloroform).

An authentic sample of this material, prepared in 46% yield as described by Hilbert and Jansen,[2] admixed with **2c** obtained by the acetylation of **2b** does not depress the melting point, and the infrared absorption spectra of the two samples are identical.

4-Ethoxy-1-[2,3,4-tri-*O*-acetyl-6-*O*-(3,4,6-tri-*O*-acetyl-2-benzamido-2-deoxy-β-D-glucosyl)-β-D-glucosyl]-2(1*H*)-pyrimidinone (4)

A vigorously stirred mixture of 0.47 g. (1.1 mmoles) of the nucleoside **2b**, 0.80 g. of silver carbonate, 3 g. of Drierite (predried at 200° for 12 hr.), and 1.43 g. (3.1 mmoles) of the bromide[3] (**3**) in 10 ml. of azeotropically dried benzene is refluxed in the dark for 3 hr., and then stirred at room temperature for a further 10 hr. The mixture is diluted with chloroform, and filtered, and the solid is washed with chloroform. The filtrate and washing are combined, and evaporated to dryness under diminished pressure, giving a gum which, on trituration with ether, solidifies. The solid is dissolved in methanol, the solution is treated with activated carbon, the suspension is filtered, and the filtrate is concentrated to a small volume; the resulting crystals are recrystallized once more from methanol; yield 90 mg. (9%, based on **2b**) of analytically pure product, m.p. 287–288°, $[\alpha]_D^{26}$ −6.5° (*c* 1, chloroform), λ_{max}^{EtOH} 274 nm. (ϵ 4,970).

No attempt has yet been made to improve the yield.

REFERENCES

(1) D. W. Reynolds and W. L. Evans, *Org. Syn.*, Coll. Vol. **3**, 434 (1955).
(2) G. E. Hilbert and E. F. Jansen, *J. Am. Chem. Soc.*, **58**, 60 (1936).
(3) C. L. Stevens and P. Blumbergs, *J. Org. Chem.*, **30**, 2723 (1965).

[102] 5′-*O*-Tritylthymidine

Selective Tritylation of a Pyrimidine Nucleoside

H. RANDALL MUNSON, JR.

DEPARTMENT OF CHEMISTRY, GEORGETOWN UNIVERSITY,
WASHINGTON, D.C. 20007

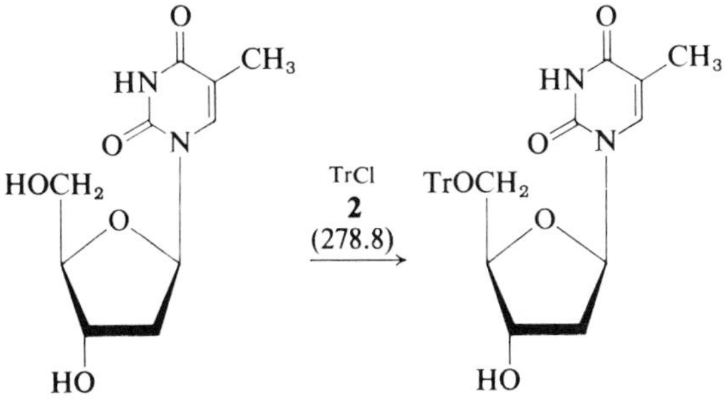

HOCH$_2$ ... **1** (242.2) $\xrightarrow[(278.8)]{TrCl\ \mathbf{2}}$ TrOCH$_2$ ... **3** (484.5)

where Tr is $(C_6H_5)_3C$.

INTRODUCTION

The title compound (**3**) is a useful intermediate, especially for preparing derivatives of thymidine (**1**) that are acylated at only the 3-hydroxyl group of the carbohydrate moiety (Sec. III [118]). The original preparation[1] of **3** required one week for completion of reaction at room temperature, because, at that time, thymidine was such a rare compound that it was decided not to use a higher temperature. The synthesis described herein[2] is accomplished at an elevated temperature in a short period of time to give 5′-*O*-tritylthymidine (**3**) in 86% yield. An alternative procedure is given in Sec. III [118]; however, the product is there converted directly into 3′-*O*-acetylthymidine, without isolation and purification of **3**.

PROCEDURE

5′-*O*-Tritylthymidine (3)

A solution of 5.0 g. (20.6 mmoles) of thymidine (**1**) and 7.0 g. (25.1 mmoles) of chlorotriphenylmethane (**2**)[3] in 100 ml. of anhydrous pyridine[a] is placed in a preheated (100°) glycerol bath, and heating under an air

[a] Pyridine is refluxed overnight with barium oxide, followed by distillation. The dry pyridine is stored over anhydrous calcium sulfate.

condenser (drying tube) is conducted at 100°, with stirring, for 30 min. The reaction mixture is cooled to room temperature, and is slowly poured into 1.5 l. of vigorously stirred ice–water. The precipitate is filtered off, washed with water until free from pyridine, and dried overnight in a vacuum desiccator containing phosphorus pentaoxide. The white solid is crystallized from acetone–benzene to give 8.57 g. (86%) of **3**, m.p. 128–130° (lit.[1] 128°).

REFERENCES

(1) P. A. Levene and R. S. Tipson, *J. Biol. Chem.*, **109**, 623 (1935).
(2) J. P. Horwitz, J. A. Urbański, and J. Chua, *J. Org. Chem.*, **27**, 3300 (1962).
(3) W. E. Bachmann, *Org. Syn.*, **23**, 100 (1943).

[103] 1-(2-Amino-2-deoxy-β-D-glucopyranosyl)thymine

The N-*Trifluoroacetyl Protecting Group in Synthesis of a Pyrimidine Nucleoside of a 2-Amino Sugar by the Trimethylsilyl Fusion Method*

M. L. WOLFROM, H. B. BHAT, and P. McWAIN

DEPARTMENT OF CHEMISTRY, THE OHIO STATE UNIVERSITY, COLUMBUS, OHIO 43210

INTRODUCTION

A number of *N*-protecting groups have been utilized in polypeptide syntheses, and one of the more convenient ones is the trifluoroacetyl group introduced in the amino acid series[1]; it has been utilized by Wolfrom and Bhat[2] in a pyrimidine nucleoside synthesis with 2-amino-2-deoxy-D-glucopyranose. 1,3,4,6-Tetra-*O*-acetyl-2-amino-2-deoxy-β-D-glucose hydrochloride (**4**) is prepared by the method of Bergmann and Zervas,[3] who converted 2-amino-2-deoxy-D-glucose hydrochloride (**1**), through the Schiff base (**3**) with *p*-methoxybenzaldehyde (**2**, *p*-anisaldehyde), into the tetraacetate (**4**). Compound **4** is transformed to the trifluoroacetamido derivative (**5**) and this can be converted into 3,4,6-tri-*O*-acetyl-2-deoxy-2-(trifluoroacetamido)-α-D-glucosyl bromide (**6**), which has been prepared[4] by an alternative method in somewhat lower yield. This bromide (**6**) can be transformed into the acylated nucleoside (**8**) by controlled fusion[5] with 5-methyl-2,4-bis(trimethylsiloxy)pyrimidine (**7**).[6] The *N*-trifluoroacetyl group is readily removed by mild treatment with either an acid or a base; in the present procedure, it is removed, together with the acetyl groups, by treatment with methanolic ammonia, to yield 1-(2-amino-2-deoxy-β-D-glucopyranosyl)thymine (**9**). The anomeric nature of **9** was established by the nuclear magnetic resonance spectrum of its tetraacetate.[7] The trifluoroacetyl group is an essentially nonparticipating group, and this property is advantageous in controlling the reactivity of the protected amino sugar in nucleoside syntheses.

PROCEDURE

2-Deoxy-2-(*p*-methoxybenzylidene)amino-β-D-glucopyranose[3] (3)

A solution of 15.6 g. (72.6 mmoles) of 2-amino-2-deoxy-D-glucose hydrochloride (**1**) in 73 ml. of *N* sodium hydroxide is treated, under

1 (215.6) + **2** (136.1) → **3** (297.3) $\xrightarrow{\text{1. } Ac_2O,\ C_5H_5N;\ \text{2. HCl}}$ **4** (383.8) → **5** (443.3) → **6** (464.2) $\xrightarrow{\textbf{7}\ (270.4)}$ **8** (509.4) → **9** (287.3)

where Ac is CH_3CO.

vigorous stirring, with 1.0 g. of *p*-anisaldehyde (**2**). A crystalline product forms rapidly. The mixture is refrigerated overnight and filtered, and the crystalline product is successively washed with cold water and ether–alcohol (1:4 v/v); yield 17.3 to 19.4 g. (80–90%), m.p. 166° (dec.).

1,3,4,6-Tetra-*O*-acetyl-2-amino-2-deoxy-β-D-glucose Hydrochloride[3] (4)

Dried **3** (15.0 g., 50.4 mmoles) is slowly dissolved in a precooled solution of 45 ml. of acetic anhydride in 81 ml. of dry pyridine while the temperature is maintained below 40° by occasional cooling in an ice–water bath. On complete dissolution, the mixture is kept overnight at room temperature, and is then stirred into a mixture of ice and water. Crystallization occurs on standing at 0°; after 1 hr., the material is collected, and washed with cold water; yield 21.1 g. (90%) of 1,3,4,6-tetra-*O*-acetyl-2-deoxy-2-(*p*-methoxybenzylidene)amino-β-D-glucose, m.p. 188°, $[\alpha]_D^{21}$ +98.6° (chloroform).

1,3,4,6-Tetra-*O*-acetyl-2-deoxy-2-(*p*-methoxybenzylidene)amino-β-D-glucose (46.4 g., 99.8 mmoles) is dissolved in acetone, the solution is heated to boiling, and a molar proportion of hydrochloric acid (20 ml. of 5 *N* hydrochloric acid) is added with vigorous stirring. The mixture forms a gelatinous mass; this is cooled, an equal volume of ether is added, and the mixture is cooled to 0°. The precipitate is collected on a filter, and washed with ether to yield 34.5 g. (90%) of **4**, dec. 230° without melting, $[\alpha]_D^{21}$ +29.7° (water).[3]

1,3,4,6-Tetra-*O*-acetyl-2-deoxy-2-(trifluoroacetamido)-β-D-glucose[2] (5)

1,3,4,6-Tetra-*O*-acetyl-2-amino-2-deoxy-β-D-glucose hydrochloride (**4**) (20 g., 52.2 mmoles) is suspended in dichloromethane (200 ml.) and pyridine (20 ml.), and trifluoroacetic anhydride (20 ml.) is added, with cooling and stirring. After stirring the solution for 20 min., it is washed with ice–water, dried (sodium sulfate), and concentrated to a small volume. Addition of ether affords crystals which are removed by filtration to give 21.5 g. (90%) of **5**, m.p. 167°, $[\alpha]_D^{22}$ −13° (*c* 2.4, chloroform).

3,4,6-Tri-*O*-acetyl-2-deoxy-2-(trifluoroacetamido)-α-D-glucosyl Bromide[2] (6)

To a suspension of 1,3,4,6-tetra-*O*-acetyl-2-deoxy-2-(trifluoroacetamido)-β-D-glucose (**5**) (3.0 g., 6.8 mmoles) in dichloromethane (3 ml.) is added acetic acid (3 ml.) almost saturated at 0° with hydrogen bromide. After the mixture has been kept for 3 hr. at room temperature, the solvent is removed, the residual sirup is dissolved in ether, and the solution is evaporated to a sirup. The bromide is crystallized from ether–hexane to yield 2.8 g. (92%) of **6**, m.p. 96°, $[\alpha]_D^{21}$ +126° (*c* 2.92, chloroform).

1-[3,4,6-Tri-*O*-acetyl-2-deoxy-2-(trifluoroacetamido)-β-D-glucosyl]thymine[2] (8)

The above bromide (**6**) (3.5 g., 7.5 mmoles) is intimately mixed with 5-methyl-2,4-bis(trimethylsiloxy)pyrimidine[6] (**7**) (Sec. III [119]) (3.5 g., 12.9 mmoles) in a flask which is then evacuated with a water pump. The flask is now closed, and the mixture is slowly heated to fusion (140°) and kept molten for 30 min.; the mixture is cooled, and evacuated to dryness. The dark-colored residue is treated with 80% methanol, and the mixture is evaporated to dryness. The residue is extracted with hot chloroform, and the extract is filtered, washed with water, and dried. The sirup obtained on removal of the solvent is crystallized from methanol–water to yield 3.1 g. (80%) of **8**, m.p. 236°, $[\alpha]_D^{22}$ $-47°$ (*c* 2.50, chloroform).

1-(2-Amino-2-deoxy-β-D-glucopyranosyl)thymine[2] (9)

The acetylated derivative (**8**) (540 mg., 1.1 mmoles) is dissolved in methanol (20 ml.), and the solution is almost saturated at 0° with ammonia. After being kept at room temperature for 5 days, the solution is evaporated to dryness, and the residue is crystallized from methanol to yield 245 mg. (80%) of **9**, m.p. 240–242°, $[\alpha]_D^{26}$ $+5.4°$ (*c* 2.44, water).

REFERENCES

(1) F. Weygand and E. Scendes, *Angew. Chem.*, **64**, 136 (1952).
(2) M. L. Wolfrom and H. B. Bhat, *Chem. Commun.*, **1966**, 146; *J. Org. Chem.*, **32**, 1821 (1967).
(3) M. Bergmann and L. Zervas, *Ber.*, **64**, 975 (1931); D. Horton, *J. Org. Chem.*, **29**, 1776 (1964).
(4) R. G. Strachan, W. V. Ruyle, T. Y. Shen, and R. Hirschmann, *J. Org. Chem.*, **31**, 507 (1966).
(5) T. Nishimura, B. Shimizu, and I. Iwai, *Chem. Pharm. Bull.* (Tokyo), **12**, 1471 (1964); *Chem. Abstr.*, **62**, 9223 (1965).
(6) L. Birkofer and A. Ritter, *Angew. Chem.*, **71**, 372 (1959); L. Birkofer, P. Richter, and A. Ritter, *Chem. Ber.*, **93**, 2804 (1960); E. Wittenburg, *Z. Chem.*, **4**, 303 (1964).
(7) M. L. Wolfrom and M. W. Winkley, to be published.

[104] 1-(2-Deoxy-β-D-*arabino*-hexopyranosyl)thymine [(2-Deoxy-β-D-glucosyl)thymine]

The Hilbert–Johnson Synthesis Applied to Preparation of 2-Deoxyaldopyranosyl Nucleosides

B. R. AARONOFF, R. D. BABSON, and A. J. ZAMBITO

MERCK & CO., INC., RAHWAY, NEW JERSEY 07065

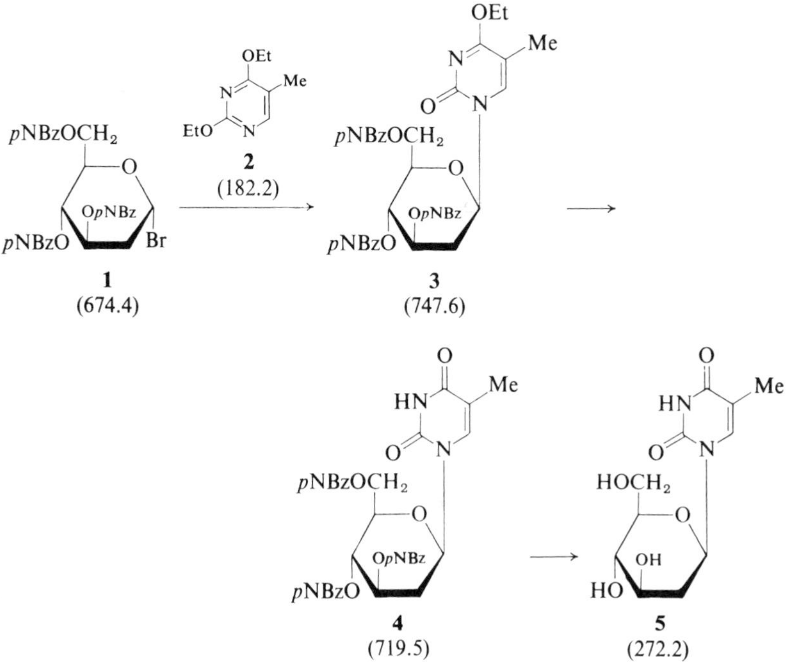

where *p*NBz is *p*-nitrobenzoyl.

INTRODUCTION

(2-Deoxy-β-D-glucosyl)thymine[1] (**5**) has been found to inhibit the pyrimidine phosphorylase that cleaves both uridine and thymidine, but it does not inhibit thymidine phosphorylase, which is specific for the (2-deoxy-D-ribosyl)diketopyrimidine structure.[2–4] Its preparation is another example of the use of 2-deoxy-poly-*O*-*p*-nitrobenzoylglycosyl halides in the synthesis of pyrimidine 2′-deoxynucleosides by the Hilbert–Johnson procedure.[5] In this instance, intermediate **4** is deacylated in methanol containing methoxide ion.

PROCEDURE

1-(2-Deoxy-3,4,6-tri-*O*-*p*-nitrobenzoyl-β-D-*arabino*-hexosyl)-4-ethoxy-5-methyl-2(1*H*)-pyrimidinone[1] (3)

A mixture of 80 g. (0.12 mole) of 2-deoxy-3,4,6-tri-*O*-*p*-nitrobenzoyl-α-D-*arabino*-hexosyl bromide (**1**) (Sec. III [117]) and 88 g. (0.48 mole) of 2,4-diethoxy-5-methylpyrimidine (**2**) (Sec. II [30]) in 1 l. of dichloromethane is evaporated to dryness under diminished pressure at 25°.[a] The resulting, intimate mixture is heated at 45–50°/0.1 mm. Hg for 4.5 hr. The melt is cooled, and successively triturated in a Waring Blendor with two 500-ml. portions of ether[b] and three 350-ml. portions of benzene, and the product is air-dried to give 67 g. (74.5%) of **3**, m.p. 230–240°.

1-(2-Deoxy-3,4,6-tri-*O*-*p*-nitrobenzoyl-β-D-*arabino*-hexosyl)thymine[1] (4)

The crude, protected nucleoside (**3**) (67 g., 0.09 mole) is dissolved in 600 ml. of 1:3 (w/w) hydrogen chloride–methanol, and the resulting solution is stirred for 18 hr. The product that separates is filtered off, and dried for 2 days over potassium hydroxide in a vacuum desiccator, giving 47 g. (73%) of **4**, m.p. 240–244° (softening, 165°).

1-(2-Deoxy-β-D-*arabino*-hexopyranosyl)thymine[1] (5)

The protecting *p*-nitrobenzoyl groups are removed by dissolving 47 g. (0.065 mole) of **4** in 2.76 l. of methanol containing 4.5 g. of sodium methoxide, and keeping the solution for 18 hr. The pH of the mixture is adjusted to 6 by the addition of glacial acetic acid, and the solution is evaporated to dryness at 40° under diminished pressure. The residue is dissolved in 2 l. of water, and the solution is washed with three 800-ml. portions of chloroform. The aqueous layer is evaporated to dryness at 40° under diminished pressure, and the residue is extracted with three 500-ml. portions of boiling isopropyl alcohol. To the cooled, filtered extract is added 3 ml. of concentrated hydrochloric acid, and the solution is concentrated to *ca.* 150 ml., and kept at 0–5° for 18 hr. The product is filtered off, washed with ether, and air-dried, yielding 7 g. (39%) of **5**, m.p. 228–230°, $[\alpha]_D^{25}$ +7.2° (*c* 1.0, water).

Fractional recrystallization of the product is conducted by dissolving it in boiling isopropyl alcohol (80 ml./g., plus 10% w/w of Darco G 60).

[a] In preparations of this size and larger, better mixing of the reactants is achieved by this method than by manual blending. This technique results in higher yields of **3**.

[b] Pyrimidine **2** may be recovered from the ether filtrate.

The filtered solution is kept undisturbed while being cooled to room temperature, and is then kept at 0–5° overnight. The supernatant liquor is decanted, and the crystals (45–50%) are collected, washed with ether, and air-dried, m.p. 230–232°, $[\alpha]_D^{25}$ +6.2° (*c* 1.0, water) (both unchanged by further recrystallization). The mother liquors are evaporated to dryness, and the resulting solid is recrystallized as described above, to give an additional crop of crystals having the same melting point and specific rotation as the first crop. The total yield from four crops is 5.8 g. (17.5% from **1**), m.p. 230–232°, $[\alpha]_D^{25}$ +6.2° (*c* 1.0, water).

REFERENCES

(1) W. W. Zorbach and G. J. Durr, Jr., *J. Org. Chem.*, **27**, 1474 (1962).
(2) M. Zimmerman, *Biochem. Biophys. Res. Commun.*, **16**, 600 (1964).
(3) P. Langen and G. Etzold, *Biochem. Z.*, **339**, 190 (1963).
(4) T. A. Krenitsky, J. W. Mellors, and R. K. Barclay, *J. Biol. Chem.*, **240**, 1281 (1964).
(5) G. E. Hilbert and T. B. Johnson, *J. Am. Chem. Soc.*, **52**, 4489 (1930).

[105] 1-(2-Deoxy-β-D-*arabino*-hexopyranosyl)thymine and its α-D Anomer

[Anomeric (2-Deoxy-D-glucopyranosyl)thymines]

2-Deoxyaldopyranosyl Nucleosides from O-Acetylglycals of Hexoses by the "Mercuri" Method

G. ETZOLD and P. LANGEN

INSTITUTE OF BIOCHEMISTRY, GERMAN ACADEMY OF SCIENCES, BERLIN, BERLIN–BUCH, GERMANY (D.D.R.)

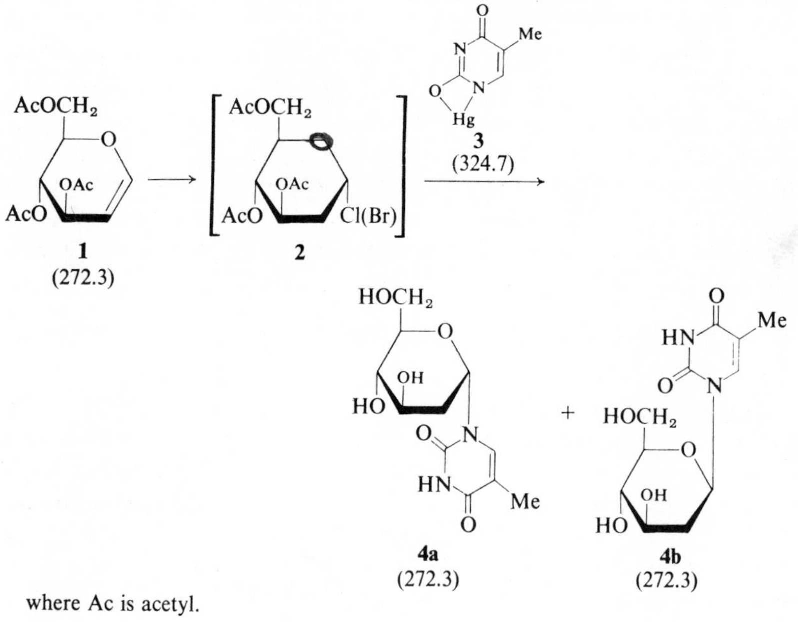

1 (272.3)

2

3 (324.7)

4a (272.3)

4b (272.3)

where Ac is acetyl.

INTRODUCTION

(2-Deoxy-β-D-glucopyranosyl)thymine (**4b**),[1] first synthesized by the Hilbert–Johnson procedure,[1a] is, both *in vitro*[1c,2] and *in vivo*,[3] a powerful inhibitor of uridine-deoxyuridine phosphorylase (EC 2.4.2.3). The α-D anomer (**4a**) is less active.[1c] The preparation described here is an example of the synthesis of anomeric pairs of pyrimidine nucleosides by the coupling of mercury derivatives of pyrimidines with poly-*O*-acetyl-2-

deoxyhexosyl halides, formed as intermediates by the addition of hydrogen halide to tri-*O*-acetylglycals. The mixture of **4a** (67%) and **4b** (33%) thus obtained can be used for effective inhibition of the above-mentioned enzyme.[1c,3] The anomers may be separated by column chromatography on silica gel.

PROCEDURE

Monothyminylmercury[4] (3)

To 50 ml. of acetic anhydride and 0.5 ml. of pyridine in a 100-ml., round-bottomed flask is added 12.6 g. (0.1 mole) of thymine. The mixture is refluxed, with stirring, until all the thymine dissolves (10–15 min.), and refluxing is continued for 15 min. The brown mixture is allowed to cool to room temperature, and the crystalline 1-acetylthymine is filtered off and washed with ice-cold acetic anhydride and ether; yield 14.5 g. (86%), m.p. 196°.

To a warm solution of 6.37 g. (20 mmoles) of mercuric acetate in 200 ml. of methanol is added 3.36 g. (20 mmoles) of 1-acetylthymine, with stirring, and the suspension is refluxed, with stirring, for 2 hr. It is allowed to cool, and is kept overnight, with stirring maintained. The fine, white precipitate of monothyminylmercury (**3**) is separated by repeated suction-filtration, using the same filter, until the filtrate is clear. The amorphous product is washed with methanol, and dried at 80° overnight; yield 6.4 g. (quantitative), m.p. $>360°$.

1-(2-Deoxy-β-D-*arabino*-hexopyranosyl)thymine and α-D anomer[1c] (4b and 4a)

Into a solution of 5.4 g. (20 mmoles) of 3,4,6-tri-*O*-acetyl-D-glucal[5] (**1**) in 20 ml. of sodium-dried benzene in a 100-ml., round-bottomed flask (fitted with a gas-inlet tube and cooled in ice) is passed a moderately rapid stream of dry hydrogen chloride or hydrogen bromide. After saturation of the solution (15–20 min.), the gas-stream is sharply diminished, but continued for at least 60 min. The solution is then evaporated to dryness at 30°, and the sirupy residue is diluted with 15 ml. of dry benzene and re-evaporated. This procedure is twice repeated, and the crude, sirupy 3,4,6-tri-*O*-acetyl-2-deoxy-D-*arabino*-hexosyl halide (**2**) is dissolved in 10 ml. of dry benzene.

In a 50 ml., round-bottomed flask, a suspension of 2.27 g. (7 mmoles) of monothyminylmercury (**3**) in 20 ml. of *N*,*N*-dimethylformamide and 7 ml. of anhydrous toluene is azeotropically dried by distilling off the toluene at atmospheric pressure, with magnetic stirring. The condenser is replaced by a calcium chloride tube, and the mixture is allowed to cool to room temperature. The benzene solution of the halide (**2**) is added rapidly, in one portion, to the vigorously stirred suspension. With slight warming, all of the material dissolves to an almost clear solution. With exclusion of moisture, stirring is continued for 3 hr., and the small amount of insoluble material is filtered off. The filtrate is mixed with 40 ml. of chloroform and thoroughly shaken with 70 ml. of 15% aqueous potassium iodide.[a] The chloroform layer is dried with sodium sulfate, and evaporated to dryness at 30°. The sirupy residue is extracted with three 25-ml. portions of dry, boiling cyclohexane (the cyclohexane that remains after the last decantation is removed under diminished pressure) and the extracts are discarded.

For deacetylation, the brown sirupy mixture of the acetylated nucleosides is dissolved in 20 ml. of absolute methanol, and *M* methanolic sodium methoxide is added until the solution gives a positive test with phenolphthalein paper. The solution is then refluxed on a water bath, and more *M* methanolic sodium methoxide is added until an alkaline reaction to phenolphthalein remains unchanged during reflux for 10 min. After a total of about 30 min. of refluxing, the mixture is diluted with 20 ml. of water, and the sodium ions are removed by adding Amberlite IR-120 (H^+) ion-exchange resin, in portions, until the solution is neutral. The resin is filtered off, and the solution of the nucleoside is evaporated to a sirup under diminished pressure.

A chromatographic tube (2.5 × 50 cm.) is uniformly packed by pressing down small portions (height, 1–2 cm.) of 80 g. of Celite 535,[b] thoroughly premixed in a porcelain mortar with 30 ml. of the lower phase of 4:1:2 ethyl acetate–isopropyl alcohol–water (solvent system A). The crude nucleoside sirup is diluted to a volume of about 7 ml. with the same phase, and thoroughly mixed with 14 g. of the Celite, and the mixture is pressed onto the top of the column. The chromatogram is developed with the upper phase of solvent system A, under slight pressure (0.5 ml./min.; 10 ml./fraction). The nucleoside content of the eluate may be examined by measuring the ultraviolet absorption at 260 nm., or by paper[c] or thin-layer[d] chromatography. The fractions (about 25–60) containing both

[a] To remove the mercury salts.

[b] Fresh, or used, Celite 535 is digested with 6 *N* hydrochloric acid at room temperature overnight, filtered off by suction, and washed free of acid with water. It is given a final wash with acetone, and dried in an oven at 100° overnight.

[c] About 0.03 ml. per fraction per spot; descending development on Schleicher & Schüll paper No. 2043b; solvent mixture 8:5:5 ethyl acetate–isopropyl alcohol–25% aqueous ammonia; $R_{thymine}$ of **4a** 0.77 and of **4b** 0.70 (detected by ultraviolet light).

[d] About 0.01 ml. per fraction per spot, on a 0.25-mm. layer of silica gel HF_{254} (Merck) according to Stahl; solvent mixture: 5:1:1 2-butanone–isopropyl alcohol–25% aqueous ammonia; R_f (two runs of 15 cm. distance) of **4a**, 0.24, and of **4b**, 0.20 (detection by ultraviolet light).

anomers are evaporated under diminished pressure. The residue is re-evaporated with two 5-ml. portions of isopropyl alcohol, and dissolved in a small portion of hot isopropyl alcohol. When kept in air at room temperature for 1 to 2 days, the nucleoside mixture, consisting of 67% of **4a** and 33% of **4b**, gradually crystallizes. It is mixed with about 5 ml. of 2:1 ethyl acetate–isopropyl alcohol and filtered; yield 580 mg. (33%, based on **3**), m.p. 160–170°, $[\alpha]_D^{23}$ +62° (*c* 1.74, water).

A slurry of 500 g. of silica gel (0.2–0.5 mm., Merck) in a homogeneous mixture of 100:8:5:5 2-butanone–isopropyl alcohol–25% aqueous ammonia–water (solvent system B) is poured into a chromatographic tube (3.8 × 100 cm.). The solvent is drained from the bottom, until its level is 5 cm. above the bed of silica gel. A solution of 300 mg. of the nucleoside mixture in 3 ml. of warm methanol is adsorbed onto 3 g. of silica gel and, after being briefly dried at 50°, the mixture is placed on the column. The adsorbed gas is removed by cautious stirring of this small zone with a glass rod, and glass wool is placed on the surface. The chromatogram is developed with solvent system B (1 ml. per min.; 20 ml. per fraction)[e]. The nucleoside distribution is examined by thin-layer chromatography, using 0.1 ml. per spot.[d,f] Fractions 290–360 contain **4a**, 361–386 contain both anomers, and 387–465 contain **4b**.

1-(2-Deoxy-α-D-*arabino*-hexopyranosyl)thymine (4a)

Fractions 290–360 from the preceding chromatogram are combined and evaporated under diminished pressure at 30°. The brown, liquid residue is re-evaporated with 10 ml. of absolute ethanol, and dissolved in a mixture of 10 ml. of chloroform and 10 ml. of water. The aqueous layer is thoroughly washed with two 5-ml. portions of chloroform, and the combined chloroform layers are extracted once with 5 ml. of water. The aqueous phases are combined and evaporated under diminished pressure to a sirup which is dissolved in 0.5 ml. of the lower phase of solvent system A; the solution is then mixed with 1 g. of Celite. The mixture is pressed down onto the top of a column (2 × 50 cm.), packed with 60 g. of Celite mixed with 18 ml. of the lower phase of solvent A, and the column is developed with the upper phase, as described above. The nucleoside-containing fractions (about 19–28) are evaporated under diminished pressure, and the residue is re-evaporated with two 3-ml. portions of absolute ethanol. After the addition of a few drops of isopropyl alcohol, **4a** crystallizes within a few hours.[g]

[e] At the beginning, the column becomes yellow, but the colored products are eluted before the nucleosides appear.

[f] Direct examination of the eluate by ultraviolet absorption measurement is not possible, because of the high extinction coefficient of 2-butanone in the same wavelength range as that of the nucleosides.

[g] Scratching may help induce crystallization.

It is suspended in ethyl acetate, and the suspension is filtered; yield 155 mg., m.p. 182°. After recrystallization from isopropyl alcohol,[g] it has m.p. 182–183° (softening at 179°), $[\alpha]_D^{20}$ +90° (*c* 1.63, water), $\lambda_{max}^{H_2O}$ 266 nm., $\epsilon_{260}^{H_2O}$ 9,100.

1-(2-Deoxy-β-D-*arabino*-hexopyranosyl)thymine (4b)

Fractions 387–465 are processed, including purification by means of a Celite column (the resulting fractions 24–33 are used), as with **4a**. On scratching, **4b** readily crystallizes; yield 40 mg., m.p. 231°. After recrystallization from absolute ethanol, it has m.p. 232° (subliming at 228°), $[\alpha]_D^{20}$ +4° (*c* 0.691, water), $\lambda_{max}^{H_2O}$ 266 nm., $\epsilon_{260}^{H_2O}$ 8,900.

REFERENCES

(1) (a) W. W. Zorbach and G. J. Durr, *J. Org. Chem.*, **27**, 1474 (1962); (b) J. J. K. Novák and F. Šorm, Czech. Pat. 107,594 (1963); *Chem. Abstracts*, **60**, 5627 (1964); (c) G. Etzold and P. Langen, *Chem. Ber.*, **98**, 1988 (1965).

(2) (a) P. Langen and G. Etzold, *Biochem. Z.*, **339**, 190 (1963); (b) M. Zimmerman, *Biochem. Biophys. Res. Commun.*, **16**, 600 (1964).

(3) P. Langen and G. Etzold, *Mol. Pharmacol.*, **2**, 89 (1966).

(4) M. Hoffer, *Chem. Ber.*, **93**, 2777 (1960).

(5) W. Roth and W. Pigman, *Methods Carbohydrate Chem.*, **2**, 405 (1963).

[106] 1-(2-Deoxy-β-D-*erythro*-pentopyranosyl)-thymine and its α-D Anomer [Anomeric (2-Deoxy-D-ribopyranosyl)thymines]

2-Deoxyaldopyranosyl Nucleosides from O-*Acetylglycals of Pentoses by the "Mercuri" Method*

G. ETZOLD and P. LANGEN

INSTITUTE OF BIOCHEMISTRY, GERMAN ACADEMY OF SCIENCES, BERLIN, BERLIN–BUCH, GERMANY (D.D.R.)

1 (150.1) → **2** (318.3) → **3** (200.2) →

4 (236.7) —**5** (324.7)→ acetates of **6a** and **6b** —NH_3→ **6a** (242.2) + **6b** (242.2)

where Ac is acetyl.

INTRODUCTION

(2-Deoxy-β-D-ribopyranosyl)thymine[1] (**6b**), the pyranoid isomer of thymidine, is, in contrast to the latter, resistant to pyrimidine phosphorylases.[2] Its preparation demonstrates the synthesis of anomeric 2-deoxypentopyranosyl pyrimidines by the "mercuri" method, using 3,4-di-*O*-acetyl-D-glycals as starting materials for the non-isolated 3,4-di-*O*-acetyl-2-deoxy-D-pentopyranosyl halide intermediates.[a] The mixture of anomers (acetates of **6a** and **6b**) is deacetylated by ammonolysis, and the anomers are separated by column chromatography on Celite.

PROCEDURE

1,2,3,4-Tetra-*O*-acetyl-α-D-arabinose[6] (2)

Into a 250-ml., 3-necked, round-bottomed flask, equipped with a stirrer, reflux condenser, and a thermometer, and containing 10 g. of anhydrous sodium acetate and 130 ml. of acetic anhydride, is added 37.5 g. (0.25 mole) of β-D-arabinose (**1**). The mixture is slowly heated, with stirring, to 50–70°. When the reaction begins, the flask is immediately cooled with ice–water, and the temperature is maintained at about 80° until the sugar has entirely dissolved. The solution is kept for 30 min. at 100° and, after being cooled, it is poured into 1 l. of ice–water. The sirupy tetraacetate (**2**) crystallizes within a few hours. It is filtered off and washed with water; yield 35–40 g.[b] (44–50%), m.p. 94–96°. Recrystallization from ethanol gives pure **2**, m.p. 98–100°, $[\alpha]_D^{20}$ −43.6° (*c* 2.0, chloroform).

3,4-Di-*O*-acetyl-D-arabinal[7] (3)

To a mixture of 40 g. (126 mmoles) of 1,2,3,4-tetra-*O*-acetyl-α-D-arabinose (**2**), 60 ml. of acetic acid, 3.5 ml. of acetic anhydride, and 6 g. of red phosphorus, cooled in an ice-bath, 36 g. (11.4 ml.) of bromine is added dropwise, with stirring, the temperature being kept below 10° (30–35 min.). Water (7.2 ml.) is added dropwise, at a temperature below 10° (10 min.). The mixture containing crude tri-*O*-acetyl-D-arabinopyranosyl bromide is stirred for 1 hr. at room temperature.

In the meantime, a solution of 68.5 g. of sodium acetate trihydrate and 140 ml. of acetic acid in 200 ml. of water is well cooled with an ice–salt mixture. With vigorous stirring, 83.5 g. of zinc dust and a solution of 3.6 g. of copper sulfate pentahydrate in 14 ml. of water are added. With stirring, the glycosyl halide is added in several portions, the temperature being

[a] By the same method, starting with 3,4-di-*O*-acetyl-L-arabinal and 3,4-di-*O*-acetyl-D-xylal, the anomers of (2-deoxy-L-*erythro*-pentopyranosyl)thymine[3] and of (2-deoxy-D-*threo*-pentopyranosyl)thymine[4] have been prepared. Both β-D anomers proved to be powerful inhibitors of uridine-deoxyuridine phosphorylase (EC 2.4.2.3)[2,5]

[b] The product may be used in the next stage without further purification.

kept at −5 to 0° (15 min.). After being stirred for 3 hr. at 0°, the mixture is filtered, and the zinc is washed with 50 ml. of 50% aqueous acetic acid. The filtrate is diluted with 500 ml. of ice–water, and extracted with four 150-ml. portions of chloroform. The combined chloroform extracts are washed successively with water, saturated aqueous sodium hydrogen carbonate, and water (to remove any acetic acid present). The chloroform solution is dried with sodium sulfate overnight in a refrigerator, and evaporated to a sirup under diminished pressure at 40°; the sirup is fractionally distilled, to give **3** as a thin sirup; yield, about 15 g. (60%), b.p. 88–90°/1 mm. Hg, $[\alpha]_D^{20}$ +265° (*c* 2.0, chloroform).

1-(2-Deoxy-β-D-*erythro*-pentopyranosyl)thymine (6b)

3,4-Di-*O*-acetyl-D-arabinal (**3**) (7 g., 35 mmoles) is dissolved in 30 ml. of anhydrous benzene, the solution is cooled in ice, and a moderately rapid stream of dry hydrogen chloride is passed in. After saturation (about 15 min.) of the solution, it is kept at 0° for 15 min. and then evaporated to dryness under diminished pressure at 30°. The residue is diluted with 15 ml. of dry benzene and the solution is re-evaporated to a small volume. This procedure is twice repeated, and the crude, sirupy 3,4-di-*O*-acetyl-2-deoxy-D-*erythro*-pentosyl chloride (**4**) is dissolved in 10 ml. of dry benzene.

As described in the preparation of 1-(2-deoxy-D-glucosyl)thymine (Sec. III [105]), the benzene solution of the halide (**4**) is added rapidly, in one portion, to a suspension of 5.2 g. (16 mmoles) of monothyminylmercury (**5**) (Sec. III [105]) in 40 ml. of *N*,*N*-dimethylformamide and 15 ml. of toluene (azeotropically dried) in a 100-ml. flask. The mixture is stirred for 3 hr. and filtered. The filtrate is diluted with 70 ml. of chloroform, and shaken with 170 ml. of 15% aqueous potassium iodide. The chloroform layer is dried with sodium sulfate, and evaporated to dryness under diminished pressure.

For deacetylation of the mixture of protected nucleosides, the residue is dissolved in 60 ml. of absolute methanol presaturated with ammonia, and the solution is kept at room temperature overnight in a stoppered flask. The solvent is removed under diminished pressure at 40°, and the sirup is treated with 3 ml. of the lower phase of 12:1:6 ethyl acetate–isopropyl alcohol–water. The solution is mixed with 10 g. of Celite 535 and pressed down on a column (5.2 × 100 cm.) that has been packed with a mixture of 1,000 g. of Celite 535 and 300 ml. of the lower phase of the solvent system.[c] The column is developed with the upper phase of the solvent mixture (1 ml. per min., 20-ml. fractions). The nucleoside content is estimated by measurement of the ultraviolet absorption or by thin-layer

[c] See Celite column chromatography of 1-(2-deoxy-D-glucosyl)thymine, Sec. III [105], for details.

chromatography (0.1 ml. per spot).[d] Fractions 101–165 (second main peak) are combined, evaporated to dryness, and twice re-evaporated with a small volume of absolute ethanol, giving colorless crystals of chromatographically pure **6b** which are suspended in ethyl acetate and filtered off; yield 605 mg. (16%, based on **5**), m.p. 222–224° (recrystallization from isopropyl alcohol does not change the m.p.), $[\alpha]_D^{20}$ +27° (*c* 1.08, water), $\lambda_{max}^{H_2O}$ 265 nm. (ϵ 10,300), $\epsilon_{260}^{H_2O}$ 9,700.

1-(2-Deoxy-α-D-*erythro*-pentopyranosyl)thymine (6a)

Fractions 220–325 are evaporated under diminished pressure, and the sirupy residue is re-evaporated with two 5-ml. portions of absolute ethanol. On the addition of 1 ml. of isopropyl alcohol, and with scratching, **6a** crystallizes. It is suspended in 10 ml. of ethyl acetate and filtered off; yield 585 mg. (15%, based on **5**), m.p. 226° (unchanged after recrystallization from isopropyl alcohol), $[\alpha]_D^{20}$ −46° (*c* 1.2, water), $\lambda_{max}^{H_2O}$ 265 nm. (ϵ 10,100), $\epsilon_{260}^{H_2O}$ 9,600.

REFERENCES

(1) G. Etzold and P. Langen, *Naturwissenschaften*, **53**, 178 (1966).
(2) G. Etzold, B. Preussel, and P. Langen, *Mol. Pharmacol.*, **4**, 20 (1968).
(3) G. Etzold, R. Hintsche, and P. Langen, *Chem. Ber.*, **101**, 226 (1968).
(4) E. Wittenburg, G. Etzold, and P. Langen, *Chem. Ber.*, **101**, 494 (1968).
(5) G. Etzold, B. Preussel, R. Hintsche, and P. Langen, *Acta Biol. Med. Ger.*, **20**, 437 (1968).
(6) J. Kuszmann and L. Vargha, *Rev. Chim. Acad. Rep. Populaire Roumaine*, **7**, 1025 (1962).
(7) L. Vargha and J. Kuszmann, *Chem. Ber.*, **96**, 411 (1963).

[d] On silica gel HF_{254} (Merck) according to Stahl (0.25-mm. layer); solvent mixture, 47:3 water-saturated ethyl acetate–methanol, detection by ultraviolet light, R_f of **6b**, 0.34, and of **6a**, 0.23.

[107] 1-(3-Deoxy-α-D-*erythro*-pentofuranosyl)-thymine and 1-(3-Deoxy-β-D-*erythro*-pentofuranosyl)-thymine

[1-(3-Deoxy-α-D-ribofuranosyl)thymine and its β-D Anomer]

The Hilbert–Johnson Synthesis Applied to Preparation of the Anomers of Pyrimidine 3-Deoxyaldofuranosyl Nucleosides

RUTH F. NUTT and EDWARD WALTON

MERCK, SHARP AND DOHME RESEARCH LABORATORIES, DIVISION OF MERCK & CO., INC., RAHWAY, NEW JERSEY 07065

$HOCH_2$ O OMe → **1** (146.1)

$HOCH_2$ O OMe, OH → **2** (148.2)

*p*NBzOCH_2 O OMe, *p*NBzO → **3** (446.4)

*p*NBzOCH_2 O Br, *p*NBzO — **4** (495.3)

+ **5** (154.2) (OMe, Me, N, MeO, N) →

R′OCH_2 … **6**, R = OMe, R′ = *p*NBz (554.5) + **7**, R = OMe, R′ = *p*NBz (554.5)

8, R = OMe, R′ = H (242.2) — **9**, R = OMe, R′ = H (242.2)

10, R = OH, R′ = H (228.2) — **11**, R = OH, R′ = H (228.2)

where *p*NBz is *p*-nitrobenzoyl.

INTRODUCTION

The preparation of 1-(3-deoxy-α-D-*erythro*-pentofuranosyl)thymine and its β-D anomer[1] is an example of the utility of crystalline 3-deoxy-2,5-di-*O*-*p*-nitrobenzoyl-β-D-*erythro*-pentofuranosyl bromide[1] (**4**) in the synthesis of both anomers of pyrimidine 3′-deoxynucleosides by the Hilbert–Johnson procedure.[2] The intermediate poly-*O*-acylglycosylpyrimidinones are deblocked in two steps. The acyl groups are first removed from the sugar moiety with methanolic sodium methoxide, and then the 4-*O*-methyl group is removed with methanolic hydrogen chloride. Crystalline, characterized intermediates are obtained at all stages in the synthesis.

PROCEDURE

Methyl 3-Deoxy-2,5-di-*O*-*p*-nitrobenzoyl-β-D-*erythro*-pentoside (3)

A solution of 15 g. (0.11 mole) of methyl 2,3-anhydro-β-D-ribofuranoside[3] (**1**) in 750 ml. of ethanol is shaken at 80° with 1 tablespoonful of Raney nickel catalyst[a] and hydrogen at 2.8 kg./cm.^{2}. The uptake of hydrogen ceases after about 6 hr. The mixture is filtered, and the catalyst is washed with hot ethanol. The filtrate and washings are combined and concentrated,[b] and two 15-ml. portions of toluene are distilled from the sirup (to remove traces of ethanol). Methyl 3-deoxy-β-D-*erythro*-pentofuranoside[3] (**2**) (15.8 g.) is obtained as a colorless sirup.

A solution of 2.0 g. (13.5 mmoles) of **2** in 50 ml. of dry (barium oxide) pyridine is cooled in an ice bath and treated with 7.5 g. (40.5 mmoles) of *p*-nitrobenzoyl chloride. The mixture is stirred at 25° for 20 hr., and then cooled in an ice bath and treated with 1 ml. of water. After being stirred for 30 min., the pasty mass is concentrated to 20 ml. and diluted with 100 ml. of chloroform. The chloroform solution is washed with three 50-ml. portions of saturated aqueous sodium hydrogen carbonate and 50 ml. of water, dried, and concentrated to a sirup (7 g.) which is crystallized from 10 ml. of benzene by adding petroleum ether; 4.6 g. of product, m.p. 106–109°, is obtained. Recrystallization from benzene by adding petroleum ether gives 4.4 g. (75%) of **3**, m.p. 108–110°, $[\alpha]_D$ −33°, $[\alpha]_{578}$ −35° (*c* 1.0, chloroform); $\tau(CDCl_3)$[c] 4.9 (singlet, C-1 proton).

3-Deoxy-2,5-di-*O*-*p*-nitrobenzoyl-β-D-*erythro*-pentosyl Bromide (4)

A warm solution of 3.8 g. (8.51 mmoles) of methyl 3-deoxy-2,5-di-*O*-*p*-nitrobenzoyl-β-D-*erythro*-pentoside (**3**) in 16 ml. of acetic acid is cooled

[a] Raney nickel similar to the W-3 form described by A. A. Pavlic and H. Adkins, *J. Am. Chem. Soc.*, **68**, 1471 (1946).

[b] All concentrations are conducted under diminished pressure in a rotary evaporator.

[c] Nuclear magnetic resonance spectra are obtained on a Varian Associates Model A-60 spectrometer.

to 10° and immediately treated with 1 ml. of acetyl bromide and 16 ml. of a cold solution of 33% (w/w) hydrogen bromide in acetic acid. The solution is kept at 10° for 20 min., during which time a solid is precipitated. Thin-layer chromatography (tlc) on alumina with 1:1 (v/v) benzene–chloroform shows zones (developed with iodine vapor) at R_f 0.2 (glycosyl halide, **4**), 0.6 (by-product), and 0.7 (starting material, **3**). Disappearance of the spot at R_f 0.7 after about 10 min. indicates that the reaction is complete. The reaction mixture is evaporated to dryness, and three 20-ml. portions of dry toluene are added, and removed under diminished pressure (to remove traces of hydrogen bromide and acetic acid). The crystalline residue (m.p. 118–124°) is recrystallized from 20 ml. of dichloromethane plus 40 ml. of ether, giving 3.4 g. (81%) of 3-deoxy-2,5-di-*O*-*p*-nitrobenzoyl-β-D-*erythro*-pentosyl bromide (**4**), m.p. 128–131°, $[\alpha]_D$ −50°, $[\alpha]_{578}$ −53° (*c* 1.18, dichloromethane), $\lambda_{max}^{CH_2Cl_2}$ 261 nm. (ε 28,900); τ($CDCl_3$) 3.39 (doublet, C-1 proton, $J_{1,2}$ 0.8 Hz).

1-(3-Deoxy-2,5-di-*O*-*p*-nitrobenzoyl-β-D-*erythro*-pentosyl)-4-methoxy-5-methyl-2(1*H*)-pyrimidinone (6) and 1-(3-Deoxy-2,5-di-*O*-*p*-nitrobenzoyl-α-D-*erythro*-pentosyl)-4-methoxy-5-methyl-2(1*H*)-pyrimidinone (7)

A solution of 4.47 g. (28.8 mmoles) of 2,4-dimethoxy-5-methylpyrimidine[4] (**5**) (Sec. II [30]) in 240 ml. of dry (molecular sieves) dichloromethane is stirred at 25° with 6.67 g. (13.5 mmoles) of the bromide **4**. The course of the reaction is followed by tlc on alumina with chloroform. The plates are developed with iodine vapor, and show zones for starting pyrimidine (**5**) at R_f 0.9, β-D anomer (**6**) at R_f 0.6, α-D anomer (**7**) at R_f 0.4, and halide **4** at R_f 0.2. The intermediate 1-(3-deoxy-2,5-di-*O*-*p*-nitrobenzoyl-D-*erythro*-pentosyl)-2,4-dimethoxy-5-methylpyrimidinium bromide is observable as a zone at the origin. After 72 hr., the reaction mixture is concentrated, and the residual solid is leached with about 200 ml. of ether to remove most of the excess pyrimidine (**5**). The ether-insoluble solid (7.0 g.) is chromatographed[d] on 70 g. of acid-washed alumina (Merck) with chloroform as the solvent. After elution of a small amount of **5**, fractions containing almost pure β-D anomer (**6**) are obtained. Recrystallization of this material from benzene–petroleum ether gives a total of 2.1 g. (28%) of purified **6**, m.p. 164–170°, $[\alpha]_D$ −26°, $[\alpha]_{578}$ −27° (*c* 0.67, chloroform); λ_{max}^{Nujol} 5.81 and 5.86 μm. (ester); λ_{max}^{MeOH} 261 nm. (ε 28,200); τ($CDCl_3$) 3.79 (doublet, C-1′ proton, $J_{1',2'}$ 1.3 Hz).

Further elution of the column with chloroform gives several fractions containing both **7** and **6**, followed by fractions containing almost pure

[d] The ratio of the height to diameter of the column is about 1:1. A fritted-glass Büchner funnel of medium porosity may be used.

α-D anomer (**7**). Recrystallization of this material from chloroform–methanol gives a total of 1.3 g. (18%) of pure **7**, m.p. 218–219°, $[\alpha]_D$ −209°, $[\alpha]_{578}$ −222° (*c* 0.23, chloroform); λ_{max} 260 nm. (ε 26,200); τ(CDCl$_3$) 3.41 (doublet, C-1′ proton, $J_{1',2'}$ 3.8 Hz).

1-(3-Deoxy-β-D-*erythro*-pentofuranosyl)-4-methoxy-5-methyl-2(1*H*)-pyrimidinone (8)

In apparatus predried at 110°, a mixture of 1.54 g. (2.78 mmoles) of the protected nucleoside (**6**) with 34 ml. of dry (molecular sieves) methanol is treated with a solution prepared from 100 mg. (4.35 mg.-atoms) of sodium and 3 ml. of dry methanol, and the mixture is refluxed for 1 hr. The reaction solution is evaporated to dryness, about 50 ml. of water is added, and the insoluble methyl *p*-nitrobenzoate is removed and well washed with water. The filtrate and washings are combined, and stirred with 15 g. of moist Dowex-50 W-A4 (H^+) for 10 min. The resin and precipitated *p*-nitrobenzoic acid are removed and washed with water, and the filtrate and washings are combined, and washed with three 50-ml. portions of ether. The water layer is filtered and concentrated to dryness, and the residue (640 mg., m.p. 190–196°) is recrystallized from methanol; 444 mg. (62%) of pure **8** is obtained, m.p. 196–198° (transition, 180–190°), $[\alpha]_D$ +25°, $[\alpha]_{578}$ +27° (*c* 0.77, water); $\lambda_{max}^{H_2O}$ nm. (ε × 10^{-3}): 280 (6.6), 203 (18.6), 215 (infl.) (12.0).

1-(3-Deoxy-β-D-*erythro*-pentofuranosyl)thymine (10)

A suspension of 395 mg. (1.54 mmoles) of 4-methoxy derivative **8** in 15 ml. of methanol is treated with 1.5 ml. of 30.6% (w/w) hydrogen chloride in methanol, and the solution is kept at 25° in a stoppered flask. After 6 days, no further change in the ultraviolet absorption spectrum is observable. The solution is evaporated to dryness, and one portion of methanol and three successive portions of benzene are added to and distilled from the residue. Crystallization of the residue from 1 ml. of methanol plus 3 ml. of ether gives 300 mg. (81%) of **10**, which melts at 96–100° (resolidifies, and remelts[e] at 155–157°); $[\alpha]_D$ +1.4°, $[\alpha]_{578}$ +2.3° (*c* 0.44, water); $\lambda_{max}^{H_2O}$ nm. (ε × 10^{-3}): pH 1, 269 (9.5); pH 7, 269 (9.6); pH 13, 268 (7.0); τ(D$_2$O) 4.19 (doublet, C-1′ proton, $J_{1',2'}$ 1.8 Hz).

1-(3-Deoxy-α-D-*erythro*-pentofuranosyl)-4-methoxy-5-methyl-2(1*H*)-pyrimidinone (9)

By the method described for the preparation of the β-D anomer (**8**), 960 mg. (1.73 mmoles) of the pyrimidinone derivative (**7**) gives 292 mg.

[e] If the compound is dried to constant weight at 56°, the lower melting point, caused by solvent of crystallization, is not observed.

(60%) of **9**, m.p. 185–187°, $[\alpha]_D$ −157°, $[\alpha]_{578}$ −166° (*c* 0.22, water); $\lambda_{max}^{H_2O}$ nm. ($\epsilon \times 10^{-3}$): 281 (6.6), 204 (18.9).

1-(3-Deoxy-α-D-*erythro*-pentofuranosyl)thymine (11)

By the procedure described for the synthesis of the β-D anomer (**10**), 279 mg. (1.1 mmoles) of the 4-methoxy derivative (**9**) gives 200 mg. (76%) of **11**; m.p. 188–191°, $[\alpha]_D$ −112°, $[\alpha]_{578}$ −118° (*c* 0.17, water); $\lambda_{max}^{H_2O}$ nm. ($\epsilon \times 10^{-3}$): pH 1, 269 (9.9); pH 7, 269 (10.0); pH 13, 268 (7.6); $\tau(D_2O)$ 3.93 (doublet, C-1′ proton, $J_{1',2'}$ 3.9 Hz).

REFERENCES

(1) E. Walton, F. W. Holly, G. E. Boxer, and R. F. Nutt, *J. Org. Chem.*, **31**, 1163 (1966).
(2) G. E. Hilbert and T. B. Johnson, *J. Am. Chem. Soc.*, **52**, 4489 (1930).
(3) (a) C. D. Anderson, L. Goodman, and B. R. Baker, *J. Am. Chem. Soc.*, **80**, 5247 (1958); (b) E. Walton, F. W. Holly, G. E. Boxer, R. F. Nutt, and S. R. Jenkins, *J. Med. Chem.*, **8**, 659 (1965).
(4) W. Schmidt-Nickels and T. B. Johnson, *J. Am. Chem. Soc.*, **52**, 4511 (1930).

[108] 1-(2,3-Dideoxy-β-D-*glycero*-pent-2-enofuranosyl)thymine [1] (3′-Deoxy-2′-thymidinene)

Base-catalyzed Elimination Reaction Applied to the Synthesis of Pyrimidine 2′,3′-Unsaturated Nucleosides

JEROME P. HORWITZ and JONATHAN CHUA

DETROIT INSTITUTE OF CANCER RESEARCH DIVISION OF THE MICHIGAN CANCER FOUNDATION, DETROIT, MICHIGAN 48201

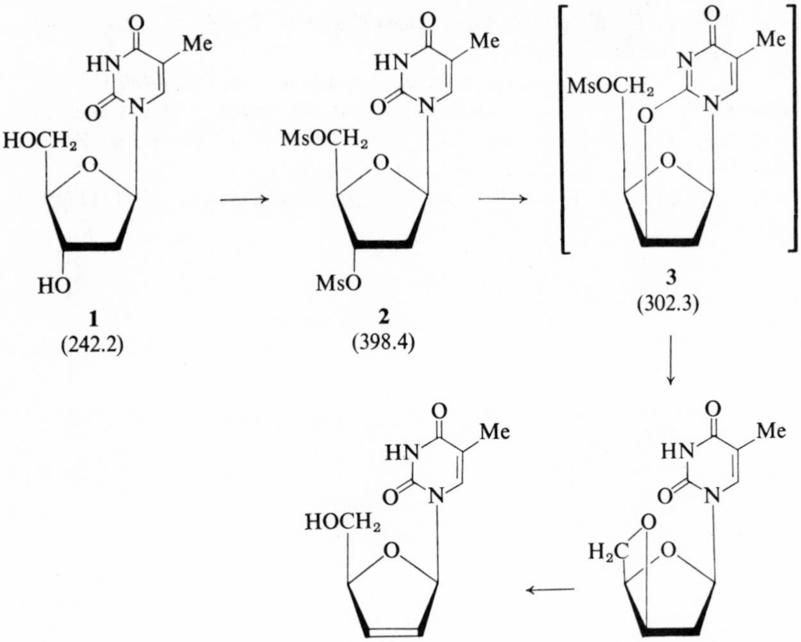

1 (242.2) → **2** (398.4) → [**3**] (302.3) → **4** (224.2) → **5** (224.2)

where Ms is methylsulfonyl.

INTRODUCTION

The facile conversion of 1-(3,5-anhydro-2-deoxy-β-D-*threo*-pentosyl)-pyrimidines into corresponding 2′,3′-unsaturated nucleosides by the action of potassium *tert*-butoxide in methyl sulfoxide provides a convenient route to a series of potentially important intermediates for the synthesis of modified pyrimidine nucleosides.[1] The procedure here outlined is also applicable to the synthesis of 2′,3′-unsaturated derivatives of uracil,[1] 4-thiothymine,[1] and cytosine.[2]

The requisite 3′,5′-anhydronucleosides (**4**) are readily obtained by the action of sodium hydroxide on the corresponding pyrimidine 2′-deoxy-3′,5′-di-*O*-(methylsulfonyl)nucleoside[3] (**2**); this gives **4** *via* an intermediate 2,3′-anhydronucleoside[4] (**3**).

PROCEDURE

3′,5′-Di-*O*-(methylsulfonyl)thymidine[3] (2)[a]

To a solution of 2.42 g. (10 mmoles) of thymidine (**1**) in 30 ml. of dry pyridine at 0° is added 2.3 ml. of methanesulfonyl chloride, and the solution is kept overnight at 0°. Approximately 2 ml. of water is added to the reaction mixture, which is kept in a refrigerator for an additional hr. and then poured, with vigorous stirring, into *ca.* 300 ml. of ice–water. The granular product is collected, washed with generous quantities of water, and air-dried, giving 3.9 g. (98%) of product, m.p. 162–165° (dec.). On recrystallization from 90% ethanol, compound **2** separates as colorless needles; yield 3.4 g. (85%), m.p. 170–171° (dec.) [lit.[3] 168–169° (dec.)].

1-(3,5-Anhydro-2-deoxy-β-D-*threo*-pentofuranosyl)thymine[4] (4)

A solution of 3.74 g. (9.4 mmoles) of 3′,5′-di-*O*-(methylsulfonyl)-thymidine (**2**) in 250 ml. of water containing 28 ml. of *N* sodium hydroxide is refluxed for 2 hr., cooled to room temperature, neutralized (phenol-phthalein) with 9.2 ml. of *N* hydrochloric acid, and evaporated to dryness under diminished pressure, the last traces of moisture being removed by adding and evaporating absolute ethanol. The residue is triturated with five 25-ml. portions of hot acetone, the suspension is filtered, and the filtrate is evaporated to dryness under diminished pressure. The residue is dissolved in ethanol, and the solution is decolorized (Norit) and concentrated on a hot plate to *ca.* 20 ml. The product, which separates as a mat of colorless needles, is collected and air-dried, giving 1.55 g. (74%) of **4**, m.p. 190–193°. A second recrystallization from the same solvent gives material of m.p. 193–194°, $[\alpha]_D^{24}$ −127° (*c* 1, water) which is dried at $100°/10^{-2}$ mm. Hg prior to use in the next step.

1-(2,3-Dideoxy-β-D-*glycero*-pent-2-enofuranosyl)thymine[4] (5)

A solution of 0.36 g. (1.6 mmoles) of **4** in 10 ml. of dry methyl sulfoxide[b] containing 0.36 g. (3.2 mmoles) of potassium *tert*-butoxide,[c] protected

[a] This procedure is virtually identical with that described in reference 3.
[b] Methyl sulfoxide (J. T. Baker, reagent grade) is distilled from calcium hydride.
[c] Potassium *tert*-butoxide is obtained from M. S. A. Corp. and is used as received.

from moisture, is stirred magnetically at room temperature for 2 hr. The mixture is neutralized to *ca.* pH 7 (test paper) with ethanolic acetic acid, and the solution is evaporated to dryness[d] at 50°/10^{-2} mm. Hg. The residue is triturated with five 20-ml. portions of hot acetone, the salts are removed by filtration, and the filtrate is evaporated to dryness under diminished pressure. The residue is dissolved in 25 ml. of absolute ethanol, and the solution is decolorized (Norit), concentrated to *ca.* 5 ml. by boiling, and diluted with 25 ml. of benzene. The volume of the mixture is again diminished to *ca.* 5 ml. by boiling, and the process is repeated until the solution becomes turbid. The unsaturated nucleoside separates as a colorless, granular solid; yield 0.32 g. (89%), m.p. 156–160°. A second recrystallization from the same solvent system gives 0.285 g. (79%) of pure **5**, m.p. 165–166° (with resolidification of the melt, and no further evidence of a change of state to 200°), $[\alpha]_D^{25}$ −42° (*c* 0.69, water).

REFERENCES

(1) J. P. Horwitz, J. Chua, M. A. Da Rooge, M. Noel, and I. L. Klundt, *J. Org. Chem.*, **31**, 205 (1966).
(2) J. P. Horwitz, J. Chua, J. T. Donatti, and M. Noel, *J. Org. Chem.*, **32**, 817 (1967).
(3) A. M. Michelson and A. R. Todd, *J. Chem. Soc.*, **1955**, 816.
(4) J. P. Horwitz, J. Chua, J. A. Urbanski, and M. Noel, *J. Org. Chem.*, **28**, 942 (1963).

[d] Failure to remove the last traces of methyl sulfoxide at this point leads to subsequent difficulty in the crystallization of **5**.

[109] 2-(2-Deoxy-β-D-*arabino*-hexopyranosyl)-*as*-triazine-3,5(2*H*,4*H*)-dione [1-(2-Deoxy-β-D-glucosyl)-6-azauracil]

The Modified Hilbert–Johnson Synthesis of a 6-Azauracil Nucleoside

G. J. DURR

DEPARTMENT OF CHEMISTRY, LE MOYNE COLLEGE, SYRACUSE, NEW YORK 13214

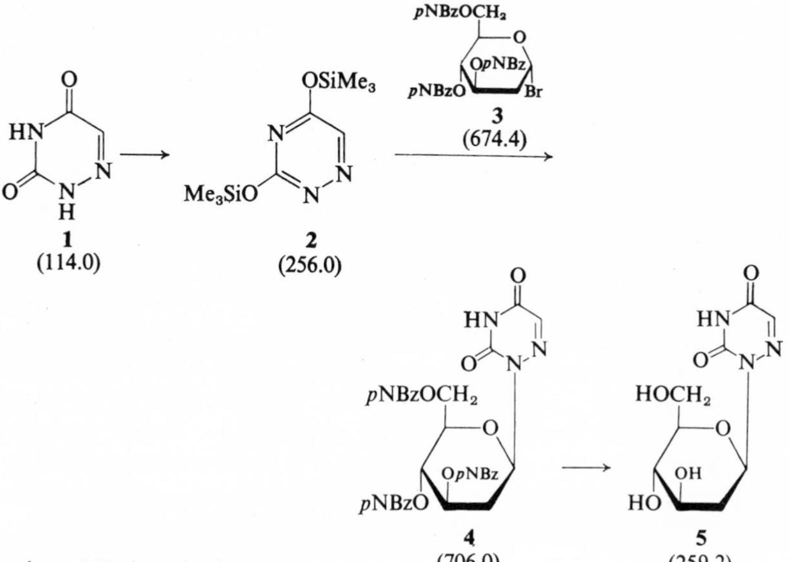

1 (114.0) → 2 (256.0) —[3 (674.4)]→ 4 (706.0) → 5 (259.2)

where *p*NBz is *p*-nitrobenzoyl.

INTRODUCTION

6-Azauracil (**1**) is an anticancer agent[1] and (2-deoxy-D-glucosyl)thymine (Sec. III [104]) is an inhibitor of a pyrimidine phosphorylase.[2] Thus, the title compound[3] (**5**) incorporates structural features of both of these interesting antimetabolites. Syntheses of 6-azauracil nucleosides have been complicated by the formation of the undesired N-3 isomer.[4] The preparation of **5** is an example of the application of the trimethylsilyl method[5] to the synthesis of a 6-azapyrimidine nucleoside. This method has the advantage that the "natural," N-1 nucleosides are readily isolated in fair yields.

PROCEDURE

3,5-Bis(trimethylsiloxy)-*as*-triazine (2)

A magnetically stirred mixture of 3.9 g. (34 mmoles) of 6-azauracil[6] (**1**) and 30 ml. of hexamethyldisilazane is refluxed until a clear solution results and the temperature rises to 141° (1 hr.). The excess hexamethyldisilazane is removed at 20 mm. Hg, and the residue is purified by distillation at 142–144°/25 mm. Hg; yield 7.9 g. (90%), m.p.[a] *ca.* 35°.

2-(2-Deoxy-3,4,6-tri-*O*-*p*-nitrobenzoyl-β-D-*arabino*-hexosyl)-*as*-triazine-3,5(2*H*,4*H*)-dione (4)

To a magnetically stirred melt of 7.5 g. (29.3 mmoles) of 3,5-bis(trimethylsiloxy)-*as*-triazine (**2**) at 75° is added 3.6 g. (5.35 mmoles) of 2-deoxy-3,4,6-tri-*O*-*p*-nitrobenzoyl-α-D-*arabino*-hexosyl bromide (**3**) (Sec. III [117]). On being stirred, the mixture becomes clear, and a gas (Me_3SiBr?) is evolved. The temperature is raised to 85°, and then maintained at that temperature overnight, at which time the reaction mixture is solid. The mixture is extracted with three 250-ml. portions of chloroform, each containing 2 ml. of absolute ethanol[b]; the residue is 6-azauracil (**1**), 2.7 g., m.p. 270–271°. The chloroform extracts are combined, and evaporated to an oil at 60°/20 mm. Hg. On digestion with 200 ml. of boiling absolute ethanol, the oil slowly solidifies, and the solid, crude, protected nucleoside (**4**) (2.3 g., m.p. 235–250°) is removed by filtering the hot mixture. The product (**4**) is recrystallized from glacial acetic acid–water, yielding 990 mg., m.p. 241–242°, which then crystallizes and remelts at 287.5–288.5°. An additional 340 mg. (m.p. 235–237°, crystallizing and then melting at 286–288°) is obtained from the mother liquors, giving a total yield of 35%. The pure product may also be obtained as an interconvertible, polymorphous form, having m.p. 250–252°.

2-(2-Deoxy-β-D-*arabino*-hexopyranosyl)-*as*-triazine-3,5(2*H*,4*H*)-dione (5)

A mixture of 1.17 g. (1.66 mmoles) of *pure*, protected nucleoside (**4**) and 40 ml. of absolute methanol is treated with 4 ml. of *M* sodium methoxide in absolute methanol. The resulting, clear solution is stirred overnight at room temperature, and evaporated to dryness under diminished pressure at 60°. The resulting oil is dissolved in 25 ml. of water, the pH of the solution is adjusted to 1 with concentrated hydrochloric acid, and the solution is washed with two 25-ml. portions of chloroform (to remove the methyl *p*-nitrobenzoate). The aqueous layer is then evaporated under

[a] The product (**2**) is extremely sensitive to atmospheric moisture, which causes its hydrolysis to **1**.

[b] The ethanol is added to cleave the remaining trimethylsiloxy groups from the *as*-triazine ring.

diminished pressure to an oil. The residue is extracted with 100 ml. of hot acetone, the extract is filtered to remove sodium chloride, and the filtrate is concentrated under diminished pressure to 10 ml. On addition of 10 ml. of ethyl acetate, 130 mg. of **5** is obtained. An additional 210 mg. of **5** is recovered from the mother liquors, giving a total yield of 79%, m.p.[c] 214–215.5°, $[\alpha]_D^{24}$ +0.1° (*c* 0.50, water), $\lambda_{max}^{0.1\,N\,HCl}$ 259 nm. (log ϵ 3.74), $\lambda_{min}^{0.1\,N\,HCl}$ 227 nm., $\lambda_{max}^{pH\,7\,buffer}$ 254 nm. (log ϵ 3.82), $\lambda_{min}^{pH\,7\,buffer}$ 220 nm., $\lambda_{max}^{0.1\,N\,NaOH}$ 252 nm. (log ϵ 3.79), $\lambda_{max}^{95\%\,EtOH}$ 260 nm. (log ϵ 3.87), $\lambda_{min}^{95\%\,EtOH}$ 235 nm.

REFERENCES

(1) J. Skoda, *Progr. Nucleic Acid Res.*, **3**, 197 (1963); G. B. Elion and G. H. Hitchings, *Advan. Chemotherapy*, **2**, 91 (1965).
(2) W. W. Zorbach and G. J. Durr, *J. Org. Chem.*, **27**, 1474 (1962); G. Etzold and P. Langen, *Chem. Ber.*, **98**, 1988 (1965); M. Zimmerman, *Biochem. Biophys. Res. Commun.*, **16**, 600 (1964).
(3) G. J. Durr, J. F. Keiser, and P. A. Ierardi, *J. Heterocyclic Chem.*, **4**, 291 (1967).
(4) J. Gut, *Advan. Heterocyclic Chem.*, **1**, 204 (1963).
(5) T. Nishimura and I. Iwai, *Chem. Pharm. Bull.* (Tokyo), **12**, 353 (1963).
(6) J. Gut, *Collection Czech. Chem. Commun.*, **23**, 1588 (1958).

[c] Compound **5** also exists in an amorphous form, m.p. 114–117° after drying; the material is analytically pure, and may be converted into its crystalline form by crystallization from ethyl acetate.

[110] 2-(2-Deoxy-β-D-*erythro*-pentofuranosyl)-6-methyl-*as*-triazine-3,5(2*H*,4*H*)-dione (6-Azathymidine)

The "Mercuri" Procedure Applied to Preparation of 6-Aza Analogs of Pyrimidine Nucleosides

M. PRYSTAŠ and F. ŠORM

INSTITUTE OF ORGANIC CHEMISTRY AND BIOCHEMISTRY,
CZECHOSLOVAK ACADEMY OF SCIENCES, PRAGUE, CZECHOSLOVAKIA

INTRODUCTION

6-Azathymidine [2-(2-deoxy-β-D-*erythro*-pentofuranosyl)-6-methyl-*as*-triazine-3,5(2*H*,4*H*)-dione, **8**] is[1–4] an inhibitor of the synthesis of nucleic acid,[5,6] and a much more potent antagonist of thymine and thymidine than is 6-azathymine [6-methyl-*as*-triazine-3,5(2*H*,4*H*)-dione].[5] Its biosynthesis and bacteriostatic properties have been described by Prusoff and Welch.[1] The synthesis of 6-azathymidine[3,4] makes use of condensation of the protected intermediate (**4**) with a protected 2-deoxy-D-*erythro*-pentofuranosyl chloride[7] (**5**). The condensation product (**6**) is hydrogenolyzed over 10% palladium-on-active-charcoal catalyst, to yield only the β-D anomer (**7**), in contrast to the two anomers obtained[4] from 3-(diphenylmethyl)-6-azauracil [4-(diphenylmethyl)-*as*-triazine-3,5(2*H*,4*H*)-dione]. This method has been used for the preparation of 6-azauridine [2-β-D-ribofuranosyl-*as*-triazine-3,5(2*H*,4*H*)-dione] and its 5-methyl derivative,[8] 1-β-D-xylofuranosyl-6-azauracil [2-β-D-xylofuranosyl-*as*-triazine-3,5(2*H*,4*H*)-dione],[9] and the anomeric 2′-deoxy-5-(trifluoromethyl)-6-azauridines [2-(2-deoxy-D-*erythro*-pentofuranosyl)-6-(trifluoromethyl)-*as*-triazine-3,5(2*H*,4*H*)-diones].[10]

PROCEDURE

1-(Trifluoroacetyl)-6-azathymine [6-Methyl-2-(trifluoroacetyl)-*as*-triazine-3,5(2*H*,4*H*)-dione, 2][4,8]

A mixture of 2.54 g. (20 mmoles) of 6-azathymine[11] (**1**) (Sec. II [38]) and 6.50 g. (35 mmoles) of trifluoroacetic anhydride is boiled until the former dissolves completely (90 min.). The resulting solution is heated for 70 min. at 55°, diluted with 50 ml. of toluene, and evaporated to dryness under diminished pressure. The residual oil is dissolved in 50 ml. of

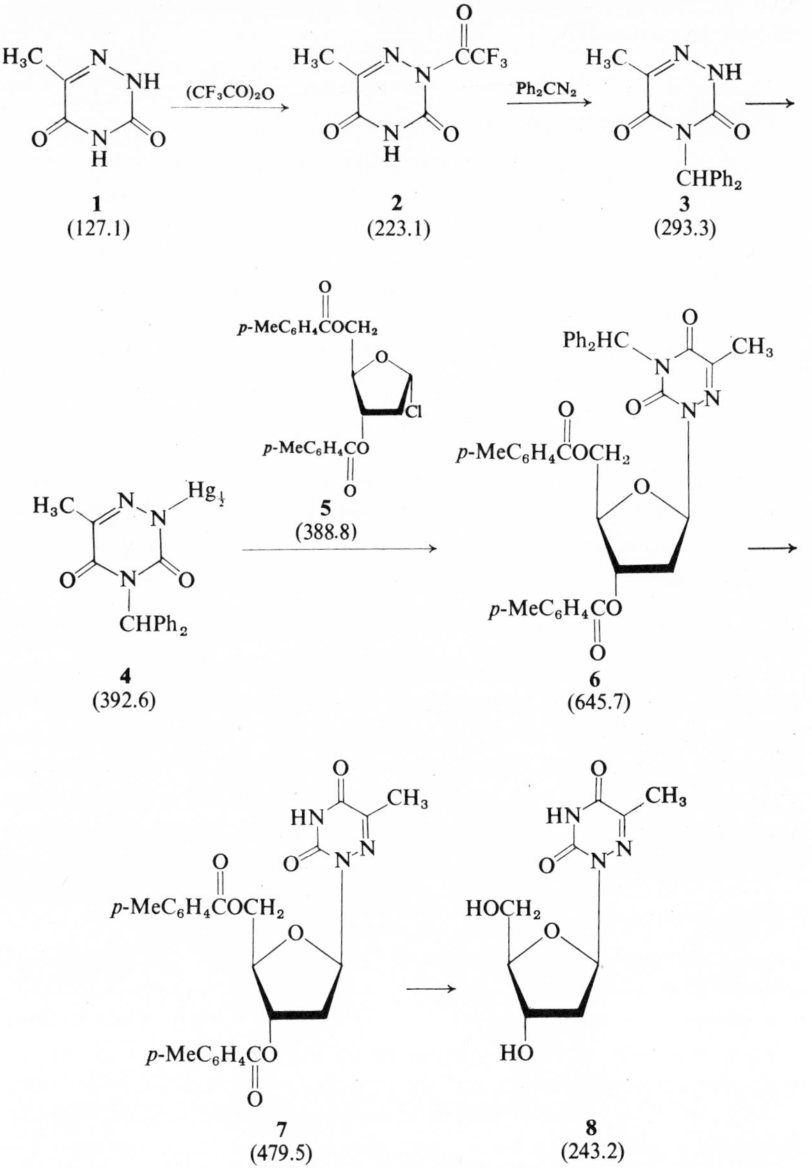

$(CF_3CO)_2O$ Ph_2CN_2

1 (127.1) **2** (223.1) **3** (293.3)

5 (388.8)

4 (392.6) **6** (645.7)

7 (479.5) **8** (243.2)

toluene, and the solution is evaporated under diminished pressure, giving a crystalline product which is dissolved in 50 ml. of hot toluene; the solution is evaporated to dryness under diminished pressure, and the procedure is repeated three times. The crude trifluoroacetyl derivative (**2**), m.p. 102°, is obtained in almost quantitative yield; on recrystallization from hot toluene, m.p. 104–106°.

3-(Diphenylmethyl)-6-azathymine [4-(Diphenylmethyl)-6-methyl-*as*-triazine-3,5(2*H*,4*H*)-dione, 3][4,8]

A solution of 4.46 g. (20 mmoles) of the crude **2** in 70 ml. of hot toluene is cooled and treated, with cooling, with a solution of 8 g. of diphenyldiazomethane in 70 ml. of anhydrous toluene; after 25 min., the evolution of nitrogen ceases. The mixture is decolorized by addition of several drops of 3 *N* hydrochloric acid, diluted with 30 ml. of chloroform, and washed with three 100-ml. portions of water at 60°. [The aqueous washings are combined, decolorized with activated carbon, filtered, and evaporated to dryness under diminished pressure. On recrystallization of the crystalline residue from water, there is obtained 0.7 g. of unchanged 6-azathymine (**1**).] The organic layer is concentrated (without being dried) under diminished pressure to 15 ml., and the solution is allowed to cool and kept at room temperature for 10 hr., depositing crystals of **3** that are collected by suction, and washed with cold toluene; yield 4.0 g. (68%), m.p. 170–171°. The melting point is unchanged on recrystallization from toluene.

Mercury Salt (4) of 3-(Diphenylmethyl)-6-azathymine[8] (3)

A solution of 2.933 g. (10 mmoles) of **3** in 20 ml. of 0.5 *N* potassium hydroxide and 20 ml. of ethanol is added dropwise, during 15 min., to a solution of 1.358 g. (5 mmoles) of mercuric chloride in 100 ml. of hot water, and the mixture is kept at room temperature for 20 hr. The resulting precipitate is collected with suction, and washed with 300 ml. of lukewarm water. The mercury salt (**4**) is dried at 70°/0.2 mm. Hg for 6 hr.; yield, 3.80 g. (97%).

3-(Diphenylmethyl)-3′,5′-di-*O*-*p*-toluoyl-6-azathymidine (6)

An azeotropically dried solution of 15.7 g. (20 mmoles) of **4** in 120 ml. of toluene is treated with 17.94 g. (46 mmoles) of 2-deoxy-3,5-di-*O*-*p*-toluoyl-D-*erythro*-pentosyl chloride[7] (**5**) (Sec. V [158]). The resulting solution gradually deposits a black precipitate. The mixture is kept at room temperature for 3 days, and the precipitate is filtered off, and washed with a solution of 3 g. of potassium hydrogen carbonate in 50 ml. of water. The same aqueous solution is used to wash the toluene filtrate.[a] The toluene layer is successively washed with 100 ml. of water, a solution of 20 g. of potassium iodide in 80 ml. of water, and two 100-ml. portions of

[a] The combined aqueous washings may be treated with active carbon, and acidified (Congo Red) with 4 *N* hydrochloric acid to afford 4.46 g. of *p*-toluic acid, m.p. 180–181°.

water, dried (anhydrous sodium sulfate), and evaporated to dryness under diminished pressure. The residue (18 g.) is chromatographed on a column of 900 g. of neutral alumina (Brockmann, activity III). The column is successively washed with 1.8 l. of carbon tetrachloride (fractions 1–9), 3.0 l. of benzene (fractions 10–24), and 2.4 l. of 5:1 (v/v) benzene–ethyl acetate (fractions 25–36), a 200-ml. fraction being collected every 25 min. The benzene fractions 15–19 are combined, and evaporated to dryness under diminished pressure. The residue should show a single spot (R_f 0.25) when chromatographed on a thin layer of loose, neutral alumina (Brockmann, activity III) in benzene. The amorphous ester (**6**) is dried at 40°/0.2 mm. Hg for 10 hr.; yield 11.2 g. (43%). [Fractions 28–31 are combined, and evaporated to dryness under diminished pressure; the residue is crystallized from toluene, to give 4.8 g. of **3**, m.p. 170–172°.]

3′,5′-Di-*O*-*p*-toluoyl-6-azathymidine (7)

A solution of 9.7 g. (15 mmoles) of crude ester (**6**) in 200 ml. of dry *p*-dioxane is evaporated to dryness under diminished pressure, and the residue is dissolved in a mixture of 50 ml. of dry *p*-dioxane and 200 ml. of dry butyl alcohol. The compound (**6**) is hydrogenolyzed at room temperature over 4 g. of 10% palladium-on-active-carbon catalyst for 5 hr. The suspension is filtered, the filtrate is evaporated to dryness under diminished pressure, and the resulting oil is dissolved in 100 ml. of boiling ethanol. The solution is cooled, and the resulting, crude ester (**7**), m.p. 173–175°, is recrystallized from 150 ml. of hot ethanol; yield 5.25 g., m.p. 175–176°, $[\alpha]^{25}_{5461}$ $-70.3°$, $[\alpha]^{25}_{D}$ $-59.3°$ (*c* 0.75, pyridine).

The mother liquors are combined, and evaporated to dryness under diminished pressure, and the residue is chromatographed in benzene–ethyl acetate on 100 g. of silica gel (previously deactivated by the addition of 5% of water). The fractions obtained by elution with 4:1 benzene–ethyl acetate (250 ml.) are combined, and evaporated to dryness under diminished pressure. On crystallization from 20 ml. of hot ethanol, the residue (1 g.) affords an additional 0.82 g. of ester (**7**); total yield 6.07 g. (85%), m.p. 173–175°.

6-Azathymidine (8)

A mixture of 479 mg. (1 mmole) of the di-*p*-toluate (**7**) in 10 ml. of absolute methanol with 5 ml. of 0.4 *M* sodium methoxide in methanol is kept at room temperature for 10 hr. The solution is rendered neutral by the addition of 1.5 g. of Dowex-50W (H^+) ion-exchange resin, the suspension is filtered, and the filtrate is evaporated to dryness under diminished pressure. The residue is dissolved in 10 ml. of cold water, and the solution is washed with two 20-ml. portions of chloroform, and freeze-dried. The glassy residue is then dried over sulfuric acid in a vacuum desiccator at 20 mm. Hg for 10 hr.; yield 238 mg. (98%), $[\alpha]_{5461}^{25}$ −95.5°, $[\alpha]_{D}^{25}$ −76.7° (*c* 0.5, pyridine); $\lambda_{max}^{pH\,1.9}$ 264 nm. (log ϵ 3.78), $\lambda_{min}^{pH\,1.9}$ 238 nm. (log ϵ 3.56).

REFERENCES

(1) W. H. Prusoff and A. D. Welch, *J. Biol. Chem.*, **218**, 929 (1956).
(2) R. H. Hall and R. Haselkorn, *J. Am. Chem. Soc.*, **80**, 1138 (1958).
(3) M. Prystaš and F. Šorm, *Rev. Chim.* (Bucharest), **7**, 1181 (1962).
(4) J. Pliml, M. Prystaš, and F. Šorm, *Collection Czech. Chem. Commun.*, **28**, 2588 (1963).
(5) W. H. Prusoff, L. G. Lajtha, and A. D. Welch, *Biochim. Biophys. Acta*, **20**, 209 (1956).
(6) W. H. Prusoff, *Biochem. Pharmacol.*, **2**, 221 (1959).
(7) M. Hoffer, *Chem. Ber.*, **93**, 2777 (1960).
(8) M. Prystaš and F. Šorm, *Collection Czech. Chem. Commun.*, **27**, 1578 (1962).
(9) T. Tkaczynski, J. Šmejkal, and F. Šorm, *Collection Czech. Chem. Commun.*, **29**, 1736 (1964).
(10) M. P. Mertes, S. E. Saheb, and D. Miller, *J. Med. Chem.*, **9**, 876 (1966).
(11) J. Gut, *Collection Czech. Chem. Commun.*, **23**, 1588 (1958).

[111] 2-(2-Deoxy-β-D-*erythro*-pentofuranosyl)-6-(trifluoromethyl)-*as*-triazine-3,5(2*H*,4*H*)-dione

Synthesis of a 6-Azapyrimidine Nucleoside by the Trimethylsilyl Ether Method

T. Y. SHEN, W. V. RUYLE, and R. L. BUGIANESI

MERCK, SHARP & DOHME RESEARCH LABORATORIES,
DIVISION OF MERCK AND CO., INC., RAHWAY, NEW JERSEY 07065

1 (181.1) → **2** (313.1) → [**3** (450.6)] → **4** (595.4) → **5** (297.2)

where *p*NBz is *p*-nitrobenzoyl.

INTRODUCTION

The compound[1] reported here incorporates two structural modifications of the naturally occurring nucleoside thymidine, namely, the replacement of the ring C-6 methine group by nitrogen, and replacement of the C-5 methyl group by a trifluoromethyl group. The synthetic procedure illustrates a general method, using bis(trimethylsilyl) ethers of *as*-triazines in the preparation of 5-substituted 6-azapyrimidine nucleosides.[1] Independent syntheses of the compound have been reported.[2,3]

PROCEDURE

6-(Trifluoromethyl)-3,5-bis(trimethylsiloxy)-*as*-triazine[2] (2)

A suspension of 7.75 g. (42.8 mmoles) of 6-(trifluoromethyl)-*as*-triazine-3,5(2*H*,4*H*)-dione (**1**) (Sec. II [43]) in 18 ml. of hexamethyldisilazane containing 0.5 ml. of chlorotrimethylsilane, in a 50-ml., round-bottomed flask fitted with a condenser and a drying tube, is refluxed by heating in an oil bath at 150–170°; ammonium chloride collects in the condenser. When dissolution occurs, the excess hexamethyldisilazane and chlorotrimethylsilane are distilled off at 30°/1 mm. Hg (oil bath at 100°). The residual oil crystallizes on being kept at 0° overnight, giving 10 g. (*ca.* 100%) of **2**, m.p. 35°.

2-(2-Deoxy-3,5-di-*O*-*p*-nitrobenzoyl-β-D-*erythro*-pentosyl)-6-(trifluoromethyl)-*as*-triazine-3,5(2*H*,4*H*)-dione (4)

A mixture of 3 g. (6.67 mmoles) of 2-deoxy-3,5-di-*O*-*p*-nitrobenzoyl-D-*erythro*-pentosyl chloride[4] (**3**) (Sec. III [60]) and 4 g. (12.5 mmoles) of 6-(trifluoromethyl)-3,5-bis(trimethylsiloxy)-*as*-triazine (**2**) is fused for 30 min. at 150°/25 mm. Hg, and then cooled. Dichloromethane (25 ml.) is added to the dark mixture, followed by the addition of 5 ml. of methanol, which cleaves the silyl ether groups. The precipitate that forms is filtered off, and recrystallized from methanol–acetone to give 2.0 g. (51%) of compound **4**, m.p. 209–211°.

2-(2-Deoxy-β-D-*erythro*-pentofuranosyl)-6-(trifluoromethyl)-*as*-triazine-3,5-(2*H*,4*H*)-dione (5)

A solution of 700 mg. (1.2 mmoles) of 2-(2-deoxy-3,5-di-*O*-*p*-nitrobenzoyl-β-D-*erythro*-pentosyl)-6-(trifluoromethyl)-*as*-triazine-3,5(2*H*,4*H*)-dione (**4**) and 1.75 ml. of diisopropylamine in 400 ml. of methanol is refluxed for 15 min. The methanol is evaporated, and the residue is partitioned between water and chloroform. The aqueous layer is washed with chloroform, and this is then extracted with small volumes of water. The combined aqueous layers are treated with 2 ml. of Dowex-50 X4 (H^+) ion-exchange resin until the solution is neutral, the resin is filtered off, and the filtrate is evaporated to dryness. The crude sirup (500 mg.) is chromatographed on silica gel (prepacked in dichloromethane), and eluted with 1:10 methanol–dichloromethane. The product is obtained as a glass, which slowly crystallizes on being kept at room temperature. The crude, crystalline product is recrystallized by dissolving it in a small volume of acetone, slowly adding ether to incipient turbidity, and keeping the solution at room temperature, to give 100 mg. (20%) of pure **5**, m.p. 152–154°, $[\alpha]_D^{25}$ −59° (*c* 1.0, water), $\lambda_{max}^{0.1\,N\,HCl}$ 268 nm. (ε 5,880); $\lambda_{max}^{pH\,11}$ 263 nm. (ε 5,610).

REFERENCES

(1) T. Y. Shen, W. V. Ruyle, and R. L. Bugianesi, *J. Heterocyclic Chem.*, **2**, 495 (1965).

(2) M. P. Mertes and S. E. Saheb, *J. Heterocyclic Chem.*, **2**, 491 (1965).

(3) A. Dipple and C. Heidelberger, *J. Med. Chem.*, **9**, 715 (1966).

(4) R. K. Ness, D. L. MacDonald, and H. G. Fletcher, Jr., *J. Org. Chem.*, **26**, 2895 (1961).

[112] 5-Amino-2-β-D-ribofuranosyl-*as*-triazin-3(2*H*)-one (6-Azacytidine)

*5-Chloro-2-(2,3,5-tri-*O*-acyl-β-D-ribosyl)-*as*-triazin-3(2*H*)-ones as Intermediates in the Synthesis of 5-Amino-substituted Derivatives of 2-β-D-Ribofuranosyl-*as*-triazin-3(2*H*)-ones*

J. ŽEMLIČKA

INSTITUTE OF ORGANIC CHEMISTRY AND BIOCHEMISTRY, CZECHOSLOVAK ACADEMY OF SCIENCES, PRAGUE, CZECHOSLOVAKIA

$\xrightarrow[\text{[Me}_2\text{N=CH—Cl]}^+\text{Cl}^-]{\text{SOCl}_2}$ $\xrightarrow{\text{NH}_3}$

1, R = Bz (557.5)
2, R = Ac (371.3)

3, R = Bz (576.0)
4, R = Ac (389.8)

$\xrightarrow[\text{MeOH}]{\text{NH}_3}$

5, R = Bz (556.6)
6, R = Ac (370.4)

7 (244.2)

INTRODUCTION

6-Azacytidine [5-amino-2-β-D-ribofuranosyl-*as*-triazin-3(2*H*)-one, **7**] is known as an antimetabolite possessing a cytostatic[1] and an abortive[2] activity. In its preparation, 2′,3′,5′-tri-*O*-benzoyl-6-azauridine[3] (**1**) or 2′,3′,5′-tri-*O*-acetyl-6-azauridine[3,4] (**2**) is used as the starting material. Reaction of either of these with phosphorus pentasulfide in pyridine[5] affords the corresponding 2′,3′,5′-tri-*O*-acyl-4-thio-6-azauridines,[3] from which 6-azacytidine (**7**) is obtained by ammonolysis at an elevated temperature and pressure.[3] Another method,[6] involving use of 5-chloro-2-β-D-ribofuranosyl-*as*-triazin-3(2*H*)-one 2′,3′,5′-tribenzoate (**3**) or 2′,3′,5′-tri-acetate (**4**) is more advantageous. Compounds **3** and **4** are prepared by treatment of 2′,3′,5′-tri-*O*-benzoyl-6-azauridine (**1**) and 2′,3′,5′-tri-*O*-acetyl-6-azauridine (**2**), respectively, with thionyl chloride in the presence of a catalytic amount of (chloromethylene)dimethylammonium chloride.[7] Ammonolysis of the chloro derivatives **3** and **4** affords 2′,3′,5′-tri-*O*-benzoyl-6-azacytidine (**5**) and 2′,3′,5′-tri-*O*-acetyl-6-azacytidine (**6**), respectively.[6] Removal of the *O*-acyl groups from compound **5** or **6** to give 6-azacytidine (**7**) is accomplished by the action of methanolic ammonia.[6] 6-Azacytidine (**7**) may also be obtained by coupling of 5-(methylthio)-*as*-triazin-3(2*H*)-one with 2,3,5-tri-*O*-benzoyl-D-ribosyl chloride and subsequent ammonolysis.[8]

PROCEDURE

6-Azacytidine [5-Amino-2-β-D-ribofuranosyl-*as*-triazin-3(2*H*)-one, 7]

(*A*) From 2-(2,3,5-Tri-*O*-benzoyl-β-D-ribosyl)-*as*-triazine-3,5(2*H*,4*H*)-dione (1)

A suspension of 76.7 g. (0.138 mole) of 2′,3′,5′-tri-*O*-benzoyl-6-azauridine (**1**) in 100 ml. of dry chloroform,[a] in a 500-ml., round-bottomed flask equipped with a reflux condenser,[b] is treated successively with 111 ml. (1.38 mole) of thionyl chloride[c] and 4.2 ml. (0.054 mole) of *N*,*N*-dimethylformamide,[d] and the mixture is refluxed for 14.5 hr. The mixture is cooled, and evaporated to dryness under diminished pressure, and the resulting crude intermediate **3** is dissolved in 200 ml. of hot benzene. The solution is cooled, and added dropwise during 75 min., with stirring, and cooling

[a] Commercial chloroform is treated with 10% (by weight) of phosphorus pentaoxide, and the chloroform is distilled off.

[b] Vigorous evolution of hydrogen chloride and sulfur dioxide occurs during the reaction, which should, therefore, be performed in a hood; the reflux condenser should be equipped with a calcium chloride drying tube.

[c] Thionyl chloride is purified by distillation from a mixture with linseed oil.

[d] In this manner, (chloromethylene)dimethylammonium chloride is generated *in situ*.[7] Commercial *N*,*N*-dimethylformamide is treated with 10% (by weight) of phosphorus pentaoxide, and the amide is distilled off.

with ice, to 100 ml. of a saturated solution of ammonia in methanol[e] in a 500-ml., three-necked flask equipped with a 500-ml. dropping funnel, a stirrer, and a potassium hydroxide drying tube. A precipitate separates, and is collected with suction; the filtrate is evaporated to dryness under diminished pressure. The two portions of solid are combined, and washed with a total of 2.0 l. of water (until no chloride ions are detectable in the filtrate), and then with three 100-ml. portions of methanol. After being dried under diminished pressure, the crude **5** weighs 72.2 g. (94%). Compound **5** is then heated in a 500-ml. autoclave at 100° (bath temperature) with 250 ml. of methanolic ammonia[e] for 24 hr. The reaction mixture is cooled, and filtered, the filtrate is evaporated to dryness under diminished pressure, and the solid residue is refluxed briefly with 300 ml. of 2:1 benzene–ethanol. The hot suspension is filtered, and the material on the filter is re-extracted in the same way with 150 ml. of the hot solvent-mixture. The resulting material is dissolved in 60 ml. of hot water, the solution is filtered with activated carbon, and the filtrate is diluted with 60 ml. of ethanol, to deposit pure 6-azacytidine (**7**), which is collected with suction and dried under diminished pressure; yield 18.1 g. (54%), m.p. 222–224° (dec.)[f]; $\lambda_{max}^{H_2O}$ 260 nm. (ϵ_{max} 8,400); R_f 0.08, in butyl alcohol saturated with water, on Whatman No. 1 paper (descending technique). On cooling the mother liquors to −15°, an additional crop (7.26 g., 22%) of 6-azacytidine (**7**) is obtained.

(*B*) From 2-(2,3,5-Tri-*O*-acetyl-β-D-ribosyl)-*as*-triazine-3,5(2*H*,4*H*)-dione (2)

In a 250-ml., three-necked, round-bottomed flask equipped with a stirrer and a reflux condenser is placed a mixture of 9.25 g. (0.025 mole) of 2′,3′,5′-tri-*O*-acetyl-6-azauridine (**2**), 20 ml. (0.28 mole) of thionyl chloride,[b] and 0.6 ml. of *N*,*N*-dimethylformamide.[c] The mixture is refluxed,[d] with stirring, for 26.5 hr., cooled, and evaporated to dryness under diminished pressure, and the resulting crude sirupy **4** is dissolved in 50 ml. of chloroform. The solution is added dropwise during 15 min., with stirring and cooling with ice, to 30 ml. (0.15 mole) of 5.05 *M* ethanolic ammonia[g] in a 250-ml., three-necked, round-bottomed flask (equipped with a stirrer and a 100-ml. dropping funnel protected by a potassium hydroxide drying tube). The precipitate is filtered off through a thin layer of Hyflo Super Cel and activated carbon, and washed thoroughly with

[e] Methanol is saturated with ammonia gas while being cooled with ice.

[f] In some preparations, a product melting at 217° (dec.) is obtained, chemically identical with the material melting at 222–224°, as may be shown by paper chromatography, ultraviolet spectrophotometry, and analysis.

[g] Prepared in analogy to methanolic ammonia (*cf.* footnote *e*).

chloroform. The filtrate and washings are combined, and evaporated under diminished pressure, giving crude, sirupy **6**, which is dissolved in 30 ml. of methanol presaturated at 0° with ammonia.[e] The solution is kept at room temperature for 3 days, and deposits 3.47 g. (57%) of pure 6-azacytidine (**7**) which is of the same quality as that obtained by procedure *A*. The mother liquors are evaporated to dryness under diminished pressure, and the residue is briefly refluxed with 100 ml. of chloroform, which is then decanted. The residue is briefly refluxed with 20 ml. of methanol, to afford an additional 1.63 g. (27%) of 6-azacytidine (**7**), of satisfactory purity. Recrystallization may be performed from aqueous ethanol as described in procedure *A*.

REFERENCES

(1) F. Šorm and J. Veselý, *Experientia*, **17**, 355 (1961).
(2) K. Čerey, J. Elis, and H. Rašková, *Biochem. Pharmacol.*, **14**, 1549 (1965).
(3) V. Černěckij, S. Chládek, F. Šorm, and J. Smrt, *Collection Czech. Chem. Commun.*, **26**, 87 (1962).
(4) J. Beránek and J. Pitha, *Collection Czech. Chem. Commun.*, **29**, 625 (1964).
(5) J. J. Fox, D. Van Praag, I. Wempen, I. L. Doerr, L. Cheong, J. E. Knoll, M. I. Eidinoff, A. Bendich, and G. B. Brown, *J. Am. Chem. Soc.*, **81**, 178 (1959).
(6) J. Žemlička and F. Šorm, *Collection Czech. Chem. Commun.*, **30**, 2052 (1965).
(7) H. H. Bosshard, R. Mory, M. Schmid, and H. Zollinger, *Helv. Chim. Acta*, **42**, 1653 (1959).
(8) Y. Mizuno, M. Ikehara, and K. A. Watanabe, *Chem. Pharm. Bull.* (Tokyo), **10**, 653 (1962).

[113] 1-(3-Amino-3-deoxy-β-D-glucopyranosyl)uracil*

Application of the Baer–Fischer Synthesis of
3-Deoxy-3-nitroaldohexopyranosides to Nucleosides

HERBERT A. FRIEDMAN and JACK J. FOX

DIVISION OF BIOLOGICAL CHEMISTRY, SLOAN-KETTERING INSTITUTE FOR CANCER RESEARCH, SLOAN-KETTERING DIVISION OF CORNELL UNIVERSITY MEDICAL COLLEGE, NEW YORK, NEW YORK 10021

1 (244.2) → 2 (242.2) → 3 (302.2) → 4 (273.3)

INTRODUCTION

Several 3-amino-3-deoxy sugar derivatives exhibit antibiotic properties.[1] Nucleosides containing the 3-amino-3-deoxy-D-ribosyl moiety have been

* The authors acknowledge the support of the United States Public Health Service (Grant No. CA-08748).

found in Nature,[2] but nucleosides containing a 3-amino-3-deoxyaldohexosyl moiety have not as yet been isolated from natural sources. The first chemical synthesis of 3-amino-3-deoxyhexosyl nucleosides was reported by Baker *et al.*,[3] who condensed suitably protected 3-amino-3-deoxy-aldohexopyranoses (of the *allo* and *altro* configurations) with mercury salts of certain purines.

Described here is the first application[4] of the Baer–Fischer[5] "periodate–nitromethane" procedure to the synthesis of a pyrimidine nucleoside of a 3-amino-3-deoxyaldohexose from uridine (**1**), which is readily available. Oxidation of uridine with sodium metaperiodate affords the[a] "dialdehyde" (**2**); this is condensed with nitromethane in aqueous alkali to give **3**. Catalytic reduction of the nucleoside (**3**) of the 3-deoxy-3-nitro-β-D-glucopyranose gives 1-(3-amino-3-deoxy-β-D-glucopyranosyl)uracil (**4**). Similar procedures have been employed[7] for the synthesis of other pyrimidine or purine nucleosides of 3-amino-3-deoxy sugars.

PROCEDURE

1-(3-Deoxy-3-nitro-β-D-glucopyranosyl)uracil (3)[b]

To a stirred, ice-cold solution of 10.7 g. (0.05 mole) of sodium metaperiodate in 80 ml. of water is added 12.2 g. (0.05 mole) of uridine (**1**) in small portions. After completion of the addition, the solution is stirred at room temperature for 20 min.[c] and then kept overnight in a refrigerator. The precipitate (A) of inorganic salts is filtered off, and the filtrate is poured into 800 ml. of vigorously stirred ethanol, giving a heavy, white precipitate (B). Precipitates A and B are combined, and well washed with ethanol, and the solid is discarded. The filtrate and washings are combined, and evaporated *in vacuo* to yield the "dialdehyde" (**2**) as a thick sirup.

To a solution of the "dialdehyde" and nitromethane (3.0 ml., 0.056 mole) in 80 ml. of water is added 50 ml. of *N* sodium hydroxide dropwise, with vigorous stirring. After being kept for 4 hr., 50 ml. of Dowex-50 (H^+) is added, and the mixture is stirred for 30 min. The precipitate (consisting of resin plus product) is filtered off[d] and extracted with methanol (to dissolve the nucleoside), and the resin is removed by filtration. Evaporation of the

[a] Spectroscopic evidence[6] indicates that the carbonyl groups of the dialdehydes from sugars exist in a hydrated form.

[b] This procedure is a modification of that given in Ref. 4.

[c] If a test for excess periodate with starch–iodide paper is negative at this point, just enough of the oxidant should be added to give a positive test. This ensures complete oxidation of the uridine.

[d] The filtrate contains some 3-deoxy-3-nitro-β-D-glucosyl nucleoside (**3**), together with smaller amounts of other hexosyl isomers.[8] Concentration of this filtrate *in vacuo* to a small volume and storage of the solution overnight yields *ca.* 1 g. of **3**.

filtrate to dryness gives pale-yellow crystals which are washed with a small quantity of ice-cold water.

The yield of colorless, crystalline nucleoside (**3**) is 10.5 g. (69%), m.p. 175–177° (lit.,[4] 175–176°), unchanged on recrystallization from water; $[\alpha]_D^{25}$ +33° (*c* 0.75, methanol), $\lambda_{max}^{H_2O}$ 256.5 nm. (ϵ_{max} 9,200).

1-(3-Amino-3-deoxy-β-D-glucopyranosyl)uracil[4] (4)

To a solution of 15 g. (0.05 mole) of crystalline **3** in 100 ml. of 50% aqueous methanol is added 50 g. (wet weight) of activated Raney nickel.[e] The mixture is hydrogenated in a Parr hydrogenator at an initial pressure of 3 atmospheres. The reaction is complete when the uptake of hydrogen ceases.[f] The catalyst is removed by gravity filtration,[g] and washed with 700 ml. of 50% aqueous methanol. The filtrate and washings are evaporated to dryness *in vacuo* (bath temperature below 40°). The resulting sirup is dissolved in 30 ml. of hot water, and the solution is decolorized with charcoal and cooled, giving colorless, fine needles (11.0 g., 80%), m.p. 166–167° (sinters), 179–182° (effervescence), $[\alpha]_D^{25}$ +33° (*c* 0.88, water); $\lambda_{max}^{pH\,7}$ 257 nm., $\lambda_{max}^{pH\,1}$ 258 nm., $\lambda_{max}^{pH\,13}$ 259 nm.

REFERENCES

(1) For comprehensive reviews, see A. B. Foster and D. Horton, *Advan. Carbohydrate Chem.*, **14**, 432 (1959); J. D. Dutcher, *ibid.*, **18**, 259 (1963).
(2) See chapter by J. J. Fox, K. A. Watanabe, and A. Bloch, in *Progr. Nucleic Acid Res.*, **5**, 251 (1966).
(3) B. R. Baker, J. P. Joseph, and R. E. Schaub, U. S. Patent 2,852,505 (Sept. 16, 1958); *Chem. Abstracts*, **53**, 8175d (1959).
(4) K. A. Watanabe and J. J. Fox, *Chem. Pharm. Bull.* (Tokyo), **12**, 975 (1964); K. A. Watanabe, J. Beránek, H. A. Friedman, and J. J. Fox, *J. Org. Chem.*, **30**, 2735 (1965).
(5) H. H. Baer and H. O. L. Fischer, *J. Am. Chem. Soc.*, **81**, 5184 (1959); *Proc. Natl. Acad. Sci. U. S.*, **44**, 991 (1958).
(6) J. W. Rowen, F. H. Forziati, and R. E. Reeves, *J. Am. Chem. Soc.*, **73**, 4484 (1951); C. D. Hurd, P. J. Baker, Jr., R. P. Holysz, and W. H. Saunders, Jr., *J. Org. Chem.*, **18**, 186 (1953).

[e] Raney Catalyst Co., Inc., Chattanooga, Tennessee.

[f] The progress of the reduction may be checked by thin-layer chromatography on silica gel with 5:1 chloroform–methanol as the solvent. The amino compound (**4**) does not migrate, whereas the nitro compound (**3**) does.

[g] Caution is required, to avoid letting the catalyst become dry. Activated Raney nickel is highly pyrophoric.

(7) J. Beránek, H. A. Friedman, K. A. Watanabe, and J. J. Fox, *J. Heterocyclic Chem.*, **2**, 188 (1965); F. W. Lichtenthaler, H. P. Albrecht, and G. Olferman, *Angew. Chem.*, **77**, 131 (1965); F. W. Lichtenthaler and H. Zinke, *ibid.*, **78**, 774 (1966); F. W. Lichtenthaler and H. P. Albrecht, *Chem. Ber.*, **99**, 575 (1966).
(8) K. A. Watanabe and J. J. Fox, *J. Org. Chem.*, **31**, 211 (1966).

[114] 1-(3-Amino-3-deoxy-3-*C*-methyl-β-D-glucopyranosyl)uracil

The Dialdehyde–Nitroethane Cyclization Applied to Dialdehydes from Nucleosides

FRIEDER W. LICHTENTHALER and HORST ZINKE

INSTITUT FÜR ORGANISCHE CHEMIE, TECHNISCHE HOCHSCHULE DARMSTADT, 61 DARMSTADT, GERMANY

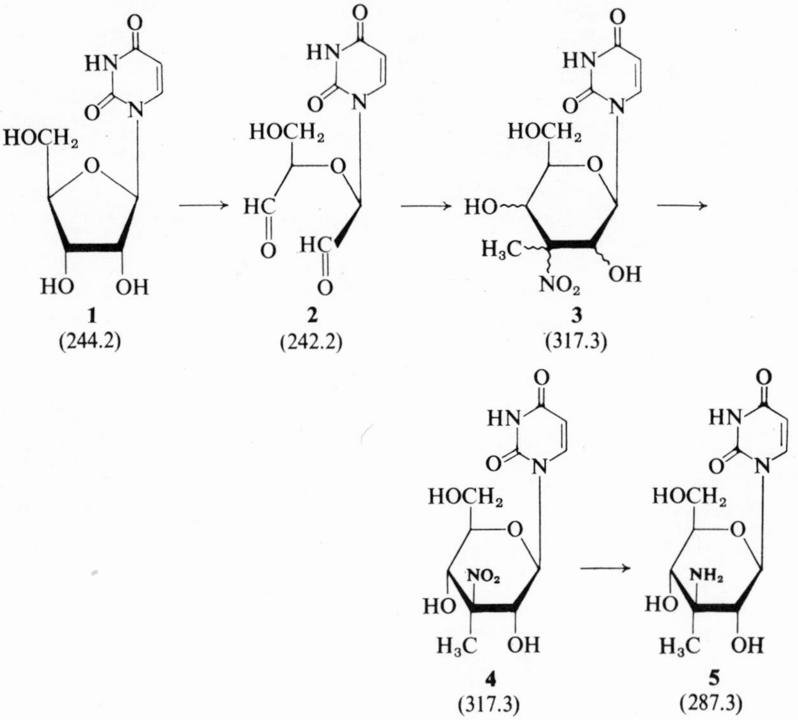

1 (244.2) → **2** (242.2) → **3** (317.3) → **4** (317.3) → **5** (287.3)

INTRODUCTION

Cyclization of a 1,4- or 1,5-dialdehyde with nitroethane offers a convenient method for introducing a nitrogen-containing function, together with a *C*-methyl branch, into five- or six-membered rings.[1] When applied to dialdehydes (*e.g.*, **2**) from nucleosides, 3-*C*′-methyl branched nucleosides are obtained, having a nitro or amino group at the branching point.[2]

PROCEDURE

1-(3-Deoxy-3-*C*-methyl-3-nitro-β-D-glucopyranosyl)uracil (4)

To a magnetically stirred, ice-cooled solution of 21.4 g. of sodium metaperiodate (0.10 mole) in 300 ml. of water is added 24.4 g. (0.1 mole) of uridine (**1**) in small portions during 15 min. Stirring is continued for 6 hr. at room temperature, and the solution is concentrated *in vacuo* at 35° to about 100 ml. Addition of methanol (300 ml.) precipitates most of the sodium iodate formed; this is removed by filtration, and washed with methanol (100 ml.). The filtrate and washings are combined, and evaporated to dryness under diminished pressure below 40°. The residue is dissolved in 200 ml. of methanol, and a small amount of inorganic material is removed by filtration. The resulting solution of the dialdehyde (**2**) is diluted with 400 ml. of methanol, and nitroethane (1.11 ml., 0.10 mole) is added, followed by dropwise addition, with vigorous stirring and ice-cooling, of 0.1 *M* sodium methoxide in methanol (100 ml.). The mixture is allowed to warm to room temperature, and is kept for 24 hr. Following deionization of the solution with a strongly acidic, ion-exchange resin ("Merck I"), which is filtered off and washed with 600 ml. of methanol, the filtrate and washings are combined, and concentrated under diminished pressure at 35° to about 200 ml. The product, which separates after standing for 2 days at room temperature,[a] is filtered off; weight 9.7 g. The mother liquor is then concentrated to about 100 ml., resulting in a second crop (4.1 g.). From the filtrate, a third fraction (3.64 g.) is obtained after concentration to 50 ml. Total yield: 17.45 g. of the isomeric mixture **3**, consisting mainly of the D-*gluco* compound (**4**).

The crude product is dissolved in boiling methanol (45 ml. per g. of **3**) and the solution is concentrated to one-tenth of its volume under diminished pressure at 35°. The crystals that have separated after 2 days[a] are filtered off. Another recrystallization by the same procedure yields chromatographically[b] pure product **4**; yield 9.02 g. (28%, based on **1**), m.p. 226–240° (dec.), $[\alpha]_D^{20}$ +25.5° (*c* 1, water).

1-(3-Amino-3-deoxy-3-*C*-methyl-β-D-glucopyranosyl)uracil (5)

To a prehydrogenated suspension of 2 ml. of freshly prepared Raney nickel[3] in 20 ml. of water is added a solution of 7.92 g. (25 mmoles) of **4**

[a] Crystallization of the isomeric mixture **3** proceeds rather slowly, requiring 1–2 days for completion.

[b] Tlc on Kieselgel PF_{254} (E. Merck AG., Darmstadt) (0.2-mm. layers), using 55:16:5 butyl acetate–acetic acid–water; detection by ultraviolet light.

in 400 ml. of 1:1 methanol–water, and the hydrogenation is continued. After 15 hr., 1.8 l. of hydrogen has been absorbed. The catalyst is removed, and washed thoroughly with 1:1 methanol–water, and the filtrate and washings are combined, and concentrated *in vacuo* at 35° to about 100 ml. Crystals, which separate slowly, are collected by suction filtration after 24 hr. (2.1 g.). Two further crops are obtained from the mother liquor by lessening the volume to about 50 and then 20 ml. The crops are combined (5.10 g.), and recrystallized from water, giving 4.73 g. (62%) of the chromatographically[c] pure aminonucleoside (**5**) as the monohydrate, m.p. 146–148°, $[\alpha]_D^{20}$ +39° (*c* 1, water).

REFERENCES

(1) S. W. Gunner, W. G. Overend, and N. R. Williams, *Chem. Ind.* (London), **1964**, 1523; H. H. Baer and G. U. Rao, *Ann.*, **686**, 210 (1965); F. W. Lichtenthaler, H. Leinert, and H. K. Yahya, *Z. Naturforsch.*, **21b**, 1004 (1966); F. W. Lichtenthaler and P. Emig, *Tetrahedron Letters*, **1967**, 577; F. W. Lichtenthaler and H. K. Yahya, *Carbohyd. Res.*, **5**, 485 (1967).

(2) F. W. Lichtenthaler and H. Zinke, *Angew. Chem.*, **78**, 774 (1966); *Angew. Chem. Intern. Ed. Engl.*, **5**, 737 (1966).

(3) S. Nishimura, *Bull. Chem. Soc. Japan*, **32**, 61 (1959).

[c] Tlc as described in footnote *b* with 4:1 ethanol–conc. ammonia as the running phase; detection by ultraviolet light or with ninhydrin.

[115] 1-β-D-Arabinofuranosyluracils*

A General Method for the Conversion of β-D-Ribofuranosyluracils into their β-D-Arabinofuranosyl Analogs

IRIS WEMPEN and JACK J. FOX

DIVISION OF BIOLOGICAL CHEMISTRY, SLOAN-KETTERING INSTITUTE FOR CANCER RESEARCH, SLOAN-KETTERING DIVISION OF CORNELL UNIVERSITY MEDICAL COLLEGE, NEW YORK, NEW YORK 10021

INTRODUCTION

This synthesis[1] demonstrates a facile method for the inversion of the hydroxyl group on C-2 of the D-ribofuranosyl moiety of ribonucleosides to give the D-*arabino* configuration; it is generally applicable to uracil ribonucleosides. By this method, the 1-β-D-ribofuranosyl derivatives of uracil, 5-fluorouracil, and thymine are converted into the corresponding 1-β-D-arabinofuranosyl derivatives of uracil[2] (Sec. III [119]), 5-fluorouracil,[3,4] and thymine (Sec. III [119]).[5,6] The thionocarbonate intermediate (**3**) is not isolated because, under these reaction conditions, compounds **2** (R = H, F, or Me) are converted directly into 2,2′-anhydronucleosides (**4**). All of the D-arabinofuranosyl compounds (**6**) may also be converted into their cytosine analogs (see example in Section III [96]). 1-β-D-Arabinofuranosyl-5-fluorouracil exhibits[3] antitumor activity against Leukemia B82 and Sarcoma 180.

PROCEDURE

5-Fluoro-1-(5-*O*-trityl-β-D-ribofuranosyl)uracil[3] **(2, R = F)**

5-Fluorouridine[3,7] (**1**, R = F) (8.3 g., 0.032 mole) is treated with 9.8 g. (0.035 mole)[a] of chlorotriphenylmethane in 80 ml. of pyridine (distilled from powdered, anhydrous barium oxide). The amber solution is kept at 0–5° overnight, and is then heated in a boiling-water bath for 2 hr., and cooled. The solution is poured, in a thin stream, into 1 l. of well stirred ice–water, whereupon a reddish gum is precipitated. After 15 min., the aqueous, supernatant liquor is decanted from the gum, which is then treated with fresh ice–water. The second wash is decanted and discarded, the residue is dissolved in chloroform, and the solution is washed with water, and dried with anhydrous sodium sulfate. The suspension is filtered by

* The authors acknowledge the support of the United States Public Health Service (Grant No. CA-08748).

[a] To avoid formation of any ditrityl derivative, a larger excess of chlorotriphenylmethane should not be used.

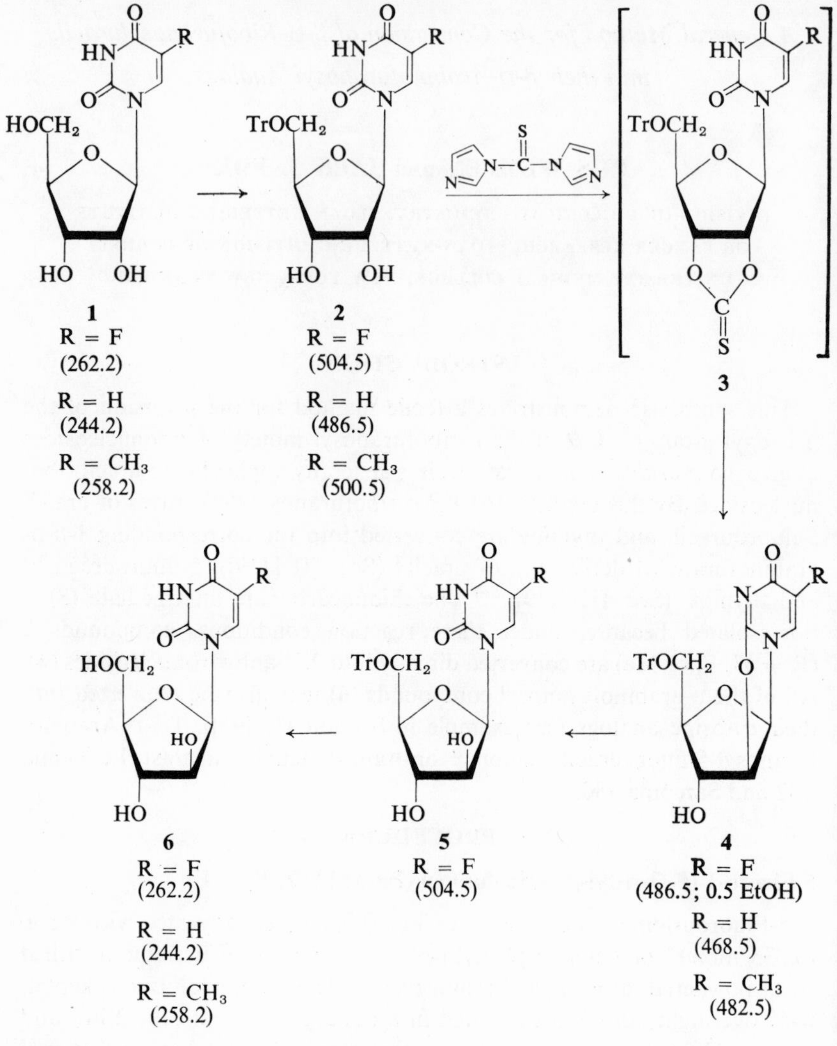

1
R = F (262.2)
R = H (244.2)
R = CH_3 (258.2)

2
R = F (504.5)
R = H (486.5)
R = CH_3 (500.5)

3

4
R = F (486.5; 0.5 EtOH)
R = H (468.5)
R = CH_3 (482.5)

5
R = F (504.5)

6
R = F (262.2)
R = H (244.2)
R = CH_3 (258.2)

where Tr is triphenylmethyl.

suction, the desiccant is washed with a small amount of chloroform, and the filtrate and washings are combined and evaporated to dryness under diminished pressure. The resulting sirup is dried by adding and evaporating several portions of absolute ethanol, is then dissolved in the minimal volume of hot methanol, and the solution is allowed to cool slowly. Slow crystallization occurs, and is aided by scratching the inside walls of the flask. The suspension is thoroughly cooled, and filtered by suction, giving clusters of prisms; yield 10 g. (65%), m.p. 202–204°. This product is usually of satisfactory purity, but it may be recrystallized from ethanol.

Bis(imidazol-1-yl)thione[1,8]

A solution of predried imidazole (59 g., 0.87 mole) in 700 ml. of dry benzene is heated to *ca.* 60°, and thoroughly flushed with a slow stream of dry nitrogen, preferably by use of a gas sparger. To the efficiently stirred solution is added dropwise (in a well ventilated hood) a solution of thiophosgene[b] (25 g., 0.22 mole) in dry benzene, the temperature being as high as the exothermic reaction permits. A continuous, slow stream of nitrogen is passed in during the addition and the following reflux period. (Omission of the nitrogen results in a 50% diminution of yield.) After addition of the thiophosgene is complete, the reaction mixture is refluxed for 5 hr., and cooled. Precipitation of a solid usually starts at the onset of the reflux period. The orange-red mixture is kept overnight at room temperature under an atmosphere of nitrogen, and the precipitate (imidazole hydrochloride) is removed by rapid filtration on a large Büchner funnel, preferably under a stream of nitrogen, and thoroughly extracted with dry benzene until any occluded yellow solid is dissolved. The benzene extract is filtered, and the filtrate is evaporated *in vacuo* to a thin sirup, which is treated with *ca.* 50 ml. of dry tetrahydrofuran and re-evaporated. This procedure is repeated until the residual benzene has been replaced by tetrahydrofuran. During this treatment, the sirup solidifies to a bright-yellow, granular solid. The suspension is thoroughly chilled, and the solid is removed by filtration and successively washed with small portions of cold tetrahydrofuran containing increasing proportions of dry ether and then with dry ether. The residual solvent is removed in a vacuum desiccator, and the solid, m.p. 100–102°, is stored in a refrigerator under an atmosphere of nitrogen. A second crop is obtained by keeping the tetrahydrofuran–ethyl ether filtrate (under nitrogen) at 0° overnight; total yield 50–55 g. (80–90%); stored under these conditions, the product is usable for several weeks. The ultraviolet absorption spectrum of bis(imidazol-1-yl)thione in tetrahydrofuran shows a strong maximum at 293 nm.

2,2′-Anhydro-[5-fluoro-1-(5-*O*-trityl-β-D-arabinofuranosyl)uracil][1] (4, R = F)

In a 2-l., 3-necked, round-bottomed flask (equipped with a stirrer, thermometer, and reflux condenser protected by a drying tube) is placed a suspension of 25 g. (0.05 mole) of dry **2** (R = F) in 1 l. of toluene (distilled from sodium), and the internal temperature is raised to *ca.* 80°.

[b] Available from the Chemicals Procurement Laboratories, Inc., 18–17 130th Street, College Point, New York.

To the well stirred suspension, a solution of 9.4 g. (0.053 mole) of bis(imidazol-1-yl)thione in dry toluene is added in one portion, and the reaction mixture is heated until it is refluxing. After *ca.* 10 min., dissolution occurs, and usually, within an additional 10 min., precipitation begins, accompanied by fading of the yellow-orange color. After being refluxed for 1 hr., the reaction mixture is thoroughly cooled, and the precipitate is filtered off by suction and thoroughly washed with dry toluene. The crude product is recrystallized from boiling methanol, the color being removed with decolorizing carbon. The yield of pure **4** (R = F) is 19–20 g., m.p. 200–203°. An additional 2–3 g. of product, m.p.[c] 180–182°, may be obtained by evaporation of the original toluene filtrate, and crystallization of the residual sirup from hot methanol with liberal use of decolorizing carbon. The total yield of **4** is *ca.* 90%.

2,2′-Anhydro-[1-(5-*O*-trityl-β-D-arabinofuranosyl)uracil][1] **(4, R = H)**

A suspension of 5.8 g. (0.012 mole) of dry 5′-*O*-trityluridine[9] (Sec. III [136]) (**2**, R = H) in 250 ml. of anhydrous toluene is treated with a toluene solution of 2.3 g. (0.013 mole) of bis(imidazol-1-yl)thione exactly as described for the 5-fluoro analog (**4**, R = F). Recrystallization of the crude solid from methanol affords a highly crystalline product, 3.3 g., m.p. 217–219°. After partial evaporation of the methanol filtrate, a second crop is obtained, which, after recrystallization from methanol, weighs 1.5 g., m.p. 206–209°. (The melting behavior of this compound is apparently dependent on the crystal form, as the high- and the low-melting materials are spectrophotometrically and chromatographically identical.) The total yield of recrystallized product is 4.8 g. (85%).

2,2′-Anhydro-[1-(5-*O*-trityl-β-D-arabinofuranosyl)thymine][10] **(4, R = CH_3)**

A suspension of 4.5 g. (9 mmoles) of the 5′-*O*-tritylnucleoside[5] (**2**, R = CH_3) in 100 ml. of dry toluene is treated with a solution of 1.8 g. of bis(imidazol-1-yl)thione in toluene. The reaction mixture is processed by the procedure described for the 5-fluoro analog (**4**, R = F). Recrystallization of the combined, crude solids from ethanol yields micaceous plates, 2.9 g. (67%), m.p. 225–230°.

5-Fluoro-1-(5-*O*-trityl-β-D-arabinofuranosyl)uracil[1] **(5, R = F)**

To a solution of compound **4** (R = F) (12.0 g., 0.025 mole) in 1 l. of 50% ethanol is added 70 ml. of *N* sodium hydroxide, and the reaction

[c] The difference in the melting points of the two crops of product has been definitely proved to result from isomorphic crystal structures.

mixture is kept at room temperature for 1 hr.,[d] during which, the ultraviolet absorption maximum (in 50% ethanol) shifts to 268 nm. The solution is rendered neutral with 2 *N* acetic acid, and the alcohol is evaporated off under diminished pressure. The resulting, heavy suspension is filtered by suction, the precipitate is thoroughly washed with water, and the damp solid is recrystallized from hot, 50% ethanol. Pure **5** crystallizes as micaceous flakes; yield 12–13 g. (*ca.* 92%), m.p. 172–174° (efferv.).

1-β-D-Arabinofuranosyl-5-fluorouracil[1] (**6**, R = F)

An ethereal suspension of **5** (12.0 g., 0.024 mole) is treated with *ca.* 50 ml. of ether presaturated at 0° with anhydrous hydrogen chloride. The flask is tightly stoppered, and the reaction mixture is stirred magnetically at 0° for 2 hr. The yellowish suspension is filtered by suction, and the solid is thoroughly triturated with dry ether, dissolved in the minimal volume of hot ethanol, and the solution allowed to cool slowly. The resulting crystals are filtered off, and washed with dry ether; yield 5–6 g. (*ca.* 92%), m.p. 187–188°, $[\alpha]_D^{24}$ +128° (*c* 0.21, water), λ_{max}^{pH5-6} 269.5 nm. (ϵ_{max} 9,170); λ_{min} 234 nm. (ϵ_{min} 1,780). The absence of starting material (**1**, R = F) from the product may be demonstrated by paper electrophoresis in borate buffer (pH 9.2) at 900 volts.

1-β-D-Arabinofuranosyluracil[1] (**6**, R = H)

Compound **4** (R = H) may be converted into the unsubstituted β-D-arabinofuranosyl nucleoside (**6**, R = H), by cleavage of the 2,2′-anhydro bond and acid hydrolysis of the trityl ether by the procedures used for conversion of **4** (R = F) into **6** (R = F). The crude product is recrystallized from hot, moist methanol; yield of pure **6** (R = H) *ca.* 65–70% [based on **4** (R = H)], m.p. 222–224°.

1-β-D-Arabinofuranosylthymine[5,6] (**6**, R = CH_3)

Cleavage of the 2,2′-anhydro bond of **4** (R = CH_3), and subsequent detritylation, are achieved as described for the conversion of **4** (R = F) into **6** (R = F). The crude solid is recrystallized from aqueous ethanol (25%); yield of pure **6** (R = CH_3), *ca.* 81% [based on **4** (R = CH_3)], m.p. 248–250°.

REFERENCES

(1) J. J. Fox, N. Miller, and I. Wempen, *J. Med. Chem.*, **9**, 101 (1966).

(2) Alternative syntheses of 1-β-D-arabinofuranosyluracil are described by (a) D. M. Brown, A. R. Todd, and S. Varadarajan, *J. Chem. Soc.*, **1956**, 2388; and (b) J. F. Codington, R. Fecher, and J. J. Fox, *J. Am. Chem. Soc.*, **82**, 2794 (1960).

[d] Prolonged treatment with base must be avoided, as this causes a side-reaction which eventually results in complete loss of the product.[11]

(3) N. C. Yung, J. H. Burchenal, R. Fecher, R. Duschinsky, and J. J. Fox, *J. Am. Chem. Soc.*, **83**, 4060 (1961).

(4) Alternative syntheses of 1-β-D-arabinofuranosyl-5-fluorouracil are described by E. J. Reist, J. H. Osiecki, L. Goodman, and B. R. Baker, *J. Am. Chem. Soc.*, **83**, 2208 (1961); and F. Keller, N. Sugisaka, A. R. Tyrrill, L. H. Brown, J. E. Bunker, and I. J. Botvinick, *J. Org. Chem.*, **31**, 3842 (1966).

(5) J. J. Fox, N. Yung, and A. Bendich, *J. Am. Chem. Soc.*, **79**, 2775 (1957).

(6) An alternative synthesis of 1-β-D-arabinofuranosylthymine has been described by T. Y. Shen, H. M. Lewis, and W. V. Ruyle, *J. Org. Chem.*, **30**, 835 (1965).

(7) R. Duschinsky, E. Pleven, E. Malbica, and C. Heidelberger, *Abstracts Papers Am. Chem. Soc. Meeting*, **132**, 19C (1957).

(8) This procedure is a modification of that described by W. Ried and B. M. Beck, *Ann.*, **646**, 96 (1961).

(9) (a) P. A. Levene and R. S. Tipson, *J. Biol. Chem.*, **104**, 385 (1934); (b) J. F. Codington, I. L. Doerr, and J. J. Fox, *J. Org. Chem.*, **29**, 558 (1964).

(10) I. L. Doerr, J. F. Codington, and J. J. Fox, *J. Med. Chem.*, **10**, 247 (1967).

(11) J. J. Fox, N. Miller, and R. Cushley, *Tetrahedron Letters*, **1966**, 4927.

[116] 1-β-D-Arabinofuranosyl-2-thiouracil and 1-β-D-Arabinofuranosyl-2-thiocytosine

Reaction of a 2,2′-Anhydronucleoside with Hydrogen Sulfide

W. V. RUYLE and T. Y. SHEN

MERCK, SHARP & DOHME RESEARCH LABORATORIES,
DIVISION OF MERCK AND CO., INC., RAHWAY, NEW JERSEY 07065

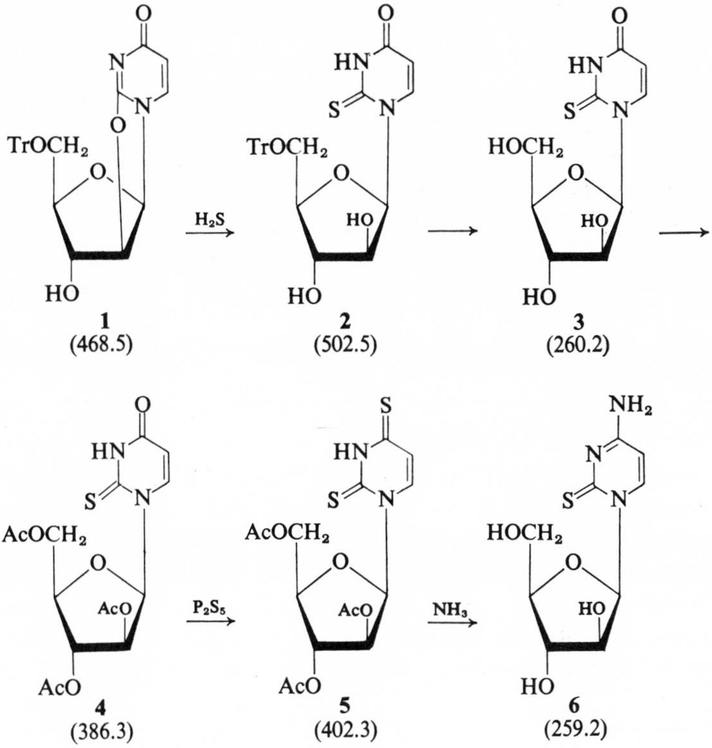

where Tr is triphenylmethyl (trityl).

INTRODUCTION

Thiopyrimidine nucleosides are of interest owing to their occurrence as constituents of certain transfer ribonucleic acids (tRNA). A simple synthesis of D-arabinosyl-2-thiopyrimidines is achieved by opening of the anhydro ring of compound **1** with hydrogen sulfide.[1,2] Conversion of the 2-thionucleoside (**3**) into the 2,4-dithio compound (**5**), followed by selective amination and ammonolysis, yields the 2-thiocytosine nucleoside (**6**). Preparation of an analogous series of compounds in the D-ribose series has been reported by Ueda and coworkers.[3] Other methods for the synthesis of 2-thiopyrimidine nucleosides have been described by Shaw and co-workers.[4,5]

PROCEDURE

2-Thio-1-(5-*O*-trityl-β-D-arabinofuranosyl)uracil (2)

A stream of hydrogen sulfide gas is passed into a solution of 8.0 g. (17 mmoles) of 2,2′-anhydro-[1-(5-*O*-trityl-β-D-arabinofuranosyl)uracil][1,6] (**1**) (Sec. III [115]) in 75 ml. of dry *N*,*N*-dimethylformamide and 5.6 ml. of triethylamine. The temperature of the mixture is gradually raised to 95° during 1.5 hr. and then to 115° during the next 4.5 hr. After being kept at room temperature overnight, the reaction mixture is poured into 300 ml. of water, and the product is extracted with two 100-ml. portions of ethyl acetate.[a] The combined extracts are washed with water, dried with magnesium sulfate, and evaporated to dryness under diminished pressure, to give 9.0 g. of a foamy glass. Chromatography on 450 g. of silica gel,[b] and elution with 1:25 methanol–dichloromethane, yields 7.8 g. (90%) of an amorphous, pale-yellow solid (single peak). The product shows a single spot on a silica gel, thin-layer chromatogram (1:9 methanol–dichloromethane).

1-β-D-Arabinofuranosyl-2-thiouracil (3)

A mixture of 7.8 g. (15.5 mmoles) of 2-thio-1-(5-*O*-trityl-β-D-arabinofuranosyl)uracil (**2**) and 80 ml. of 80% (v/v) aqueous acetic acid is heated on a steam bath for 20 min. The mixture is evaporated to dryness under diminished pressure, and the residue is partitioned between water and ether. Evaporation of the aqueous phase yields 4.0 g. of a crystalline solid which, on trituration with ethanol, yields 3.2 g. (80%) of **3**, m.p. 199–204°. Recrystallization from ethanol gives analytically pure **3**, m.p. 203–205°, $[\alpha]_{589}^{25}$ +110° (*c* 1.0, water), $\lambda_{max}^{H_2O}$ 276 nm. (ε 14,700); $\lambda_{min}^{H_2O}$ 245 nm. (ε 4,500); $\lambda_{max}^{0.1\,N\,NaOH}$ 270 nm. (ε 14,000) and 241 nm. (ε 21,700); and $\lambda_{min}^{0.1\,N\,NaOH}$ 262 nm. (ε 13,600).

2-Thio-1-(2,3,5-tri-*O*-acetyl-β-D-arabinosyl)uracil (4)

A solution of 2.5 g. (9.6 mmoles) of compound **3** in 4 ml. of pyridine and 20 ml. of acetic anhydride is heated on a steam bath for 1 hr. The

[a] The addition of saturated aqueous sodium chloride aids in dispersing the emulsion that forms.

[b] This chromatography was omitted in a subsequent preparation, with no diminution in the overall yield of **3**.

mixture is evaporated under diminished pressure to an oil, and 20-ml. portions of ethanol and toluene are successively added and evaporated off. Recrystallization of the product from ethanol yields 3.31 g. (89%) of **4**, m.p. 140.5–141.5°.

2,4-Dithio-1-(2,3,5-tri-*O*-acetyl-β-D-arabinosyl)uracil (5)

A mixture of 3.2 g. of 2-thio-1-(2,3,5-tri-*O*-acetyl-β-D-arabinosyl)uracil (**4**), 55 ml. of dry pyridine, and 7.4 g. of phosphorus pentasulfide is refluxed for 3.5 hr. The mixture is cooled, and poured into 300 ml. of cold water, the suspension is stirred for 20 min., and the solid product is filtered off and well washed with water. The moist solid is dissolved in 25 ml. of pyridine, and the solution is warmed on a steam bath until evolution of hydrogen sulfide ceases. The product is precipitated by the gradual addition of 200 ml. of cold water, the suspension is filtered, and the precipitate is washed and dried at 110°/0.1 mm. Hg; yield of crude **5**, 3.26 g. (98%), m.p. 145–147°. Recrystallization from toluene–hexane gives 3.01 g. (91%) of bright-yellow crystals, m.p. 146–147°, $\lambda_{max}^{H_2O}$ 281 nm. (ϵ 20,400) (with inflections at 360, 340, 300, and 198 nm.).

1-β-D-Arabinofuranosyl-2-thiocytosine (6)

A solution of 2.0 g. of 2,4-dithio-1-(2,3,5-tri-*O*-acetyl-β-D-arabinosyl)-uracil (**5**) in 100 ml. of methanol is saturated at 0° with anhydrous ammonia, and heated at 100° for 3 hr. in a pressure bomb fitted with a glass liner.[c] The reaction mixture is evaporated under diminished pressure to a gum, and most of the acetamide is removed by sublimation at 60°/0.1 mm. Hg. The residue is chromatographed on 100 g. of silica gel; elution of the column with dichloromethane containing methanol, in increasing concentrations of 2–25%, gives fractions containing acetamide and a series of brown gums. The desired product is then eluted with 3:7 methanol–dichloromethane; yield 386 mg. (30%), m.p. 175–180° (dec.). Recrystallization from methanol–isopropyl alcohol gives analytically pure **6**, m.p. 180–182° (dec.), $\lambda_{max}^{pH\,1}$ 278 nm. (ϵ 17,800) and 230 nm. (ϵ 16,800); $\lambda_{min}^{pH\,1}$ 252 nm. (ϵ 7,900); and $\lambda_{max}^{pH\,13}$ 252.5 nm. (ϵ 22,900).

REFERENCES

(1) For leading references, see W. V. Ruyle and T. Y. Shen, *J. Med. Chem.*, **10**, 331 (1967).

[c] The optimal conditions of temperature and time for this conversion have not yet been established. It has been found, for example, that none of the desired product could be isolated when the mixture was heated for 40 hr.

(2) D. M. Brown, D. B. Parihar, A. R. Todd, and S. Varadarajan, *J. Chem. Soc.*, **1958**, 3028. These workers converted a 2,5′-anhydronucleoside into a 2-thionucleoside by reaction with hydrogen sulfide.
(3) T. Ueda, Y. Iida, K. Ikeda, and Y. Mizuno, *Chem. Pharm. Bull.* (Tokyo), **14**, 666 (1966).
(4) G. Shaw and R. N. Warrener, *J. Chem. Soc.*, **1958**, 153.
(5) G. Shaw, R. N. Warrener, M. H. Maguire, and R. K. Ralph, *J. Chem. Soc.*, **1958**, 2294.
(6) J. J. Fox, N. Miller, and I. Wempen, *J. Med. Chem.*, **9**, 101 (1966).

[117] 1-(2-Deoxy-β-D-*arabino*-hexopyranosyl)uracil [(2-Deoxy-D-glucosyl)uracil]

The Hilbert–Johnson Synthesis Applied to Preparation of 2-Deoxyaldopyranosyl Nucleosides

W. WERNER ZORBACH and H. RANDALL MUNSON JR.

DEPARTMENT OF CHEMISTRY, GEORGETOWN UNIVERSITY, WASHINGTON, D. C. 20007

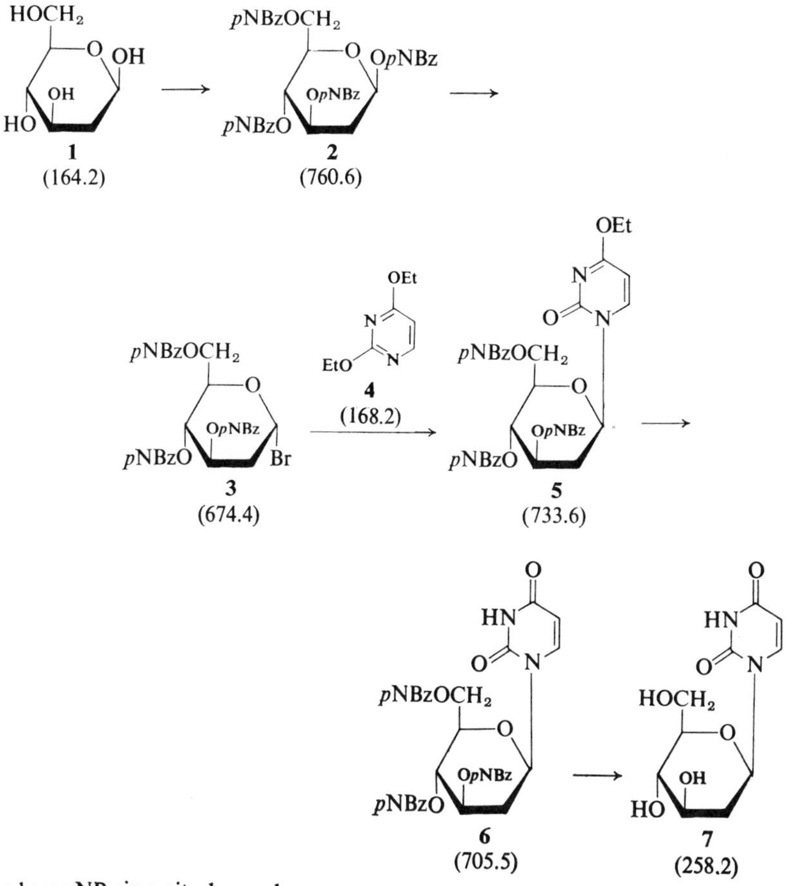

1 (164.2) → **2** (760.6) → **3** (674.4) + **4** (168.2) → **5** (733.6) → **6** (705.5) → **7** (258.2)

where *p*NBz is *p*-nitrobenzoyl.

INTRODUCTION

(2-Deoxy-D-glucosyl)uracil[1] (**7**) is an inhibitor of a nonspecific, pyrimidine phosphorylase.[2] Its preparation is an example of the utility of stable, crystalline 2-deoxy-poly-*O*-*p*-nitrobenzoylglycosyl halides[3] in the direct synthesis of pyrimidine 2′-deoxynucleosides[1c] by the Hilbert–Johnson procedure.[4] Deacylation of the penultimate compound (**6**) is, in this case, performed by ammonolysis.

PROCEDURE

2-Deoxy-1,3,4,6-tetra-*O*-*p*-nitrobenzoyl-β-D-*arabino*-hexose[5] (2)

To 50 ml. of absolute pyridine in a 100-ml. flask is added 7.5 g. of pure *p*-nitrobenzoyl chloride, and the mixture is warmed, with stirring, until the latter dissolves; it is then cooled to 0°, whereupon part of the chloride separates in finely divided form. With efficient stirring, 1.64 g. (10 mmoles) of dry 2-deoxy-β-D-*arabino*-hexopyranose (**1**) is added in small portions during 30 min. The suspension is stirred for 1 hr. at 0°, the flask is stoppered, and the mixture is kept in a refrigerator for 3 days. The flask and contents are allowed to warm to room temperature, the unreacted *p*-nitrobenzoyl chloride is decomposed by the careful addition, with stirring, of an excess of a saturated aqueous solution of sodium hydrogen carbonate, and the resulting mixture is slowly added, with stirring, to 1 l. of ice–water. After being stirred for 1 hr. at room temperature, the separated material is filtered off, washed thoroughly with water, and dried over calcium chloride in a vacuum desiccator. The crude product is dissolved in hot nitromethane, and the solution is decolorized with Darco G-60 and filtered. The filtrate is allowed to cool, and the separated material is recrystallized three times from nitromethane. It is dried for 15 hr. at 150°/0.1 mm. Hg, giving 5.68 g. (75%) of the tetrakis-*p*-nitrobenzoate (**2**), m.p. 213–214°, $[\alpha]_D^{23}$ −9° (*c* 1, nitromethane).

2-Deoxy-3,4,6-tri-*O*-*p*-nitrobenzoyl-α-D-*arabino*-hexosyl Bromide[5] (3)

To a magnetically stirred suspension of 7.6 g. (10 mmoles) of 2-deoxy-1,3,4,6-tetra-*O*-*p*-nitrobenzoyl-β-D-*arabino*-hexose (**2**) in 50 ml. of dry dichloromethane is added 67 ml. of a saturated solution of hydrogen bromide in dichloromethane (containing approximately 0.3 meq. of hydrogen bromide per ml.). The flask is stoppered, and stirring is continued for 2 hr., during which time the nitrobenzoate (**2**) slowly dissolves. The *p*-nitrobenzoic acid, which separates in almost quantitative yield, is removed by suction filtration on a sintered-glass funnel, and rinsed with a little dry dichloromethane. The filtrate is evaporated to dryness at 25°, and the sirupy residue is dissolved in 75 ml. of dry dichloromethane. To the latter is added 75 ml. of dry ether, the flask is stoppered, and the mixture is kept at room temperature for 15 hr. The resulting crystals are removed by suction, and washed with a little dry ether; yield 5.86 g. (87%), m.p. 145–148° (dec.), $[\alpha]_D^{20}$ +88° (*c* 3.8, dichloromethane).

1-(2-Deoxy-3,4,6-tri-*O*-*p*-nitrobenzoyl-β-D-*arabino*-hexosyl)-4-ethoxy-2(1*H*)-pyrimidinone (5)

To 10.8 g. (69.3 mmoles) of 2,4-diethoxypyrimidine (**4**) (Sec. II [30]) is added 2.23 g. of the bromide **3**. The mixture is stirred magnetically, with rigorous exclusion of moisture, for 3 hr., after which time it solidifies. It is allowed to stand overnight at room temperature, and the solid is crushed and then thoroughly stirred in 250 ml. of dry ether. The insoluble material is filtered off, washed with four 10-ml. portions of ether, and dissolved in 200 ml. of dry dichloromethane. The volume of the solution is diminished to about 90 ml. by boiling, and 40 ml. of ether is added. The solution is kept in a refrigerator overnight, and the crystals are filtered off. The crystalline, protected nucleoside (**5**) is washed four times with small portions of ether; yield 1.41 g. (58%), m.p. 271–272.5°, $[\alpha]_D^{20}$ −7° (*c* 0.5, dichloromethane).

1-(2-Deoxy-3,4,6-tri-*O*-*p*-nitrobenzoyl-β-D-*arabino*-hexosyl)uracil (6)

To a solution of 1.41 g. (1.92 mmoles) of the protected nucleoside (**5**) in 100 ml. of dichloromethane is added 25 ml. of 20% (w/w) hydrogen chloride–methanol. The solution is kept overnight at room temperature, the solvent is evaporated under diminished pressure at 40°, and the residue is re-evaporated three times with 10-ml. portions of absolute ethanol. The residue is dissolved in 130 ml. of dry dichloromethane, and the solution is stirred with a little Celite 535, and filtered. The filtrate is diminished in volume to about 100 ml. by boiling, and 90 ml. of dry ether is added. The solution is kept in a refrigerator overnight, and the crystalline, de-ethylated nucleoside[a] is filtered off; yield 1.26 g. (94%). One recrystallization from dichloromethane gives pure **6**, m.p. 143–150°, crystallizing between 168° and 173°, and melting at 260–263°, $[\alpha]_D^{24}$ −8° (*c* 0.5, dichloromethane).

1-(2-Deoxy-β-D-*arabino*-hexopyranosyl)uracil (7)

A suspension of 1.26 g. (1.78 mmoles) of the de-ethylated nucleoside (**6**) in 100 ml. of absolute methanol saturated with ammonia is stirred overnight in a stoppered flask. The solvent is removed at 40° under diminished pressure, and the dry residue is mixed with 50 ml. of water. After being extracted with ten 50-ml. portions of ether,[b] the aqueous layer is stirred

[a] The product is sufficiently pure for use in the deacylation procedure which follows.

[b] To remove the *p*-nitrobenzamide completely.

with Darco G-60 and Rexyn 300 mixed anion–cation exchange resin (Fisher Scientific Co.), the mixture is filtered, and the filtrate is evaporated under diminished pressure at 40°. The residue is re-evaporated several times with small amounts of absolute ethanol, giving hygroscopic material. Crystallization from ethyl acetate–ether is very slow and incomplete; yield 176 mg. (38%), m.p. 168–169° and also 196–197.5°, $[\alpha]_D^{24}$ +6° (*c* 0.5, water), λ_{max}^{MeOH} 260 nm. (log ϵ 3.98).

REFERENCES

(1) (a) J. J. Fox, L. F. Cavalieri, and N. Chang, *J. Am. Chem. Soc.*, **75**, 4315 (1963); (b) G. Etzold and P. Langen, *Chem. Ber.*, **98**, 1988 (1965); (c) W. W. Zorbach, H. R. Munson, and K. V. Bhat, *J. Org. Chem.*, **30**, 3955 (1965).

(2) Private communication from Dr. M. Zimmerman of Merck, Sharp and Dohme Research Laboratories, Rahway, N. J.

(3) W. W. Zorbach and T. A. Payne, Jr., *J. Am. Chem. Soc.*, **80**, 5564 (1958); **81**, 1519 (1959); **82**, 4979 (1960).

(4) G. E. Hilbert and T. B. Johnson, *J. Am. Chem. Soc.*, **52**, 4489 (1930).

(5) W. W. Zorbach and G. Pietsch, *Ann.*, **655**, 26 (1962).

[118] 1-(5-Deoxy-β-D-*erythro*-pent-4-enofuranosyl)-uracil and 1-(2,5-Dideoxy-β-D-*glycero*-pent-4-enofuranosyl)thymine (5′-Deoxy-4′-uridinene and 5′-Deoxy-4′-thymidinene)

Synthesis of 4′,5′-Unsaturated Nucleosides[1]

J. P. H. VERHEYDEN and J. G. MOFFATT

INSTITUTE OF MOLECULAR BIOLOGY, SYNTEX RESEARCH, PALO ALTO, CALIFORNIA 94304

INTRODUCTION

A 4′,5′-unsaturated pentofuranosyl moiety is present[2] in the nucleoside antibiotic angustmycin A. Pyrimidine ribo- and 2′-deoxyribo-nucleosides (**7a** and **7b**) containing this structural feature may be prepared by the reaction of the corresponding acetylated 5′-deoxy-5′-iodonucleosides (**5a** and **5b**) with silver fluoride in pyridine[3,4] followed by hydrolysis of the acetyl group(s). The replacement of the 5′- hydroxyl group in **3a** and **3b** by iodine provides further examples of the general method described elsewhere in this Volume (Sec. III [126]).

PROCEDURE

2′,3′-Di-*O*-acetyluridine (3a)

Uridine (**1a**) (10.0 g., 41 mmoles) and chlorotriphenylmethane (11.4 g., 41 mmoles) are dissolved in anhydrous pyridine (100 ml.), and the solution is heated at 100° for 3 hr.[a] The mixture is cooled to room temperature, acetic anhydride (80 ml., 840 mmoles) is added, and after 20 hr., the solution is slowly added to vigorously stirred ice–water (800 ml.), giving a gummy precipitate that is dissolved in acetone (50 ml.), and reprecipitated by adding the solution to 800 ml. of ice–water in a blender. The resulting, light-tan powder is collected, washed with water, and dried *in vacuo*, giving 21.8 g. (96%) of almost pure 2′,3′-di-*O*-acetyl-5′-*O*-trityluridine (**2a**).[5] Thin-layer chromatography with 3:2 dichloromethane–ethyl acetate shows a single, major product contaminated only by traces of two impurities. The product (**2a**) (20.0 g., 36 mmoles) is dissolved, without further purification, in 80% acetic acid (90 ml.) and boiled for 7 min. The mixture is then evaporated to dryness under diminished pressure, and the resulting

[a] For tritylation of uridine, see Sec. III [136].

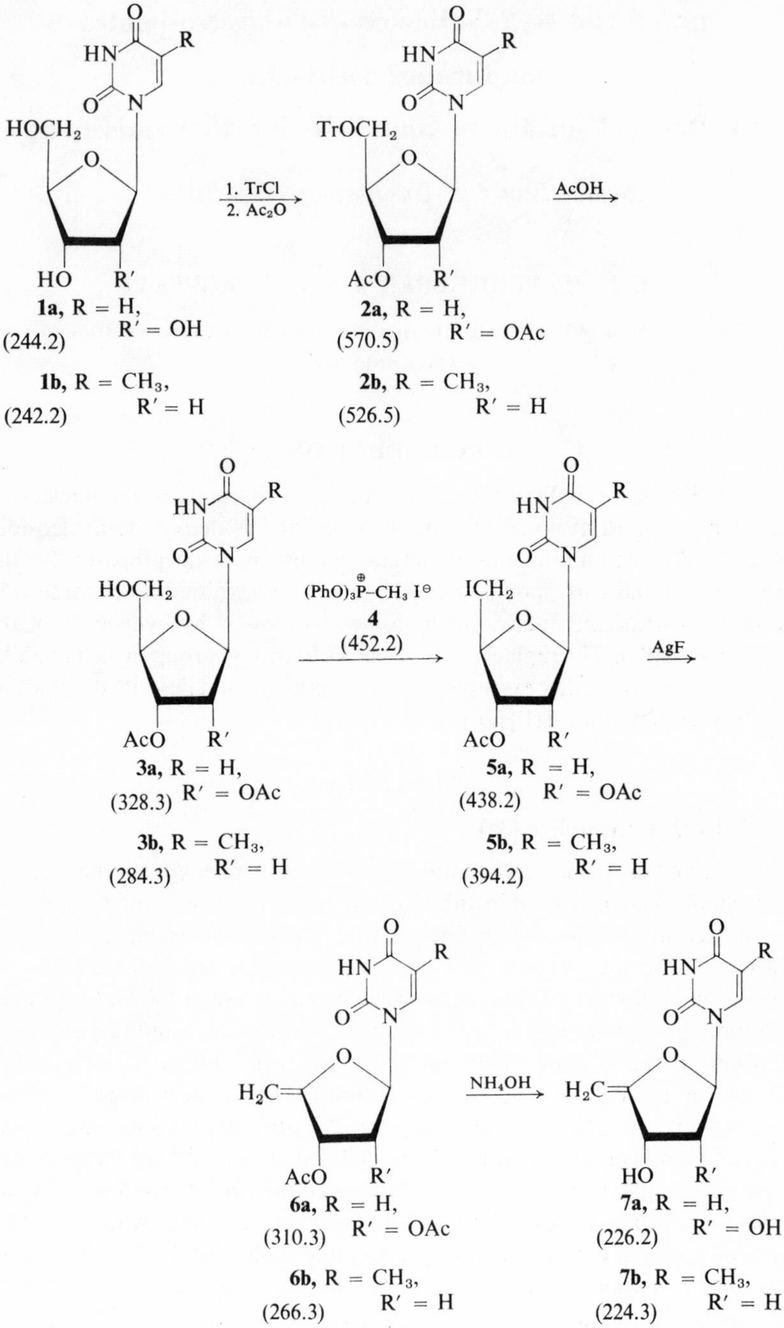

1. TrCl
2. Ac_2O

AcOH

1a, R = H, R′ = OH (244.2)

1b, R = CH_3, R′ = H (242.2)

2a, R = H, R′ = OAc (570.5)

2b, R = CH_3, R′ = H (526.5)

$(PhO)_3\overset{\oplus}{P}-CH_3\ I^{\ominus}$
4
(452.2)

AgF

3a, R = H, R′ = OAc (328.3)

3b, R = CH_3, R′ = H (284.3)

5a, R = H, R′ = OAc (438.2)

5b, R = CH_3, R′ = H (394.2)

NH_4OH

6a, R = H, R′ = OAc (310.3)

6b, R = CH_3, R′ = H (266.3)

7a, R = H, R′ = OH (226.2)

7b, R = CH_3, R′ = H (224.3)

where Ac is acetyl and Ph is phenyl.

sirup is partitioned between water (200 ml.) and benzene (70 ml.). The aqueous phase is washed twice with benzene, and concentrated under diminished pressure to 100 ml. It is then extracted with six 30-ml. portions of ethyl acetate, and the extracts are dried, and evaporated to dryness. The resulting froth is dried under high vacuum giving 4.8 g. (40%) of 2′,3′-di-*O*-acetyluridine (**3a**) that is homogeneous by thin-layer chromatography, employing 19:1 chloroform–isopropyl alcohol; its n.m.r spectrum is consistent with that of the assigned structure. The product may be crystallized from acetone–pentane; m.p.[6] 142–143°.

2′,3′-Di-*O*-acetyl-5′-deoxy-5′-iodouridine[1] (5a)

A solution of compound **3a** (6.4 g., 19.5 mmoles) and methyltriphenoxyphosphonium iodide (**4**) (Sec. III [126]) (13 g., 30 mmoles) in anhydrous *N*,*N*-dimethylformamide (100 ml.) is kept at room temperature for 2 hr. (thin-layer chromatography in 1:1 carbon tetrachloride–acetone shows that the reaction is complete). Methanol (10 ml.) is added, and the solution is evaporated to dryness under diminished pressure. The resulting sirup is dissolved in chloroform (100 ml.), and the solution is washed successively with aqueous sodium hydrogen carbonate, aqueous sodium thiosulfate, and water, dried (sodium sulfate), and evaporated to dryness. The residue is dissolved in chloroform (20 ml.), and hexane is slowly added, to give 7.2 g. (84%) of pure **5a**, m.p. 162–164°.

1-(2,3-Di-*O*-acetyl-5-deoxy-β-D-*erythro*-pent-4-enosyl)uracil (6a)

A mixture of compound **5a** (876 mg., 2 mmoles) and silver fluoride[b] (600 mg., 4.8 mmoles) in pyridine (20 ml.) is shaken in the dark at room temperature for 4 days. The mixture is then filtered through a Celite pad, and the filtrate is evaporated almost to dryness under diminished pressure. Ethyl acetate (50 ml.) and then water (50 ml.) are added, and the mixture is thoroughly shaken. A precipitate is removed by filtration and discarded, and the organic phase is evaporated to dryness, giving a brown sirup (0.80 g.) that is purified by preparative, thin-layer chromatography on a glass plate (20 × 100 cm.) coated with a 1.3-mm. layer of Merck silica gel HF, with ethyl acetate as the solvent. Elution of the main, ultraviolet-absorbing zone with acetone gives 525 mg. (85%) of analytically pure **6a**, characterized by n.m.r. spectroscopy,[1] but not yet obtained crystalline.

1-(5-Deoxy-β-D-*erythro*-pent-4-enofuranosyl)uracil (7a)

The diacetate (**6a**) (835 mg., 2.7 mmoles) is dissolved in a mixture of methanol (8 ml.) and concentrated ammonium hydroxide (8 ml.), and the

[b] The quality of the silver fluoride is important. A suitable grade is obtainable from the Research Inorganic Chemical Co., 11686 Sheldon, Sun Valley, California.

solution is kept at room temperature for 1 hr. Evaporation to dryness under diminished pressure gives a crystalline residue that is freed from ammonium acetate by preparative, thin-layer chromatography with 1:1 acetone–ethyl acetate; this gives a single, ultraviolet-absorbing zone. Elution of this zone with acetone, and evaporation of the solution gives 500 mg. (87%) of pure, crystalline (**7a**), which is recrystallized from acetone, m.p. 169–170°.

3′-*O*-Acetylthymidine[7,8] (3b)

Thymidine (**1b**) (9.76 g., 40 mmoles) and chlorotriphenylmethane (11.2 g., 40 mmoles) are dissolved in anhydrous pyridine (100 ml.) and heated at 100° for 2 hr. (see Sec. III [102]) (thin-layer chromatography with 9:1 ethyl acetate–methanol shows greater than 95% conversion into a single product). The mixture is cooled to room temperature, and acetic anhydride (40 ml., 420 mmoles) is added and, after being stored for 23 hr., the solution is slowly added to ice–water (2 l.) with vigorous agitation in a blender. The resulting pale, cream-colored precipitate is collected by filtration, thoroughly washed with water, and dried in a vacuum desiccator giving 21.0 g. (100%) of 3′-*O*-acetyl-5′-*O*-tritylthymidine (**2b**), shown by thin-layer chromatography with 7:3 carbon tetrachloride–acetone to be at least 95% pure, and contaminated only by traces of three fast-moving, tritylated products. The product (**2b**), crystallized from benzene–hexane, has m.p. 105° but, as prepared, is sufficiently pure for direct use.

The product (21.0 g., 40 mmoles) is dissolved in boiling 80% acetic acid (100 ml.) and the solution is refluxed for 10 min., cooled, and added, with stirring, to 1.5 l. of ice–water. The precipitated triphenylmethanol is removed by filtration, and the aqueous filtrate is evaporated to dryness under diminished pressure. After being dried overnight under high vacuum, the crude product, shown by thin-layer chromatography with 9:1 ethyl acetate–methanol to contain one major product and three impurities in trace amounts, weighs 9.03 g. (79%). Crystallization from ethyl acetate–acetone gives pure **3b**; yield 7.39 g. (65%), m.p. 176–177° (lit.[8] m.p. 176°).

3′-*O*-Acetyl-5′-deoxy-5′-iodothymidine (5b)

A solution of 3′-*O*-acetylthymidine (**3b**) (4.26 g., 15 mmoles) and methyltriphenoxyphosphonium iodide (**4**) (10.0 g., 22 mmoles) in anhydrous *N*,*N*-dimethylformamide (50 ml.) is kept at room temperature for 75 min. (thin-layer chromatography with 9:1 ethyl acetate–methanol indicates completion of the reaction). The solution is then evaporated to dryness under diminished pressure, the residue is dissolved in chloroform (50 ml.), and the solution is successively washed with aqueous sodium thiosulfate (20 ml.)

and two portions of water, dried, and evaporated to dryness. The residue is crystallized from chloroform–hexane, giving 4.91 g. (83%) of **5b**, m.p. 134–134.5° (lit.[8] m.p. 131°).

1-(3-*O*-Acetyl-2,5-dideoxy-β-D-*glycero*-pent-4-enosyl)thymine (6b)

A mixture of **5b** (1.92 g., 4.9 mmoles) and silver fluoride (1.50 g., 12 mmoles) in anhydrous pyridine is shaken at room temperature for 3 days, and processed as described for **6a**. The crude product is purified by preparative, thin-layer chromatography on two preparative silica plates by use of 2:1 carbon tetrachloride–acetone. Elution of the main ultraviolet-absorbing zone with acetone gives 0.85 g. (66%)[c] of **6b** as an analytically pure sirup.

1-(2,5-Dideoxy-β-D-*glycero*-pent-4-enofuranosyl)thymine (7b)

The acetylated product (**6b**) (0.85 g., 3.2 mmoles) is dissolved in a mixture of methanol (8 ml.) and concentrated ammonium hydroxide (8 ml.), and the solution is kept at room temperature for 1 hr., and evaporated to dryness. The product crystallizes from methanol, giving 580 mg. (85%) of **7b** as needles, m.p. 208–210° (immediately resolidifying, and then melting at 270°).

REFERENCES

(1) J. P. H. Verheyden and J. G. Moffatt, *J. Am. Chem. Soc.*, **88,** 5684 (1966).
(2) H. Hoeksema, G. Slomp, and E. E. van Tamelen, *Tetrahedron Letters*, **1964**, 1787.
(3) B. Helferich and E. Himmen, *Ber.*, **61,** 1825 (1928).
(4) L. Hough and B. Otter, *Chem. Commun.*, **1966**, 173.
(5) P. A. Levene and R. S. Tipson, *J. Biol. Chem.*, **104,** 385 (1934).
(6) G. W. Kenner, A. R. Todd, R. F. Webb, and F. J. Weymouth, *J. Chem. Soc.*, **1954**, 2288.
(7) J. P. Horwitz, J. A. Urbanski, and J. Chua, *J. Org. Chem.*, **27,** 3300 (1962).
(8) A. M. Michelson and A. R. Todd, *J. Chem. Soc.*, **1953**, 951.

[c] On a smaller scale, a 78% yield is obtained.

[119] Anomeric Pentofuranosyluracils and Pentofuranosylthymines

The Synthesis of Anomeric Pyrimidine Nucleosides by the Trimethylsilyl Method

ISSEI IWAI, TAKUZO NISHIMURA, and BUNJI SHIMIZU

RESEARCH DEPARTMENT, SANKYO CO., TOKYO, JAPAN

INTRODUCTION

β-D-Arabinofuranosylpyrimidines have been isolated from sponge,[1] and β-D-arabinofuranosylcytosine has been found to inhibit growth of various tumors in mice.[2] In D- or L-arabinofuranosyl- or D- or L-lyxofuranosylpyrimidines, the β anomers have the 1′,2′-*cis* configuration. However, it has been rather difficult to synthesize the 1′,2′-*cis* nucleosides starting from poly-*O*-acylglycosyl halides. The anomeric pyrimidine nucleosides (**14**–**19**) are obtained by condensing trimethylsilyl derivatives of pyrimidines (**3** or **4**) with the poly-*O*-acylglycosyl halides (**5**–**7**) and then removing the protecting groups.[3–5]

PROCEDURE

2,4-Bis(trimethylsiloxy)pyrimidine[3] (3)

To a suspension of 11.3 g. (0.10 mole) of dry, powdered uracil (**1**) and 21.0 g. (0.19 mole) of chlorotrimethylsilane in 100 ml. of dry *p*-dioxane is added, dropwise, a solution of 19.5 g. of triethylamine in dry *p*-dioxane, with stirring, under anhydrous conditions at room temperature, and stirring is continued for 7 hr. The precipitated mixture of triethylamine hydrochloride and uracil is filtered off and washed with three 20-ml. portions of dry *p*-dioxane. The filtrate and washings are combined, and evaporated to dryness under diminished pressure. The viscous oil resulting is distilled at 116°/12 mm. Hg, to give 15.7 g. (72%, based on the uracil consumed) of a colorless oil (**3**), m.p. 31–33°, $\lambda_{max}^{p\text{-dioxane}}$ 258 nm. (By treatment of the still residue and the filter cake with water, 1.6 g. of uracil is recovered.)

5-Methyl-2,4-bis(trimethylsiloxy)pyrimidine[3] (4)

To a suspension of 12.6 g. (0.10 mole) of dry, powdered thymine (**2**) and 21.7 g. of chlorotrimethylsilane[a] in 300 ml. of dry benzene is added dropwise, with mechanical stirring, a solution of 20.2 g. of triethylamine in benzene. By a procedure similar to that given for the preparation of **3**,

[a] Hexamethyldisilazane may be used[6] instead of chlorotrimethylsilane–triethylamine for the preparation of **4**.

Me_3SiCl, Et_3N →; 1. R′X, 2. aq. EtOH →

1, R = H (112.1)
2, R = CH_3 (126.1)

3, R = H (256.4)
4, R = CH_3 (270.4)

5, R′ = 2,3,5-tri-*O*-benzoyl-D-ribosyl, X = Cl (480.9)
6, R′ = 2,3,5-tri-*O*-benzoyl-D-arabinosyl, X = Br (525.3)
7, R′ = 2,3,5-tri-*O*-acetyl-D-lyxosyl, X = Cl (294.7)

MeONa–MeOH →

8a, R = H, R′ = α-D-ribo$(Bz)_3$ (556.5)
8b, R = H, R′ = β-D-ribo$(Bz)_3$ (556.5)
9, R = CH_3, R′ = D-ribo$(Bz)_3$ (570.5)
10a, R = H, R′ = α-D-arabino$(Bz)_3$ (556.5)
10b, R = H, R′ = β-D-arabino$(Bz)_3$ (556.5)
11a, R = CH_3, R′ = α-D-arabino$(Bz)_3$ (570.5)
11b, R = CH_3, R′ = β-D-arabino$(Bz)_3$ (570.5)
12, R = H, R′ = D-lyxo$(Ac)_3$ (370.3)
13, R = CH_3, R′ = D-lyxo$(Ac)_3$ (384.3)

14a, R = H, R″ = α-D-ribo (244.2)
14b, R = H, R″ = β-D-ribo (244.2)
15a, R = CH_3, R″ = α-D-ribo (258.2)
15b, R = CH_3, R″ = β-D-ribo (258.2)
16a, R = H, R″ = α-D-arabino (244.2)
16b, R = H, R″ = β-D-arabino (244.2)
17a, R = CH_3, R″ = α-D-arabino (258.2)
17b, R = CH_3, R″ = β-D-arabino (258.2)
18a, R = H, R″ = α-D-lyxo (244.2)
18b, R = H, R″ = β-D-lyxo (244.2)
19a, R = CH_3, R″ = α-D-lyxo (258.2)
19b, R = CH_3, R″ = β-D-lyxo (258.2)

22.2 g. (89%, based on thymine consumed) of oily product is obtained by distillation at 123–125°/13 mm. Hg, and the product crystallizes on being kept at room temperature, m.p. 63–65°, $\lambda_{max}^{p\text{-dioxane}}$ 266 nm. (From the still residue, 1.0 g. of thymine is recovered.)

The Anomers of 1-(2,3,5-Tri-*O*-benzoyl-D-ribosyl)uracil[4] (8a, 8b)

A mixture of **3** (2.56 g., 0.01 mole) with the amount of **5** obtained from the reaction of 5.04 g. of 1-*O*-acetyl-2,3,5-tri-*O*-benzoyl-D-ribose[b] with dry, ethereal hydrogen chloride,[7] is heated at 190° for 40 min. The mixture is cooled, and dissolved in aqueous ethanol, and the solution is evaporated under diminished pressure to a brown gum which is treated with 80 ml. of hot benzene. Insoluble **1** (0.38 g.) is recovered by filtration. The filtrate is kept in the cold overnight, to furnish 1.3 g. of compound **8b** as needles, m.p. 144–145°. Much product remains in the mother liquor, which is placed on a column (2.5 cm. i.d. × 20 cm.) packed with silica gel; the column is washed with benzene, and then eluted with chloroform. Those fractions containing nucleoside derivatives are combined and evaporated, and the residue is treated with benzene, to yield 0.8 g. of crystals. The crude, crystalline material is purified by fractional recrystallization. By treatment with warm benzene, sparingly soluble clusters of crystals (*A*) are obtained; these are removed by filtration. Evaporation and cooling of the filtrate gives needles, which are recrystallized from benzene to afford an additional 1.0 g. of **8b** (Sec. III [122] and [135]), m.p. 144–145°, $[\alpha]_D^{29}$ $-48°$ (*c* 1.9, chloroform). The total yield of **8b** is 2.3 g. (58%, based on reacted **3**).

The sparingly benzene-soluble clusters (*A*) are recrystallized from benzene–ethanol. Pure **8a** (0.35 g., 9%) is obtained, m.p. 203–205°, $[\alpha]_D^{29}$ $-83°$ (*c* 1.9, chloroform).

1-α-D-Ribofuranosyluracil[4] (14a)

1-(2,3,5-Tri-*O*-benzoyl-α-D-ribosyl)uracil (**8a**) (400 mg., 0.72 mmole) is suspended in 100 ml. of absolute methanol presaturated with dry ammonia at 0°, and the suspension is kept in a refrigerator for 2 days. The crystals gradually dissolve, giving a clear solution; this is evaporated to dryness under diminished pressure, and the residue is dissolved in water. The solution is repeatedly extracted with chloroform (in order to remove benzamide), and evaporated to dryness under diminished pressure. The residue is dissolved in about 10 ml. of absolute ethanol, and a large volume of petroleum ether is added to the solution. The resulting precipitate is filtered off, and dried, giving 130 mg. (74%) of **14a**; $[\alpha]_D^{25}$ $-68°$ (*c* 1.0, water); $\lambda_{max}^{H_2O}$ 264 nm. (ϵ 10,050) (pH 2.9–6.2), 263 nm. (pH 11).

Uridine[4] (14b)

A solution of **8b** (560 mg., 1 mmole) and sodium methoxide (200 mg.) in 3 ml. of absolute methanol is refluxed for about 30 min., and cooled.

[b] Cyclo Chemical Corporation, Los Angeles, Calif. 90001.

Sodium ions are removed from the mixture by treatment with Dowex-50 (H^+) ion-exchange resin, and the solution is evaporated to dryness. The residue is twice recrystallized from 99% ethanol, to give 210 mg. (86%) of **14b**, m.p. 165–166°, $[\alpha]_D^{30}$ +4.6° (*c* 5.3, water), $\lambda_{max}^{H_2O}$ 262 nm. (ϵ 10,300).

The Anomers of 1-D-Ribofuranosylthymine[4] (15a, 15b)

A mixture of **4** (2.70 g., 0.01 mole) with **5**, prepared by treating 5.04 g. of 1-*O*-acetyl-2,3,5-tri-*O*-benzoyl-D-ribose[b] with 100 ml. of dry ethereal hydrogen chloride,[7] is dissolved in 50 ml. of dry benzene. After evaporation of the solvent, the residue is heated at 190° for 45 min. The reaction mixture is treated with about 50 ml. of hot, aqueous ethanol. After evaporation of the solvent, the residue is dissolved in benzene, unreacted thymine (0.25 g.) is recovered by filtration, to the filtrate is added a large volume of hexane, and the supernatant liquor is decanted; this procedure is repeated twice. The gummy residue (**9**) is dissolved in 80 ml. of absolute methanol containing 0.2 g. of sodium methoxide, and the solution is refluxed for 1 hr., cooled, and evaporated to dryness under diminished pressure. The residue is dissolved in water, and the solution is repeatedly washed with ether and then applied to the top of a column (3.5 cm. i.d. × 8 cm.) of Dowex-50 (H^+) ion-exchange resin. The column is eluted with water, and the ultraviolet-absorbing fractions are evaporated to dryness; the residue is treated with absolute ethanol, to give 1.05 g. (50%) of a crystalline mixture of the anomers of 1-D-ribofuranosylthymine. Fractional recrystallization from ethanol separates[8] **15a** from **15b**.

Thus, 0.36 g. of 1-α-D-ribofuranosylthymine is obtained, m.p. 174–175°, $[\alpha]_D^{26}$ −52.3° (*c* 1.60, water); $\lambda_{max}^{H_2O}$ 267 nm. (ϵ 10,000) (pH 3.7–6.2), and 266 nm. (pH 10.6–12.0).

Furthermore, 0.64 g. of 1-β-D-ribofuranosylthymine is obtained, m.p. 183–184.5°, $[\alpha]_D^{27}$ −10.0° (*c* 4.0, water), $\lambda_{max}^{H_2O}$ 267 nm. (ϵ 9,700) (pH 3.9–6.2), and 266 nm. (pH 10.6–12.1).

The Anomers of 1-(2,3,5-Tri-*O*-benzoyl-D-arabinosyl)uracil[5] (10a, 10b)

A mixture of 1.28 g. (5 mmoles) of **3** with 2.63 g. (5.1 mmoles) of[9] **6** is heated at 190° for 40 min., the resulting brownish gum is dissolved in aqueous ethanol, the solution is evaporated to dryness, the residue is mixed with benzene, and the suspension is filtered. To the filtrate is added a large volume of petroleum ether, and the resulting precipitate is filtered off and chromatographed on silica gel (40 g.). The column is developed with the following solvent systems: fraction 1, benzene (2 l.); fraction 2, 4:1 benzene–chloroform (500 ml.); fraction 3, 7:3 benzene–chloroform (500 ml.); fraction 4, 3:2 benzene–chloroform (500 ml.); fraction 5,

11:9 benzene–chloroform (1.5 l.); and fraction 6, 1:1 benzene–chloroform (1 l.). Fractions 5 and 6 are combined, and evaporated, to yield 1.38 g. of a resin that, on trituration with ethanol, gives a crystalline product; recrystallization from ethanol affords 0.39 g. (14%, based on **3**) of compound **10b**, m.p. 202–203°, $[\alpha]_D^{26}$ +72.0° (*c* 2.0, chloroform).

After removal of **10b**, the mother liquor is evaporated, to yield 0.97 g. (35%, based on **3**) of crude 1-(2,3,5-tri-*O*-benzoyl-α-D-arabinosyl)uracil **(10a)**.

1-α-D-Arabinofuranosyluracil[5] (16a)

To a solution of 25 mg. of sodium methoxide in 35 ml. of absolute methanol is added 500 mg. (0.9 mmole) of **10a**. The solution is refluxed for 45 min., and evaporated to dryness, and the residue is dissolved in water. The solution is washed with three portions of ether, and then treated with small portions of Dowex-50 (H^+) ion-exchange resin. The resulting solution is adsorbed on a column of Dowex-1 (OH^-) ion-exchange resin. After being well washed with water, the column is eluted with water saturated with carbon dioxide. The eluates that show ultraviolet absorption are combined, and evaporated to dryness, to give 207 mg. (94%) of **16a**, $\lambda_{max}^{H_2O}$ nm. (pH): 263 (3.2), 263 (7.0), 263 (9.55), and 265 (14.0).

1-β-D-Arabinofuranosyluracil[5] (16b)

1-(2,3,5-Tri-*O*-benzoyl-β-D-arabinosyl)uracil (200 mg., 0.36 mmole) is debenzoylated with sodium methoxide in refluxing methanol, and the resulting methyl benzoate and sodium ions are removed. The crude product is recrystallized from methanol, to give 79 mg. (90%) of 1-β-D-arabinofuranosyluracil **(16b)** (Sec. III [115]), m.p. 225–227.5°, $[\alpha]_D^{27}$ +126° (*c* 1.0, water), $\lambda_{max}^{H_2O}$ nm. (ε; pH): 263 (11,100; 3.2), 263 (11,100; 7.0), 263 (9,300; 9.6), 267 (9,300; 14.0). The melting point and optical rotation show good agreement with the values reported for an authentic sample prepared from natural sources.[10]

The Anomers of 1-D-Arabinofuranosylthymine[5] (17a, 17b)

5-Methyl-2,4-bis(trimethylsiloxy)pyrimidine (**4**) (1.15 g., 4.3 mmoles) is condensed with **6** (2.19 g., 4.3 mmoles)[9] by the procedure used for preparing the anomers of **10**. After chromatographic purification on silica gel, 1.13 g. (47%) of a mixture of **11a** and **11b** is obtained. Treatment of the mixture (500 mg.) with sodium methoxide in methanol, followed by the removal of sodium ions with Dowex-50 ion-exchange resin gives 0.25 g. of a 2.7:1.0 mixture of the α and β anomers of 1-D-arabinofuranosylthymine.

This mixture is triturated with ethanol, to give the crude β-D anomer (Sec. III [115]) as crystals. Recrystallization from aqueous ethanol affords 40 mg. (18%, based on the mixture of **11a** and **11b**) of **17b**, m.p. 246–248°, $[\alpha]_D^{28}$ +90.0° (*c* 0.8, pyridine), +98.0° (*c* 0.1, water),[10] $\lambda_{max}^{H_2O}$ nm. (ϵ; pH): 269 (10,300; 3.2), 269 (10,300; 7.0), 269 (9,300; 9.6), and 272 (9,000; 14.0).

The mother liquor of **17b** is evaporated to dryness, and the residue is purified on a column of Dowex-1 (OH^-) ion-exchange resin, as described for the preparation of **16a**, to afford **17a**; yield 140 mg. (62%, based on the mixture of **11a** and **11b**), $[\alpha]_D^{27}$ +81.0° (*c* 2.6, pyridine), +39.0° (*c* 2.0, water), $\lambda_{max}^{H_2O}$ nm. (pH): 268 (3.2), 268 (7.0), 268 (9.5), and 270 (14.0).

The Anomers of 1-D-Lyxofuranosyluracil[5] (18a, 18b)

2,3,5-Tri-*O*-acetyl-D-lyxosyl chloride (**7**) (0.46 g., 1.6 mmoles)[11] is condensed with an equimolar amount of **3** by heating at 190° for 40 min. The mixture is cooled, treated with petroleum ether, and chromatographed on silica gel with benzene–chloroform as the eluant, to yield 0.32 g. (54%) of a resinous product (**12**) that is deacetylated with sodium methoxide in methanol. After removal of methyl acetate by evaporation, and of sodium ions as described for **16a** and **16b**, the product is dissolved in methanol, and then ethanol is added. From 260 mg. of **12**, 66 mg. (39%) of crystalline α-D anomer is obtained, m.p. 201–203°, $[\alpha]_D^{27}$ +62.5° (*c* 1.0, water).

The mother liquor is concentrated, and kept in a refrigerator, to afford the crude β-D anomer (**18b**). Recrystallization from methanol–ethanol yields 19 mg. (11%) of pure **18b**, m.p. 200.5–202°, $[\alpha]_D^{27}$ +107° (*c* 1.0, water), $\lambda_{max}^{H_2O}$ nm. (ϵ; pH): 263 (10,500; 3.2), 263 (10,500; 7.0), 264 (10,300; 9.5), and 266 (9,000; 14.0). Its physical constants are in good agreement with those reported.[12]

The Anomers of 1-D-Lyxofuranosylthymine[5] (19a, 19b)

Condensation of 2,3,5-tri-*O*-acetyl-D-lyxosyl chloride (**7**) (0.55 g., 1.89 mmoles) with **4** (1.89 mmoles) is conducted as described for **15** or **17**, to give a mixture of the anomers of 1-(2,3,5-tri-*O*-acetyl-D-lyxosyl)thymine (0.30 g., 41%). Hydrolysis of the triacetate (0.23 g.), followed by purification of the product on a column of Dowex-1 (OH^-) gives 0.118 g. (77%) of a mixture of **19a** and **19b**. The mixture is dissolved in the minimal volume of hot absolute ethanol, and the solution is kept at room temperature, to give the crude α-D anomer (**19a**) as colorless needles. After removal of the needles, the crude β-D anomer (**19b**) is obtained by keeping the filtrate in a refrigerator overnight. Additional amounts of **19a** and **19b** are obtained by evaporation of the mother liquor to dryness and fractional recrystallization of the residue as described for **18a** and **18b**. The anomers thus obtained are separately recrystallized three times from absolute ethanol.

α-D Anomer (**19a**): yield 50 mg. (32%, based on the triacetate), m.p. 203–204.5°, $[\alpha]_D^{27}$ +80.0° (*c* 1.2, water), $\lambda_{max}^{H_2O}$ nm. (ε; pH): 268 (10,500; 3.2), 268 (10,500; 7.0), 269 (10,100; 9.5), and 270 (8,200; 14.0).

β-D Anomer (**19b**): yield 25 mg. (16%, based on the triacetate), m.p. 184–186°, $\lambda_{max}^{H_2O}$ nm. (ε; pH): 269 (10,000; 3.2), 269 (10,000; 7.0), 270 (10,000; 9.5), and 270.5 (8,500; 14.0).

REFERENCES

(1) W. Bergmann and R. J. Feeney, *J. Am. Chem. Soc.*, **72**, 2809 (1950); *J. Org. Chem.*, **16**, 981 (1951).

(2) J. S. Evans, E. A. Musser, G. D. Mengel, K. R. Forsbald, and J. H. Hunter, *Proc. Soc. Exptl. Biol. Med.*, **106**, 350 (1961).

(3) T. Nishimura and I. Iwai, *Chem. Pharm. Bull.* (Tokyo), **12**, 352 (1964).

(4) T. Nishimura, B. Shimizu, and I. Iwai, *Chem. Pharm. Bull.* (Tokyo), **12**, 1471 (1964).

(5) T. Nishimura and B. Shimizu, *Chem. Pharm. Bull.* (Tokyo), **13**, 803 (1965).

(6) E. Wittenburg, *Chem. Ber.*, **99**, 2380 (1966).

(7) N. Yung and J. J. Fox, *Methods Carbohyd. Chem.*, **2**, 109 (1963).

(8) J. Farkaš, L. Kaplan, and J. J. Fox, *J. Org. Chem.*, **29**, 1469 (1964).

(9) H. G. Fletcher, Jr., *Methods Carbohyd. Chem.*, **2**, 228 (1963).

(10) W. Bergmann and D. C. Burke, *J. Org. Chem.*, **20**, 1501 (1955).

(11) H. Zinner and H. Brandner, *Chem. Ber.*, **89**, 1507 (1956).

(12) R. Fecher, J. F. Codington, and J. J. Fox, *J. Am. Chem. Soc.*, **83**, 1889 (1961).

[120] 1-(β-D-*erythro*-Pentofuranosyl-3-ulose)uracil and 1-(β-D-*erythro*-Pentofuranosyl-2-ulose)uracil ("2′-Ketouridine" and "3′-Ketouridine")

Oxidation of a Secondary Hydroxyl Group of the Sugar Moiety of a Nucleoside

A. F. COOK and J. G. MOFFATT

INSTITUTE OF MOLECULAR BIOLOGY, SYNTEX RESEARCH, PALO ALTO, CALIFORNIA 94304

1 (728.8) $\xrightarrow[\text{DCC, H}^+]{\text{Me}_2\text{SO}}$ **2** (726.8) $\xrightarrow[\text{CHCl}_3]{\text{HCl}}$ **3** (242.2)

4 (728.8) $\xrightarrow[\text{DCC, H}^+]{\text{Me}_2\text{SO}}$ **5** (726.8) $\xrightarrow[\text{CHCl}_3]{\text{HCl}}$ **6** (242.2)

where Tr is triphenylmethyl.

INTRODUCTION

Early attempts to oxidize the 3′-hydroxyl group in 5′-substituted thymidine derivatives led only to β-elimination of thymine, without the detection of any of the intermediate pentos-3-ulose derivative. Oxidation[1] of 2′,5′-di-*O*-trityluridine (**1**) and of 3′,5′-di-*O*-trityluridine (**4**) with methyl sulfoxide and dicyclohexylcarbodiimide,[2] however, gives the corresponding pentosulose derivatives (**2** and **5**), from which the trityl groups may be removed, affording free "3′-ketouridine" (**3**) and "2′-ketouridine" (**6**). The oxidation may also be performed with methyl sulfoxide–acetic anhydride or methyl sulfoxide–phosphorus pentaoxide.[1]

PROCEDURE

2′,5′-Di-*O*-trityluridine[1,3] (1) and 3′,5′-Di-*O*-trityluridine[1,4] (4)

A mixture of uridine (15 g., 61.5 mmoles) and chlorotriphenylmethane (51.4 g., 183 mmoles) in pyridine (150 ml.) is kept at room temperature overnight and then at 100° for 4 hr. The solution is cooled, and poured into vigorously stirred ice–water (1 l.), and the gummy precipitate is dissolved in chloroform. The chloroform solution is washed successively with 5% aqueous cadmium chloride and water, dried (sodium sulfate), and evaporated *in vacuo* to a yellow sirup. Crystallization from benzene–ether gives 13.6 g. (30%) of compound[3] **1**, which is homogeneous by thin-layer chromatography[a] with 1:1 chloroform–ethyl acetate, and has m.p. 217–220°. Chromatography of the mother liquors on a column of 1.2 kg. of Merck silicic acid with 1:1 chloroform–ethyl acetate gives a small amount of 2′,3′,5′-tri-*O*-trityluridine (m.p. 286–288°, from ethyl acetate), followed by a further 3.8 g. of crystalline **1** (total yield, 39%) and then 8.40 g. of chromatographically homogeneous (but amorphous) 3′,5′-di-*O*-trityluridine[4] (**4**).

2′,5′-Di-*O*-trityl-3′-ketouridine[1] (2)

Compound **1** (3.02 g., 4 mmoles) is dissolved in a mixture of anhydrous methyl sulfoxide[b] (15 ml.) and benzene (15 ml.) containing dicyclohexylcarbodiimide (2.48 g., 12 mmoles) and pyridine (0.32 ml., 4 mmoles). Trifluoroacetic acid (0.16 ml., 2 mmoles) is added, and the mixture is kept at room temperature overnight. Oxalic acid (1.3 g., 12 mmoles) is then added (to decompose excess carbodiimide), and, after 30 min., chloroform (50 ml.) and water (50 ml.) are added and dicyclohexylurea is removed by filtration. The chloroform layer is successively washed with *N* sodium hydrogen carbonate and water (twice), dried (sodium sulfate) and evaporated to dryness *in vacuo*. The residue is dissolved in hot methanol, and allowed to cool very slowly, giving 1.40 g. (46%) of **2**, m.p. 146–148°.

[a] Thin-layer chromatography is performed on layers (0.25 mm. thick) of Merck silica gel GF, and preparative thin-layer chromatography on glass plates (20 × 100 cm.) coated with a 1.3-mm. layer of Merck silica gel HF.

[b] Distilled, and stored over Linde Molecular Sieve Type 4A.

Chromatography of the mother liquors on four preparative, thin-layer plates with 10:1 chloroform–ethyl acetate permits separation of a further 0.89 g. (total yield, 66%) of pure, crystalline **2** (from some slower-moving, unreacted **1**).

3′-Ketouridine[1] (3)

A solution of **2** (1.228 g., 1.69 mmoles) in anhydrous chloroform (20 ml.) is cooled to 0°, and a fresh solution of hydrogen chloride in chloroform (9.3 ml. of 0.40 *M*, 3.72 mmoles) is added dropwise, with stirring, during 20 min. The mixture is then stirred for 1 hr., ether (50 ml.) is added, and the white precipitate is collected by centrifugation. It is then washed with six 25-ml. portions of ether, and dried *in vacuo* over potassium hydroxide, to give 404 mg. (99%) of **3** as an analytically pure, amorphous white solid that melts at 130–135°, and has not yet been crystallized. It is homogeneous by paper chromatography with 5:2:3 butyl alcohol–acetic acid–water, R_f 0.41 (just ahead of uridine). On paper electrophoresis[5] in *M* boric acid at pH 6.0, it has a mobility of 1.2 (relative to 1.0 for uridine).

3′,5′-Di-*O*-trityl-2′-ketouridine[1] (5)

Compound **4** (728 mg., 1 mmole) is oxidized in anhydrous methyl sulfoxide (10 ml.) and benzene (10 ml.) containing dicyclohexylcarbodi-imide (0.62 g.), pyridine (0.075 ml.), and trifluoroacetic acid (0.04 ml.) exactly as described for the oxidation of **1**. The washed chloroform solution is evaporated to dryness, giving an amorphous product (1.02 g.) that has not yet been obtained crystalline. It is purified by preparative, thin-layer chromatography on three 20 × 100 cm. plates, with two consecutive developments with 9:1 chloroform–ethyl acetate, which cleanly separates **5** from some slower-moving, unreacted starting material (**4**). The faster-moving, ultraviolet-absorbing zone is eluted with acetone, and the extracts are combined and evaporated, affording 455 mg. (63%) of homogeneous, analytically pure, noncrystalline **5**, fully characterized by physical methods[1] and by borohydride reduction to a mixture of **4** and 1-(2,5-di-*O*-trityl-β-D-arabinofuranosyl)uracil in the ratio of 1:4.

2′-Ketouridine[1] (6)

Compound **5** (0.60 g., 0.82 mmole) is treated at 0° for 1 hr. with 2.2 equiv. of hydrogen chloride in anhydrous chloroform as described for the detritylation of **2**. The resulting, white precipitate is thoroughly washed with ether, and dried *in vacuo* over potassium hydroxide, giving 144 mg. (72%) of chromatographically and analytically pure **6**, which moves just ahead of uridine by borate electrophoresis at pH 6.0. The product is amorphous (m.p. 186–189°); further purification has not been attempted because of its lability above pH 8. Borohydride reduction gives uridine and 1-β-D-arabinofuranosyluracil in the ratio of 1:9.

REFERENCES

(1) A. F. Cook and J. G. Moffatt, *J. Am. Chem. Soc.*, **89**, 2697 (1967).
(2) K. E. Pfitzner and J. G. Moffatt, *J. Am. Chem. Soc.*, **87**, 5661 (1965).
(3) N. C. Yung and J. J. Fox, *J. Am. Chem. Soc.*, **83**, 3060 (1961).
(4) J. Žemlička, *Collection Czech. Chem. Commun.*, **29**, 1734 (1964).
(5) J. F. Codington, R. Fecher, and J. J. Fox, *J. Am. Chem. Soc.*, **82**, 2794 (1960).

[121] 5-β-D-Ribofuranosyluracil (Pseudouridine)

Use of a Pyrimidinyllithium in Synthesis of C-Glycosyl Derivatives

D. M. BROWN and M. G. BURDON

UNIVERSITY CHEMICAL LABORATORY, CAMBRIDGE, AND SCHOOL OF MOLECULAR SCIENCES, UNIVERSITY OF WARWICK, ENGLAND

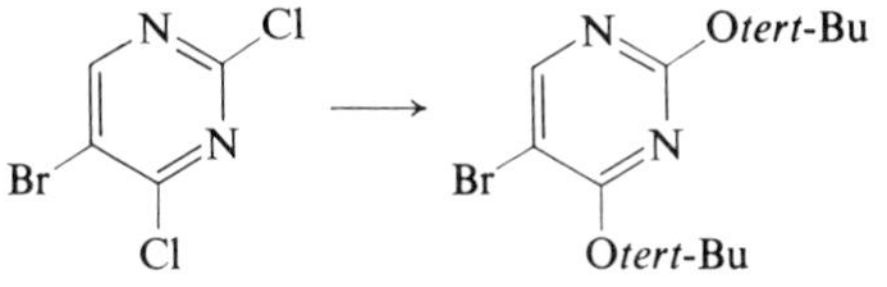

Br, N, Cl, N, Cl → Br, N, O*tert*-Bu, N, O*tert*-Bu

1 (227.9) **2** (303.2)

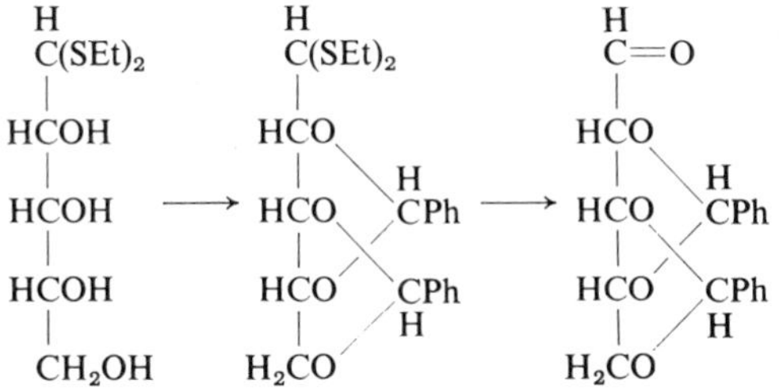

H C(SEt)$_2$ HCOH HCOH HCOH CH$_2$OH → H C(SEt)$_2$ HCO HCO HCO H$_2$CO H CPh CPh H → H C=O HCO HCO HCO H$_2$CO H CPh CPh H

3 (256.4) **4** (432.6) **5** (326.3)

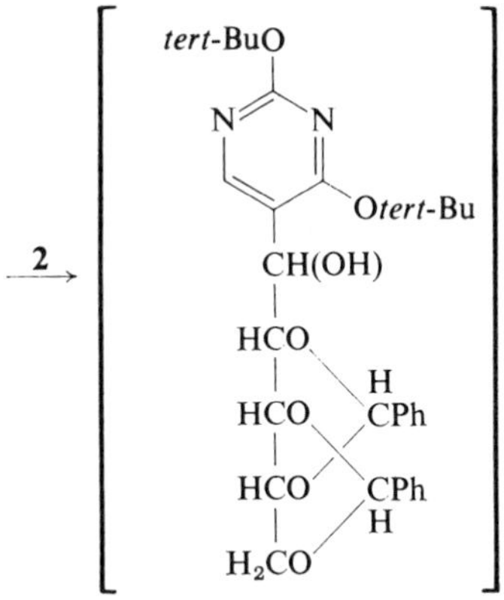

2 → [*tert*-BuO N N O*tert*-Bu CH(OH) HCO HCO HCO H$_2$CO H CPh CPh H] → O HN NH O HOCH$_2$ O HO OH

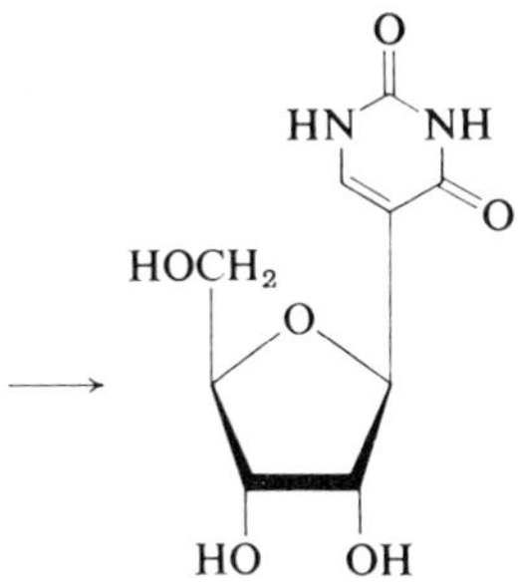

6 (550.6) **7** (244.2)

INTRODUCTION

5-β-D-Ribofuranosyluracil (pseudouridine, **7**) is a nucleoside that is unusual in having a *C*-glycosyl linkage. It is present in transfer ribonucleic acid, and it may be obtained from this source[1] or from human urine.[2] The synthesis involves the conversion of a 5-bromopyrimidine into the corresponding 5-lithio compound, followed by condensation with a derivative (**5**) of *aldehydo*-D-ribose. Removal of the protecting groups, and ring-closure to a mixture of the α- and β-D-ribofuranosyl derivatives, are effected by treatment with acid, and the anomers are separated chromatographically.[3]

PROCEDURE

5-Bromo-2,4-di-*tert*-butoxypyrimidine (2)

Dry *tert*-butyl alcohol (350 ml.) is slowly added, under an atmosphere of nitrogen, to a stirred suspension of 9.3 g. of sodium hydride (50% dispersion in oil) in 100 ml. of petroleum ether (b.p. 80–100°). After the initial, vigorous reaction has subsided, the mixture is refluxed for 10 min., and the white precipitate almost entirely dissolves. The suspension is cooled, 18 g. (0.079 mole) of 5-bromo-2,4-dichloropyrimidine[4] (**1**) is added dropwise during 10 min. at such a rate that gentle refluxing is just maintained, and the solution is refluxed for 2 hr., cooled, and evaporated to dryness under diminished pressure. The residues from two such preparations are combined, 250 ml. of water is added, and the aqueous solution is rapidly extracted with four 250-ml. portions of ether. The extracts are combined, washed with water, dried (sodium sulfate), and evaporated to dryness, and the residue is distilled through a 15-cm. Vigreux column. The fraction having b.p. 94–102°/0.7 mm. Hg is collected; yield 29.3 g. (61%), m.p. 58°. Refractionation gives material having m.p. 63–64°.

D-Ribose Diethyl Dithioacetal[5] (3)

A mixture of 25 g. (0.17 mole) of D-ribose, 25 ml. of concentrated hydrochloric acid, and 25 ml. of ethanethiol, initially at 0°, is shaken, with cooling, until homogeneous (20 min.). Water (200 ml.) is added, and the solution is poured onto a column (600 ml.) of Amberlite IR-45 (OH^-) cation-exchanger. The product is eluted with 800 ml. of water, and the total effluent, which must remain neutral, is evaporated to dryness under diminished pressure, giving a thick sirup that crystallizes on being kept at room temperature. The crystals are dried over phosphorus pentaoxide, and dissolved in the minimal volume of hot chloroform, and three volumes of benzene are added, affording crystals of **3**; yield 28 g. (64%), m.p. 82–83°.

2,4:3,5-Di-*O*-benzylidene-D-ribose Diethyl Dithioacetal[6] (4)

A mixture of 2.56 g. (0.01 mole) of the dithioacetal (**3**) and 10 ml. of freshly distilled benzaldehyde is warmed until it is homogeneous, and is then cooled to 0°. Dry hydrogen chloride is passed into the solution until the initial precipitate has almost vanished (2–3 min.). Ethanol (40 ml.) is added, the solution is cooled to 0°, and after 1 hr., the crop of crystalline **4** is collected; yield 3.2 g. (74%). It may be recrystallized from aqueous methanol, and then has m.p. 120–122°, but the unrecrystallized material is satisfactory for the next step. Attempts to increase the scale of this reaction have resulted in lower yields.

2,4:3,5-Di-*O*-benzylidene-D-ribose[7] (5)

Compound **4** (6.5 g., 0.015 mole), 10.5 g. of mercuric chloride, 10.5 g. of yellow mercuric oxide, 150 ml. of acetone, and 10 ml. of water are mixed, and the mixture is refluxed, with vigorous mechanical stirring, for 6 hr. The hot suspension is filtered, the solids are washed with 100 ml. of acetone, and the filtrate and washings are combined, and evaporated to dryness in the presence of 4.5 g. of yellow mercuric oxide. The residue is extracted with four portions of hot chloroform, and the chloroform extracts are combined, washed successively with aqueous potassium iodide and water, dried (sodium sulfate), and evaporated to a glass that crystallizes from aqueous acetone. This product, essentially the hydrate, is dehydrated by being dried at 100° *in vacuo* for 6 hr., and then has m.p. 167–168°. It should show a strong infrared band at 1,700 $cm.^{-1}$, but no absorption characteristic of hydroxyl groups at 3,500 $cm.^{-1}$.

5-β-D-Ribofuranosyluracil (7)

It is essential that the apparatus be rigorously dried, and that moisture be excluded throughout the reaction by use of drying tubes. Freshly sublimed **2** (0.91 g., 3 mmoles) is dissolved in 20 ml. of dry tetrahydrofuran in a 100-ml., two-necked flask. The solution is magnetically stirred under an atmosphere of oxygen-free nitrogen, and is cooled to −70°. A solution of butyllithium (3 mmoles) in ether, prestandardized by double titration,[8] is added, and stirring is continued for 10 min. A solution of **5** (0.98 g., 3 mmoles) (predried at 100°/0.1 mm. Hg for 3 hr.) in 20 ml. of dry tetrahydrofuran is added dropwise during 30 min., and the solution is stirred for 2 hr. at −70°, and then allowed to warm to room temperature overnight. Water (100 ml.) is added, and the solution is extracted with three portions of ether. The extracts are combined, dried, and evaporated to dryness, and the residual gum is kept at 80°/10 μm. Hg for several hours (to sublime off unchanged **2**). The product (**6**) may be isolated by crystallization at this stage, but it is more convenient to avoid crystallization.

The gum is dissolved in 10 ml. of methanol, concentrated hydrochloric acid (1 ml.) is added, and the solution is heated at 60° for 2 min., and evaporated to dryness under diminished pressure. The residue[a] is dissolved in 20 ml. of water, and the solution is washed with 100 ml. of ether. The aqueous phase is made approximately 0.5 *N* in ammonia, and is applied to a column (2 × 18 cm.) of Dowex-1 (Cl^-). Elution is effected with a linear-gradient system, in which 5 l. of a solution 0.005 *M* with respect to ammonium chloride, ammonium hydroxide, and sodium tetraborate is replaced by 5 l. of 0.02 *M* ammonium chloride.[1] The elution is followed by measuring the optical absorbance at 260 nm. The last peak contains the product, pseudouridine. The α-D anomer is present in the penultimate peak.[1] The solution containing the product is made 0.5 *M* in ammonia, applied to a column similar to that described, and, after the column has been washed with water, the product is eluted with 0.1 *N* acetic acid. The effluent is evaporated to dryness under diminished pressure, and the residue is evaporated several times with methanol, to give **7** as a gum that crystallizes from methanol; yield 131 mg. (18%), m.p. 220–222°, λ_{max} (pH 7) 262 nm. (ϵ 7,900); λ_{max} (pH 12) 287 nm. (ϵ 7,800).

The α-D anomer may be isolated in the same way; yield 58 mg. (8%), m.p. 207–210°.

REFERENCES

(1) W. E. Cohn, *J. Biol. Chem.*, **235**, 1488 (1960).

(2) W. E. Cohn, V. Kurkov, and R. W. Chambers, *Biochem. Prepn.*, **10**, 135 (1963).

(3) D. M. Brown, M. G. Burdon, and R. P. Slatcher, *Chem. Commun.*, **1965**, 77.

(4) J. Chesterfield, J. F. W. McOmie, and E. R. Sayer, *J. Chem. Soc.*, **1955**, 3478.

(5) H. Zinner, *Chem. Ber.*, **83**, 275 (1950); **86**, 495 (1953).

(6) H. Zinner and E. Wittenburg, *Chem. Ber.*, **94**, 1298 (1961).

(7) H. Zinner and H. Schmandke, *Chem. Ber.*, **94**, 1304 (1961).

(8) H. Gilman and J. W. Morton, *Org. Reactions*, **8**, 258 (1954).

[a] If this residue is dissolved in 5 ml. of methanol, pseudouridine (**7**) crystallizes directly. One or two recrystallizations are necessary in order to separate it completely from the α-D anomer. A yield of *ca.* 10% of the theoretical is obtained.

[122] Uridine, Cytidine, and Their 5-Fluoro Derivatives

The Hilbert–Johnson Stereospecific Procedure Applied to Preparation of D-*Ribosylpyrimidines*

M. PRYSTAŠ and F. ŠORM

INSTITUTE OF ORGANIC CHEMISTRY AND BIOCHEMISTRY, CZECHOSLOVAK ACADEMY OF SCIENCES, PRAGUE, CZECHOSLOVAKIA

BzOCH$_2$ O Cl BzO OBz
2
(480.9)

N OMe N R OMe
1a, R = H
(140.1)
1b, R = F
(158.1)

OMe R N O N BzOCH$_2$ O BzO OBz
3a, R = H
(570.5)
3b, R = F
(588.5)

O HN R O N BzOCH$_2$ O BzO OBz
4a, R = H
(556.5)
4b, R = F
(574.5)

NH$_2$ N R O N HOCH$_2$ O HO OH
5a, R = H
(243.2)
5b, R = F
(261.2)

INTRODUCTION

5-Fluoro derivatives of uridine and cytidine exhibit a marked anticancer activity.[1] The ester **3b** is a suitable intermediate for their synthesis, and is obtained as a single product by reaction of base **1b** with a protected D-ribofuranosyl chloride (**2**) in acetonitrile.[2] The analogous reaction of 2,4-dimethoxypyrimidine (**1a**) with the glycosyl halide **2** is fully stereospecific, and affords the β-D anomer (**3a**) exclusively.[3]

PROCEDURE

4-Methoxy-1-(2,3,5-tri-*O*-benzoyl-β-D-ribosyl)-2(1*H*)-pyrimidinone (3a)

A mixture of 280.2 mg. (2.0 mmoles) of 2,4-dimethoxypyrimidine (**1a**) (Sec. II [30]), 20.2 ml. of a 0.1 *M* solution of 2,3,5-tri-*O*-benzoyl-D-ribosyl chloride (**2**) (Sec. III [81]) in acetonitrile, and 1 g. of molecular sieves is refluxed for 6 hr., cooled, kept at room temperature for 2 days and filtered. The precipitate is washed with acetonitrile, and dissolved in 20 ml. of boiling chloroform, and the solution is evaporated to dryness under diminished pressure; the residue is crystallized from ethanol, to give 605 mg. of the ester (**3a**), m.p. 199–199.5°. The mother liquors are combined with the acetonitrile filtrate, and evaporated to dryness under diminished pressure; the residue is chromatographed on a thin layer (1 plate, 18 × 48 cm.) of loose alumina (Brockmann activity, II–III) in 7:2 benzene–ethyl acetate. The zone of R_f 0.35 affords an additional 218 mg. of the ester (**3a**); total yield 823 mg. (72%), m.p. 200–201° (from ethanol).

2′,3′,5′-Tri-*O*-benzoyluridine (4a)

A solution of 570 mg. (1 mmole) of the tribenzoate (**3a**) in 100 ml. of chloroform is saturated at 0° with gaseous hydrogen chloride, kept at room temperature for 12 hr., and evaporated to dryness under diminished pressure. Co-evaporation of the residue with toluene affords an almost quantitative yield of uridine tribenzoate (**4a**) (Sec. III [119] and [135]), m.p. 146–148° (from toluene).

Cytidine (5a)

A mixture of 570 mg. (1.0 mmole) of the 4-methoxy derivative (**3a**) and 60 ml. of methanolic ammonia (presaturated at 0°) is heated in a pressure vessel at 100° for 15 hr., cooled, and evaporated to dryness under diminished pressure. The residue is co-evaporated with two 30-ml. portions of water, and dissolved in 10 ml. of water, the aqueous solution is washed with chloroform, and filtered with activated carbon, and the filtrate is evaporated to dryness under diminished pressure. The colorless residue is treated with a solution of 50 mg. of sulfuric acid in 10 ml. of ethanol. The crystals that separate are collected, washed with ethanol, and dried for 5 hr. at 120°/0.1 mm. Hg. The yield of cytidine sulfate[4] (**5a**) is 242 mg. (82%), m.p. 225°.

5-Fluoro-2,4-dimethoxypyrimidine[5] (1b)

A mixture of 1.67 g. (10 mmoles) of 2,4-dichloro-5-fluoropyrimidine[6] (Sec. I [34]) and 25 ml. of 1.2 *M* sodium methoxide in methanol is refluxed for 20 min., allowed to cool, diluted with 50 ml. of dry ether, and neutralized with gaseous carbon dioxide. The resulting suspension is filtered, the precipitate is washed with 30 ml. of dry ether, and the filtrates are combined, and evaporated to dryness under diminished pressure (bath temperature, 23°). The crystalline residue is purified by sublimation at 75°/15 mm. Hg to afford 1.52 g. (96%) of the pure fluoro derivative (**1b**), m.p. 49–50°.

5-Fluoro-4-methoxy-1-(2,3,5-tri-*O*-benzoyl-β-D-ribosyl)-2(1*H*)-pyrimidinone (3b)

A mixture of 316 mg. (2 mmoles) of the 5-fluoropyrimidine derivative (**1b**), 51 ml. of a 0.04 *M* solution of the protected glycosyl halide (**2**) in acetonitrile, and 2 g. of molecular sieves is refluxed for 15 hr. The suspension is filtered, the filtrate is evaporated to dryness under diminished pressure, and the residue is chromatographed on a column of 100 g. of neutral alumina (Brockmann activity, II–III). The column is successively eluted with 400 ml. of benzene, and 1.2 l. of an exponential gradient (fractions 1–60) obtained from 1.2 l. of 14:5 benzene–ethyl acetate and 1.0 l. of benzene, 20-ml. fractions being taken at 10-min. intervals. The chromatographically homogeneous fractions 27–36 (R_f 0.4, as determined by chromatography on a thin layer of alumina of the above quality in 3:1 benzene–ethyl acetate) are combined, and evaporated to dryness under diminished pressure, and the residue is crystallized from absolute ethanol to give 576 mg. (49%) of the ester (**3b**), m.p. 195–197°. A single recrystallization from absolute ethanol gives pure **3b**, m.p. 197–198°, $[\alpha]_D^{25}$ +2.4° (*c* 0.8, pyridine).

2′,3′,5′-Tri-*O*-benzoyl-5-fluorouridine (4b)

A solution of 588 mg. (1.0 mmole) of the ester (**3b**) in 50 ml. of anhydrous chloroform presaturated at 0° with dry, gaseous hydrogen chloride is kept at room temperature for 8 hr., and evaporated to dryness under diminished pressure. The residue is co-evaporated with three 20-ml. portions of anhydrous toluene, and recrystallized from anhydrous toluene, to give 568 mg. (99%) of pure tribenzoate (**4b**), m.p. 213–214°, $[\alpha]_D^{25}$ −73° (*c* 0.9, chloroform).

5-Fluorocytidine (5b)

A mixture of 588 mg. (1.0 mmole) of the tribenzoate (**3b**) and 50 ml. of methanol presaturated at 0° with gaseous ammonia is heated in a pressure vessel at 90° for 16 hr., cooled, and evaporated to dryness under diminished pressure. The residue is dissolved in 10 ml. of water, the aqueous solution is filtered with activated carbon, and the filtrate is freeze-dried; the residue is triturated with hot ethanol, to give 204 mg. (78%) of 5-fluorocytidine (**5b**), m.p. 193°.

REFERENCES

(1) C. Heidelberger, N. K. Chaudhuri, P. Dannenberg, D. Mooren, L. Griesbach, R. Duschinsky, R. J. Schnitzer, E. Pleven, and J. Scheiner, *Nature*, **179**, 663 (1957); C. Heidelberger, L. Griesbach, B. J. Montag, D. Mooren, O. Cruz, R. J. Schnitzer, and E. Grunberg, *Cancer Res.*, **18**, 305 (1958); C. Heidelberger, L. Griesbach, O. Cruz, R. J. Schnitzer, and E. Grunberg, *Proc. Soc. Exptl. Biol. Med.*, **97**, 470 (1958); N. C. Yung, J. H. Burchenal, R. Fecher, R. Duschinsky, and J. J. Fox, *J. Am. Chem. Soc.*, **83**, 4060 (1961).

(2) M. Prystaš and F. Šorm, *Collection Czech. Chem. Commun.*, **29**, 2956 (1964).

(3) M. Prystaš and F. Šorm, *Collection Czech. Chem. Commun.*, **31**, 1035 (1966).

(4) G. A. Howard, B. Lythgoe, and A. R. Todd, *J. Chem. Soc.*, **1947**, 1052.

(5) M. Prystaš and F. Šorm, *Collection Czech. Chem. Commun.*, **30**, 1900 (1965).

(6) F. Hoffmann–La Roche and Co., Brit. Pat. 877, 318 (1960); *Chem. Abstracts*, **56**, 8724 (1962).

[123] 5-Aminouridine

Amination of a Pyrimidine Nucleoside

D. W. VISSER

DEPARTMENT OF BIOCHEMISTRY, UNIVERSITY OF SOUTHERN CALIFORNIA, LOS ANGELES, CALIFORNIA 90033

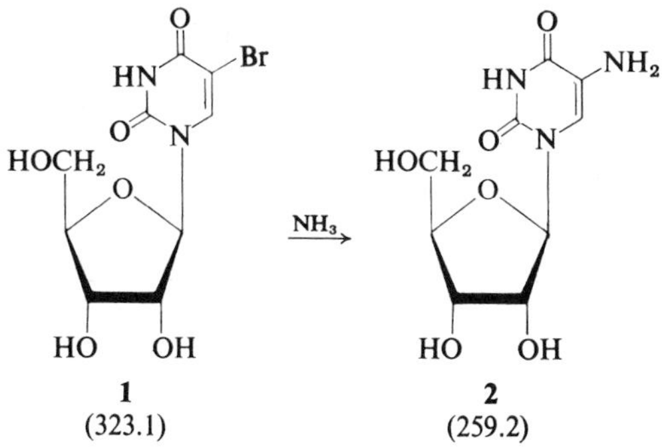

1 (323.1) $\xrightarrow{NH_3}$ 2 (259.2)

INTRODUCTION

5-Aminouridine (**2**) has a wide range of biological effects. Compound **2** inhibits the growth of bacteria,[1] fungi,[2] viruses,[3] protozoa,[4] and tumors.[5] The analog inhibits incorporation of carbamylaspartate into pyrimidines of both ribonucleic acid and 2′-deoxyribonucleic acid in rat liver,[6] and of phosphate-^{32}P into phospholipids and nucleotides of ribonucleic acids of rat-liver slices and hepatoma.[7]

Compound **2** is metabolized to the 5′-monophosphate, 5′-pyrophosphate, 5′-triphosphate, and 5′-(glycosyl pyrophosphates), and is incorporated into ribonucleic acid by Ehrlich ascites cells.[8] The 5′-monophosphate of **2** is a potent inhibitor of orotidine 5′-phosphate decarboxylase.[8] Viruses grown in the presence of the nucleoside analog are hypersensitized to nitrous acid.[9] Preparation[8,10] of **2** and its 5′-monophosphate[9] has been reported.

PROCEDURE

5-Aminouridine (2)

To 10.0 g. (0.03 mole) of 5-bromouridine[11] (**1**) in a stainless-steel container is added 50 ml. of dry, liquid ammonia. The container is sealed, and kept at 55° for 48 hr. The ammonia is then allowed to evaporate, and the last traces of ammonia are removed over concentrated sulfuric acid in a vacuum desiccator. The brown residue is dissolved in 200 ml. of 0.1 *N* hydrochloric acid, and the solution is passed through a column (3 × 30 cm.) of Dowex 50-X8 (H^+) ion-exchange resin. The resin is eluted with water to remove any unreacted **1**, until the optical absorbance ($A_{280\ nm.}$) of the eluate is zero. The hydrochloride of **2** is eluted with 0.5 *N* hydrochloric acid. Fractions containing the product, as determined by the absorbance ($A_{260\ nm.}$), are combined, and flash-evaporated to 50 ml. Addition of 150 ml. of absolute ethanol causes immediate formation of colorless crystals of the hydrochloride of **2**. Recrystallization is effected from aqueous ethanol.

The hydrochloride is dissolved in the minimal volume of hot water, and the pH is adjusted to 7.0 with concentrated ammonium hydroxide. Crystallization of **2** occurs on adding three volumes of hot, absolute ethanol, and cooling; yield 5.8 g. (72%), m.p. 214–216°; $\lambda_{max}^{pH\ 1}$ 265 nm. (ϵ 9,300), $\lambda_{max}^{pH\ 7}$ 294 nm. (ϵ 7,400), and $\lambda_{max}^{pH\ 12}$ 289 nm. (ϵ 6,200). A_{280}/A_{260} 0.61 at pH 1.0, and 2.04 at pH 12. Compound **2** gives a yellow color with *p*-(dimethylamino)benzaldehyde in glacial acetic acid.[12]

REFERENCES

(1) R. E. Beltz and D. W. Visser, *J. Biol. Chem.*, **226**, 1035 (1957); E. C. Theil and S. Zamenhof, *ibid.*, **238**, 3058 (1963); D. B. Dunn and J. D. Smith, *Biochem. J.*, **68**, 627 (1958).

(2) M. Roberts and D. W. Visser, *J. Biol. Chem.*, **194**, 695 (1952).

(3) D. W. Visser, D. L. Lagerborg, and H. E. Pearson, *Proc. Soc. Exptl. Biol. Med.*, **79**, 571 (1952).

(4) G. W. Kidder and B. C. Dewey, *J. Biol. Chem.*, **178**, 383 (1949).

(5) D. W. Visser, in C. P. Rhoads (Ed.), *Antimetabolites and Cancer*, American Association for the Advancement of Science, Washington, D. C., 1955, p. 47.

(6) W. C. Werkheiser and D. W. Visser, *Cancer Res.*, **15**, 644 (1955); M. L. Eidinoff, J. E. Knoll, B. J. Marano, and D. Klein, *ibid.*, **19**, 738 (1959).

(7) W. C. Werkheiser, R. J. Winzler, and D. W. Visser, *Cancer Res.*, **15**, 641 (1955).

(8) D. A. Smith, P. Roy-Burman, and D. W. Visser, *Biochim. Biophys. Ac*
119, 221 (1966).

(9) L. Thiry, *Virology*, **28**, 543 (1966).

(10) M. Roberts and D. W. Visser, *J. Am. Chem. Soc.*, **74**, 668 (1952); I. Wempen, I. L. Doerr, L. Kaplan, and J. J. Fox, *ibid.*, **82**, 1624 (1960).

(11) P. A. Levene and F. B. LaForge, *Ber.*, **45**, 615 (1912); T. K. Fukuhara and D. W. Visser, *J. Biol. Chem.*, **190**, 95 (1951).

(12) S. N. Chakravarti and M. S. Roy, *Analyst*, **62**, 603 (1937).

[124] 5-Bromo-2′-deoxyuridine and 5-Amino-2′-deoxyuridine

Substitution Reactions of the Pyrimidine Ring of a Nucleoside

D. W. VISSER

DEPARTMENT OF BIOCHEMISTRY, UNIVERSITY OF SOUTHERN CALIFORNIA, LOS ANGELES, CALIFORNIA 90033

HOCH$_2$ HN O O N HO **1** (228.2) — Ac_2O, Br_2 → AcOCH$_2$ HN O Br O N AcO **2** (391.2) — NH_3 → HOCH$_2$ HN O Br O N HO **3** (307.1); — NH_3 → HOCH$_2$ HN O NH_2 O N HO **4** (243.2)

where Ac is acetyl.

INTRODUCTION

The thymidine-like behavior of 5-bromo-2′-deoxyuridine (**3**) and the effects of incorporation of **3** into 2′-deoxyribonucleic acid (DNA) of bacteria, viruses, plants, and animals have been studied extensively, and summarized in several reviews.[1] Incorporation of 5-bromouracil into DNA usually causes mutagenesis, which has been interpreted as being the result of anomalous base-pairing. DNA containing 5-bromouracil is markedly more sensitive than naturally occurring DNA to ultraviolet light, X-rays, visible light, and decay of internally incorporated phosphate-^{32}P. The procedure given here for the synthesis of **3** and its diacetate (**2**) is a modification of the original preparations.[2,3]

The inhibitory effect of 5-amino-2′-deoxyuridine (**4**) on bacterial growth[4] has been attributed, in part, to the fact that it inhibits the formation and utilization of thymidylate.[3] Compound **4** partially replaces thymidine as a growth requirement in a thymine-requiring strain of *Escherichia coli*.[5] 5-Aminouracil[6] and compound[7] **4** undergo minor incorporation into the DNA of bacteria. The preparation of **4** is a modification[3] of the method described by Beltz and Visser.[4]

PROCEDURE

3′,5′-Di-*O*-acetyl-5-bromo-2′-deoxyuridine (2)

A suspension of 5 g. (21.9 mmoles) of 2′-deoxyuridine (**1**) in 30 ml. of acetic anhydride is heated until dissolution occurs. A solution of 3.8 g. (23.8 mmoles) of bromine in 3 ml. of glacial acetic acid is added, with cooling, to maintain a temperature of 25°. After being kept overnight in the cold, the solution is evaporated to a thick sirup at 30°/1 mm. Hg. The slightly yellow sirup is kept overnight at 0.5 mm. Hg in a desiccator containing potassium hydroxide. On addition of about 30 ml. of anhydrous ethanol, the diacetate (**2**) crystallizes slowly; yield 7.42 g. (86.6%), m.p. 153–155°.

5-Bromo-2′-deoxyuridine (3)

Compound **2** (7.4 g.) is dissolved in 85 ml. of anhydrous methanol containing 10% of anhydrous ammonia, and the solution is kept at room temperature for two days. The solvent is evaporated at 20°, and the residue is dissolved in the minimal volume of warm, anhydrous ethanol. Crystallization of **3** occurs on cooling; yield 5.45 g. (81%, based on **1**), m.p. 192–193°; $\lambda_{max}^{pH\,1}$ 278 nm. (ϵ 8,500), and $\lambda_{max}^{pH\,12}$ 278 nm. (ϵ 5,800). A_{280}/A_{260} 1.86 at pH 1 and 1.61 at pH 12.

5-Amino-2′-deoxyuridine (4)

Compound **2** (7.4 g., 18.9 mmoles) is converted directly into **4** by treatment with 25 ml. of liquid ammonia in a stainless-steel container at 50° for 24 hr. After removal of the ethanol and ammonia, an aqueous solution of the residue is adsorbed on Dowex-50 X8 (H^+), and subsequently eluted with dilute hydrochloric acid. Fractions containing the product, as determined by A_{260}, are combined, and concentrated to about 40 ml. at 35°.

Crystallization of the hydrochloride occurs on addition of anhydrous ethanol (3 ml./g.). One recrystallization yields 3.47 g. (56.6%, based on **1**), m.p. 185° (dec.); $\lambda_{max}^{pH\,1}$ 265 nm. (ϵ 9,300), and $\lambda_{max}^{pH\,12}$ 289 nm. (ϵ 6,200); A_{280}/A_{260} 0.61 at pH 1 and 2.04 at pH 12.

The free base is obtained by adjusting the pH of an aqueous solution of the hydrochloride to 7 with ammonium hydroxide, and crystallizing from aqueous alcohol; m.p. 195–197°.

REFERENCES

(1) R. W. Brockman and E. P. Anderson, *Ann. Rev. Biochem.*, **32**, 490 (1962); R. E. Handschumacher and A. D. Welch, in E. Chargaff and J. N. Davidson (Eds.), *The Nucleic Acids*, Vol. III, Academic Press Inc., New York, N. Y., 1960, p. 453; R. W. Brockman *Advan. Cancer Res.*, **7**, 129 (1963); W. Szybalski, *The Molecular Basis of Neoplasia*, University of Texas Press, Austin, Texas, 1962, p. 147; R. W. Brockman and E. P. Anderson, in J. H. Quastel and R. M. Hochster (Eds.), *Metabolic Inhibitors*, Vol. I, Academic Press Inc., New York, N. Y., 1963, p. 239; D. R. Krieg, *Progr. Nucleic Acid Res.*, **2**, 125 (1963); A. Wacker, *ibid.*, **1**, 369 (1963).

(2) R. E. Beltz and D. W. Visser, *J. Am. Chem. Soc.*, **77**, 736 (1955).

(3) M. Friedland and D. W. Visser, *Biochim. Biophys. Acta*, **51**, 148 (1961).

(4) R. E. Beltz and D. W. Visser, *J. Biol. Chem.*, **226**, 1035 (1956).

(5) D. W. Visser, S. Kabat, and M. Lieb, *Biochim. Biophys. Acta*, **76**, 463 (1963).

(6) A. Wacker, S. Kirschfeld, D. Hartman, and D. Weinblum, *J. Mol. Biol.*, **2**, 69 (1960).

(7) S. Kabat and D. W. Visser, *Biochim. Biophys. Acta*, **82**, 680 (1964).

[125] 2′-Deoxy-5-hydroxyuridine

Hydroxylation via *Addition of Aqueous Bromine*

EUGENE G. PODREBARAC and C. C. CHENG

MIDWEST RESEARCH INSTITUTE, KANSAS CITY, MISSOURI 64110

1. Br_2–H_2O
2. pyridine

1 (228.2) → **2** (244.2)

INTRODUCTION

The increasing attention being paid to the biological importance of 5-substituted pyrimidine nucleosides and nucleotides has necessitated a re-evaluation of the existing synthetic methods and a search for more-convenient methods for the preparation of many of the compounds in this category. Recorded methods for the preparation of 2′-deoxy-5-hydroxyuridine (**2**) include (*a*) the addition of aqueous bromine to 2′-deoxyuridine (**1**), followed by treatment of the resulting dihydro intermediate with lead oxide, resulting in a yield[1] of 30%, and (*b*) treatment of 5-bromo-2′-deoxyuridine with sodium hydrogen carbonate, giving a yield[2] of 28%. The present procedure is a modification of Ueda's synthesis of 5-hydroxyuridine[3] from uridine, adapted to the 2′-deoxyuridine series by conducting the reaction at room temperature.

PROCEDURE

2′-Deoxy-5-hydroxyuridine (2)

To a stirred solution of 5.0 g. (0.112 mole) of **1** in 200 ml. of water at room temperature, aqueous bromine is added, dropwise, until a light-yellow color persists (*ca.* 1.5 ml. of bromine). Air is then bubbled through the solution until it becomes colorless. To the clear solution is added 150 ml. of pyridine in small portions so that the temperature of the solution does not exceed 25°. The resulting solution is kept at room temperature overnight, and evaporated under diminished pressure below 30° to a viscous sirup. This is dissolved in *ca.* 200 ml. of absolute ethanol at 50°, and the solution is filtered. The filtrate is concentrated to *ca.* half its original volume, and refrigerated overnight. The analytically pure solid that separates is filtered off, and dried at 25°/1 mm. Hg for 15 hr.; yield 2.5 g. (47%), m.p.[a] 199–201°; $\lambda_{max}^{pH\,1}$ 279 nm. (ϵ 8,100); $\lambda_{max}^{pH\,11}$ 239 nm. (ϵ 7,100) and 304 nm. (ϵ 6,600).

REFERENCES

(1) R. E. Beltz and D. W. Visser, *J. Am. Chem. Soc.*, **77**, 736 (1955).
(2) T. Y. Shen, J. F. McPherson, and B. O. Linn, *J. Med. Chem.*, **9**, 366 (1966).
(3) T. Ueda, *Chem. Pharm. Bull.* (Tokyo), **8**, 455 (1960).

[a] Previous investigators have reported melting points for **2** in the range of 208–211°. Our analytically pure product, in different preparations, always melted at 199–201°.

[126] 5′-Deoxy-5′-iodouridine and 3′-Deoxy-3′-iodo-5′-*O*-*p*-nitrobenzoylthymidine[1]

Direct Replacement of a Hydroxyl Group of a Nucleoside by Iodine

J. P. H. VERHEYDEN and J. G. MOFFATT

INSTITUTE OF MOLECULAR BIOLOGY, SYNTEX RESEARCH, PALO ALTO, CALIFORNIA 94304

$(PhO)_3\overset{\oplus}{P}-CH_3\ I^{\ominus}$ **2** (452.2)

1 (284.3) → **3** (394.2) → H^+ → **4** (354.1)

5 (242.2) → **6** (391.3) → **2** → **7** (501.2)

where Ph is phenyl and *p*NBz is *p*-nitrobenzoyl.

INTRODUCTION

Replacement of the primary hydroxyl group in pyrimidine nucleosides has previously been accomplished by a two-step procedure involving initial *p*-toluenesulfonylation followed by nucleophilic displacement of the *p*-tolylsulfonyloxy group by iodide ion.[2] A more rapid and efficient procedure consists of reaction of the protected nucleoside (*e.g.*, **1**) with methyltriphenoxyphosphonium iodide (**2**) in *N,N*-dimethylformamide to give the 5′-deoxy-5′-iodonucleoside (*e.g.*, **3**) directly. The method may also be applied to replacement of the 3′-hydroxyl group of 5′-protected pyrimidine 2′-deoxynucleosides by iodine, to give the corresponding 3′-iodo derivative with retention of configuration.[1]

PROCEDURE

Methyltriphenoxyphosphonium Iodide[3] (2)

Triphenyl phosphite (31 g., 0.1 mole) and methyl iodide (21 g., 0.15 mole) are refluxed together for 36 hr., during which time the temperature of the liquid rises to 115°. The solution is cooled, and anhydrous ether (200 ml.) is gradually added, causing the separation of pale-brown crystals of compound **2**; this is repeatedly washed by decantation with fresh portions of anhydrous ether until the washings are essentially colorless (usually, 6–8 times). The crystalline product is then dried, and stored *in vacuo*. The material so obtained is quite stable when stored *in vacuo*, and is entirely suitable for direct use. Recrystallization from acetone–ether[3] is not recommended.

5′-Deoxy-5′-iodo-2′,3′-*O*-isopropylideneuridine[1] (3)

A solution of 2′,3′-*O*-isopropylideneuridine (**1**) (Sec. III [132]) (1.80 g., 6.3 mmoles)[a] and methyltriphenoxyphosphonium iodide (**2**) (5.7 g., 12.6 mmoles) in anhydrous *N,N*-dimethylformamide (20 ml.)[b] is kept at room temperature for 3 hr., and then evaporated to dryness *in vacuo*. The residue is dissolved in chloroform (100 ml.), and the solution is successively washed (twice) with aqueous sodium thiosulfate and water, dried (sodium sulfate), and evaporated to dryness. The resulting oil is dissolved in hot chloroform (30 ml.) and, on slow addition of hexane, 2.13 g. (85%) of compound **3** separates as colorless crystals, m.p. 166–167° (lit.[2] 164°).

5′-Deoxy-5′-iodouridine[1] (4)

A solution of compound **3** (500 mg., 1.3 mmoles) in 80% acetic acid (10 ml.) is heated at 100° for 90 min., and evaporated to dryness *in vacuo*; traces of solvent are removed by addition and evaporation of methanol. The residue is crystallized from ethanol, giving 383 mg. (85%) of compound **4**, m.p. 183–185° (lit.[4] 182–183°).

[a] Commercially available from Aldrich Chemical Company, Milwaukee, Wisconsin.
[b] Dried by distillation, and stored over Linde Molecular Sieves, Type 4A.

5′-*O*-*p*-Nitrobenzoylthymidine[5] (6)

Recrystallized *p*-nitrobenzoyl chloride (1.05 g., 5.5 mmoles) is added in portions during 30 min. to a solution of thymidine (**5**) (1.21 g., 5 mmoles) in pyridine (10 ml.), and the yellow solution is kept at room temperature overnight. The mixture is then added to 400 ml. of well stirred ice–water, and the resulting precipitate is collected, washed with water, and dried. The crude product readily dissolves in chloroform (20 ml.), from which 0.94 g. (48%) of compound **6** immediately crystallizes; m.p. 180–181°, unchanged by recrystallization from ethyl acetate. A further quantity (7%) is obtainable by chromatography of the mother liquors on a column of silicic acid with acetone.

3′-Deoxy-3′-iodo-5′-*O*-*p*-nitrobenzoylthymidine (7)

A solution of 5′-*O*-*p*-nitrobenzoylthymidine (**6**) (225 mg., 0.58 mmole) and methyltriphenoxyphosphonium iodide (**2**, 500 mg., 1.1 mmole) in *N*,*N*-dimethylformamide is kept at room temperature for 24 hr., and then evaporated to dryness. The residue is dissolved in chloroform (25 ml.), and the solution is successively washed with aqueous sodium thiosulfate and water (twice), dried (sodium sulfate), and evaporated to dryness. The residue is chromatographed on a column of silicic acid (60 g.) by use of a gradient of ethyl acetate in chloroform (0 to 50%). The fractions containing the major, ultraviolet-absorbing product (R_f 0.5 on thin-layer chromatograms with 1:1 chloroform–ethyl acetate) are pooled, and evaporated to dryness, giving 240 mg. (82%) of chromatographically homogeneous **7**. Recrystallization from chloroform–hexane affords 200 mg. (70%) of pure **7**, m.p. 154–156°.

REFERENCES

(1) J. P. H. Verheyden and J. G. Moffatt, *J. Am. Chem. Soc.*, **86**, 2093 (1964).
(2) P. A. Levene and R. S. Tipson, *J. Biol. Chem.*, **106**, 113 (1934).
(3) S. R. Landauer and H. N. Rydon, *J. Chem. Soc.*, **1953**, 2224.
(4) D. M. Brown, A. R. Todd, and S. Varadarajan, *J. Chem. Soc.*, **1957**, 868.
(5) K. E. Pfitzner and J. G. Moffatt, *J. Am. Chem. Soc.*, **87**, 5611 (1965).

[127] 2′-Deoxy-5-(methylamino)uridine and 2′-Deoxy-5-(dimethylamino)uridine

Methylamination of a Pyrimidine Nucleoside

D. W. VISSER

DEPARTMENT OF BIOCHEMISTRY, UNIVERSITY OF SOUTHERN CALIFORNIA, LOS ANGELES, CALIFORNIA 90033

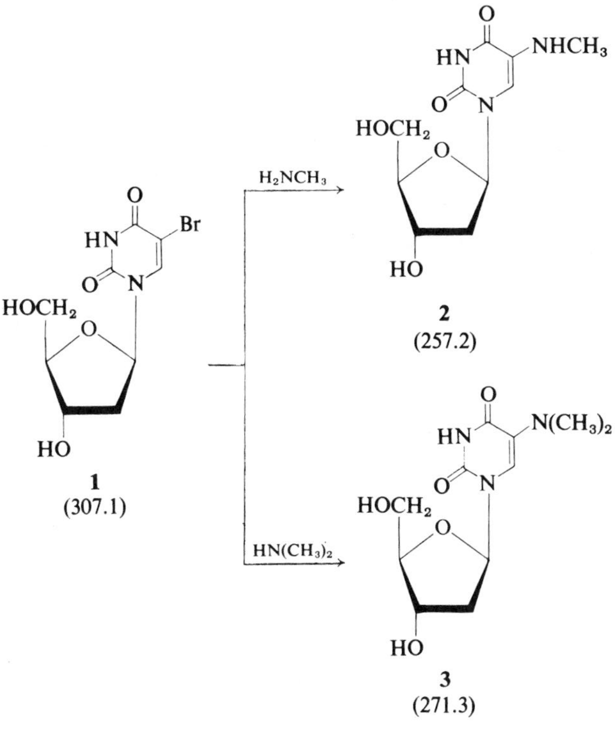

H_2NCH_3 → **2** (257.2)

1 (307.1)

$HN(CH_3)_2$ → **3** (271.3)

INTRODUCTION

2′-Deoxy-5-(methylamino)uridine[1] (**2**) enhances cell growth and division in a thymineless strain of *Escherichia coli* K_{12}. 2′-Deoxy-5-(dimethylamino)uridine (**3**) has no effect on cell growth or division under the same conditions.[1] In both tissue-culture and rabbit-eye assays, **2** produces a potent, highly specific suppression of *Herpes simplex* virus propagation, whereas **3** is ineffective.[2,3] Compound **2** has an extremely low cytotoxicity,[3] and is ineffective against several other 2′-deoxyribonucleic acid viruses.[3] It is cleaved by human-spleen thymidine phosphorylase *in vitro* at one quarter the rate for thymidine.[3] It is also a very weak inhibitor of incorporation of hypoxanthine and orotic acid into nucleic acids in an ascites-cell system, and has no effect on the synthesis of protein.[3]

PROCEDURE

2′-Deoxy-5-(methylamino)uridine[1] (2)

A mixture of thoroughly dry 5-bromo-2′-deoxyuridine[4,5] (**1**) (Sec. III [124]) (5 g., 16.3 mmoles) and 18 ml. of methylamine is heated in a stainless-steel container at 80° for 18 hr. After removal of the amine from the yellow-colored mixture, the pH of an aqueous solution of the product is adjusted to 2 with concentrated hydrochloric acid, and the solution is passed through a column (2 × 20 cm.) of Dowex-50 (H^+) ion-exchange resin, which is then washed with water until unreacted **1** is removed, as determined by its optical absorbance ($A_{280\ nm.}$). Compound **2** is eluted from the column with *N* ammonium hydroxide; the fractions containing **2**, as determined by their absorbance ($A_{280\ nm.}$), are evaporated to dryness at 40° under diminished pressure. The product crystallizes as colorless needles from aqueous ethanol; yield 1.82 g. (43%), m.p. 178.5–179°; lit.[3] $[\alpha]_D^{25}$ +48°; lit.[3] $\lambda_{max}^{pH\ 1}$ 204 nm. (ϵ 9,400), and 266 nm. (ϵ 9,300); lit.[3] $\lambda_{max}^{pH\ 12}$ 218 nm. (ϵ 18,600), and 293 nm. (ϵ 5,800); lit.[3] pK_1 2.55, and pK_2 9.30.

2′-Deoxy-5-(dimethylamino)uridine[1] (3)

The procedure for the preparation of **3** from 5 g. (16.3 mmoles) of **1** and dimethylamine is similar to that described for the preparation of **2**, yielding 2.4 g. (54%) of **3** as colorless needles, m.p. 188–190°; lit.[3] $\lambda_{max}^{pH\ 1}$ 267 nm. (ϵ 9,500); $\lambda_{max}^{pH\ 12}$ 220 nm. (ϵ 13,000), and 284 nm. (ϵ 6,200).

REFERENCES

(1) D. W. Visser, S. Kabat and M. Lieb, *Biochim. Biophys. Acta*, **76**, 463 (1963).
(2) M. M. Nemes and M. R. Hilleman, *Proc. Soc. Exptl. Biol. Med.*, **119**, 515 (1965).
(3) T. Y. Shen, J. F. McPherson, and B. O. Linn, *J. Med. Chem.*, **9**, 366 (1966).
(4) R. E. Beltz and D. W. Visser, *J. Am. Chem. Soc.*, **77**, 736 (1955).
(5) M. Friedland and D. W. Visser, *Biochim. Biophys. Acta*, **51**, 148 (1961).

[128] 2′,3′-Dideoxyuridine

Deoxygenation of a Hydroxyl Group of the Sugar Moiety of a Nucleoside

K. E. PFITZNER and J. G. MOFFATT

INSTITUTE OF MOLECULAR BIOLOGY, SYNTEX RESEARCH, PALO ALTO, CALIFORNIA 94304

$HOCH_2$ Ur, HO — **1** (228.2) $\xrightarrow{TrCl}$ TrOCH$_2$ Ur, HO — **2** (470.5) $\xrightarrow{MsCl}$ TrOCH$_2$ Ur, MsO — **3** (548.6) $\xrightarrow{NaI}$

TrOCH$_2$ Ur, I — **4** (580.4) $\xrightarrow{HOAc}$ HOCH$_2$ Ur, I — **5** (338.1) $\xrightarrow{H_2/Pd}$ HOCH$_2$ Ur — **6** (212.2)

where Ms is methylsulfonyl, Tr is triphenylmethyl, and Ur is [uracil-1-yl: HN, O, O, N]

INTRODUCTION

2′,3′-Dideoxyuridine (**6**), as its triphosphate, is a potential inhibitor of the biosynthesis of 2′-deoxyribonucleic acid. Its synthesis provides examples of the selective protection of primary hydroxyl groups by tritylation, displacement of a secondary methylsulfonyloxy group (with participation of the 2-oxo function of the uracil ring *via* the 2,2′-anhydronucleoside) by iodide, and hydrogenolysis of an iodo group. The procedure[1] is essentially similar to that used previously for the synthesis of 3′-deoxythymidine.[2]

PROCEDURE

2′-Deoxy-5′-*O*-trityluridine[1] (2)

2′-Deoxyuridine (**1**) (4.56 g., 20 mmoles) and chlorotriphenylmethane (6.67 g., 24 mmoles) are heated together in anhydrous pyridine (50 ml.) at 100° for 45 min. The solution is then cooled and gradually added to 1 l. of ice–water with vigorous stirring; after 30 min., the precipitated solid is collected by filtration and washed with water. The crude product is dried, and crystallized twice from acetone–benzene, giving 5.40 g. of **2**; m.p. 204–205°. A further 1.64 g. of pure product (total yield, 74%) is obtained by chromatography of the mother liquors on a column of Davison silicic acid (grade 923) with 19:1 chloroform–acetone.

2′-Deoxy-3′-*O*-(methylsulfonyl)-5′-*O*-trityluridine[1] (3)

Distilled methanesulfonyl chloride (5 ml., 64 mmoles) is added dropwise to an ice-cooled, stirred solution of **2** (5.47 g., 11.6 mmoles) in anhydrous pyridine (25 ml.). After the solution has been kept at 4° overnight, water (2 ml.) is added, and the solution is slowly poured into 2 l. of vigorously stirred ice–water. The resulting precipitate is collected, washed with water, dissolved in hot ethanol (100 ml.), and decolorized with 1 g. of carbon. The ethanolic suspension is then filtered, and added to 1 l. of ice–water, and the resulting, amorphous precipitate is collected and dried *in vacuo*. Compound **3** so obtained (6.3 g.) is homogeneous by thin-layer chromatography on silicic acid with 1:1 chloroform–ethyl acetate, but has not yet been obtained crystalline.

2′,3′-Dideoxy-3′-iodo-5′-*O*-trityluridine[1,a] (4)

Compound **3** (5.48 g., 10 mmoles) is dissolved in 1,2-dimethoxyethane[b] (100 ml.), sodium iodide (15 g., 0.1 mole) is added, and the solution is refluxed for 5 hr. The solvent is evaporated off *in vacuo*, the residue is dissolved in dichloromethane, and the solution is successively washed with 5% sodium thiosulfate and water, and dried with sodium sulfate. The solution is evaporated to dryness, and the residue is chromatographed on a column (300 g.) of Davison silicic acid with 3:1 dichloromethane–ether, giving chromatographically homogeneous **4**, which is crystallized from methanol; yield 2.65 g. (46%), m.p. 138–140°. Continued elution gives a further 2.26 g. of unreacted **3**, which is retreated as above with sodium iodide in 1,2-dimethoxyethane, to give a

[a] An alternative method[3] for the direct iodination of the glycosyl residue in pyrimidine nucleosides is described elsewhere in this Volume (See J. P. H. Verheyden and J. G. Moffatt, Sec. III [126].)

[b] Purified by distillation from lithium aluminum hydride.

further 1.0 g. (total yield, 64%) of **4**. An analytical sample has m.p. 142–143°.

2′,3′-Dideoxy-3′-iodouridine[1] (5)

A solution of compound **4** (2.08 g., 3.58 mmoles) in 80% acetic acid (100 ml.) is heated at 100° for 45 min. The solvent is then evaporated *in vacuo*, and residual acetic acid is removed by repeated addition and evaporation of methanol. The residue is crystallized from 1:1 dichloromethane–ethyl acetate, giving compound **5** (0.58 g., 45%), m.p. 161–162° (dec.). Chromatography of the mother liquors on a column of 200 g. of silicic acid with ethyl acetate gives a further 0.30 g. of **5**. The products are combined, and recrystallized from acetone, to give 0.79 g. (62%) of analytically pure **5**.

2′,3′-Dideoxyuridine[1] (6)

Compound **5** (0.73 g., 2.15 mmoles) is dissolved in 50% aqueous methanol (100 ml.) containing triethylamine (0.7 ml., 5 mmoles) and 10% palladium-on-carbon catalyst (200 mg.). The mixture is vigorously stirred in an atmosphere of hydrogen at room temperature. Hydrogen uptake is complete in 15 min., and, after filtration of the catalyst, the filtrate is evaporated to dryness *in vacuo*. The residue is dissolved in aqueous methanol, and the solution is passed through a column (10 ml.) of Dowex-2 (HCO_3^-) ion-exchange resin. The resin is washed free from ultraviolet-absorbing products with water, and the eluates are combined and evaporated to dryness. Crystallization of the residue from acetone–hexane gives 375 mg. (83%) of **6**, m.p. 115°. This product is contaminated by 4% of uracil, which is difficult to remove by recrystallization. A completely pure sample is obtained by preparative paper chromatography with 7:1:2 isopropyl alcohol–concentrated ammonium hydroxide–water; after crystallization from acetone–ether, m.p. 116–117°.

REFERENCES

(1) K. E. Pfitzner and J. G. Moffatt, *J. Org. Chem.*, **29**, 1508 (1964).
(2) A. M. Michelson and A. R. Todd, *J. Chem. Soc.*, **1955**, 816.
(3) J. P. H. Verheyden and J. G. Moffatt, *J. Am. Chem. Soc.*, **86**, 2093 (1964).

[129] 2′,3′-*O*-(Ethoxymethylene)uridine

2′,3′-O-(Ethoxymethylene) Derivatives of Ribonucleosides as Specifically Substituted Intermediates in the Preparation of Ribo-oligonucleotides Possessing the Naturally Occurring 3′ → 5′ Internucleotide Linkage

J. ŽEMLIČKA

INSTITUTE OF ORGANIC CHEMISTRY AND BIOCHEMISTRY,
CZECHOSLOVAK ACADEMY OF SCIENCES, PRAGUE, CZECHOSLOVAKIA

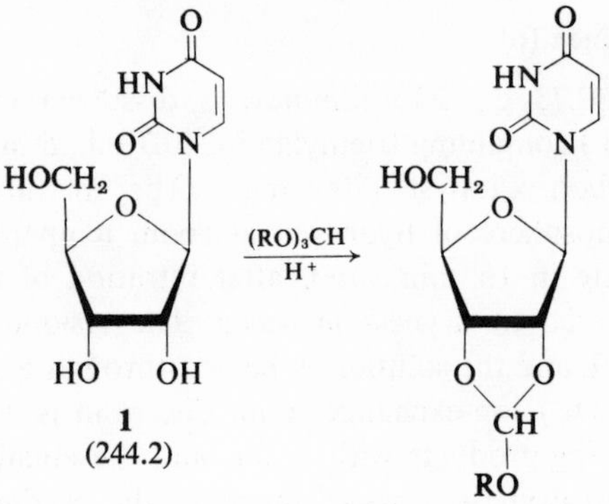

1
(244.2)

2
2a, R = Et
(300.3)

INTRODUCTION

Ribonucleoside 2′,3′-*O*-(alkoxymethylene) derivatives[1–5] (**2**) are important intermediates in the preparation of ribo-oligonucleotides possessing the naturally occurring 3′ → 5′ internucleotide linkage,[4–10] as well as of some other nucleoside derivatives.[11] The intermediates **2** may readily be prepared by reaction of the corresponding ribonucleosides with trimethyl[2] or triethyl orthoformate[1,3–5] in the presence of an acid catalyst (anhydrous hydrogen chloride,[1,4,5] *p*-toluenesulfonic acid,[2] or trichloroacetic acid[3]), usually in *N*,*N*-dimethylformamide as the solvent. With hydrogen chloride as the catalyst, the 2′,3′-*O*-(ethoxymethylene) derivatives of uridine[1,12] (Sec. IV [155]), inosine,[8] adenosine,[4] cytidine,[5] guanosine,[5] 5,6-dihydrouridine,[9] and 6-azacytidine[10] may be prepared. Ribonucleosides containing amino groups attached to the heterocyclic moiety require more than 1 equivalent of anhydrous hydrogen chloride in the reaction with the orthoformate.[4,5]

PROCEDURE

2′,3′-*O*-(Ethoxymethylene)uridine (2a)

A solution of 2.44 g. (0.01 mole) of uridine (**1**) in 30 ml. of *N*,*N*-dimethylformamide[a] is placed in a 100-ml., round-bottomed flask and treated successively with 5.6 ml. (0.04 mole) of ethyl orthoformate and 0.7 ml. (3.3 mmoles) of a 4.7 *M* solution of anhydrous hydrogen chloride in *N*,*N*-dimethylformamide.[b] The mixture is kept at room temperature for 20 hr.,[c] and then shaken with 4.2 g. (0.05 mole) of sodium hydrogen carbonate at room temperature for 30 min. The solid is filtered off, and washed with 10 ml. of *N*,*N*-dimethylformamide, and the filtrates are combined, and evaporated below 1 mm. Hg (bath temperature, 35–40°).[d] The residual sirup is co-evaporated under analogous conditions with 100 ml. of *p*-dioxane.[e,f] A solution of the residue in 50 ml. of *p*-dioxane is then filtered through a thin layer of Hyflo Super Cel to remove the small turbidity. The clear filtrate is mixed with 50 ml. of water,[g] and the mixture is kept at room temperature for 2 hr., and evaporated at 20° (bath temperature) below 1 mm. Hg.[d] The residue is co-evaporated with 100 ml. of *p*-dioxane,[f] to afford a quantitative yield of sirupy **2a**, which may be used directly as the nucleoside component in the synthesis of ribo-oligonucleotides.

Analytically pure 2′,3′-*O*-(ethoxymethylene)uridine (**2a**) is prepared as follows. The crude **2a** is dissolved in 10 ml. of chloroform, and the resulting solution is added, with stirring, to 200 ml. of ether. A further

[a] Commercial *N*,*N*-dimethylformamide is dried by vacuum distillation from 10% (by weight) of phosphorus pentaoxide, and is stored over molecular sieves.

[b] The reagent is prepared by passing dry hydrogen chloride into *N*,*N*-dimethylformamide cooled in ice. When a precipitate (probably *N*,*N*-dimethylformamide hydrochloride) separates during the saturation, it is dissolved by the addition of a further portion of *N*,*N*-dimethylformamide.

[c] The course of the reaction is checked by descending chromatography on Whatman No. 1 paper in 7:1:2 (v/v) isopropyl alcohol–concentrated ammonium hydroxide–water. The reaction is complete when no uridine (R_f 0.47) is present in the sample of the reaction mixture. Prior to chromatography, the sample of the reaction mixture is made alkaline with triethylamine.

[d] The receiver is cooled in Dry Ice–ethanol. The top of the condenser is fitted with a dropping funnel, to permit introduction of solvents into the evacuated flask.

[e] Commercial *p*-dioxane is dried by distillation from sodium.

[f] *p*-Dioxane is gradually added in portions of about 20 ml.

[g] The addition of water is necessary, in order to convert the by-product (R_f 0.84 in 7:1:2 (v/v) isopropyl alcohol–concentrated ammonium hydroxide–water) into 2′,3′-*O*-(ethoxymethylene)uridine (**2a**). The by-product is probably formed by reaction of the orthoformate with the free 5′-hydroxyl group of **2a**.

crop is obtained by the addition of an equal volume of light petroleum.[h] The product is collected[i] with suction, washed with 50 ml. of light petroleum, and dried under diminished pressure; yield, 2.3 g. (77%) of **2a**, as an amorphous, hygroscopic, white powder; R_f 0.72 in 7:1:2 (v/v) isopropyl alcohol–concentrated ammonium hydroxide–water.

REFERENCES

(1) J. Žemlička, *Chem. Ind.* (London), **1964**, 581.
(2) M. Jarman and C. B. Reese, *Chem. Ind.* (London), **1964**, 1493.
(3) F. Eckstein and F. Cramer, *Chem. Ber.*, **98**, 995 (1965).
(4) S. Chládek, J. Žemlička, and F. Šorm, *Collection Czech. Chem. Commun.*, **31**, 1785 (1966).
(5) J. Žemlička, S. Chládek, A. Holý, and J. Smrt, *Collection Czech. Chem. Commun.*, **31**, 3198 (1966).
(6) J. Smrt, *Collection Czech. Chem. Commun.*, **29**, 2049 (1964).
(7) B. E. Griffin, C. B. Reese, G. F. Stephenson, and D. R. Trentham, *Tetrahedron Letters*, **1966**, 4349.
(8) A. Holý and K. H. Scheit, *Chem. Ber.*, **99**, 3778 (1966).
(9) J. Smrt, *Collection Czech. Chem. Commun.*, **32**, 198 (1967).
(10) A. Holý, J. Smrt, and F. Šorm, *Collection Czech. Chem. Commun.*, **32**, 2980 (1967).
(11) N. J. Leonard and R. A. Laursen, *Biochemistry*, **4**, 365 (1965); A. Holý, N. C. Spassovska, and J. Smrt, *Collection Czech. Chem. Commun.*, **29**, 2567 (1964); A. Holý, J. Smrt, and F. Šorm, *ibid.*, **30**, 1635 (1965); A. Holý, *ibid.*, **30**, 1635 (1965).
(12) S. Chládek and J. Žemlička, *Collection Czech. Chem. Commun.*, **33**, 232 (1968).

[h] The fraction boiling at 40–60° is used. Light petroleum is added as long as any **2a** is precipitated.

[i] Filtration of **2a** must be performed with exclusion of atmospheric moisture. From the results of elementary analysis, the product thus obtained is pure. However, as shown by paper chromatography in 7:1:2 (v/v) isopropyl alcohol–concentrated ammonium hydroxide–water, traces of a by-product (R_f 0.52) are present.

[130] 5-Fluorouridine*

The Mercuri Method of Synthesis of Pyrimidine Nucleosides

IRIS WEMPEN and JACK J. FOX

DIVISION OF BIOLOGICAL CHEMISTRY, SLOAN-KETTERING INSTITUTE FOR CANCER RESEARCH, SLOAN-KETTERING DIVISION OF CORNELL UNIVERSITY MEDICAL COLLEGE, NEW YORK, NEW YORK 10021

5-Fluorouracil-Hg 1 (328.6)

2 (481.0) → 3 (574.5) → 4 (262.2)

where Bz is benzoyl.

INTRODUCTION

The preparation of 5-fluorouridine[1,2] (**4**) is an illustration of the use of the mercuri procedure[3] for the synthesis of a pyrimidine nucleoside that is still commercially unavailable. The nucleoside serves (Section III [115] and [96]) as a chemical precursor for the synthesis of the biochemically interesting 1-β-D-arabinofuranosyl derivatives of 5-fluorouracil and 5-fluorocytosine.

PROCEDURE

5-Fluorouracil-mercury[1,4] (1)

A mixture of mercuric acetate (31.9 g., 0.1 mole) and 600 ml. of methanol is stirred and boiled under reflux until complete dissolution occurs, and a hot solution of 13.0 g. (0.1 mole) of 5-fluorouracil[5] in 250 ml. of water is then added; immediate precipitation of 5-fluorouracil-mercury (**1**) occurs. Stirring is continued while the suspension is allowed to cool to room temperature, and the product is removed by filtration, and dried in a desiccator; yield 33 g. (100%), no m.p. below 360°.

* The authors acknowledge the support of the United States Public Health Service (Grant No. CA-08748).

5-Fluoro-1-(2,3,5-tri-*O*-benzoyl-β-D-ribosyl)uracil[1] (3)

The sirupy halide (**2**) is prepared from 10.1 g. (0.02 mole) of[a] 1-*O*-acetyl-2,3,5-tri-*O*-benzoyl-D-ribose[6] by the usual procedure,[7] and is dissolved in dry benzene. The solution is added to a well stirred, refluxing suspension of thoroughly dried 5-fluorouracil-mercury (**1**) (3.3 g., 0.01 mole) in 350 ml. of dry toluene (distilled from sodium), giving a homogeneous, slightly yellow solution. After the reaction mixture has been refluxed for 30 min., it is filtered, and the filtrate is evaporated under diminished pressure, giving a glass which is dissolved in 100 ml. of ethyl acetate. The solution is successively washed with two 50-ml. portions of aqueous potassium iodide (30 g./70 ml. of water) and water, dried (anhydrous sodium sulfate), and filtered. The filtrate is concentrated *in vacuo*, giving a glass which is treated with 25 ml. of warm chloroform, whereupon crystallization occurs. The suspension is cooled and filtered, and the crystals are washed with a small amount of cold chloroform and then with ether. A second crop is obtained by reprocessing the mother liquor. The total yield of **3** is 4.3 g. (74%), m.p. 207–209°.

5-Fluorouridine[1,2] (4)

A suspension of the tribenzoate (**3**) (4.5 g., 0.74 mole) in 100 ml. of 1:1 ethanol–water is stirred magnetically, and treated with 20 ml. of *N* sodium hydroxide. After 2 hr., the light-yellow solution should have a pH of 10–12. The solution is rendered neutral with glacial acetic acid, and the ethanol is removed with a rotary evaporator. The resulting, aqueous solution is treated batchwise with Dowex-50 (H^+) ion-exchange resin. The benzoic acid precipitated together with the resin is removed by filtration, and the filtrate is treated with fresh resin. This process is repeated until the aqueous filtrate gives a negative flame-test for sodium. The filtrate is washed three times with chloroform (to remove residual benzoic acid) and evaporated *in vacuo* to a semi-solid mass which is freed of traces of water by repeated addition and evaporation of toluene. The solid is then dissolved in the minimal volume of hot ethanol, and the solution is allowed to cool slowly, giving crystals of **4** (1.7 g., 88%), m.p. 180–182°. The product is of sufficient purity to be used directly in further syntheses. Ultraviolet absorption data: in *N* hydrochloric acid, λ_{max} at 269 nm. (ϵ_{max} 8,950); λ_{min} at 234 nm. (ϵ_{min} 1,680); pK_a 7.57 (determined spectrophotometrically).

REFERENCES

(1) N. C. Yung, J. H. Burchenal, R. Fecher, R. Duschinsky, and J. J. Fox, *J. Am. Chem. Soc.*, **83**, 4060 (1961).

[a] This sugar derivative is available from Calbiochem., Box 54282, Los Angeles, California 90054.

(2) R. Duschinsky, E. Pleven, E. Malbica, and C. Heidelberger, *Abstracts Papers Am. Chem. Soc. Meeting*, **132**, 19c (1957).
(3) For a discussion of this procedure, see J. J. Fox and I. Wempen, *Advan. Carbohydrate Chem.*, **14**, 283 (1959).
(4) M. Hoffer, R. Duschinsky, J. J. Fox, and N. Yung, *J. Am. Chem. Soc.*, **81**, 4112 (1959).
(5) R. Duschinsky, E. Pleven, and C. Heidelberger, *J. Am. Chem. Soc.*, **79**, 4559 (1957).
(6) (a) R. K. Ness, H. W. Diehl, and H. G. Fletcher, Jr., *J. Am. Chem. Soc.*, **76**, 763 (1954); (b) H. M. Kissman, C. Pidacks, and B. R. Baker, *ibid.*, **77**, 18 (1955).
(7) For a detailed description of the synthesis of halide **2**, see I. Wempen and J. J. Fox, *Methods Enzymol., Nucleic Acids*, **12A**, 59 (1967); N. C. Yung and J. J. Fox, *Methods Carbohydrate Chem.*, **2**, 108 (1963). See also, Sec. III [81].

[131] 5-Hydroxyuridine

Substitution Reactions of the Pyrimidine Ring of a Nucleoside

D. W. VISSER

DEPARTMENT OF BIOCHEMISTRY, UNIVERSITY OF SOUTHERN CALIFORNIA,
LOS ANGELES, CALIFORNIA 90033

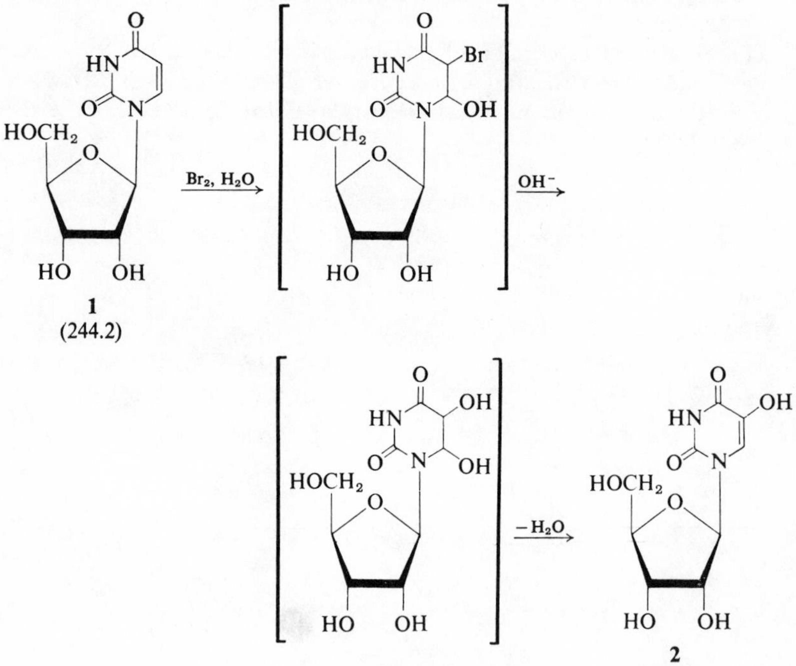

INTRODUCTION

5-Hydroxyuridine (**2**) undergoes most of the reactions of uridine, including conversion into 5-hydroxyuridine 5′-(glycosyl pyrophosphate), and minor incorporation into ribonucleic acid.[1] Compound **2** produces various effects on metabolic processes.[2,3] The nucleoside analog inhibits the growth of molds,[4] bacteria,[5] tumors,[3,6] and the propagation of viruses.[7] The 5-hydroxyuracil moiety may be classified as a "minor base" constituent of ribonucleic acid, inasmuch as **2** has recently been isolated from yeast ribonucleic acid and from commercial uridine.[8]

The preparation of **2** has been reported[4,9,10]; the procedure described here is a modification of Ueda's method.[10] The modification involves use of a lower reaction temperature, resulting in less decomposition and higher yields, and provides the mild conditions necessary for the preparation of the nucleoside analog pyrophosphate and triphosphate from the corresponding phosphates of **1** (see Section IV [151]).

PROCEDURE

5-Hydroxyuridine (2)

Bromine and bromine-water are added, with stirring, to 5.0 g. (16.4 mmoles) of **1** in 30 ml. of water until a light-yellow color persists. Excess bromine is removed by aeration, and the volume is adjusted to 55 ml. Freshly distilled pyridine (6 ml.) is added (pH 5.7), and the mixture is kept for 20 hr. at 37°. The solvents are removed from the yellow solution at 35° under diminished pressure. Removal of the residual pyridine and water (which inhibit crystallization of the product) is effected by the addition and distillation of absolute ethanol. Hot ethanol is now added, and **2** crystallizes on cooling. One recrystallization from aqueous ethanol yields colorless crystals of **2**; yield 3.63 g. (60.5%), m.p. 238–240° (rapid heating); $\lambda_{max}^{pH\,1}$ 281 nm. (ϵ 8,300), $\lambda_{max}^{pH\,12}$ 305 nm.; $pK_2 = 7.8$; A_{280}/A_{260} 1.8 at pH 1.0 and 0.97 at pH 12. Compound **2** gives a blue color with ferric chloride,[11] and is unstable in alkali.

REFERENCES

(1) D. A. Smith and D. W. Visser, *J. Biol. Chem.*, **240**, 446 (1965).

(2) R. Ben-Ishai and B. E. Volcani, *Biochim. Biophys. Acta*, **21**, 265 (1956); S. Spiegelman, in O. H. Gaebler (Ed.), *Enzymes: Units of Biological Structure and Function*, Academic Press, New York, N. Y., 1954, p. 67; S. Spiegelman, in D. W. McElroy and B. Glass (Eds.), *A Symposium on the Chemical Basis of Heredity*, Johns Hopkins Press, Baltimore, Maryland, 1957, p. 232; A. I. Aronson and S. Spiegelman, *Biochim. Biophys. Acta*, **53**, 84 (1961); C. O. Doudney and F. L. Haas, *Proc. Natl. Acad. Sci. U. S.*, **44**, 390 (1958); **45**, 709 (1959); *Nature*, **184**, 114 (1959).

(3) D. W. Visser, in C. P. Rhoads (Ed.), *Antimetabolites and Cancer*, American Association for the Advancement of Science, Washington, D.C., 1955, p. 47.

(4) M. Roberts and D. W. Visser, *J. Biol. Chem.*, **194**, 695 (1952).

(5) I. J. Slotnick, D. W. Visser, and S. C. Rittenberg, *J. Biol. Chem.*, **203**, 647 (1953).

(6) K. Sugiura and H. J. Creech, *Ann. N. Y. Acad. Sci.*, **63**, 962 (1955).

(7) D. W. Visser, D. L. Lagerborg, and H. E. Pearson, *Proc. Soc. Exptl. Biol. Med.*, **79**, 571 (1952); H. E. Pearson, D. L. Lagerborg, and D. W. Visser, *ibid.*, **93**, 61 (1956); A. Shug, H. R. Mahler, and D. Fraser, *Biochim. Biophys. Acta*, **42**, 255 (1960).
(8) A. W. Lis and W. E. Passarge, *Arch. Biochem. Biophys.*, **114**, 593 (1966).
(9) P. A. Levene and F. B. LaForge, *Ber.*, **45**, 616 (1912).
(10) T. Ueda, *Chem. Pharm. Bull.* (Tokyo), **8**, 455 (1960).
(11) D. Davidson and O. Baudisch, *J. Biol. Chem.*, **64**, 619 (1925).

[132] 2′,3′-*O*-Isopropylideneuridine

An Intermediate in the Preparation of 5′-Substituted Uridines

R. STUART TIPSON

KENSINGTON, MARYLAND 20795

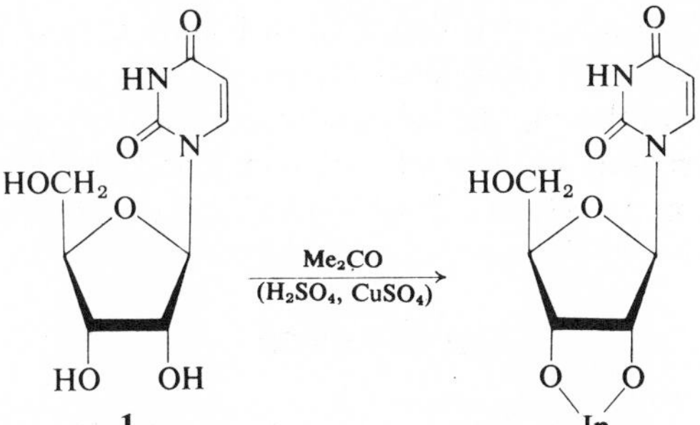

$$\mathbf{1}\ (244.2) \xrightarrow[(H_2SO_4,\ CuSO_4)]{Me_2CO} \mathbf{2}\ (284.3)$$

INTRODUCTION

2′,3′-*O*-Isopropylideneuridine[1] (**2**) is a useful intermediate in the preparation[1] of uridine 5′-phosphate (5′-uridylic acid) and other 5′-substituted uridines. The method given here is extremely simple, gives a practically quantitative yield, and, except for uridine (**1**), involves the use of inexpensive reagents.

PROCEDURE

2′,3′-*O*-Isopropylideneuridine[1] (2)

Dry, finely powdered uridine (**1**) (10 g., 0.041 mole) is suspended in 250 ml. of acetone (analytical grade) in a 500-ml., Pyrex bottle, and then 20 g. of anhydrous copper(II) sulfate and 0.25 ml. of concentrated sulfuric acid are added. The bottle is tightly corked, and the suspension is shaken at room temperature or in a warm-room at 37° for at least 48 hr. The mixture is now filtered, with suction, through hard filter-paper, and the bottle, the cork, and the copper sulfate are well washed with five 25-ml. portions of pure acetone.

The clear filtrate and washings are combined, and transferred to a 500-ml., Pyrex bottle containing 10 g. of dry, calcium hydroxide powder, the bottle is tightly corked, and the suspension is shaken at room temperature for 1 hr., to neutralize the sulfuric acid. The mixture is filtered, with suction, through hard filter-paper precoated with a layer of 2 g. of dry, calcium hydroxide powder,[a] and the bottle, the cork, and the insoluble material are well washed with five 25-ml. portions of pure acetone. The filtrate and washings are then combined, and evaporated to dryness under diminished pressure. The product invariably crystallizes out during the evaporation, giving a colorless, crystalline mass; yield of **2**, 11.64 g. (99.9%). It is readily recrystallized from absolute methanol or in the following way. The product is dissolved in boiling acetone (1 g./12 ml.), the solution is cooled to room temperature, and pentane is added to faint opalescence. On nucleating the solution with a trace of compound **2**, and keeping it in a refrigerator overnight, the compound crystallizes in rosettes of fine needles, m.p. 159–160°, $[\alpha]_D^{27}$ −15.8° (*c* 1.01, absolute methanol).

REFERENCE

(1) P. A. Levene and R. S. Tipson, *J. Biol. Chem.*, **106**, 113 (1934).

[a] The calcium hydroxide is placed in a test tube, 10 ml. of acetone is added, and the suspension is vigorously shaken and poured over the filter paper (which is under suction).

[133] 2′,3′-*O*-Isopropylidene-3-methyluridine

Methylation of Uracil Derivatives with Diazomethane

W. SZER and D. SHUGAR

INSTITUTE OF BIOCHEMISTRY AND BIOPHYSICS, POLISH ACADEMY OF SCIENCES, WARSZAWA, POLAND

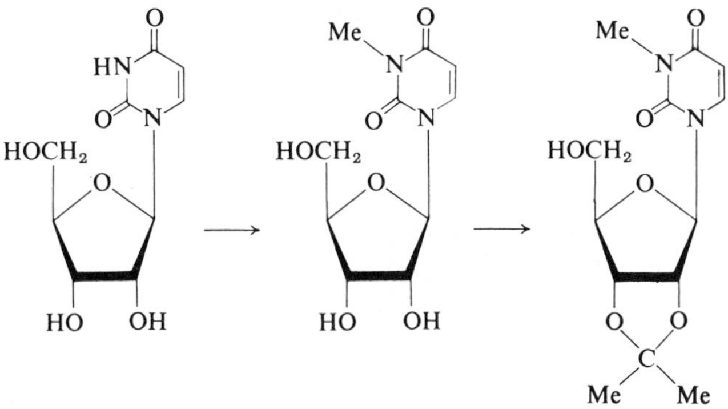

1	2	3
(244.2)	(258.2)	(218.3)

INTRODUCTION

3-Methyluridine (**2**) has been isolated from soluble ribonucleic acid (sRNA), but whether it is a normal constituent, or the product of deamination of 3-methylcytidine,[1] has not yet been established. The 2′:3′-cyclic phosphate and the polynucleotides of **2** are completely resistant to pancreatic ribonuclease.[2] Polynucleotides of **2** do not exhibit secondary structure or form complexes with poly(adenylic acid).[3] The synthesis of **2** is typical of the action of diazomethane on 1-substituted uracil derivatives.[4] *N*-Methylation of a uridine derivative with diazomethane was introduced[5] in 1934. The introduction of the acid-labile, 2′,3′-*O*-isopropylidene group[6] is a typical procedure for protection of 2′,3′-*cis*-glycols in ribonucleosides; however this procedure is not applicable to cytidine and its derivatives. In such instances, the method of Chambers *et al.*[7] may be applied. The use of a benzylidene protecting group is equally advantageous.[8]

It is important to note that 3-methyluridine, like other 1,3-disubstituted derivatives of uracil, is unstable in alkaline media.[3] The procedure described here is also applicable to the preparation of 3,5-dimethyluridine.[9]

PROCEDURE

3-Methyluridine (2)

To a solution of 2 g. (8.2 mmoles) of uridine (**1**) in 50 ml. of methanol is slowly added an ethereal solution of diazomethane prepared from 10 g. of 1-methyl-3-nitro-1-nitrosoguanidine.[a] Decolorization of the solution and evolution of nitrogen are at first rapid. An excess of the reagent is added, until the yellow color persists, and the solution is kept at room temperature. The course of the reaction is followed by taking an aliquot of the reaction mixture and measuring its optical absorbance at 262 nm. in aqueous medium at pH 7 and pH 12. After about 30 min., the optical absorbance is the same at both pH values, indicating completion of the reaction. Evaporation of solvent under diminished pressure gives a sirup. Paper chromatography (Whatman No. 1 paper, ascending) with 7:1:2 (v/v) isopropyl alcohol–concentrated ammonia (*d*, 0.88)–water (solvent A) should show one major product (**2**) having R_f 0.78, and a minor component, R_f 0.83, presumably the 2′-methyl ether[10] of **2**. The products are chromatographed on 100 g. of potato starch with water-saturated butyl alcohol, 5-ml. fractions being analyzed for absorption at 262 nm. A minor peak, presumably the 2′-methyl ether of **2**, appears at about 70 ml. The main peak, containing **2**, is eluted in the range of 140–240 ml.; the fractions are combined, and evaporated, to give a sirup that, on being kept overnight, crystallizes; yield of **2**, 1.5 g. (71%). After recrystallization from hot methanol, and from ethyl acetate, m.p. 119–120°, $[\alpha]_D^{26}$ −20.1° (water); at pH 1–12, λ_{max} 262 nm. (ϵ_{max} 9.11 × 10^3).

2′,3′-*O*-Isopropylidene-3-methyluridine (3)

Dry, finely powdered **2** (1.05 g., 4 mmoles) and 2 g. of anhydrous cupric sulfate are suspended in 25 ml. of dry acetone. Concentrated sulfuric acid (0.025 ml.) is added, and the suspension is shaken at 37° for 48 hr. The mixture is filtered, and the residue is washed with 10 ml. of acetone. The filtrate and washings are combined, and shaken with dry calcium hydroxide for 1 hr. The mixture is filtered, the insoluble matter is washed with 10 ml. of acetone, and the filtrate and washings are combined, and evaporated to dryness under diminished pressure. The residue is crystallized from 2 ml. of ethanol plus 0.5 ml. of ether, the solution being kept overnight at 0°, to give 0.88 g. of **3**; a further 0.19 g. is obtained from the mother liquor; total yield, 1.07 g. (90%). Recrystallization from the

[a] 1-Methyl-3-nitro-1-nitrosoguanidine (10 g.) is covered with 100 ml. of ether, the mixture is chilled, and 40 ml. of 40% aqueous potassium hydroxide is slowly added. On completion of the reaction, the ethereal solution of diazomethane is decanted.

same solvent mixture gives a product having m.p. 182–183°, and R_f 0.86 (solvent A).[b]

REFERENCES

(1) (a) R. H. Hall, *Biochem. Biophys. Res. Commun.*, **12**, 361 (1963); (b) R. H. Hall, *Biochemistry*, **4**, 661 (1965).
(2) W. Szer and D. Shugar, *Acta Biochim. Polon.*, **7**, 491 (1960).
(3) W. Szer and D. Shugar, *Acta Biochim. Polon.*, **8**, 235 (1960).
(4) H. T. Miles, *Biochim. Biophys. Acta*, **22**, 247 (1956).
(5) P. A. Levene and R. S. Tipson, *J. Biol. Chem.*, **104**, 385 (1934).
(6) P. A. Levene and R. S. Tipson, *J. Biol. Chem.*, **106**, 113 (1934).
(7) J. M. Gulland and J. Smith, *J. Chem. Soc.*, **1948**, 1527.
(8) R. W. Chambers, P. Shapiro, and V. Kurkov, *J. Am. Chem. Soc.*, **82**, 970 (1960).
(9) W. Szer, M. Swierkowski, and D. Shugar, *Acta Biochim. Polon.*, **10**, 87 (1963).
(10) W. Szer and D. Shugar, *Biokhimiya*, **26**, 840 (1961).

[b] If isolation of **2** is not envisaged (for example, when the protected nucleoside (**3**) is intended to be subsequently phosphorylated), the above reaction sequence may be reversed, in order to avoid column chromatography in the synthesis of **2**. 2′,3′-*O*-Isopropylideneuridine[6] (m.p. 161–162°) (Sec. III [132]) is first prepared, and the product is methylated as described for **2**. The final product (**3**) is then isolated simply by crystallization.

[134] 5-Methyluridine

Mannich Reaction Applied to Uridine

E. I. BUDOWSKY, V. N. SHIBAEV, and G. I. ELISEEVA

INSTITUTE FOR CHEMISTRY OF NATURAL PRODUCTS, ACADEMY OF SCIENCES OF U.S.S.R., MOSCOW, U.S.S.R.

$\xrightarrow{CH_2O,\ Et_2NH}$ (2, Et_2NCH_2) $\xrightarrow{H_2/Pt}$ (3, Me)

1 (244.2) **2** (300.3) **3** (258.2)

where R is [ribofuranosyl: $HOCH_2$, O, HO, OH]

INTRODUCTION

5-Methyluridine (1-β-D-ribofuranosylthymine, **3**) has been proved to be a minor component of ribonucleic acid.[1] Compound **3** has been used as the starting material for the synthesis of 5-methyluridine 5′-(α-D-glucopyranosyl pyrophosphate)[2] and poly(5-methyluridylic acid).[3] The field of application of these compounds appears to be in the study of the specificity of carbohydrate-metabolizing enzymes, and of secondary structure and biological properties of poly(ribonucleotides). Compound **3** has been obtained by ribosylation of thymine derivatives,[4–6] and from 5-(hydroxymethyl)uridine.[8] The latter may be prepared by the reaction of uridine with formaldehyde.[8,9] The procedure described here is based on an application of the Mannich reaction.[2]

PROCEDURE

5-(Diethylaminomethyl)uridine (2)

An aqueous solution of formaldehyde (2.4 ml. of 40%) is added, with shaking, to a mixture of uridine (**1**) (1.00 g., 4.1 mmoles), diethylamine (4 ml.), and water (14 ml.). The mixture is placed in a tube which is then sealed, and heated at 100° for 7 hr. The resulting solution is evaporated to dryness *in vacuo*, and the residue is dissolved in water (400 ml.); the pH of the solution is adjusted to 2 with formic acid, and the solution is passed through a column (2 × 30 cm.) of Dowex-50 X2 (H^+) (100–200 mesh). The column is washed with water (250 ml.) (to remove **1**) and then **2** is eluted with *N* ammonium hydroxide (200 ml.). The solution of **2** (A_{260} 13,600) is evaporated to dryness, and the residue is dried (by co-evaporation with benzene–absolute ethanol) and dissolved in methanol. After addition of acetone, 595 mg. (48%) of a yellow, hygroscopic precipitate of crude **2** is obtained. This preparation of **2** may be used without further purification for the next stage.

For purification, the precipitate is dissolved in water (100 ml.), the pH of the solution is adjusted to 2, and the solution is applied to a column (2 × 10 cm.) of Dowex-50 (NH_4^+) (100–200 mesh). The column is washed successively with water (250 ml.) and 0.05 *M* ammonium hydroxide (250 ml.). The latter eluate (which contains **2**) is evaporated to dryness. The residue is dried, dissolved in methanol, and precipitated with acetone as a white, hygroscopic powder, m.p. 174–175°; λ_{max} 267 nm., ϵ_{max} 9,000; λ_{min} 235 nm., ϵ_{min} 2,100 (pH 2); λ_{max} 267 nm., ϵ_{max} 6,400; λ_{min} 245 nm., ϵ_{min} 4,500 (pH 12).

5-Methyluridine (3)

A mixture of crude **2** (595 mg., 1.96 mmoles), water (14 ml.), and ethanol (56 ml.) is placed in a 100-ml., round-bottomed flask fitted with a reflux condenser and an inlet tube. Platinum oxide (190 mg.) is added, the mixture is heated to 70° with magnetic stirring, and hydrogen is passed in for several min. Then the top of the condenser is stoppered, and hydrogenation is conducted in the closed system at 70° for 12 hr. The mixture is cooled, and filtered, and the filtrate is evaporated to dryness *in vacuo*. The residue is dissolved in water, and the solution is acidified (pH 2) and passed through a column (2 × 10 cm.) of Dowex-50 (H^+) (100–200 mesh). The column is washed with water, and the effluent (A_{260} 7,400) is lyophilized, giving **3** as a white powder; yield 253 mg. (50%), m.p. 184–185° (after crystallization from methanol–acetone). The preparation is homogeneous by paper chromatography, R_f 0.35 (77:23 *sec*-butyl alcohol–water).

REFERENCES

(1) J. M. Littlefield and D. B. Dunn, *Biochem. J.*, **70**, 642 (1958).
(2) N. K. Kochetkov, E. I. Budowsky, V. N. Shibaev, and G. I. Eliseeva, *Izv. Akad. Nauk SSSR, Ser. Khim.*, **1966**, 1779.

(3) B. E. Griffin, A. R. Todd, and A. Rich, *Proc. Natl. Acad. Sci. U. S.*, **44,** 1123 (1958).
(4) J. J. Fox, N. Yung, J. Davoll, and G. B. Brown, *J. Am. Chem. Soc.*, **78,** 2117 (1956).
(5) J. Farkas, L. Kaplan, and J. J. Fox, *J. Org. Chem.*, **29**, 1469 (1964).
(6) T. Nishimura, B. Shimizu, and I. Iwai, *Chem. Pharm. Bull.* (Tokyo), **12,** 1471 (1964).
(7) M. Prystaš and F. Šorm, *Collection Czech. Chem. Commun.*, **31**, 1035 (1966).
(8) R. E. Cline, R. M. Fink, and K. Fink, *J. Am. Chem. Soc.*, **81**, 2521 (1959).
(9) B. R. Baker, T. J. Schwan, and D. V. Santi, *J. Med. Chem.*, **9**, 66 (1966).

[135] 2′,3′,5′-Tri-*O*-benzoyl-4-thiouridine

Selective Thiation of an Acylated Pyrimidine Nucleoside

PATRICIA E. GARRETT

DEPARTMENT OF CHEMISTRY, GEORGETOWN UNIVERSITY,
WASHINGTON, D.C. 20007

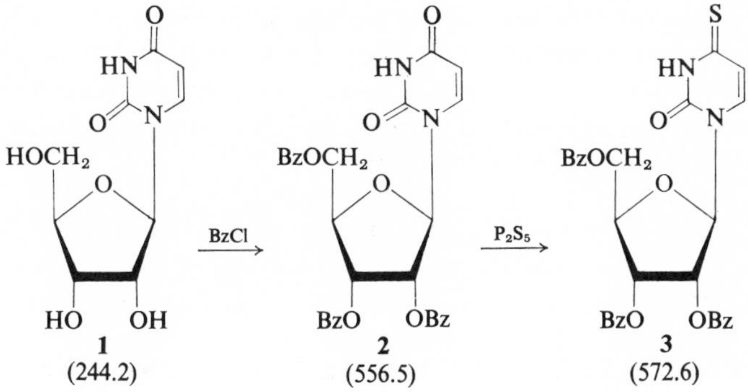

1 (244.2) $\xrightarrow{BzCl}$ 2 (556.5) $\xrightarrow{P_2S_5}$ 3 (572.6)

where Bz is benzoyl.

INTRODUCTION

The title compound (**3**) is an intermediate in the synthesis of 4-thiouridine 5′-phosphate[1] *via* 4-thiouridine (Sec. IV [154]). Its synthesis is an example of the selective thiation of C-4 of a pyrimidine nucleoside, accomplished with phosphorus pentasulfide.[2]

PROCEDURE

2′,3′,5′-Tri-*O*-benzoyluridine[2] (2)

To a solution of 80.5 g. (0.33 mole) of uridine (**1**) in 2.3 l. of anhydrous pyridine is added dropwise, with continuous stirring, 128 ml. (1.10 moles) of benzoyl chloride. The reaction mixture is heated at 55–60° for 48 hr., about 1.5 l. of pyridine is evaporated off under diminished pressure, and the mixture is filtered. The filtrate is slowly poured, with vigorous stirring, into 5 l. of ice–water, stirring is continued for 1 hr., and part of the oil that is initially formed is converted into a fluffy precipitate. The solid is filtered off, and the oil that remains in the vessel is re-treated with ice and water. The process is repeated until crystalline material no longer separates. The residual oil is dissolved in chloroform, and the solution is washed with water, dried (sodium sulfate), and evaporated to dryness, giving a residue that is combined with the precipitate and recrystallized from benzene; yield 167 g. (90%) of compound **2**, m.p. 142–143°.

2′,3′,5′-Tri-*O*-benzoyl-4-thiouridine[2] (3)

A mixture of 5.56 g. (0.01 mole) of compound **2**, 8.8 g. (0.05 mole) of phosphorus pentasulfide, and 150 ml. of pyridine (reagent grade) is refluxed with stirring for 5 hr. About one-third of the pyridine is evaporated off under diminished pressure, and the residual brown mixture is poured, with stirring, into water. The water is decanted from the resulting oil, which is then dissolved in chloroform; the solution is filtered, washed twice with water, dried (sodium sulfate), and evaporated to dryness under diminished pressure. The residue is crystallized from 150 ml. of ethanol to give 4.96 g. (87%) of yellow prisms, m.p. 128–130°.

REFERENCES

(1) N. K. Kochetkov, E. I. Budowsky, V. N. Shibaev, and M. A. Grachev, *Izv. Akad. Nauk SSSR, Ser. Khim.*, **1963**, 1592.

(2) J. J. Fox, S. Van Praag, I. Wempen, I. L. Doerr, L. Cheong, J. E. Knoll, M. L. Eidinoff, and A. Bendich, *J. Am. Chem. Soc.*, **81**, 178 (1959).

[136] 5′-*O*-Trityluridine

*An Intermediate in the Synthesis of 2′,3′-Di-*O*-substituted Uridines*

R. STUART TIPSON

KENSINGTON, MARYLAND 20795

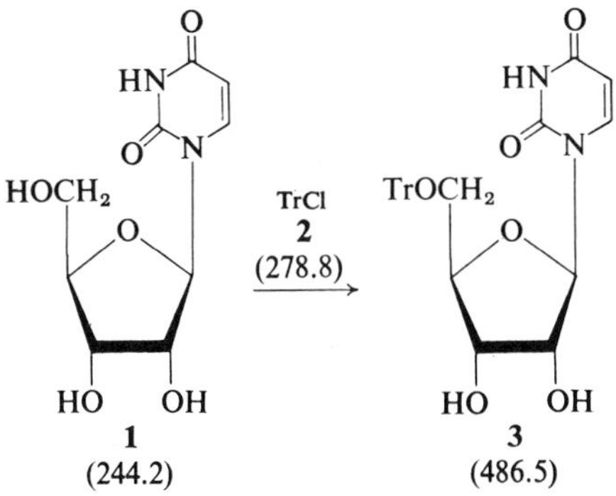

where Tr is triphenylmethyl.

INTRODUCTION

5′-*O*-Trityluridine[1] (**3**) is a key intermediate in the synthesis of 2′,3′-di-*O*-substituted derivatives of uridine, such as the 2′,3′-thionocarbonate (Sec. III [115]) and the 2′,3′-diacetate (Sec. III [118]).

PROCEDURE

5′-*O*-Trityluridine[1] (3)

A mixture of 10 g. (0.041 mole) of dry, finely powdered uridine (**1**) with 12.6 g. (0.045 mole) of pure,[a] dry chlorotriphenylmethane (**2**) is dissolved in 120 ml. of dry, redistilled pyridine. The solution is kept overnight at room temperature, with exclusion of atmospheric moisture, and is then heated, under an air condenser closed by a calcium chloride tube, in a glycerol bath at 100° during 3 hr. The solution is cooled, and poured into 1 liter of water, with vigorous stirring. The cloudy, supernatant liquor is decanted from the gummy precipitate and is filtered, and the filtrate is discarded. The gummy material is combined, well washed with water,

[a] Occasionally, commercial samples of chlorotriphenylmethane have been encountered that have undergone partial or complete hydrolysis to triphenylmethanol. To avoid use of such material, the m.p. and infrared spectrum should always be checked.

and then dissolved in acetone, and the solution is evaporated to dryness under diminished pressure. The mass is freed of pyridine by being twice heated with 200-ml. portions of boiling water, cooling the mixture, and discarding the water. The mass is now dissolved in 250 ml. of 1:1 acetone–absolute ethyl alcohol, and the solution is mixed with a little decolorizing carbon, and filtered. The filtrate is evaporated under diminished pressure to a hard mass, which is dried by dissolving it in absolute ethyl alcohol and re-evaporating to dryness. The yield of crude product is 19.0 g. (95%). This is dissolved[2] in 100 ml. of absolute ethanol, and the solution is cooled, giving colorless crystals of **3**; yield 14 g. (70%). Crude **3** is invariably contaminated with a little di-*O*-trityluridine,[3] since shown[4] to be a mixture of the 2′,5′- and 3′,5′-ditrityl ethers, which is readily soluble in cold benzene. Finely powdered, crude **3** is extracted with cold benzene (20 ml./g.), giving pure **3** (80% recovery); m.p. 200° (sharp; no foaming), $[\alpha]_D^{25}$ +8.8° (*c* 1.06, acetone), +18.8° (*c* 1.04, absolute methanol); λ_{max} 262 nm.,[2] λ_{min} 243 nm.; A_{230}/A_{260} 1.04.

REFERENCES

(1) P. A. Levene and R. S. Tipson, *J. Biol. Chem.*, **104**, 385 (1934).
(2) J. F. Codington, I. L. Doerr, and J. J. Fox, *J. Org. Chem.*, **29**, 558 (1964).
(3) P. A. Levene and R. S. Tipson, *J. Biol. Chem.*, **105**, 419 (1934).
(4) N. C. Yung and J. J. Fox, *J. Am. Chem. Soc.*, **83**, 3060 (1961).

Section IV

Nucleotides and Oligonucleotides

[137] 2′:3′-Cyclic Phosphate of Adenosine and of Uridine

Cyclization of a Nucleoside 2′(or 3′)-Morpholinophosphonic Ester Chloride to the Vicinal Hydroxyl Group by Alkaline Treatment

MORIO IKEHARA and ICHIRO TAZAWA

FACULTY OF PHARMACEUTICAL SCIENCES, HOKKAIDO UNIVERSITY, SAPPORO, JAPAN

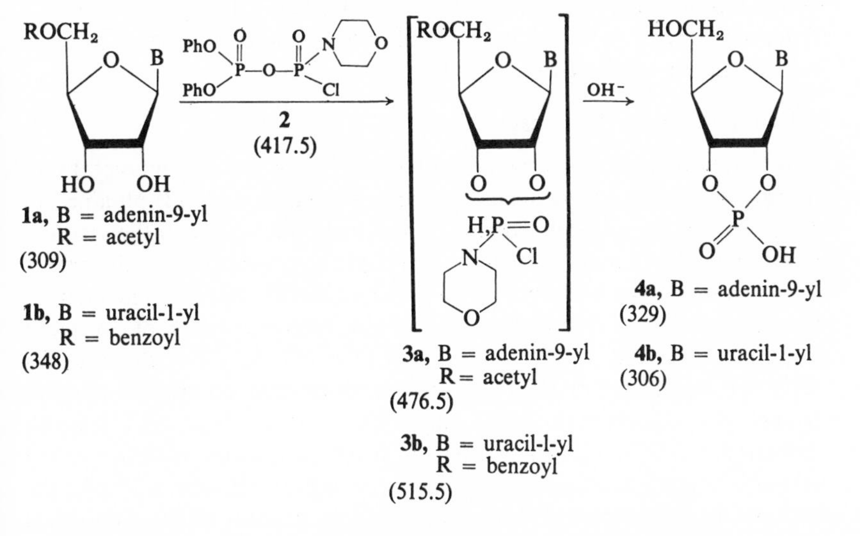

INTRODUCTION

The nucleoside 2′:3′-cyclic phosphates (**4**) are intermediates in the hydrolysis of ribonucleic acid with chemical agents and with the enzyme ribonuclease. Chemical syntheses have been reported, by several workers, of **4a** and **4b** from the corresponding nucleoside 2′(or 3′)-phosphate as the starting material.[1–3]

PROCEDURE

Morpholinophosphonic Dichloride[4]

Freshly distilled morpholine (22.4 g., 0.26 mole) is added dropwise, with stirring, to a solution of phosphoryl chloride (25.6 g., 0.17 mole) in 100 ml. of dry benzene. The temperature is kept below 20° by occasional cooling with ice–water. After the addition, the reaction mixture is kept at room temperature for 2.5 hr. with stirring, and the white precipitate is filtered off; the filtrate and washings (dry benzene) are combined, and fractionally distilled *in vacuo*. Fractions of morpholinophosphonic dichloride, b.p. 125–127°/10 mm. Hg, are collected (yield 10–15 g.) and are stored in sealed tubes.

Anhydride (2) of Diphenyl Phosphate and Morpholinophosphonic Chloride[5]

This compound (**2**) is prepared just before use, and is used *in situ*. To a *p*-dioxane (4 ml.) solution containing morpholinophosphonic dichloride (102 mg., 0.5 mmole) and diphenyl phosphate (125 mg., 0.5 mmole) is added a *p*-dioxane solution (1 ml.) of 2,6-lutidine (107 mg., 1.0 mmole). After a precipitate of lutidine hydrochloride appears, the reagent mixture is immediately employed in the following reaction.

Adenosine 2′:3′-Cyclic Phosphate[6] (4a)

A solution of 5′-*O*-acetyladenosine[7] (**1a**) (539 mg., 1 mmole) in *p*-dioxane (5 ml.) is added to the solution of the anhydride (**2**) obtained in the preceding experiment. Stirring is continued for 3 hr., and the reaction mixture is kept in a desiccator for 2 days at room temperature, the extent of phosphorylation being estimated by paper electrophoresis. The mixture is treated with 0.05 *N* ammonium hydroxide (400 ml.) and kept overnight at 30°. Amines are extracted with ether (3 × 100 ml.), and the aqueous layer is concentrated under diminished pressure to *ca.* 50 ml. The solution is applied to a column (3.5 × 80 cm.) of ECTEOLA-cellulose (Cl^-), which is then eluted with a linear gradient of 0–0.05 *M* lithium chloride in 1:4 ethanol–water. The second major peaks (containing adenosine 2′:3′-cyclic phosphate) are combined, and evaporated to dryness under diminished pressure below 20°. Lithium chloride is removed by washing with absolute methanol; yield (calculated on **1a**) *ca.* 50%; $\lambda_{max}^{H_2O}$ 260 nm.; R_f 0.46 (7:1:2 isopropyl alcohol–ammonia–water); R_f 0.38 (2:1 isopropyl alcohol–1% aqueous ammonium sulfate). Paper electrophoresis, R_{UP} 0.69[a] (0.1 *M* triethylammonium hydrogen carbonate, pH 7.5, 20 V/cm., 1 hr.).

Uridine 2′:3′-Cyclic Phosphate[6] (4b)

The reaction is conducted by essentially the same procedure as that described for **4a**, with use of the following reagents: diphenyl phosphate (75 mg., 0.3 mmole) in 2 ml. of *p*-dioxane, morpholinophosphonic dichloride (122 mg., 0.6 mmole), 2,6-lutidine (128 mg., 1.2 mmole) in 1 ml. of *p*-dioxane, and 5′-*O*-benzoyluridine (**1b**) (105 mg., 0.3 mmole) in 2 ml. of *p*-dioxane; the extent of the reaction, estimated by paper electrophoresis,

[a] R_{UP} is the distance of migration relative to that of uridine 5′-phosphate.

is 73%. Ammonium hydroxide (0.1 *N*, 40 ml.) is then added to the mixture, which is kept at room temperature for 3 hr. Extraction with ether (3 × 30 ml.), and evaporation of the ether layer gives a solution (*ca.* 20 ml.) which is applied to a column (2.2 × 34 cm.) of *O*-[2-(diethylamino)ethyl]-cellulose (HCO_3^-). The column is eluted with a linear gradient of 0–0.25 *M* triethylammonium hydrogen carbonate (pH 7.5) buffer. Evaporation of the fractions corresponding to uridine 2′:3′-cyclic phosphate gives 40 mg. (38%) of the nucleotide **4b**; $\lambda_{max}^{H_2O}$ 262 nm.; R_f 0.31 (7:1:2 isopropyl alcohol–ammonia–water), R_f 0.40 (5:2 ethanol–*M* ammonium acetate), R_f 0.43 (2:1 isopropyl alcohol–1% aqueous ammonium acetate); paper electrophoresis, R_{UP} 0.71 (0.1 *M* triethylammonium hydrogen carbonate; pH 7.5, 20 V/cm., 1 hr.).

REFERENCES

(1) C. A. Dekker and H. G. Khorana, *J. Am. Chem. Soc.*, **76**, 3522 (1954); M. Smith, J. G. Moffatt, and H. G. Khorana, *ibid.*, **80**, 6240 (1958).
(2) A. M. Michelson, *J. Chem. Soc.*, **1959**, 3655.
(3) D. M. Brown, D. I. Magrath, and A. R. Todd, *J. Chem. Soc.*, **1952**, 2708.
(4) M. Ikehara and E. Ohtsuka, *Chem. Pharm. Bull.* (Tokyo), **11**, 435 (1963).
(5) M. Ikehara and E. Ohtsuka, *Chem. Pharm. Bull.* (Tokyo), **11**, 961 (1963).
(6) M. Ikehara and I. Tazawa, *J. Org. Chem.*, **31**, 819 (1966).
(7) D. M. Brown, L. J. Haynes, and A. R. Todd, *J. Chem. Soc.*, **1950**, 3299.

[138] Adenosine 5′-(Dihydrogen Sulfatophosphate) (Anhydride of Adenosine 5′-Phosphate with Sulfuric Acid)

Use of Ethoxyacetylene to Activate Sulfuric Esters

G. R. BANKS and D. COHEN

DEPARTMENT OF CHEMISTRY, KEELE UNIVERSITY, KEELE, STAFFORDSHIRE, ENGLAND

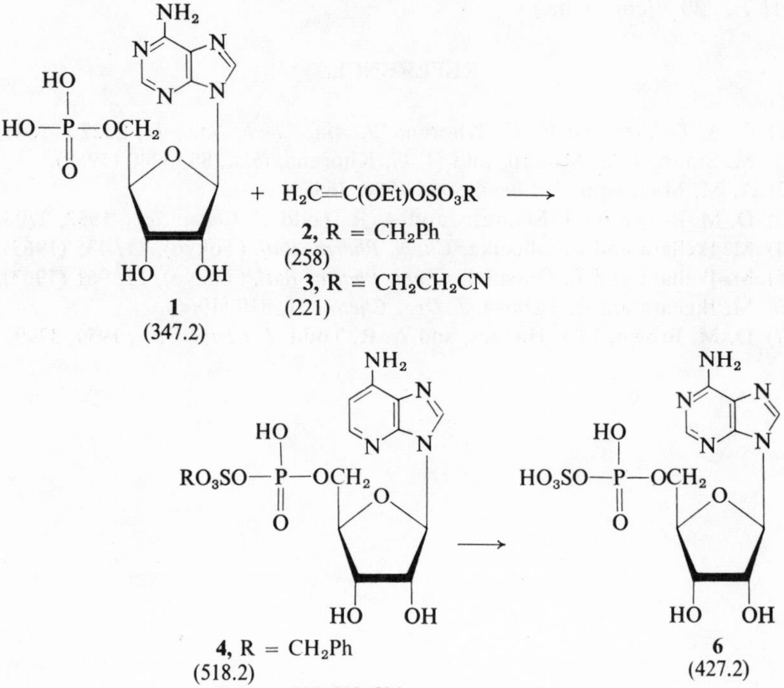

1 (347.2) + $H_2C{=}C(OEt)OSO_3R$ ⟶

2, R = CH_2Ph (258)

3, R = CH_2CH_2CN (221)

4, R = CH_2Ph (518.2)

5, R = CH_2CH_2CN (481.2)

⟶ **6** (427.2)

INTRODUCTION

Adenosine 5′-(dihydrogen sulfatophosphate) (**6**), an important analog of "active sulfate," was first synthesized by the reaction of 5′-adenylic acid (**1**) with a complex of pyridine–sulfur trioxide.[1] An independent synthesis used a condensation between 5′-adenylic acid (**1**) and inorganic sulfate, with dicyclohexylcarbodiimide as the condensing reagent.[2]

1-Ethoxyvinyl esters are now readily accessible[3] by way of direct addition of acids to ethoxyacetylene[a] and, by this route, both benzyl 1-ethoxyvinyl sulfate (**2**) and 1-ethoxyvinyl hydracrylonitrile sulfate (**3**) may be prepared and reacted *in situ* with 5′-adenylic acid to form the esterified sulfatophosphates (**4** and **5**). The protecting group of **4** is removed by hydrogenolysis, and that of **5** with aqueous sodium hydroxide.

2-Cyanoethyl sulfate (hydracrylonitrile sulfate), needed for the preparation of compound **3**, cannot be obtained analytically pure, but by conducting the reaction with **3**, rather than with **2**, a higher overall yield of compound **6** is obtained.

PROCEDURE

Benzyl 1-Ethoxyvinyl Sulfate (2)

Benzyl sulfate is prepared by a modification of Bacon and Doggart's synthesis.[5] Benzyl alcohol is treated with a solution of chlorosulfonic acid in chloroform in the presence of triethylamine, to yield an emulsion that may be separated after washing it twice with water. The chloroform layer is rendered neutral with 30% aqueous sodium hydroxide, and the resulting, colorless precipitate is recrystallized twice from 95% ethanol, to yield benzyl sodium sulfate, m.p. 194–196°.

Benzyl sodium sulfate (415 mg., 2 mmoles) is converted into the pyridinium salt by passage of an aqueous solution through a column of Amberlite IR-120 (pyridinium$^+$) ion-exchange resin, and the eluate is evaporated to dryness. The resulting gum is dissolved in dry pyridine (10 ml.), and the solution is evaporated to dryness under diminished pressure; this process is repeated. The resulting gum is dissolved in dry dichloromethane (3 ml.), and the solution is added slowly (30 min.), with stirring, to a solution of ethoxyacetylene (430 mg., 6 mmoles) in dichloromethane (2 ml.) at 0°. After 3 hr. at room temperature, the solution is evaporated under diminished pressure, to yield crude benzyl 1-ethoxyvinyl sulfate (**2**).

1-Ethoxyvinyl Hydracrylonitrile Sulfate (3)

A solution of chlorosulfonic acid (16.4 g., 0.14 mole) in dry ether (150 ml.) is cooled to −12°, mixed with hydracrylonitrile (10 g., 0.14 mole) in pyridine (22.4 g.), and the mixture is stirred. The temperature is kept below −10° for 1 hr., after which the ether is decanted from the white sludge. This residue is dissolved in the minimal volume of water, and to the

[a] Ethoxyacetylene may readily be prepared by a modification of the method used by I. N. Nazarov, Zh. A. Krasnaia, and V. P. Vinogradov, *J. Gen. Chem. USSR*, **28**, 451 (1959); see reference 3(d).

cold solution is added barium hydroxide (12.1 g., 0.14 meq., as a saturated solution).

The resultant solution is concentrated under diminished pressure to about 20 ml., and filtered. The filtrate is evaporated to dryness, and the resulting solid is recrystallized three times from aqueous ethanol.

The crude barium 2-cyanoethyl sulfate (210 mg., 1 mmole) is converted into the pyridinium salt by passage of an aqueous solution through a column of Amberlite IR-120 (pyridinium$^+$) ion-exchange resin, and the eluate is evaporated to a sirup. This is dissolved in dry pyridine, the solution is evaporated to dryness, and the process is repeated. The resulting sirup is dissolved in dry dichloromethane (4 ml.), and the solution is slowly added (30 min.) to a cold (0°), stirred solution of ethoxyacetylene (470 mg., 6.7 mmoles) in dry dichloromethane (3 ml.). The resulting solution is stirred at room temperature for 3 hr., and evaporated to dryness under diminished pressure, to yield crude 1-ethoxyvinyl hydracrylonitrile sulfate (**3**) as a sirup.

Adenosine 5′-(Dihydrogen Sulfatophosphate) (6)

(a). From Benzyl 1-Ethoxyvinyl Sulfate (2).—Benzyl 1-ethoxyvinyl sulfate (**2**) (40 mg., 0.16 mmole) is dissolved in dry dichloromethane (4 ml.), and the solution is slowly added to adenosine 5′-(pyridinium phosphate)[b] partially dissolved in a mixture of dry dichloromethane (5 ml.) and *N,N*-dimethylformamide (5 ml.). The resulting solution is kept in a stoppered flask in the dark for 5 days, and is then evaporated to dryness under diminished pressure. The residue (**4**) is dissolved in a mixture of water (20 ml.) and ethanol (1 ml.), and catalytically hydrogenolyzed with hydrogen in the presence of 10% palladium–carbon catalyst until the theoretical amount of hydrogen has been taken up. The catalyst is removed by filtration, the filtrate is concentrated to a small volume, and the solution is applied to Whatman No. 4 paper and separated chromatographically with 5:3 (v/v) isobutyric acid–0.5 *N* ammonia. Compound **6** (8.6 mg., 35%) is obtained by elution of the band having R_f 0.38 (R_f of adenylic acid, 0.53).

(b). From 1-Ethoxyvinyl Hydracrylonitrile Sulfate (3).—A solution of 1-ethoxyvinyl hydracrylonitrile sulfate (**3**) (100 mg., 0.45 mmole) in dry dichloromethane is added to a partial solution of adenosine 5′-(pyridinium phosphate) (from 21 mg., 0.06 mmole, of adenosine 5′-phosphate) in dry *N,N*-dimethylformamide (2 ml.). The resulting solution is kept in the dark at room temperature for 5 days, and evaporated to dryness under diminished pressure, and to the residue (**5**) is added 0.5 *N* sodium hy-

[b] Obtained from adenosine 5′-phosphate (**1**) (20 mg., 0.06 mmole) by ion-exchange on a column of Amberlite IR-120 (pyridinium$^+$) ion-exchange resin.

droxide (5 ml.). The solution is kept at 100° for 2 min., cooled in an ice-bath, and rendered neutral with Amberlite IR-120 (H^+) ion-exchange resin. Compound **6** (11.0 mg., 45%) is obtained as in (a), by elution after paper chromatography. An alternative chromatographic solvent system is 6:3:1 (v/v) propyl alcohol–ammonium hydroxide (*d* 0.88)–water, in which **6** has R_f 0.25 (R_f of adenylic acid, 0.16).

REFERENCES

(1) J. Baddiley, J. G. Buchanan, and R. Letters, *J. Chem. Soc.*, **1957**, 1067.

(2) P. Reichard and N. R. Ringertz, *J. Am. Chem. Soc.*, **79**, 2025 (1957).

(3) (a) H. H. Wasserman and P. S. Wharton, *J. Am. Chem. Soc.*, **82**, 661 (1960); (b) H. H. Wasserman and D. Cohen, *ibid.*, **82**, 4425 (1960); (c) G. R. Banks and D. Cohen, *Proc. Chem. Soc.*, **1963**, 83; (d) D. Cohen and H. D. Springall, in G. T. Young (ed.), *Peptides Proc. European Symp., 5th, Oxford, 1962*, 73 (1963).

(4) G. R. Banks and D. Cohen, *J. Chem. Soc.*, **1965**, 6209.

[139] Adenosine 3′(2′),5′-Diphosphate

Phosphorylation of Adenine Ribonucleoside and 2′-Deoxyribonucleoside Derivatives by Reaction with Triphenyl Phosphite and Oxidation of the Resulting Phosphorous Acid Esters

A. HOLÝ

INSTITUTE OF ORGANIC CHEMISTRY AND BIOCHEMISTRY,
CZECHOSLOVAK ACADEMY OF SCIENCES, PRAGUE, CZECHOSLOVAKIA

INTRODUCTION

Adenosine 3′(2′),5′-diphosphate (**5**) is a key intermediate for further nucleotide syntheses, *e.g.*, of coenzyme A. Its preparation consists in a nonspecific phosphorylation of nucleosides or nucleotides derived from adenine, namely, by reaction with the triphenyl ester of phosphorous acid and the subsequent oxidation of the resulting phosphites with potassium permanganate in neutral media.[1]

PROCEDURE

Adenosine 3′(2′),5′-Diphosphite (3)

A mixture of 2.67 g. (0.01 mole) of adenosine (**1**),[a] 7.80 g. (0.025 mole) of triphenyl phosphite, 15 ml. of *N,N*-dimethylformamide,[b] and 2.5 ml. of a 5.6 *M* solution of hydrogen chloride in *N,N*-dimethylformamide[c] is shaken in a 250-ml., round-bottomed flask until complete dissolution occurs. The solution is kept at room temperature for 2 days, with exclusion of atmospheric moisture, and then treated with 10 ml. of *N,N*-dimethylformamide and 6 ml. of triethylamine. The mixture is kept at 2° for 2 hr., and filtered, and the precipitate (triethylamine hydrochloride) is washed with 10 ml. of *N,N*-dimethylformamide. The filtrates are combined, shaken with 100 ml. of 1:4 concentrated ammonium hydroxide–water at room temperature for 3 hr., and the mixture is washed with three 50-ml. portions of ether. The aqueous layer is evaporated on a rotary evaporator at 40°/15 mm. Hg, and the residue is co-evaporated with three 100-ml. portions of pyridine (to remove the last traces of *N,N*-dimethylformamide). The

[a] Adenosine is dried over phosphorus pentaoxide at 100°/0.1 mm. Hg for 8 hr.

[b] Commercial *N,N*-dimethylformamide is distilled from phosphorus pentaoxide at 15 mm. Hg, and stored over molecular sieves.

[c] Dry hydrogen chloride is introduced into anhydrous *N,N*-dimethylformamide, and the resulting solution is stored in a well-stoppered bottle in a refrigerator. Its concentration is determined alkalimetrically.

HOCH$_2$ O A
HO OH
1
(267.3)

1. $(C_6H_5O)_3P$
2. OH^-

2 (37%) (331.3) (+ 2′-isomer)

3 (57%) (395.4) (+ 2′,5′-isomer)

4 (2%) (459.4)

$KMnO_4$

5 (427.4) (+ 2′,5′-isomer)

where A = (adenin-9-yl: NH_2, N, N, N, N)

residue is dissolved in 50 ml. of water, and the solution is placed on a column (4 × 70 cm.) of *O*-[2-(diethylamino)ethyl]cellulose (HCO_3^-).[d] Elution is performed with a linear gradient of triethylammonium hydrogen carbonate (pH 7.5) (2 l. of a 0.3 *M* buffer solution in the reservoir and 2 l. of water in the mixing chamber) at a flow-rate of 3 ml./min. The optical absorbance of the effluent is checked by continuous measurement of the ultraviolet absorption. Elution peak 1 contains pyridine and a small quantity of adenosine. Peak 2 (buffer concentration 0.08–0.12 *M*) contains adenosine 2′(3′)-phosphite (**2**), as shown by paper chromatography.[e] Peak 3 contains the desired reaction product (buffer concentration 0.14–0.23 *M*); these fractions are combined, and evaporated to dryness at 35°/15 mm. Hg in a rotary evaporator, and the residue is co-evaporated with two 50-ml. portions of water (to remove the volatile buffer). A solution of the residue in 25 ml. of water is then placed on a column (100 ml.) of Dowex-50 (pyridinium$^+$) (100–200 mesh),[f] and the column is eluted with 250 ml. of 10% aqueous pyridine. The eluate is treated with 20 ml. of concentrated ammonium hydroxide, and evaporated to dryness at 35°/15 mm. Hg on a rotary evaporator. The residue is co-evaporated with two 50-ml. portions of 96% ethanol, and is then dissolved in 20 ml. of methanol. The solution is treated with 0.1 ml of 30% methanolic ammonia, and the solution is added dropwise, with stirring, to 200 ml. of anhydrous ether.[g] The precipitate is immediately collected with suction, washed with 50 ml. of ether, and dried over phosphorus pentaoxide at 0.1 mm. Hg. The chromatographically and electrophoretically homogeneous[e] ammonium salt of **3** weighs 2.45 g. (57%).[h]

Adenosine 3′(2′),5′-Diphosphate (5)

A solution of 1.30 g. (0.003 mole) of the ammonium salt of adenosine 3′(2′),5′-diphosphite (**3**), 0.1 g. of sodium hydrogen carbonate, and 5 ml. of water is treated dropwise, during 2 hr., with a saturated, aqueous solution of potassium permanganate (approximately 50 ml.) until a red

[d] For the packing of a *O*-[2-(diethylamino)ethyl]cellulose column, see Sec. IV [140], footnote *j*.

[e] Descending chromatography is performed on Whatman No. 1 paper, with 7:1:2 isopropyl alcohol–concentrated ammonium hydroxide–water, the R_f values being 0.48 (**1**), 0.37 (**2**), 0.22 (**3**), 0.10 (**4**), and 0.02 (**5**). Paper electrophoresis (40 volt/cm., 45 min., Whatman paper No. 1) is performed in (*a*) 0.05 *M* triethylammonium hydrogen carbonate (pH 7.5), the mobilities being 0.50 (**2**), 1.13 (**3**), 1.55 (**4**), and 1.48 (**5**); or (*b*) 0.05 *M* sodium hydrogen citrate, the mobilities being 1.00 (**2**), 2.00 (**3**), 3.40 (**4**), and 1.85 (**5**), all based on R_f 1.00 for adenosine 2′(3′)-phosphate.

[f] For Dowex-50 (pyridinium$^+$) ion-exchange resin, see Sec. IV [140], footnote *a*.

[g] Ether is dried over sodium.

[h] A further gradient-elution affords an additional peak (buffer concentration 0.25–0.28 *M*) containing a small amount of adenosine 2′,3′,5′-triphosphite (**4**).

color persists.[i] The mixture is then kept at room temperature for 1 hr., treated with a further 1 ml. of saturated, aqueous potassium permanganate, kept at room temperature overnight, heated in a water-bath to 40°, treated with 2 ml. of methanol, and kept at 40°, with occasional stirring, until decolorization ensues. The suspension is filtered through a thin layer of Hyflo Super Cel, and the material on the filter funnel is washed with 50 ml. of warm (50–60°) water. The filtrates are combined, and passed through a column (50 ml.) of Dowex-50 (H^+) (100–200 mesh).[j] The column is eluted with 100 ml. of water, and the eluate is evaporated to dryness at 35°/15 mm. Hg on a rotary evaporator. The residue is dissolved in 5 ml. of water, and the solution is added dropwise, with stirring, to 200 ml. of 1:1 96% ethanol–acetone. The mixture is kept at 2° overnight, and the precipitate is collected with suction, washed successively with 50 ml. of 1:1 96% ethanol–acetone and 50 ml. of ether, and dried over phosphorus pentaoxide at 0.1 mm. Hg. The chromatographically and electrophoretically homogeneous[e] adenosine 3′(2′),5′-diphosphate (**5**) weighs 0.65 g. (50%).

REFERENCE

(1) A. Holý and F. Šorm, *Collection Czech. Chem. Commun.*, **31**, 1544 (1966).

[i] To determine the decolorization, a small amount of the reaction mixture is applied to a moistened filter-paper.

[j] Commercial Dowex-50 (100–200 mesh) ion-exchange resin is washed with 0.10 *M* hydrochloric acid, put into the column, and washed with redistilled water until the washings are devoid of Cl^- ions.

[140] Cytidylyl-(3′ → 5′)-adenosine

A General Method of Preparation of Diribonucleoside Monophosphates having a Known Sequence

S. CHLÁDEK

INSTITUTE OF ORGANIC CHEMISTRY AND BIOCHEMISTRY, CZECHOSLOVAK ACADEMY OF SCIENCES, PRAGUE, CZECHOSLOVAKIA

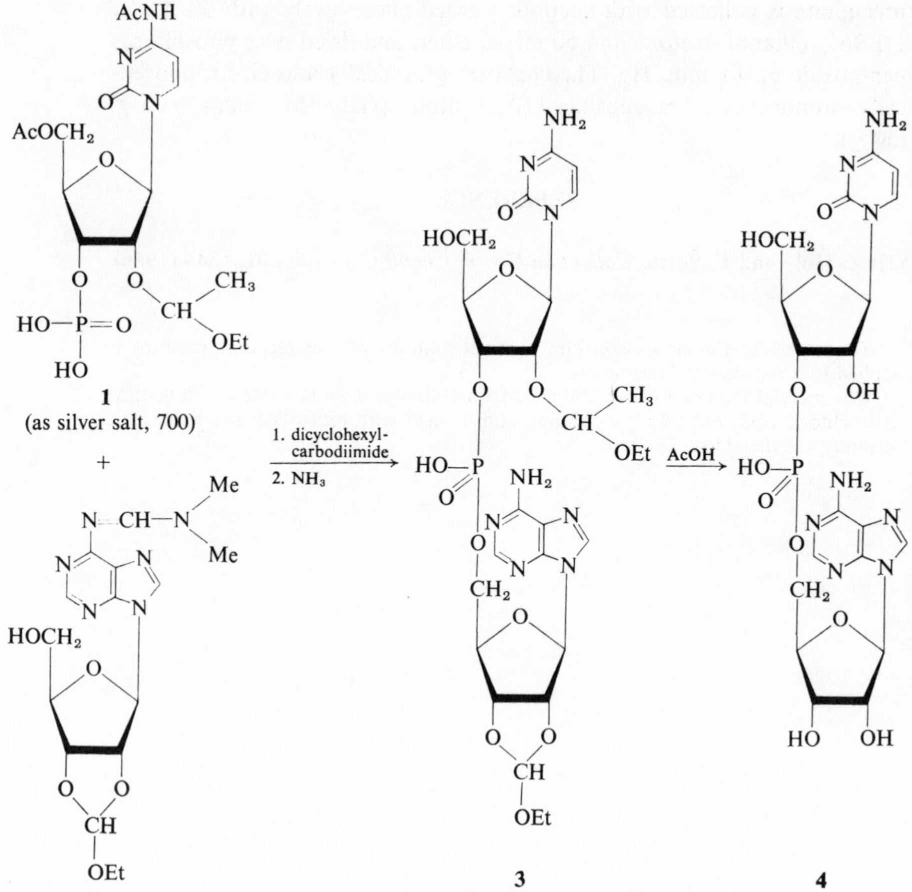

1 (as silver salt, 700)

+

2 (378.4)

1. dicyclohexylcarbodiimide
2. NH_3

3 (as triethylammonium salt) (801.7)

AcOH

4 (as ammonium salt) (589.4)

INTRODUCTION

Cytidylyl-(3′ → 5′)-adenosine (**4**) is synthesized by a general method for the preparation of oligonucleotides of the D-ribose series.[1,2] The nucleotide component started with here is a ribonucleoside 3′-phosphate. Its 2′-hydroxyl group is protected by the 1-ethoxyethyl group; the 5′-hydroxyl group and the amino group of the heterocyclic base are protected by acetyl groups.[3] The nucleoside component started with here is a ribonucleoside whose hydroxyl groups at C-2′ and C-3′ are protected by an ethoxymethylene group, and the amino group of which (on the heterocyclic moiety) is protected by a (dimethylamino)methylene group.[1] The condensation is performed by the action of dicyclohexylcarbodiimide[4] in pyridine. Removal of the protecting groups in the final stage of the synthesis is achieved by successive treatment with ammonium hydroxide and aqueous acetic acid. An alternative general method for the synthesis of oligoribonucleotides uses a different combination of protecting groups.[5]

PROCEDURE

2′-*O*-(1-Ethoxyethyl)cytidylyl-(3′ → 5′)-2′,3′-*O*-(ethoxymethylene)-adenosine (3)

The silver salt (0.70 g., 1 mmole) of *N*-acetyl-5′-*O*-acetyl-2′-*O*-(1-ethoxyethyl)cytidine 3′-phosphate (**1**) (Sec. IV [143]) is dissolved, with cooling in ice, in 20–30 ml. of 50% aqueous pyridine, and the solution is placed on a column (1 × 10 cm.) of Dowex-50 (pyridinium$^+$) ion-exchange resin[a] in a cool-room (0–4°). The elution is performed with 200 ml. of 50% aqueous pyridine, and the eluate is evaporated at 20° (bath temperature)/1 mm. Hg under continuous addition[b] of anhydrous pyridine.[c] The residue is treated with 0.8 g. (2 mmoles) of *N*-(dimethylamino) methylene-2′,3′-*O*-(ethoxymethylene)adenosine (**2**) (Sec. III [65]), and the resulting mixture is dried by five co-evaporations with anhydrous pyridine.[d] The residual material is dissolved in 10 ml. of anhydrous pyridine, and the solution is treated with 1.6 g. (8 mmoles) of dicyclohexylcarbodiimide, and kept in a tightly stoppered flask in the dark at room temperature for

[a] Dowex-50 X8 (100–200 mesh) ion-exchange resin is washed successively with water, 10% aqueous pyridine, and 100 ml. of 50% aqueous pyridine.

[b] The eluate is evaporated in a distillation apparatus equipped with a dropping funnel (for continuous addition of anhydrous pyridine).

[c] Pure pyridine is dried with sodium hydroxide, and distilled, and the distillate is stored over molecular sieves.

[d] The capillary of the vacuum distillation apparatus and the tube for releasing the vacuum are connected to a drying tower containing phosphorus pentaoxide, to ensure complete exclusion of atmospheric moisture.

4 days. The reaction mixture is then treated with 10 ml. of 50% aqueous pyridine containing 10% of triethylamine,[e] and kept at room temperature for 3 hr. Light petroleum[f] (30 ml.) is added, and the precipitate of dicyclohexylurea is filtered off, and washed with 20 ml. of 50% aqueous methanol. The filtrates are combined, and the aqueous layer is separated, washed with three 30-ml. portions of light petroleum, and concentrated to a third of the original volume[g] below 35° (bath temperature)/15 mm. Hg on a rotary evaporator. The concentrate is treated with 250 ml. of methanolic ammonia,[h] and the resulting solution is kept in a stoppered flask at room temperature for 12–15 hr. The solution is concentrated as before, and the concentrate is diluted with 60 ml. of 50% aqueous methanol,[i] and placed on a column (3.5 × 90 cm.) of *O*-[2-(diethylamino)ethyl]cellulose (HCO_3^-).[j] Elution is performed by use of a linear gradient of triethylammonium hydrogen carbonate in 50% aqueous methanol,[k] 25-ml. fractions being collected at a flow-rate[l] of 2.5 ml./min. Fractions corresponding[m] to the protected dinucleoside phosphate are evaporated to dryness below 35° (bath temperature)/15 mm. Hg on a rotary evaporator, and the residue is co-evaporated with five 20-ml. portions of 1:1 methanol–triethylamine. The yield (0.31 mmole, 31%) of the product is determined

[e] Addition of triethylamine prevents even partial removal of the labile 1-ethoxyethyl group.

[f] Light petroleum (b.p. 40–60°) is used.

[g] Care must be taken not to evaporate the solution of the substituted dinucleoside phosphate to dryness, because of the ready removal of the 1-ethoxyethyl group.

[h] Methanol saturated at 0° with dry ammonia gas is used.

[i] At this point, an additional crop of dicyclohexylurea may separate. The solution is, therefore, filtered through a thin layer of Hyflo Super Cel, and the solid is washed with 50% aqueous methanol.

[j] Commercial DEAE-cellulose (Cellex D of Calbiochem) is successively washed with 0.2 *N* sodium hydroxide, water, 0.2 *N* hydrochloric acid, water, 0.2 *N* sodium hydroxide, and water. Gaseous carbon dioxide is then introduced into the aqueous suspension of the material until the pH value is 7.5. Fine particles are removed by repeated decantation. The column is packed with a slurry of coarse particles under moderate pressure, to afford an adequate flow-rate.

[k] The linear-gradient elution is performed by use of 2 l. of 0.15 *M* triethylammonium hydrogen carbonate in 50% aqueous methanol in the reservoir and 2 l. of 50% aqueous methanol in the mixing chamber (equipped with an efficient stirrer). The buffer solution is prepared by saturating a mixture of triethylamine and water with gaseous carbon dioxide until the pH value of the resulting solution is 7.5 (the gas inlet-tube is equipped with a sintered-glass plate at the bottom). The concentration of the solution is adjusted to the value desired, by dilution with water.

[l] The ultraviolet absorption of the eluate is determined continuously by measurement of the optical absorbance at 254 nm.

[m] The first chromatographic fraction contains 2′,3′-*O*-(ethoxymethylene)adenosine, and the second fraction contains the substituted dinucleoside phosphate (**3**), which may be contaminated with a by-product in which the 1-ethoxyethyl group has been removed.

by ultraviolet spectrophotometry.[n] As is shown by chromatography in 7:1:2 isopropyl alcohol–concentrated ammonium hydroxide–water (solvent A), the product (**3**) contains a variable proportion (less than 3%) of a contaminant arising by removal of the 2′-*O*-(1-ethoxyethyl) group.

A methanolic solution of the crude product (**3**) (0.2 mmole) is applied to three sheets of Whatman No. 3MM paper, and the chromatograms are developed overnight with solvent A. The zone containing the substituted dinucleoside phosphate is eluted with 100–200 ml. of 10:1:5:2 methanol–triethylamine–water–concentrated ammonium hydroxide.[o] The eluate is evaporated below 35° (bath temperature)/15 mm. Hg on a rotary evaporator, with addition and evaporation of three to five 10-ml. portions of triethylamine, and the residue is dried at room temperature/1 mm. Hg. The triethylammonium salt of compound **3** is obtained as a colorless foam, yield 0.108 mmole.[n] R_f 0.56, on Whatman No. 1 paper in solvent A (descending technique); λ_{max} 266 nm., λ_{min} 235 nm., A_{250}/A_{260} 0.74, A_{280}/A_{260} 0.87, A_{290}/A_{260} 0.57 (0.01 *N* hydrochloric acid).

Cytidylyl-(3′ → 5′)-adenosine (4)

A solution of 0.17 mmole of the triethylammonium salt of compound **3** in 10 ml. of 80% aqueous acetic acid is kept at room temperature for 8 hr., the solution is freeze-dried, the residue is dissolved in 3.0 ml. of 5% aqueous ammonia, and the solution is applied to two sheets of Whatman No. 3MM paper.[p] The chromatograms are developed with solvent A for 20 hr. The zone corresponding to the dinucleoside phosphate (**4**) is eluted with 200 ml. of water, the eluate is freeze-dried, and the residue is dried at 10^{-4} mm. Hg; yield 0.115 mole (62%); R_f 0.21, on Whatman No. 1 paper in solvent A (descending technique); electrophoretic mobility 0.29 (relative

[n] The yield is estimated by measurement of optical absorbance of an aliquot diluted with 0.02 *N* hydrochloric acid. Compound **3** is dissolved in methanol containing 1% of triethylamine. For the dinucleoside phosphate (**4**), an aqueous solution is used.

[o] Two zones are detected on the chromatogram, namely, those of cytidylyl-(3′ → 5′)-2′,3′-*O*-(ethoxymethylene)adenosine (R_f 0.40) and 2′-*O*-(1-ethoxyethyl)-cytidylyl-(3′ → 5′)-2′,3′-*O*-(ethoxymethylene)adenosine (**3**) (R_f 0.60). The chromatograms are dried at room temperature, and the zones corresponding to compound **3** are cut out and subjected to elution by a method designed by Dr. J. Smrt (private communication). The strips are tightly wrapped round the upper part of a test tube (10-cm. diam.), and are covered, in the same way, with a sheath of polyethylene or poly(vinyl chloride) foil extending about 10 cm. beyond the bottom of the test tube and about 2 cm. above the neck. The sheath is held in place over the strips and the test tube by means of tape or soft wire. The resulting assemblage is then inverted in such a manner that the bottom of the test tube (inside the sheath) is at the top, forming a column. Finally, the eluant is poured into the upper part of the sheath, and the eluate is collected in a flask placed under the lower part of the sheath.

[p] The chromatogram should *not* be dried with hot air.

to 1.00 for cytidine 3′-phosphate).[r] The spectral properties of compound **4** are in accordance with those of compound **3**. Pancreatic ribonuclease degradation of compound **4** affords 97.5% of cytidine 3′-phosphate and adenosine in the ratio[s] 0.95:1.

REFERENCES

(1) J. Žemlička, S. Chládek, A. Holý, and J. Smrt, *Collection Czech. Chem. Commun.*, **31**, 3198 (1966).
(2) A. Holý and J. Smrt, *Collection Czech. Chem. Commun.*, **31**, 3800 (1966).
(3) S. Chládek and J. Smrt, *Chem. Ind.* (London), **1964**, 1719; J. Smrt and S. Chládek, *Collection Czech. Chem. Commun.*, **31**, 2978 (1966).
(4) P. T. Gilham and H. G. Khorana, *J. Am. Chem. Soc.*, **80**, 6212 (1958).
(5) Y. Lapidot and H. G. Khorana, *J. Am. Chem. Soc.*, **85**, 3852 (1963).

[r] Electrophoresis is performed on Whatman No. 1 paper, and 0.02 *M* disodium hydrogen phosphate (pH 7.5) is used as the buffer.

[s] Because only the naturally occurring (3′ → 5′)-diester linkage is subjected to degradation with pancreatic ribonuclease, the extent of the isomerization of the (3′ → 5′)-linkage to the (2′ → 5′)-linkage can be determined in this way. The degradation with pancreatic ribonuclease is performed with 10 μmole (approx. 21 A_{260} units) of compound **4** in 0.1 ml. of tris buffer [2-amino-2-(hydroxymethyl)-1,3,-propanediol] (pH 7.8) and 50 μg. of pancreatic ribonuclease (Calbiochem) at 37° for 3–4 hr. The reaction mixture is then applied (in a width of 2 cm.) to Whatman No. 3MM paper, and the chromatogram is developed with solvent A. The corresponding spots are eluted with 10.0 ml. of 0.01 *N* hydrochloric acid, and the optical absorbance of the eluates is measured and evaluated.

[141] 1-Methyladenosine 5′-(α-D-Glucopyranosyl Pyrophosphate)

Methylation of Adenosine Derivatives

V. N. SHIBAEV and S. M. SPIRIDONOVA

INSTITUTE FOR CHEMISTRY OF NATURAL PRODUCTS,
ACADEMY OF SCIENCES OF U.S.S.R., MOSCOW, U.S.S.R.

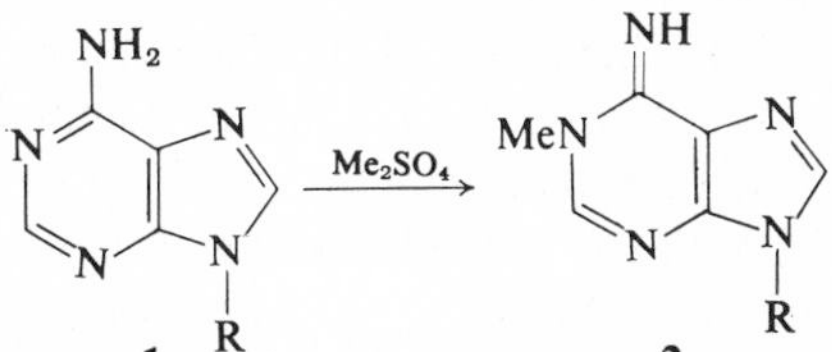

Me_2SO_4

1 (633.4) → **2** (647.4)

where R =

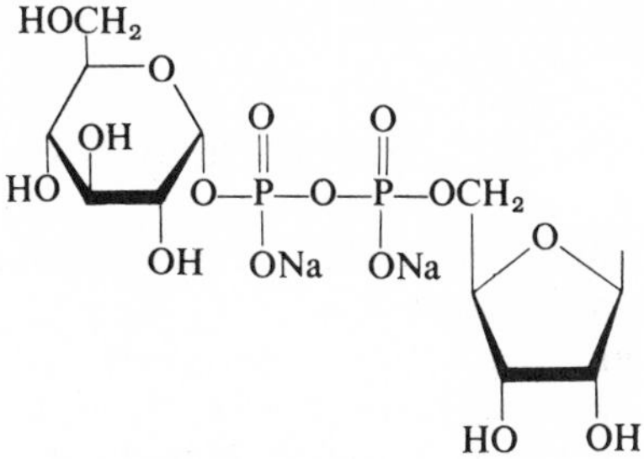

INTRODUCTION

1-Methyladenosine 5′-(α-D-glucopyranosyl disodium pyrophosphate)[1] (**2**) has been synthesized in the course of studies on the substrate specificity of carbohydrate-metabolizing enzymes. The methylation of adenosine,[2] adenosine 5′-phosphate,[3] and poly(adenylic acid)[4] has been described.

PROCEDURE

1-Methyladenosine 5′-(α-D-Glucopyranosyl Disodium Pyrophosphate) (2)

Methyl sulfate (189 mg., 1.5 mmoles) is added during 1 hr., with stirring, to a solution of adenosine 5′-(α-D-glucopyranosyl disodium pyrophosphate) (**1**) (63 mg., 0.1 mmole) in water (5 ml.); the pH of the reaction mixture is monitored with a glass electrode, and *M* sodium carbonate is added as needed, to keep the pH at 4.5–5.0. The reaction mixture is stirred for 3 hr., and analyzed by paper electrophoresis in triethylammonium hydrogen carbonate buffer of pH 7.5 (compounds **2** and **1** have mobilities of 0.45 and 0.75, respectively, compared with that of uridine 5′-phosphate as unity). If the methylation is not complete, the mixture is re-treated with methyl sulfate (1.5 mmoles) under the same conditions. The resulting solution is diluted with water to 50 ml., and passed through a column (2 × 10 cm.) of Dowex-1 X2 (Cl^-). The column is washed successively with water (100 ml.) and 0.01 *M* sodium chloride (100 ml.). Elution with 0.03 *M* sodium chloride then results in elution of **2**, A_{260} 740.

The fractions containing **2** are combined and evaporated to dryness, the residue is dissolved in water (1 ml.), and the solution is applied to a column (3 × 83 cm.) of[a] Sephadex G-10. The column is washed with water, and the eluate from 90 to 135 ml. contains compound **2**, free from inorganic salts. The resulting solution of the disodium salt (**2**) (A_{260} 630, corresponding to 68% yield) is lyophilized, giving a white powder, yield 42 mg.; λ_{max} 257 nm., ϵ_{max} 9,570; λ_{min} 235 nm., ϵ_{min} 4,400 (pH 7); λ_{max} 258 nm., ϵ_{max} 9,100; λ_{min} 235 nm., ϵ_{min} 2,700 (pH 11); $\epsilon_{300}/\epsilon_{260}$ 0.05 (pH 7), 0.21 (pH 11).

REFERENCES

(1) N. K. Kochetkov, E. I. Budowsky, V. N. Shibaev, and S. M. Spiridonova, *Khim. Prirodn. Soedin.* (in press).

(2) J. W. Jones and R. K. Robins, *J. Am. Chem. Soc.*, **85**, 193 (1964).

(3) B. E. Griffin and C. B. Reese, *Biochim. Biophys. Acta*, **68**, 185 (1963).

(4) A. M. Michelson and M. Grunberg–Manago, *Biochim. Biophys. Acta*, **91**, 92 (1964).

(5) E. I. Budowsky and V. N. Shibaev, *Vopr. Med. Khim.*, **8**, 554 (1967).

[a] In other cases, removal of inorganic salts from the solutions of nucleoside glycosyl pyrophosphates obtained after ion-exchange chromatography may be achieved effectively[5] with Sephadex G-25.

[142] 5-Amino-3-β-D-ribofuranosyl-3*H*-*v*-triazolo[4,5-*d*]pyrimidin-7(6*H*)-one 5′-phosphate (8-Azaguanosine 5′-Phosphate)

The Preparation of an Unnatural Nucleotide

JOHN A. MONTGOMERY and H. JEANETTE THOMAS

KETTERING-MEYER LABORATORY, SOUTHERN RESEARCH INSTITUTE, BIRMINGHAM, ALABAMA 35205

O, HN, N, N, H_2N, N, N, $HOCH_2$, O, HO, OH — **1** (284.2)

MeCOMe, p-$MeC_6H_4SO_3H$ →

O, HN, N, N, H_2N, N, N, $HOCH_2$, O, O, O, Me, Me — **2** (324.3)

O ‖ $NCCH_2CH_2OP(OH)_2$, $C_6H_{11}N{=}C{=}NC_6H_{11}$ →

[O, HN, N, N, H_2N, N, N, $NCCH_2CH_2OPOCH_2$, HO, O, O, O, O, Me, Me] — **3** (426.4)

1. H_2SO_4 2. LiOH 3. $Ba(OH)_2$ →

O, HN, N, N, H_2N, N, N, BaO_2POCH_2, O, O, HO, OH — **4** (499.5)

INTRODUCTION

8-Aza-5′-guanylic acid[1,2] (**4**) formed in cells by the enzymic reaction of 8-azaguanine with 5-*O*-phosphono-D-ribofuranosyl pyrophosphate, is known to be the active metabolite (of 8-azaguanine) that causes cell death.[3] This compound can conveniently be synthesized by the reaction of 2′,3′-*O*-isopropylidene-8-azaguanosine (**2**), prepared by the acid-catalyzed reaction of 8-azaguanosine (**1**) with acetone, with 2-cyanoethyl phosphate and dicyclohexylcarbodiimide,[4] followed by removal of the protecting groups.[2]

PROCEDURE

5-Amino-3-(2,3-*O*-isopropylidene-β-D-ribofuranosyl)-3*H*-*v*-triazolo[4,5-*d*]-pyrimidin-7(6*H*)-one (2)

To a suspension of 3.62 g. (12.8 mmoles) of 8-azaguanosine (**1**) in 550 ml. of dry acetone, in a 1-l., round-bottomed flask equipped with a drying tube, is added 7.30 g. (38.4 mmoles) of *p*-toluenesulfonic acid with stirring. After complete dissolution, which requires about 4 hr., the solution is stirred for 1 hr. at room temperature, concentrated to 100 ml., and poured into 300 ml. of cold water containing 70 g. of Amberlite IR-4B (OH^-) ion-exchange resin. More resin is added until pH 7 is obtained (total, 240 g. of resin). The resin is removed by filtration and washed with water, and the filtrate and washings are combined and evaporated to dryness under diminished pressure below 35°. The residue is crystallized from 75 ml. of water (charcoal treatment); yield 2.74 g. This solid is recrystallized from water; yield 2.26 g. (54%), m.p. 247° (dec.).

Barium 8-Aza-5′-guanylate [5-Amino-3-(5-*O*-phosphono-β-D-ribofuranosyl)-3*H*-*v*-triazolo[4,5-*d*]pyrimidin-7(6*H*)-one, Barium Salt] (4)

To a solution of 2.65 g. (17.5 mmoles) of 2-cyanoethyl phosphate (Sec. V [159]) in 130 ml. of anhydrous pyridine in a 250-ml., round-bottomed flask is quickly added 7.20 g. (35.0 mmoles) of dicyclohexylcarbodiimide, followed by 2.84 g. (8.75 mmoles) of 5-amino-3-(2,3-*O*-isopropylidene-β-D-ribofuranosyl)-3*H*-*v*-triazolo[4,5-*d*]pyrimidin-7(6*H*)-one (**2**); a precipitate immediately begins to form in the reaction solution. The flask is kept tightly sealed at room temperature for 2 days, and the resulting dark-red solution is then diluted with 17.5 ml. of water and kept for 1 hr. The precipitate of dicyclohexylurea (5.88 g.) is removed by filtration, and the red solution is evaporated to dryness under diminished pressure below 35°. The residue is dissolved in 100 ml. of water, and the solution is washed with five 100-ml. portions of chloroform.

The aqueous solution of 5-amino-3-(2,3-*O*-isopropylidene-β-D-ribofuranosyl)-3*H*-*v*-triazolo[4,5-*d*]pyrimidin-7(6*H*)-one 5′-(2-cyanoethyl hydrogen phosphate) (**3**) is diluted with sufficient *N* sulfuric acid to give 170 ml. of a solution 0.1 *N* in sulfuric acid. This solution is kept for 2 days at room temperature and then neutralized by the addition of 1.46 g. (8.5 mmoles) of barium hydroxide in 150 ml. of water. The precipitate of barium sulfate is removed by filtration through a Celite pad, and the filtrate, containing 5-amino-3-β-D-ribofuranosyl-3*H*-*v*-triazolo[4,5-*d*]pyrimidin-7(6*H*)-one 5′-(2-cyanoethyl hydrogen phosphate) and now at a volume of 330 ml., is diluted with sufficient 3 *N* lithium hydroxide to give 400 ml. of a solution 0.5 *N* in lithium hydroxide. The bright-yellow solution is heated in a bath at 100° for 15 min., cooled, and filtered. The cold filtrate is stirred with 315 ml. of Amberlite IR-120 (H^+) ion-exchange resin until pH 2.3 is attained, and the resin is filtered off, and washed with water. The filtrate and washings are combined, adjusted to pH 7.5 with aqueous barium hydroxide, and filtered, and the clear filtrate is diluted with 800 ml. of ethanol. The resulting white precipitate is collected by filtration; yield 3.19 g. (73%), λ_{max} in nm. (log ε): pH 1, 255 (4.11); pH 7, 255 (4.11); pH 13, 222 (4.37) and 279 (4.07).

REFERENCES

(1) J. L. Way and R. E. Parks, Jr., *J. Biol. Chem.*, **231**, 467 (1958).
(2) J. A. Montgomery and H. J. Thomas, *J. Org. Chem.*, **26**, 1926 (1961).
(3) R. W. Brockman, *Advan. Cancer Res.*, **7**, 129 (1962).
(4) P. T. Gilham and G. M. Tener, *Chem. Ind.* (London), **1959**, 542.

[143] *N*-Acetyl-5′-*O*-acetyl-2′-*O*-(1-ethoxyethyl)-cytidine 3′-Phosphate, Silver Salt

An Intermediate in the Synthesis of an Internucleotide Linkage

J. SMRT

INSTITUTE OF ORGANIC CHEMISTRY AND BIOCHEMISTRY,
CZECHOSLOVAK ACADEMY OF SCIENCES, PRAGUE, CZECHOSLOVAKIA

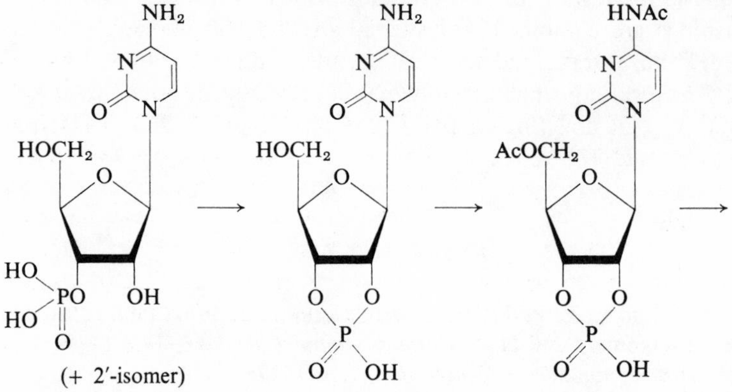

(+ 2′-isomer)
1 (323.2) → **2** (305.2) → **3** (389.2) →

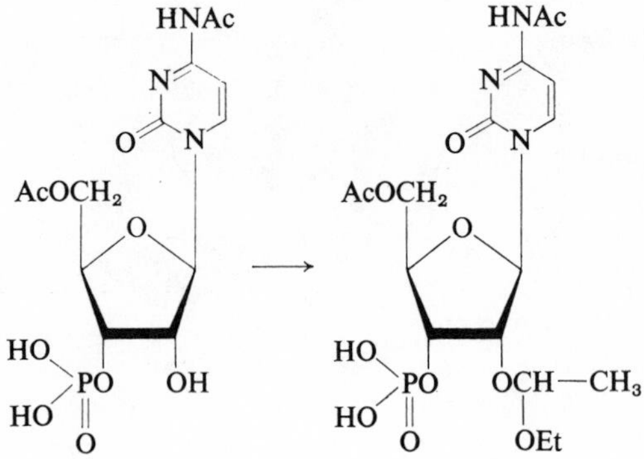

4 (407.2) → **5** (480.0)

INTRODUCTION

The pyridinium salt of *N*-acetyl-5′-*O*-acetyl-2′-*O*-(1-ethoxyethyl)cytidine 3′-phosphate[1] is an intermediate in the synthesis of oligonucleotides of the

D-ribofuranose series.[2,3] Its preparation exemplifies the use of pancreatic ribonuclease in the degradation of *N*-acetylcytidine 2′:3′-cyclic phosphate, and the application of the 1-ethoxyethyl group for the protection of the 2′-hydroxyl group in a ribonucleoside 3′-phosphate.

PROCEDURE

N-Acetyl-5′-*O*-acetylcytidine 3′-Phosphate (4), Ammonium Salt

A solution of 10 g. (0.031 mole) of cytidine 2′(3′)-phosphate [2′(3′)-cytidylic acid, **1**] in 74 ml. of 2 *N* ammonium hydroxide is successively treated with 60 ml. of *N*,*N*-dimethylformamide and a solution of 14.5 g. (0.07 mole) of dicyclohexylcarbodiimide in 60 ml. of *tert*-butyl alcohol. The resulting mixture is refluxed for 3 hr. At the end of this time, electrophoresis of a sample of the mixture on Whatman No. 1 paper in 0.05 *M* sodium dihydrogen phosphate should show a single[a] ultraviolet-absorbing spot of cytidine 2′: 3′-cyclic phosphate (**2**) (its mobility is 0.6 of that of the 2′(3′)-cytidylic acid started with). The reaction mixture is allowed to cool, and diluted with 250 ml. of water, and the mixture is shaken with 200 ml. of ether. The dicyclohexylurea that separates is filtered off, and washed with 20 ml. of water. The filtrates are combined, and the aqueous layer is washed with two 100-ml. portions of ether and evaporated to dryness at 12 mm. Hg.[b] The residue is co-evaporated with two 100-ml. portions of pyridine, to afford a mixture of ammonium and dicyclohexylguanidinium salts of cytidine 2′:3′-cyclic phosphate (**2**), containing a small amount of *N*,*N*-dimethylformamide. This mixture is shaken with a solution of 100 ml. of anhydrous pyridine[c] and 50 ml. of anhydrous *N*,*N*-dimethylformamide[d] in 100 ml. of acetic anhydride for 2 hr. At the end of this time, a clear solution is obtained; this is kept in the dark at room temperature for 15 hr., and evaporated at 12 mm. Hg (bath temperature, 45°). The residue is dried at 0.5 mm. Hg (Dry Ice–ethanol trap), and then evaporated with five 50-ml. portions of pyridine (to remove acetic anhydride). The residue is dissolved in 200 ml. of 50% aqueous pyridine, and the solution is kept at room temperature for 1 hr., and then evaporated at 12 mm. Hg, and finally at 0.5 mm. Hg. The residue is co-evaporated with six 20-ml. portions of 50% aqueous pyridine (to remove pyridinium acetate). The residual,

[a] When the electrophoresis shows the presence of unreacted starting compound, ammonia of a lower concentration had been used in the reaction. In this case, the reaction mixture is treated with an additional 3 ml. of concentrated ammonium hydroxide and 5 g. of dicyclohexylcarbodiimide, and the mixture is refluxed for 2 hr.

[b] All evaporations are performed on a rotary evaporator.

[c] Pyridine (reagent grade) is kept for several days over calcium hydride, and filtered, and the filtrate is stored over molecular sieves.

[d] *N*,*N*-Dimethylformamide (reagent grade) is distilled from 5% of phosphorus pentaoxide, and the distillate is stored over molecular sieves.

glassy *N*-acetyl-5′-*O*-acetylcytidine 2′:3′-cyclic phosphate (**3**) is dissolved in 250 ml. of 20% aqueous pyridine, 50 mg. of pancreatic ribonuclease in 5 ml. of water is added, and the mixture is kept at room temperature overnight. Electrophoresis of a sample of the resulting mixture in 0.05 *M* sodium dihydrogen phosphate should show a single, ultraviolet-absorbing spot possessing the same mobility as uridylic acid.[e] The pH of the solution is adjusted to 7.5 with dilute ammonium hydroxide, and the solution is passed through a column (1 × 5 cm.) of Dowex-50 (NH_4^+) ion-exchange resin (200 ml.). The column is eluted with 500 ml. of water, and the eluate is evaporated at 12 mm. Hg to a thick sirup which is co-evaporated with 200 ml. of ethanol, and then mixed with 100 ml. of methanol and 250 ml. of ether. The mixture is kept at room temperature for 15 hr., and the salt that separates is collected with suction, washed successively with three 50-ml. portions of 1:2 methanol–ether and one 50-ml. portion of ether, and air-dried. After additional drying at 0.1 mm. Hg, there is obtained 12.3 g. (90%) of the ammonium salt of compound **4**, R_f 0.8 in 5:2 ethanol–*M* ammonium acetate.

N-Acetyl-5′-*O*-acetyl-2′-*O*-(1-ethoxyethyl)cytidine 3′-Phosphate (5)

The ammonium salt of compound **4** (441 mg., 1 mmole), 2 ml. of anhydrous *N*,*N*-dimethylformamide, and 1.5 ml. of ethyl vinyl ether are placed in a flask fitted with a magnetic stirrer and a dropping funnel, and protected against atmospheric moisture. The suspension is cooled in Dry Ice–chloroform to −60°, and treated dropwise during 10 min. with 0.5 ml. of trifluoroacetic acid. The cooling bath is replaced by ice–water, and stirring is continued for 3 hr. The resulting, yellow solution is recooled to −60°, and treated with 2 ml. of triethylamine, and the mixture is immediately evaporated at 0.1 mm. Hg (bath temperature, 35°) to a dark sirup. A solution of the sirup in 10 ml. of methanol is poured, with stirring, into a solution of 0.5 g. of silver trifluoroacetate[f] in 40 ml. of ether. The precipitate that separates is collected by centrifugation, successively triturated with 10 ml. of methanol and 40 ml. of ether,[g] and recentrifuged. This procedure[h] is repeated twice more, and then the product is finally washed with 40 ml. of ether, and air-dried. After being dried at 0.1 mm. Hg, there

[e] When an additional spot of the cyclic phosphate (**3**) is present (its mobility is 0.6 of that of uridylic acid), an additional 50 mg. of the enzyme is added to the reaction mixture; this mixture is kept for 10 hr., and the test is repeated.

[f] Silver trifluoroacetate is prepared by dissolving silver oxide in 10% aqueous trifluoroacetic acid, evaporating the resulting neutral solution, and drying the residual salt at 0.1 mm. Hg.

[g] To the ether used in the isolation procedure should be added 0.1% of triethylamine.

[h] In the course of the isolation procedure, the silver salt of compound **5** should be protected against direct sunlight.

is obtained 461 mg. (66%) of the silver salt of compound **5**, R_f 0.91 in 5:2 ethanol–*M* ammonium acetate. The product may be stored in the dark at 4° for several weeks without any change.

REFERENCES

(1) J. Smrt and S. Chládek, *Collection Czech. Chem. Commun.*, **3**, 2978 (1966).
(2) J. Žemlička, S. Chládek, A. Holý, and J. Smrt, *Collection Czech. Chem. Commun.*, **31**, 3198 (1966).
(3) A. Holý and J. Smrt, *Collection Czech. Chem. Commun.*, **31**, 3800 (1966).

[144] 1-β-D-Ribofuranosylhydrouracil 5′-(α-D-Glucopyranosyl Pyrophosphate)

Hydrogenation of Uridine Derivatives

V. N. SHIBAEV and G. I. ELISEEVA

INSTITUTE FOR CHEMISTRY OF NATURAL PRODUCTS,
ACADEMY OF SCIENCES OF U.S.S.R., MOSCOW, U.S.S.R.

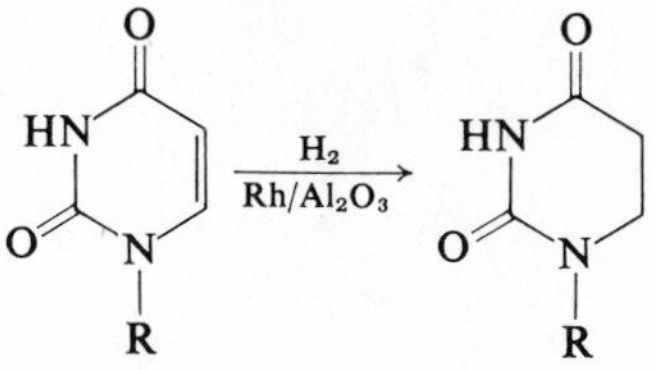

H_2, Rh/Al_2O_3

1 (610.3) → **2** (612.3)

where R =

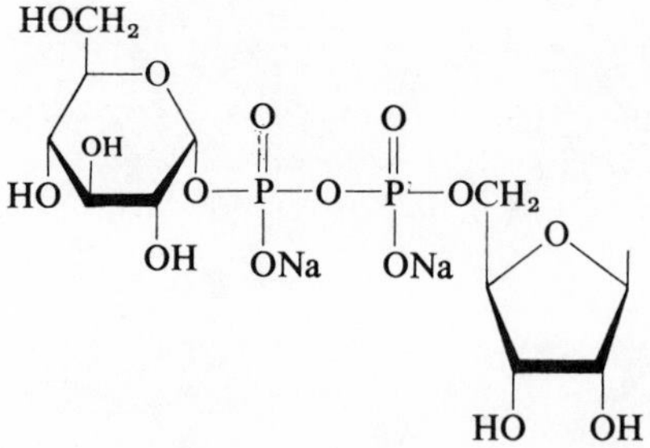

INTRODUCTION

"5,6-Dihydrouridine" 5′-(α-D-glucopyranosyl pyrophosphate)[1] (**2**) has been proved to be an active analog of uridine 5′-(α-D-glucopyranosyl pyrophosphate)[2] (**1**). The hydrogenation of uridine derivatives over rhodium-on-alumina to "5,6-dihydrouridine" derivatives has been described for nucleosides,[3,4] mononucleotides,[3] and oligonucleotides.[5]

PROCEDURE

1-β-D-Ribofuranosylhydrouracil 5′-(α-D-Glucopyranosyl Disodium Pyrophosphate) (2)

A solution of uridine 5′-(α-D-glucopyranosyl disodium pyrophosphate) (**1**) (100 mg.,[a] 0.149 mmole) in 0.005 *M* lithium acetate buffer of pH 4.0

[a] "Gee Lawson" preparation, 91% purity based on measurement of optical absorbance.

(12 ml.) is treated with 5% rhodium-on-alumina (70 mg.) and hydrogenated, with magnetic stirring, for 6–8 hr.[b] The catalyst is filtered off, and the filtrate is evaporated *in vacuo*. The residue is co-evaporated with 2:1 benzene–absolute ethanol (2 × 30 ml.), and dissolved in methanol (5 ml.). Acetone (35 ml.) and ether (5 ml.) are added, and the resulting precipitate is collected by centrifugation. This procedure is repeated thrice, the resulting precipitate is dissolved in water (5 ml.), and the solution is passed down a column (4 × 1 cm.) of Dowex-50 (Na^+) ion-exchange resin. Lyophilization of the effluent gives the disodium salt **2** (62.5 mg., 68%); $\lambda_{max}^{H_2O}$ 198 and 206 (infl.) nm., $\epsilon_{max}^{H_2O}$ 10,400 and 7,600. The product is homogeneous by paper chromatography, R_f 0.29 (5:2 ethanol–0.5 *M* ammonium acetate of pH 7.5). The chromatogram is sprayed with 0.5 *N* sodium hydroxide, heated for 15 min. at 105°, and then sprayed with a 1% solution of *p*-(dimethylamino)benzaldehyde in a solution of concentrated hydrochloric acid (10 ml.) in ethanol (90 ml.). The "dihydrouridine" derivatives give yellow spots.[5,6]

REFERENCES

(1) N. K. Kochetkov, E. I. Budowsky, V. N. Shibaev, and G. I. Eliseeva, *Izv. Akad. Nauk SSSR, Ser. Khim.*, **1965,** 914.
(2) E. I. Budowsky, T. N. Drushinina, G. I. Eliseeva, N. D. Gabrieljan, N. K. Kochetkov, M. A. Novikova, and G. L. Zhdanov, *Biochim. Biophys. Acta*, **122**, 213 (1966).
(3) W. E. Cohn and D. G. Doherty, *J. Am. Chem. Soc.*, **78**, 2863 (1956).
(4) M. Green and S. S. Cohen, *J. Biol. Chem.*, **225**, 397 (1957).
(5) C. Janion and D. Shugar, *Acta Biochim. Polon.*, **7**, 309 (1960).
(6) R. M. Fink, R. E. Cline, C. McGaughey, and K. Fink, *Anal. Chem.*, **28**, 4 (1956).

[b] The process may be monitored by measuring the decrease of A_{260}.

[145] 1-β-D-Ribofuranosylhydrouracil 5′-Phosphates (5,6-Dihydrouridine 5′-Phosphate Derivatives)

Catalytic Hydrogenation of the 5,6-Double Bond of the Pyrimidine Ring of Nucleotides

P. ROY-BURMAN

DEPARTMENT OF BIOCHEMISTRY, UNIVERSITY OF SOUTHERN CALIFORNIA, LOS ANGELES, CALIFORNIA 90033

O, HN, O, N, $ROCH_2$, O, HO, OH $\xrightarrow{H_2/Rh\ catalyst}$ O, HN, O, N, $ROCH_2$, O, HO, OH

1a, R = HO—P(=O)(HO)— (324.2) **2a**, R = HO—P(=O)(HO)— (326.2)

1b, R = HO—P(=O)(HO)—O—P(=O)(HO)— (404.1) **2b**, R = HO—P(=O)(HO)—O—P(=O)(HO)— (406.1)

1c, R = HO—P(=O)(HO)—O—P(=O)(HO)—O—P(=O)(HO)— (484.1) **2c**, R = HO—P(=O)(HO)—O—P(=O)(HO)—O—P(=O)(HO)— (486.1)

1d, R = $HOCH_2$, O, OH, HO, OH —O—P(=O)(OH)—O—P(=O)(OH)— (566.3) **2d**, R = $HOCH_2$, O, OH, HO, OH —O—P(=O)(OH)—O—P(=O)(OH)— (568.3)

INTRODUCTION

"5,6-Dihydrouridine" 5′-phosphate (**2a**) is a constituent of ribonucleic acids.[1,2] Animal tissues contain an enzyme that catalyzes the interconversion of **2a** and uridine 5′-monophosphate (**1a**).[3] Compound **2a** is enzymically phosphorylated[4,5] to the corresponding nucleoside pyrophosphate (**2b**) and triphosphate (**2c**). Enzymic incorporation of **2a**[6] and **2c**[5,7] into ribonucleic acid has been demonstrated. The functional importance of hydrouracil as a constituent of ribonucleic acid is obscure, but speculations concerning its function are based on (a) the ability of hydrouracil to open and close its ring structure reversibly[2] and (b) its failure to form hydrogen-bonded structures[8] or to direct the synthesis of polypeptides.[9] Although **2c** is not a substrate for uridine glycosyl pyrophosphate pyrophosphorylase,[5] **2d** is oxidized to "5,6-dihydrouridine" 5′-(D-glucopyranosyluronic acid pyrophosphate) by uridine glycosyl pyrophosphate dehydrogenase.[10]

PROCEDURE

The catalytic hydrogenation of uridine nucleotides is most conveniently conducted by passing hydrogen gas through a solution of the nucleotide in aqueous hydrochloric acid in the presence of a 5% rhodium-on-alumina catalyst.[11] A stoppered test tube fitted with a narrow inlet tube reaching to the bottom and an outlet tube near the top may be used as a hydrogenation apparatus.[12] Disappearance of $A_{260\,\text{nm.}}$ is used as the criterion of reduction.

1-β-D-Ribofuranosylhydrouracil 5′-Phosphate[5,11] (2a)

Uridine 5′-(disodium phosphate) pentahydrate (**1a**) (Sec. IV [151]) (1.42 g., 3.1 mmoles) is dissolved in 25 ml. of water, and the pH of the solution is adjusted to 2 with hydrochloric acid. Rhodium catalyst[a] (250 mg.) is added, and hydrogen gas is passed through the solution at 25° and atmospheric pressure. Absorbance at 260 nm. is lowered to zero within 1 to 3 hr. After removal of catalyst by centrifugation, the solution is neutralized and placed on a 2 × 40 cm. column of Dowex-1 (X8, formate) ion-exchange resin. Compound **2a** is eluted from the column by gradient elution with 9 *N* formic acid, 1 l. of water being used in the mixing chamber. The fractions containing **2a** are detected by the color reaction described by Fink *et al.*[13] Compound **1a**, if present, is completely separated from **2a** in subsequent fractions.[2,5] Fractions containing **2a** are lyophilized, dissolved in water, and converted into the sodium salt by addition of sodium hydroxide to pH 6.5. A white, amorphous powder of **2a** is obtained on

[a] A 5% rhodium-on-alumina, procured from Engelhard Industries, Inc., Newark, New Jersey.

lyophilization; yield[b] 70–85%; lit.[14] $\lambda_{max}^{pH \sim 6}$ 207 nm.[c] (ϵ 5,900), $\lambda_{max}^{pH\,13}$ 235 nm.[d] (ϵ 10,100).

In dilute alkali, the 3,4-linkage of **2a** is opened at room temperature, and *N*-carbamoyl-*N*-β-D-ribofuranosyl-β-alanine 5′-phosphate is formed. The ureido group gives a yellow color with *p*-(dimethylamino)benzaldehyde.[13] Hydrolysis of the *N*-D-ribosyl linkage of **2a** is effected by treatment with dilute acid at 100°.

1-β-D-Ribofuranosylhydrouracil 5′-Pyrophosphate (2b)

The procedure for the preparation of **2b** from 26.9 mg. (50 μmoles) of **1b** (disodium salt, pentahydrate) is similar to that described for the preparation of **2c**, and yields[b] 60% of the sodium salt of **2b** as a white, amorphous powder; $\lambda_{max}^{pH\,13}$ 235 nm.[e] The molar ratio of total phosphorus to acid-stable phosphorus of **2b** is 2. The electrophoretic mobility of **2b** is the same as that of **1b**. Compound **2b** migrates as a single compound on a paper chromatogram, as determined by color test.[13]

1-β-D-Ribofuranosylhydrouracil 5′-Triphosphate (2c)[5,7]

Uridine 5′-(dihydrogen disodium triphosphate) hexahydrate (**1c**) (31.5 mg., 50 μmoles) is dissolved in 5 ml. of cold, aqueous hydrochloric acid of pH 3.5. Rhodium catalyst[a] (10 mg.) is added, and hydrogenation is conducted as described for **2a**, except that the temperature is kept at 0°. The time required for complete reduction is about 5 hr. After removal of the catalyst, **2c** is separated from small amounts of degradation products by gradient elution on a DEAE-cellulose (HCO_3^-) column (1 × 40 cm.), with 0.2 *M* triethylammonium hydrogen carbonate (pH 7.8) and 1 liter of water in the mixing chamber. Fractions are collected at 4°. The presence of dihydrouracil-containing compounds in eluate fractions is detected by ultraviolet absorption at 230 nm. Small amounts of **2a** and **2b** are eluted prior to **2c**, which emerges as the major product. The desired fractions are collected, lyophilized, dissolved in small amounts of water, and passed through Dowex 50 (X8, Na^+) ion-exchange resin. The eluate containing the sodium salt of **2c** is relyophilized and stored at −20°; it is obtained as a dry, white powder. As determined by phosphate analysis, the yield of **2c** is 60%; $\lambda_{max}^{pH\,13}$ 235 nm.[e]

The molar ratio of total phosphorus to acid-stable phosphorus of **2c** is 3. For the determination of acid-labile phosphorus, the compound is hydrolyzed with *N* hydrochloric acid at 100° for 10 min. Its electrophoretic

[b] Calculated from phosphate analysis.
[c] Point of inflexion.
[d] Prior to decomposition. Time to half-decomposition at pH 13 and 22° is 6.5 min.
[e] Prior to decomposition.

mobility is the same as that of **1c**. As determined by color test,[13] it migrates as a single compound on a paper chromatogram developed with 5:2 (v/v) ethanol–0.5 *M* ammonium acetate of pH 7.5, and no material in the area corresponding to **2b** and **2c** is detected.

1-β-D-Ribofuranosylhydrouracil 5′-(α-D-Glucopyranosyl Pyrophosphate)[10] (2d)

Hydrogenation of 50 mg. of uridine 5′-(α-D-glucopyranosyl pyrophosphate) (**1d**), dipotassium salt, 3.5 hydrate (47.2 mg., 67 μmoles) in 10 ml. of aqueous hydrochloric acid of pH 3.8 with 25 mg. of rhodium catalyst[a] is performed at 0° as in the preparation of **2a**. After hydrogenation, and removal of the catalyst, the solution is neutralized and lyophilized, and the product is dissolved in the minimal volume of water. Compound **2d** is separated from small amounts of degradation products by paper chromatography on Whatman No. 3MM filter paper, with 5:2 (v/v) ethanol–0.5 *M* ammonium acetate of pH 7.5. The major region containing **2d**, as detected by color reagent,[13] corresponds to the area of **1d**. In this system, **2a**, **2b**, **2c**, and **2d** are separated from each other, and their R_f values are about the same as those for **1a**, **1b**, **1c**, and **1d** (namely, 0.51, 0.39, 0.33, and 0.60, respectively). Compound **2d** is eluted with cold water, and the solution is lyophilized. The resulting, white, amorphous powder is dissolved in 1 ml. of water, and the solution is desalted by passage through a column of Sephadex G-10 (1.5 × 80 cm.). Elution from the column is conducted at 4° with water. The fractions containing the product, as determined by $A_{230\ \mathrm{nm.}}$, precede those containing ammonium acetate (as determined with Nessler's reagent). The fractions containing **2d** are pooled and lyophilized, and the resulting, dry, white powder is stored at −20°. The yield[b] of **2d** is 65%; $\lambda_{max}^{pH\,13}$ 235 nm.[d]

The molar ratio of total phosphorus to acid-stable phosphorus of **2d** is 2. The D-glucose residue, released by hydrolysis for 10 min. at 100° in *N* hydrochloric acid, is identified by paper chromatography in 2:2:1 (v/v) ethyl acetate–water–pyridine.

REFERENCES

(1) R. W. Holley, J. Apgar, G. A. Everett, J. T. Madison, M. Marquisee, S. M. Merrill, J. R. Penswick, and P. A. Zamir, *Science*, **147**, 1462 (1965); H. G. Zachau, D. Dütting, and H. Feldman, *Angew. Chem., Intern. Ed.*, **5**, 422 (1966); J. T. Madison, G. A. Everett, and H. Kung, *Abstr. Cold Spring Harbor Symp. Quant. Biol.*, **31**, 43 (1966); U. L. Raj Bhandary and A. Stuart, *Federation Proc.*, **25**, 520 (1966).

(2) R. C. Huang and J. Bonner, *Proc. Natl. Acad. Sci. U. S.*, **54**, 960 (1965).
(3) L. C. Mokrasch and S. Grisolia, *Biochim. Biophys. Acta*, **33**, 444 (1959).
(4) P. Roy-Burman, S. Roy-Burman, and D. W. Visser, *Federation Proc.*, **25**, 275 (1966).
(5) P. Roy-Burman, S. Roy-Burman, and D. W. Visser, *Biochim. Biophys. Acta*, **142**, 355 (1967).
(6) D. O. Carr and S. Grisolia, *J. Biol. Chem.*, **239**, 160 (1964).
(7) P. Roy-Burman, S. Roy-Burman, and D. W. Visser, *Biochem. Biophys. Res. Commun.*, **20**, 291 (1965).
(8) P. Cerutti, H. T. Miles, and J. Frazier, *Biochem. Biophys. Res. Commun.*, **22**, 466 (1966).
(9) F. Rottman and P. Cerutti, *Proc. Natl. Acad. Sci. U. S.*, **55**, 960 (1966); J. Smrt, J. Škoda, V. Lisý, and F. Šorm, *Biochim. Biophys. Acta*, **129**, 210 (1966).
(10) P. Roy-Burman, S. Roy-Burman, and D. W. Visser, *J. Biol Chem.*, **243**, 1692 (1968).
(11) W. E. Cohn and D. G. Doherty, *J. Am. Chem. Soc.*, **78**, 2863 (1956).
(12) L. M. Joshel, *Ind. Eng. Chem. Anal. Ed.*, **15**, 590 (1943).
(13) R. M. Fink, R. E. Cline, C. McGaughey, and K. Fink, *Anal. Chem.*, **28**, 4 (1956).
(14) C. Janion and D. Shugar, *Acta Biochim. Polon.*, **7**, 309 (1960).

[146] Isocytidine 5′-Phosphate

Phosphorylation of Nucleosides with Pyrophosphoryl Chloride

N. K. KOCHETKOV, E. I. BUDOWSKY, and V. N. SHIBAEV

INSTITUTE FOR CHEMISTRY OF NATURAL PRODUCTS,
ACADEMY OF SCIENCES OF U.S.S.R., MOSCOW, U.S.S.R.

O N N NH_2 $HOCH_2$ O O O Ip **1** (283.3)

1. $P_2O_3Cl_4$ **2** (251.8) → 2. H_2O, H^+

O N N NH_2 O $(LiO)_2POCH_2$ O HO OH **3** (335.0)

INTRODUCTION

Isocytidine 5′-phosphate (**3**) has been used as the starting material for the synthesis of isocytidine 5′-(D-glucosyl pyrophosphate)[1]; this has been used for investigation of correlation of structure with biological functions for nucleoside (glycosyl pyrophosphates).[2] Phosphorylation of 2′,3′-*O*-isopropylideneisocytidine (**1**) with pyrophosphoryl chloride as described here gives better results than phosphorylation with 2-cyanoethyl phosphate and dicyclohexylcarbodiimide, described earlier.[3]

PROCEDURE

Isocytidine 5′-Phosphate (3)

Finely powdered 2′,3′-*O*-isopropylideneisocytidine[4] (**1**) (Sec. III [99]) (416.7 mg., 1.47 mmoles) is cooled in Dry Ice–acetone and treated with pyrophosphoryl chloride[5] (**2**) (0.6 ml., 920 mg., 3.68 mmoles). The mixture is allowed to warm to room temperature with stirring, and is then kept at room temperature overnight. The flask is now cooled in Dry Ice–acetone, water (0.5 ml.) is added, and the mixture is heated to room temperature with intense stirring, diluted with water (20 ml.), and kept at room temperature for 20 min.[a] To the solution is added *M* lithium hydroxide to pH 8.5, and a small amount of a precipitate (lithium phosphate) is filtered off. The filtrate is diluted with water to 500 ml., and the solution is applied to a column (3 × 19 cm.) of Dowex-1 X4 (Cl^-) (100–200 mesh). The column is washed with water (300 ml.) to remove isocytidine (A_{246} 1,246; 0.22 mmole), and eluted with 0.003 *N* hydrochloric acid. Compound **3** appears in the eluate after 3.2 l. of eluant has passed through; its elution is completed by passing a further 1.2 l. of acid solution through the column. The fractions that contain **3** (A_{246} 4,500; 0.78 mmole) are combined, rendered neutral with lithium hydroxide, and evaporated to dryness. The residue is dried by adding and evaporating benzene–absolute ethanol, and then dissolved in methanol (5 ml.), and the dilithium salt of **3** is precipitated by adding acetone (35 ml.) and ether (5 ml.). Reprecipitation is repeated twice, to give 240 mg. (50%) of the dilithium salt of **3**; λ_{max} 220 and 256 nm., ϵ_{max} 9,400 and 7,400; λ_{min} 239 nm., ϵ_{min} 5,000 (0.1 *N* hydrochloric acid); λ_{max} 255 nm., ϵ_{max} 5,800 (water); λ_{max} 226 nm., ϵ_{max} 16,000 (0.1 *N* potassium hydroxide). The isosbestic point is at 246 nm. (ϵ 5,800). The preparation of **3** is homogeneous by paper chromatography; R_f 0.17 in 5:2 ethyl alcohol–0.5 *M* ammonium acetate of pH 7.5. The electrophoretic mobility of **3** is equal to that of uridine 5′-phosphate (UMP) at pH 7.5 (0.05 *M* triethylammonium hydrogen carbonate, R_{UMP} 1.0) and lower at pH 4.0 (0.05 *M* triethylammonium acetate, R_{UMP} 0.65).

REFERENCES

(1) N. K. Kochetkov, E. I. Budowsky, and V. N. Shibaev, *Khim. Prirodn. Soedin.*, **1965**, 328.
(2) E. I. Budowsky, T. N. Drushinina, G. I. Eliseeva, N. D. Gabrieljan, N. K. Kochetkov, N. A. Novikova, V. N. Shibaev, and G. L. Zhdanov, *Biochim. Biophys. Acta*, **122**, 213 (1966).
(3) N. K. Kochetkov, E. I. Budowsky, V. N. Shibaev, G. I. Eliseeva, M. A. Grachev, and V. P. Demushkin, *Tetrahedron*, **19**, 1207 (1963).
(4) D. M. Brown, A. R. Todd, and S. Varadarajan, *J. Chem. Soc.*, **1957**, 868.
(5) H. Grunze, *Z. Anorg. Allgem. Chem.*, **296**, 63 (1958).

[a] Complete deacetonation occurs under these conditions, without splitting of the *N*-glycosyl bond. Such splitting readily occurs if the hydrolysis is conducted with hot, dilute acid.

[147] 9-β-D-ribofuranosyl-9*H*-purine-6(1*H*)-thione 5′-Phosphate (6-Mercaptopurine Ribonucleotide)

Preparation of an Unnatural, Purine Nucleotide

JOHN A. MONTGOMERY and H. JEANETTE THOMAS

KETTERING-MEYER LABORATORY, SOUTHERN RESEARCH INSTITUTE, BIRMINGHAM, ALABAMA 35205

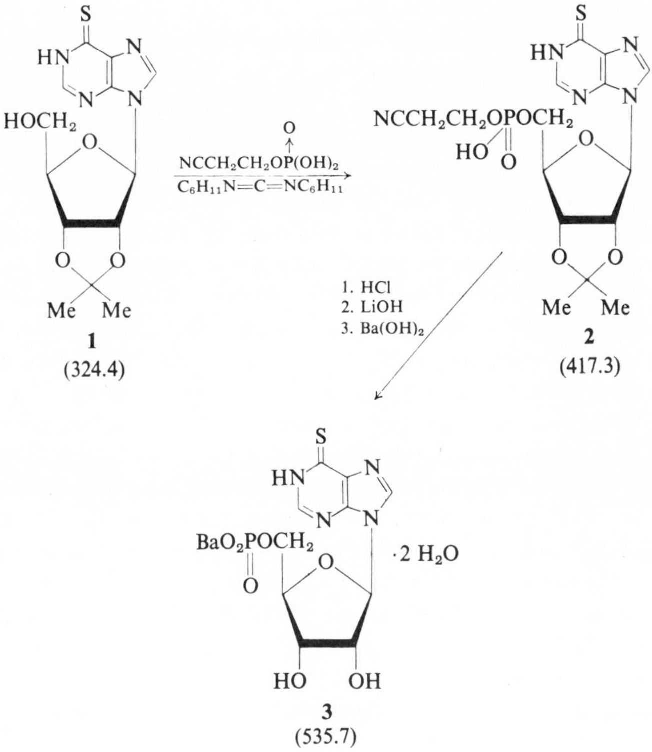

$NCCH_2CH_2OP(O)(OH)_2$, $C_6H_{11}N{=}C{=}NC_6H_{11}$

1 (324.4) → **2** (417.3)

1. HCl
2. LiOH
3. $Ba(OH)_2$

→ **3** $\cdot 2\ H_2O$ (535.7)

INTRODUCTION

6-Mercaptopurine ribonucleotide[1,2] (**3**), formed in cells by the enzymic reaction of 6-mercaptopurine with 5-*O*-phosphono-D-ribofuranosyl pyrophosphate, is known to be the active metabolite (of 6-mercaptopurine) that causes cell death.[3] The reaction of 2′,3′-*O*-isopropylidene-6-thioinosine with 2-cyanoethyl phosphate and dicyclohexylcarbodiimide,[4] followed by removal of protecting groups, provides for this compound a convenient synthesis[2] that is readily adapted to the synthesis of other nucleotides.

PROCEDURE

6-Mercaptopurine Ribonucleotide [9-(5-*O*-Phosphono-β-D-ribofuranosyl)-9*H*-purine-6(1*H*)-thione], Barium Salt (3)

To a solution of 3.16 g. (21.0 mmoles) of 2-cyanoethyl phosphate (hydracrylonitrile phosphate) (Sec. V [159]) in 150 ml. of anhydrous pyridine (in a 250-ml., round-bottomed flask) is quickly added 17.3 g. (83.8 mmoles) of dicyclohexylcarbodiimide followed by 3.39 g. (10.5 mmoles) of 9-(2,3-*O*-isopropylidene-β-D-ribofuranosyl)-9*H*-purine-6(1*H*)-thione (**1**) (Sec. III [89]), and the flask is stoppered. After about 5 min., a precipitate begins to form in the solution, which is kept at room temperature for 2 days, diluted with 21 ml. of water, and kept for 1 hr. The precipitate (dicyclohexylurea) is removed by filtration, the filtrate is evaporated to dryness under diminished pressure below 30°, and the residue is dissolved in 200 ml. of water. The solution is washed with 100 ml. of chloroform, and the aqueous solution of 9-(2,3-*O*-isopropylidene-β-D-ribofuranosyl)-9*H*-purine-6(1*H*)-thione 5′-(2-cyanoethyl hydrogen phosphate) (**2**) is diluted with *N* hydrochloric acid to give 199 ml. of a solution 0.3 *N* in hydrochloric acid. The solution is kept at room temperature for 24 hr., neutralized with the theoretical amount of 6 *N* sodium hydroxide (9.95 ml.), and the resulting aqueous solution of 9-β-D-ribofuranosyl-9*H*-purine-6(1*H*)-thione 5′-(2-cyanoethyl hydrogen phosphate), now at pH 2.2, is diluted with sufficient 3 *N* lithium hydroxide to give 292 ml. of a solution 0.5 *N* in lithium hydroxide; this is heated in a bath at 100° for 15 min., cooled, and filtered. The filtrate is stirred for 15 min. with 133 ml. of Amberlite IR-120 (H^+) ion-exchange resin, added portionwise until pH 2.2 is attained. The resin is removed by filtration and washed with water, and the filtrate and washings are combined; the pH is adjusted to 7.5 by cautious addition of aqueous barium hydroxide, and the suspension is filtered. The clear filtrate is evaporated to 260 ml., and diluted with ethanol (2 ml. per ml.) The resulting white precipitate is collected by filtration, and successively washed with ethanol and ether; yield 2.24 g. (45%); λ_{max} in nm. (log ε): pH 1, 324 (4.33); pH 7, 321 (4.36); pH 13, 311 (4.35).

REFERENCES

(1) L. N. Lukens and K. A. Herrington, *Biochim. Biophys. Acta*, **24**, 432 (1957).
(2) J. A. Montgomery and H. J. Thomas, *J. Org. Chem.*, **26**, 1926 (1961).
(3) R. W. Brockman, *Advan. Cancer Res.*, **7**, 129 (1962).
(4) P. T. Gilham and G. M. Tener, *Chem. Ind.* (London), **1959**, 542.

[148] 2-β-D-Ribofuranosyl-*as*-triazine-3,5(2*H*,4*H*)-dione 2′:3′-Cyclic Phosphate (6-Azauridine 2′:3′-Cyclic Phosphate)

Synthesis of Ribonucleoside 2′:3′-Cyclic Phosphates by Oxidative Cyclization of Ribonucleoside 2′(3′)-Phosphites

A. HOLÝ

INSTITUTE OF ORGANIC CHEMISTRY AND BIOCHEMISTRY, CZECHOSLOVAK ACADEMY OF SCIENCES, PRAGUE, CZECHOSLOVAKIA

$HOCH_2$ B O HO OH **1** (245.2)

1. $(EtO)_3P/H^+$
2. OH^-

$HOCH_2$ B O O OH O=P—H HO (+ 2′-isomer) **2** (309.3)

1. $O{=}C(CCl_3)_2$
2. OH^-

$HOCH_2$ B O O OH O=P—OH HO (+ 2′-isomer) **4** (325.3)

+

$HOCH_2$ B O O O P O OH **3** (307.3)

DCC

where B = (O, HN, N, O, N)

INTRODUCTION

Ribonucleoside 2′:3′-cyclic phosphates may be used as a source of specifically protected nucleotide derivatives,[1] or as the starting material in direct enzymic synthesis of oligoribonucleotides.[2] The present synthesis of 6-azauridine 2′:3′-cyclic phosphate (**3**) is an example of a general method[3] for the synthesis of ribonucleoside 2′:3′-cyclic phosphates that consists of (*a*) selective substitution of the *cis*-diol system of the D-ribofuranosyl group of unprotected ribonucleosides by reaction with triethyl phosphite,[4] and (*b*) subsequent oxidative cyclization of the resulting 2′(3′)-phosphites by reaction with hexachloro-2-propanone.

PROCEDURE

All evaporations are conducted at 35°/15 mm. Hg (rotary evaporator).

6-Azauridine 2′(3′)-Phosphite (2)

A suspension of 25 g. (0.10 mole) of 6-azauridine (**1**), 100 ml. of anhydrous *N,N*-dimethylformamide, 40 ml. of triethyl phosphite,[4] and 5 ml. of a 4.7 *M* solution of hydrogen chloride in *N,N*-dimethylformamide is shaken until a clear solution[a] is obtained (about 15 min.). The solution is kept at room temperature overnight, with exclusion of atmospheric moisture, and then treated with 500 ml. of 1:4 (v/v) aqueous ammonia. The resulting emulsion is shaken at room temperature for 3 hr., and washed with three 100-ml. portions of ether.[b] The aqueous layer is evaporated to dryness, and the residue is co-evaporated with three 50-ml. portions of pyridine. The partly solid residue is dissolved in 50 ml. of water, and the solution is put on a column (5 × 100 cm.) of *O*-[2-(diethylamino)ethyl]-cellulose (HCO_3^-).[c] Elution of the column is performed with a linear gradient of triethylammonium hydrogen carbonate (4 l. of water in the mixing chamber, and 4 l. of a 0.2 *M* buffer solution in the reservoir) at a flow-rate of 3–5 ml. per min. The optical absorbance of the eluate is checked by continuous measurement of the absorption. The first peak (roughly between 0.06 and 0.10 *M* salt concentration) contains unreacted

[a] The resulting solution should be strongly acidic (pH 1–2); otherwise, it is necessary to add more of the *N,N*-dimethylformamide solution of hydrogen chloride. In the case of basic ribonucleosides, a small excess of hydrogen chloride should always be present.

[b] Triethyl phosphite is highly toxic. All operations should be performed in a well ventilated hood.

[c] DEAE-cellulose Cellex D ("high capacity") is obtainable from Calbiochem. The powder is successively washed with 0.1 *N* hydrochloric acid, water, 0.1 *N* sodium hydroxide, and water. The material is then suspended in water, and the suspension is saturated with gaseous carbon dioxide. The column is packed with the adsorbent under moderate pressure, and washed with 2 *M* triethylammonium hydrogen carbonate (pH 7.5) and 5 l. of water. Regeneration is performed with 300 ml. of the same buffer solution, and the column is then washed with water.

6-azauridine (**1**). When the elution of the first peak is finished, and the optical absorbance of the effluent begins to increase, addition of the 0.2 *M* buffer solution to the mixing chamber is stopped, and the elution is continued with a constant concentration of buffer in the mixing chamber[d] until a decrease in the optical absorbance of the eluate is observed. The fractions corresponding to the second elution peak are combined, and evaporated to dryness. The residue is co-evaporated with two 50-ml. portions of water, and then dissolved in 25 ml. of water. The solution is put on a column of 200 ml. of Dowex-50 X8 (pyridinium$^+$) (100–200 mesh) ion-exchange resin,[e] and the elution is performed with 500 ml. of 10% aqueous pyridine at a flow-rate of 3 ml. per min. The effluent is treated with 50 ml. of concentrated ammonium hydroxide, and the mixture is evaporated to dryness; the residue is successively co-evaporated with one 50-ml. portion of water and three 50-ml. portions of 96% ethanol. The solid residue is mixed with the minimal volume (*ca.* 300 ml.) of methanol, the suspension is filtered with suction through a thin layer of Hyflo Super Cel, and the material on the filter is washed with 50 ml. of methanol. The filtrates are combined, and added dropwise with stirring, during 30 min., to 3 l. of anhydrous ether. The precipitate of 6-azauridine 2′(3′)-(ammonium phosphite) (**2**)[f] that separates is immediately collected with suction, washed with 100 ml. of anhydrous ether, and dried under diminished pressure over phosphorus pentaoxide; yield of chromatographically[g] homogeneous product, 24.0 g. (70%); content, 98–100%.

6-Azauridine 2′:3′-Cyclic Phosphate (3)

A suspension of 17.2 g. (0.05 mole) of the ammonium salt of **2** in 100 ml. of *N*,*N*-dimethylformamide is placed in a 500-ml., three-necked flask fitted

[d] The use of a discontinuous gradient is necessary only if the starting nucleoside is of acidic character. With neutral or basic nucleosides, unreacted nucleoside may be removed by elution with water. When the optical absorbance of the first peak decreases, elution is continued by use of a linear gradient of the buffer solution until a new increase in the optical absorbance of the eluate appears. The subsequent procedure is then identical with that given above.

[e] The commercially available material is successively washed with 0.1 *N* hydrochloric acid, water, and 50% aqueous pyridine. Before use, the column is equilibrated by washing it with 200 ml. of 10% aqueous pyridine.

[f] The ammonium salts of ribonucleoside 2′(3′)-phosphites are generally more suitable for storage than the hygroscopic triethylammonium salts (obtained directly by the column chromatography described). When the triethylammonium salt (peak 2) is used immediately in the subsequent step, its conversion into the ammonium salt is unnecessary; co-evaporation of the residue (peak 2) with 96% ethanol and drying under diminished pressure over phosphorus pentaoxide are quite sufficient.

[g] R_f of **2** is 0.35 in 7:1:2 (v/v) isopropyl alcohol–concentrated ammonium hydroxide–water on Whatman No. 1 paper; the R_f value of 6-azauridine is 0.42. In the solvent system used, the mobility of 2′(3′)-phosphites is generally lower than that of the corresponding ribonucleosides.

with a stirrer, 50-ml. dropping funnel, and calcium chloride drying-tube. The flask is immersed in ice–water, and 30 ml. of hexachloro-2-propanone[h] is added, with stirring, during 10 min. The resulting mixture is stirred at 0° for 1 hr., and filtered with suction through a thin layer of Hyflo Super Cel, and the material on the filter is washed with 10 ml. of *N,N*-dimethylformamide. The filtrates are combined, and added dropwise to 500 ml. of 1:4 (v/v) aqueous ammonia. The resulting emulsion is stirred for 3 hr., and washed with three 100-ml. portions of ether, and the aqueous layer is evaporated to dryness. The residue is dissolved in 50 ml. of water, and the solution is poured into a mixture of 100 ml. of 96% ethanol and 500 ml. of acetone. The resulting precipitate is collected, washed successively with 100 ml. of acetone and 100 ml. of ether, and dried at 20°/12 mm. Hg over phosphorus pentaoxide. Yield, 18.0 g. of the ammonium salt of **3**, contaminated with approximately 10% of the ammonium salt of 6-azauridine 2′(3′)-phosphate[i] (**4**).

A solution of this mixture of ammonium salts in 100 ml. of 2 *N* ammonium hydroxide, 90 ml. of *N,N*-dimethylformamide, and 90 ml. of *tert*-butyl alcohol is refluxed with 25 g. of dicyclohexylcarbodiimide for 6 hr.,[j] and evaporated to dryness (rotary evaporator). The residue is mixed with 500 ml. of water, and the resulting precipitate is filtered off, and washed with 100 ml. of water. The filtrates are combined, washed with three 100-ml. portions of ether, and filtered with suction through a thin layer of Hyflo Super Cel, and the material on the filter is washed with 50 ml. of water. The filtrates are combined, and evaporated to dryness, and the residue is dissolved in the minimal volume of water (*ca.* 50 ml.). The solution is applied to a column of 250 ml. of Dowex-50 (pyridinium$^+$) (100–200 mesh) ion-exchange resin, and elution is performed with 1.0 l. of 10% aqueous pyridine. The effluent is adjusted to pH 8.5–9 (pH test-paper) by addition of 4 *N* ammonium hydroxide, and the resulting solution is evaporated to dryness. The residue is co-evaporated with two 100-ml. portions of 96% ethanol, and then dissolved in 500 ml. of methanol, and

[h] Commercial hexachloro-2-propanone may be used without purification.

[i] An additional crop may be obtained by concentrating the filtrate to half its original volume, and precipitating with a solution of 20.0 g. of barium iodide in 100 ml. of ethanol. The precipitate is collected with suction, washed with acetone, and dried under diminished pressure over phosphorus pentaoxide; yield, 5.0 g. of a mixture of barium salts of compounds **3** and **4**. This mixture is converted into the ammonium salts on a column of 100 ml. of Dowex-50 (100–200 mesh) ion-exchange resin, with elution with 300 ml. of 10% aqueous pyridine. The effluent is treated with 20 ml. of concentrated ammonium hydroxide, and the solution is evaporated to dryness; the residual ammonium salts are used directly in the next step.

[j] By this procedure, 6-azauridine 2′(3′)-phosphate (**4**) is converted into 6-azauridine 2′:3′-cyclic phosphate (**3**). The course of the conversion is checked by paper chromatography in 7:1:2 (v/v) isopropyl alcohol–concentrated aqueous ammonia–water; R_f values: **4**, 0.06, and **3**, 0.40.

the solution is mixed with 0.5 g. of activated carbon; the suspension is filtered through a thin layer of Hyflo Super Cel, and the material on the filter is washed with 25 ml. of methanol. The filtrates are combined, and added dropwise, with stirring, to 2 l. of ether. The precipitate of the ammonium salt of **3** is immediately collected with suction,[k] washed with 200 ml. of ether and, while still moist with ether, dried over phosphorus pentaoxide at 20°/15 mm. Hg; yield of chromatographically homogeneous product, 13.4 g. (82%).

REFERENCES

(1) J. Smrt and F. Šorm, *Collection Czech. Chem. Commun.*, **27**, 73 (1962); J. Smrt and S. Chládek, *ibid.*, **31**, 2978 (1966); A. Holý, J. Smrt, and F. Šorm, *ibid.*, **32**, 2980 (1967).
(2) M. R. Bernfield, *J. Biol. Chem.*, **241**, 2014 (1966); D. Grünberger, A. Holý, and F. Šorm, *Collection Czech. Chem. Commun.*, **33**, 286 (1968).
(3) A. Holý and J. Smrt, *Collection Czech. Chem. Commun.*, **31**, 1528 (1966).
(4) A. Holý and F. Šorm, *Collection Czech. Chem. Commun.*, **31**, 1562 (1966).

[k] The resulting precipitate is hygroscopic if traces of solvents are present.

[149] 5′-*O*-Acetyl-2′-*O*-(tetrahydropyran-2-yl)uridine 3′-(Calcium Phosphate)

An Intermediate in the Synthesis of an Internucleotide Linkage

J. SMRT

INSTITUTE OF ORGANIC CHEMISTRY AND BIOCHEMISTRY,
CZECHOSLOVAK ACADEMY OF SCIENCES, PRAGUE, CZECHOSLOVAKIA

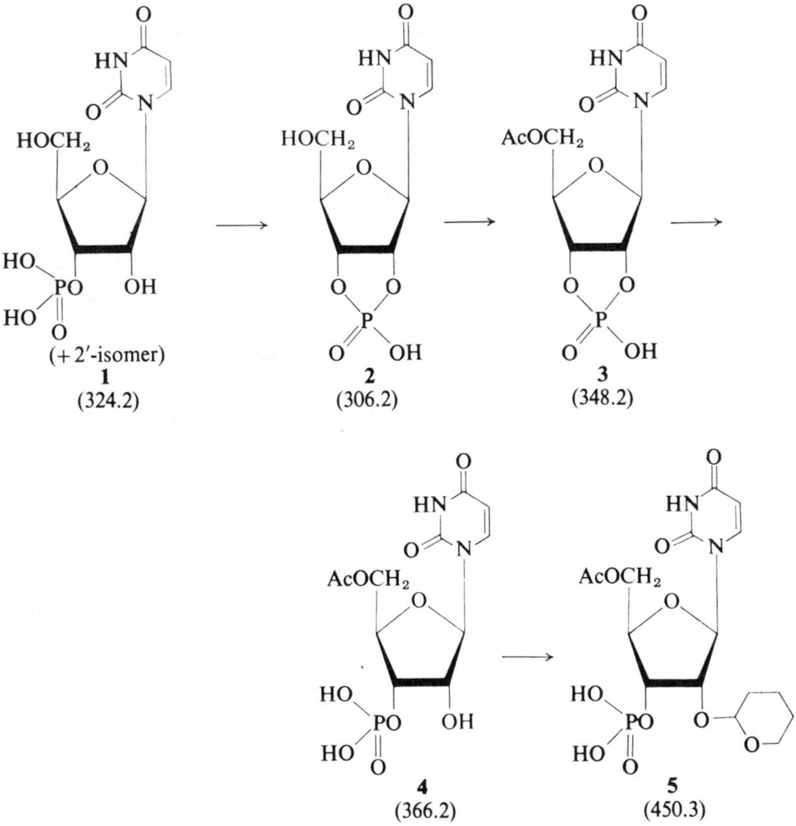

(+2′-isomer) **1** (324.2) → **2** (306.2) → **3** (348.2) → **4** (366.2) → **5** (450.3)

INTRODUCTION

5′-*O*-Acetyl-2′-*O*-(tetrahydropyran-2-yl)uridine 3′-phosphate[1] (**5**) is a readily accessible starting-compound in the synthesis of a naturally occurring inter-ribonucleotide linkage.[2] Its preparation exemplifies the preparative use of pancreatic ribonuclease in the synthesis of a ribonucleoside 3′-phosphate,[3] and the direct protection of a 2′-hydroxyl group of a ribonucleoside 3′-phosphate with a tetrahydropyranyl group.

PROCEDURE

5′-*O*-Acetyluridine 3′-(Calcium Phosphate)[1] (4)

A solution of 10 g. (0.031 mole) of uridine 2′(3′)-phosphate[a] [2′(3′)-uridylic acid, (**1**)] in 74 ml. of 2 *N* ammonia is successively treated with 60 ml. of *N,N*-dimethylformamide and a solution of 14.5 g. (0.07 mole) of dicyclohexylcarbodiimide in 60 ml. of *tert*-butyl alcohol. The resulting mixture is refluxed for 3 hr., after which, electrophoresis of a sample of the mixture on Whatman No. 1 paper in 0.05 *M* disodium hydrogen phosphate should show a single,[b] ultraviolet-absorbing spot of uridine 2′:3′-cyclic phosphate (**2**) (mobility, 0.6 of that[c] of **1**). The solution is evaporated[d] to dryness at 12 mm. Hg (bath temperature, 35°), and the residue is shaken with 100 ml. of water and 200 ml. of ether. The dicyclohexylurea that separates is filtered off, and washed with 20 ml. of water. The aqueous layer is combined with the filtrate, washed with two 100-ml. portions of ether, and evaporated to dryness (bath temperature, 35°). The residue is co-evaporated with two 100-ml. portions of pyridine, giving a residue that is a mixture of the ammonium and dicyclohexylguanidinium salts of uridine 2′:3′-cyclic phosphate (**2**); this is dissolved in 100 ml. of anhydrous pyridine[e] and 200 ml. of acetic anhydride, and the solution is kept in the dark at room temperature overnight, and evaporated to dryness at 12 mm. Hg (bath temperature, 45°). The residue is co-evaporated with four 50-ml. portions of pyridine (to remove acetic anhydride), and then dissolved in 200 ml. of 50% aqueous pyridine. After being kept for 1 hr., the mixture is evaporated (bath temperature, 35°) at 12 mm. Hg and finally at 0.5 mm. Hg. The residue (containing **3**) is co-evaporated (to remove pyridinium acetate) with five 20-ml. portions of 50% aqueous pyridine at 0.5 mm. Hg, with a trap immersed in an ethanol–Dry Ice cooling bath as the con-

[a] The sodium salt of 2′(3′)-uridylic acid is converted into the free nucleotide by passage through a column of a strongly acidic, ion-exchange resin, *e.g.*, Dowex-50 (H^+). About 6 ml. of the resin is used per mmole of the sodium salt.

[b] The reaction does not proceed to completion if ammonia of a lower concentration is used. In this case, the reaction mixture is treated with an additional 3 ml. of concentrated aqueous ammonia and 5 g. of dicyclohexylcarbodiimide, and the whole is refluxed for a further 2 hr.

[c] The reaction course may also be checked by paper chromatography in 7:1:2 isopropyl alcohol–concentrated ammonium hydroxide–water; R_f values: compound **1**, 0.1; compound **2**, 0.32.

[d] All evaporations are performed on a rotary evaporator.

[e] Commercial pyridine (analytical grade) is kept over calcium hydride for several days, and filtered, and the filtrate is stored over molecular sieves.

denser. The glassy residue is dissolved in 200 ml. of 20% aqueous pyridine, and to the solution is added 50 mg. of pancreatic ribonuclease in 5 ml. of water; the mixture is then kept at room temperature for 15 hr. At the end of this time, electrophoresis of a sample of the reaction mixture on Whatman No. 1 paper in 0.05 *M* disodium hydrogen phosphate should show a single, ultraviolet-absorbing spot of the same mobility as that of 2′(3′)-uridylic acid.[f] The resulting solution is then passed through a column (1 × 5 cm.) of 150 ml. of Dowex-50 (pyridinium form) ion-exchange resin. Elution of the column is performed with 300 ml. of 20% aqueous pyridine. The pH of the eluate is adjusted to 7.6 with triethylamine, the solution is concentrated to 100 ml., the concentrate is poured, with stirring, into a solution of 10 g. of calcium chloride in 2.0 l. of ethanol, and stirring is continued for 1 hr. The suspension is kept at room temperature overnight, and the precipitate is collected by centrifugation (3,000 r.p.m., 5 min.), and washed successively (by alternate trituration and centrifugation) with two 1.0-l. portions of ethanol and one 500-ml. portion of ether. After being dried in air and then at 0.1 mm. Hg, there is obtained 12.1 g. (85%) of the calcium salt of 5′-*O*-acetyluridine 3′-phosphate (**4**); R_f 0.56 in 5:2 ethanol–*M* ammonium acetate.

5′-*O*-Acetyl-2′-*O*-(tetrahydropyran-2-yl)uridine 3′-(Calcium Phosphate) (5)

A magnetically stirred suspension of 12 g. (0.026 mole) of the finely ground calcium salt of 5′-*O*-acetyluridine 3′-phosphate (**4**), 160 ml. of anhydrous *N,N*-dimethylformamide,[g] and 70 ml. of dihydropyran is cooled to −20°, and treated dropwise with 14.1 ml. of 5 *M* hydrogen chloride in *N,N*-dimethylformamide[h] during 15 min. under exclusion of atmospheric moisture. The cooling bath is then removed, and stirring is continued until a clear solution is obtained (about 2 hr.). After being kept at room temperature overnight, the mixture is cooled to −20°, and treated with 12 ml. of triethylamine, and the resulting thick suspension is poured into 1.5 l. of ether. The precipitate is collected on a sintered-glass funnel (Jena G3), washed with three 50-ml. portions of ether, air-dried, triturated with 200 ml. of chloroform containing 0.1% of triethylamine, and re-collected with suction. This procedure is repeated twice more, and the product is washed with 50 ml. of chloroform containing 0.1% of triethylamine. After being dried in air and then at 0.1 mm. Hg, there is obtained 12.5 g. (98%) of the

[f] If an additional spot, for compound **3**, is present, the enzyme applied was of a lower activity. In this case, the reaction mixture is kept with an additional 50 mg. of the enzyme for 10 hr., and the test is repeated.

[g] Commercial *N,N*-dimethylformamide (analytical grade) is distilled from phosphorus pentaoxide, and stored over molecular sieves.

[h] The solution is prepared by passing dry hydrogen chloride into dry *N,N*-dimethylformamide, with cooling with tap water.

calcium salt of compound **5**, R_f 0.78 in 5:2 ethanol–*M* ammonium acetate. The product may be stored at room temperature for several months without occurrence of any change.

REFERENCES

(1) J. Smrt and F. Šorm, *Collection Czech. Chem. Commun.*, **27**, 73 (1962).
(2) S. Chládek and J. Smrt, *Collection Czech. Chem. Commun.*, **29**, 214 (1964); J. Smrt, *ibid.*, **29**, 2049 (1964); J. Smrt and F. Šorm, *ibid.*, **29**, 2971 (1964).
(3) D. M. Brown, C. A. Dekker, and A. R. Todd, *J. Chem. Soc.*, **1952**, 2715.

[150] 5-Halogenouridine 5′-(Dihydrogen Phosphates)

A. M. MICHELSON

INSTITUT DE BIOLOGIE PHYSICO-CHIMIQUE, PARIS, FRANCE

1 (324.2) —(a) Cl_2 (b) Br_2 (c) I_2→ **2a**, X = Cl (358.7); **2b**, X = Br (403.1); **2c**, X = I (450.1)

INTRODUCTION

5-Chlorouridine 5′-phosphate (**2a**), 5-bromouridine 5′-phosphate (**2b**), and 5-iodouridine 5′-phosphate (**2c**) are intermediates in the synthesis of polynucleotide analogs[1] that have been used in studies of the system for the *in vitro* synthesis of proteins.[2] The syntheses given here are representative of electrophilic substitution reactions of uridine and its derivatives. The nucleotide is treated directly with the halogen to give the 5-halogeno-nucleotide.[1,3]

PROCEDURE

5-Chlorouridine 5′-Phosphate (2a)

Chlorine (107 mg., 1.5 mmoles) in 2.5 ml. of carbon tetrachloride is added to a solution of 324 mg. (1 mmole) of uridine 5′-phosphate (**1**) in 8 ml. of acetic acid, the mixture is kept at room temperature for 10 min., and the solution is then evaporated to half its original volume, heated at 100° for 5 min., and evaporated to dryness under diminished pressure; the residue is dissolved in ethanol and re-evaporated to dryness. To a concentrated solution of the product in ethanol is added 50 ml. of anhydrous ether, and the precipitated **2a** is collected by centrifugation, washed twice with ether, and dried; yield 345 mg. (96%).

5-Bromouridine 5′-Phosphate (2b)

A solution of bromine (240 mg., 1.5 mmoles) in 2 ml. of carbon tetrachloride is added to a solution of uridine 5′-phosphate (**1**) (324 mg., 1 mmole) in a mixture of 2 ml. of 0.5 *N* nitric acid and 8 ml. of *p*-dioxane, and the solution is kept at room temperature for 1 hr. The solution is evaporated to dryness under diminished pressure, the residue is dissolved in ethanol, and the solution is re-evaporated to dryness under diminished pressure; this process is repeated three times (to remove traces of water). Finally, the nucleotide is precipitated from a concentrated solution in ethanol by the addition of an excess of anhydrous ether, to give 390 mg. (89%) of **2b**.

5-Iodouridine 5′-Phosphate (2c)

To a solution of uridine 5′-phosphate (**1**) (324 mg., 1 mmole) in a mixture of 2 ml. of 0.5 *N* nitric acid and 8 ml. of *p*-dioxane is added 0.51 g. (2 mmoles) of iodine, and the mixture is kept at 100° for 1 hr. under a reflux condenser. Solvent is then removed under diminished pressure, and the residue is dried by repeated dissolution in ethanol followed by evaporation. Excess ether is then added to a concentrated solution of the residue in ethanol, to precipitate compound **2c** as a colorless powder; yield 413 mg. (88%).

REFERENCES

(1) A. M. Michelson, J. Dondon, and M. Grunberg-Manago, *Biochim. Biophys. Acta*, **55**, 529 (1962).
(2) M. Grunberg-Manago and A. M. Michelson, *Biochim. Biophys. Acta*, **80**, 431 (1964).
(3) R. Letters and A. M. Michelson, *J. Chem. Soc.*, **1962**, 71.

[151] 5-Hydroxyuridine 5′-Phosphate Derivatives

Substitution Reactions at the Pyrimidine Ring of Nucleotides

D. W. VISSER and P. ROY-BURMAN

DEPARTMENT OF BIOCHEMISTRY, UNIVERSITY OF SOUTHERN CALIFORNIA, LOS ANGELES, CALIFORNIA 90033

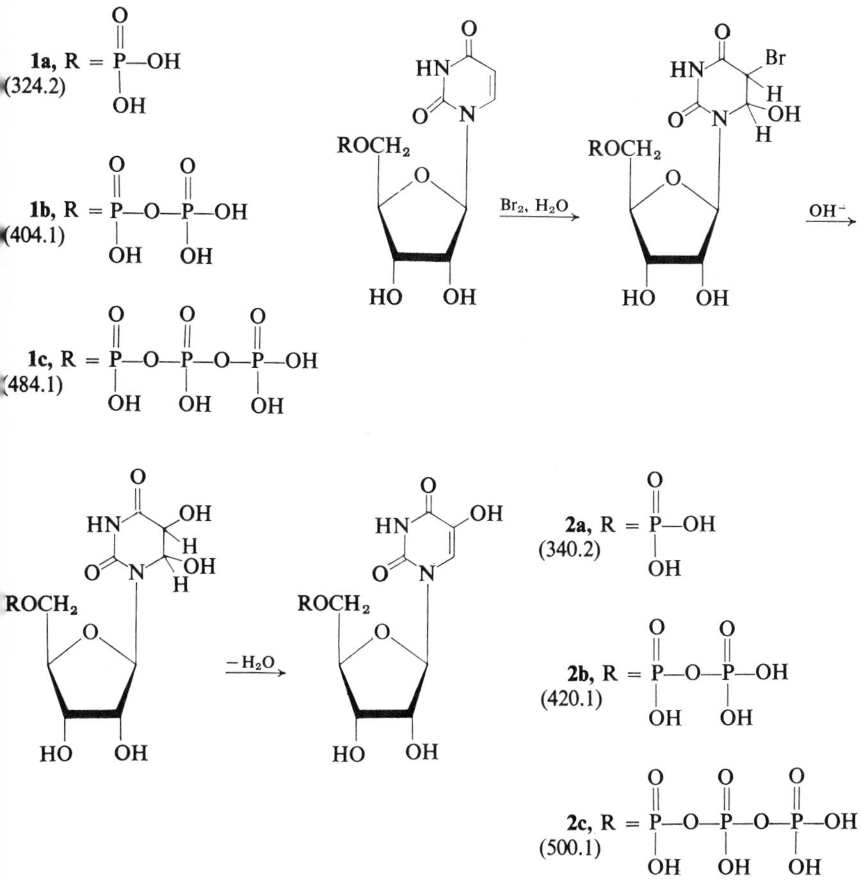

1a, R = P(=O)(OH)—OH (324.2)

1b, R = P(=O)(OH)—O—P(=O)(OH)—OH (404.1)

1c, R = P(=O)(OH)—O—P(=O)(OH)—O—P(=O)(OH)—OH (484.1)

$\xrightarrow{Br_2, H_2O}$ $\xrightarrow{OH^-}$ $\xrightarrow{-H_2O}$

2a, R = P(=O)(OH)—OH (340.2)

2b, R = P(=O)(OH)—O—P(=O)(OH)—OH (420.1)

2c, R = P(=O)(OH)—O—P(=O)(OH)—O—P(=O)(OH)—OH (500.1)

INTRODUCTION

The 5′-phosphates (**2a**, **2b**, **2c**) of 5-hydroxyuridine (Sec. III [131]) are formed in cell-free extracts of Ehrlich ascites cells[1] from the nucleoside analog. The monophosphate (**2a**) strongly inhibits orotidylic acid decarboxylase.[1] The analogous pyrophosphate (**2b**) is a substrate for polynucleotide phosphorylase[1-3] and the product of the enzymic reaction, poly(5-hydroxyuridylic acid), directs phenylalanine polymerization inefficiently[3] or not at all.[2] The nucleoside triphosphate (**2c**) is incorporated into ribonucleic acid by 2′-deoxyribonucleic acid-primed ribonucleic acid polymerase to a very limited extent.[4] Compound **2c** strongly inhibits the synthesis of ribonucleic acid, and acts as a competitive inhibitor of **1c** in the ribonucleic acid polymerase reaction.[4]

The synthetic procedures are similar to those described for the synthesis of 5-hydroxyuridine (Sec. III [131]).

PROCEDURE

5-Hydroxyuridine 5′-Phosphate[1] (2a)

Bromine and bromine-water are added to 500 mg. of uridine 5′-(disodium phosphate) (**1a**) (Sec. IV [145]), pentahydrate (1.09 mmoles) in 8 ml. of water at room temperature until a light-yellow color persists. Excess bromine is removed by aeration and the volume of the solution is adjusted to 10 ml. with water. Pyridine (5 ml.) is added, and the solution is treated as in Sec. III [131]. Alternatively, the solution may be heated on a boiling-water bath for 3 to 4 hr., with periodic addition of water (to maintain the volume) and of pyridine to keep the pH at about 6. The solution gradually turns yellow, and, within a few minutes, gives a purple color with ferric chloride.[5] At the end of the reaction period, barium hydroxide solution is added to raise the pH to 7.2, and **2a** is precipitated by addition of two volumes of ethanol. After the mixture has been kept at 4° for 4 hr., the precipitate is collected by filtration and dissolved in water, and the solution is passed through a column (1 × 40 cm.) of Dowex-1 (X8, formate) ion-exchange resin. The product is separated from the major contaminant, 5-bromouridine 5′-phosphate, by gradient elution, using 9 *N* formic acid and 1 l. of water in the mixing chamber. Fractions containing **2a** are identified by the production of a purple color with ferric chloride, and by the characteristic A_{280}/A_{260} ratio of 1.0 at pH 12 and 1.8 at pH 2.0. A smaller amount of 5-bromouridine 5′-phosphate is eluted after **2a**. The desired fractions are pooled and lyophilized, the product is dissolved in water, the pH of the solution is adjusted to 6.5 with sodium hydroxide, and the solution is lyophilized, giving a colorless, amorphous powder. The yields range from 38 to 45%, based on the assumption that ϵ_{max} of **2a** is the same as that determined for the nucleoside (Sec. III [131]); $\lambda_{max}^{pH\,2}$ 280 nm., $\lambda_{max}^{pH\,12}$ 304 nm.; A_{280}/A_{260} 1.8 at pH 2, and 1.0 at pH 12.

5-Hydroxyuridine 5′-Pyrophosphate[4] (2b)

The procedure for preparation of **2b** is similar to that described below for the preparation of **2c**. Compound **1b** (disodium salt, pentahydrate),

(430.6 mg., 0.80 mmole) yields 23% of the sodium salt of **2b** as a white, amorphorus powder; $\lambda_{max}^{pH\,2}$ 280 nm., $\lambda_{max}^{pH\,12}$ 304 nm.; A_{280}/A_{260} 1.8 at pH 2 and 1.0 at pH 12.

5-Hydroxyuridine 5′-Triphosphate[4] (2c)

Bromine and bromine-water are added to 500 mg. (0.80 mmole) of uridine 5′-(disodium triphosphate) (**1c**), hexahydrate, in 5 ml. of water at 0°, until a light-yellow color persists. Excess bromine is removed by aeration, and the volume of the solution is adjusted to 7 ml. with water. Pyridine (7 ml.) is added, and the solution is kept at 37° for 24 hr. The pH of the yellow solution is adjusted to 8.5 with 0.1 *M* ammonium hydroxide, and the solution is passed through a column (2 × 40 cm.) of DEAE-cellulose (HCO_3^-). Pyridine is removed by washing the column with water, and the products are eluted with a 0.2 *M* triethylammonium hydrogen carbonate gradient, by use of 1 l. of water in the mixing chamber. Fractions are collected at 4°. The major products, **2c** and 5-bromouridine 5′-triphosphate, emerge from the column in this order, but are usually not completely separated. Compound **2c** is recognized by its characteristic A_{280}/A_{260} ratio of 1.0 at pH 12 and 1.8 at pH 2.0. The corresponding ratios for 5-bromouridine 5′-triphosphate are 1.3 and 1.5. Usually, the first one-third of the total material (evaluated as absorbance) eluted from the column in the nucleoside triphosphate fractions is pure **2c**. Subsequent fractions contain increasing amounts of 5-bromouridine 5′-triphosphate, as ascertained by descending paper-chromatography in 5:2 (v/v) ethanol–0.5 *M* ammonium acetate of pH 7.5 (solvent A). Fractions containing mixtures of **2c** and 5-bromouridine 5′-triphosphate are combined, and the chromatographic separation on DEAE-cellulose (HCO_3^-) is repeated. The combined fractions of pure **2c** are lyophilized, and the material is dissolved in a small volume of water and converted into the sodium salt by passing the solution through a column of Dowex-50 X8 (Na^+). The effluent is lyophilized, giving a colorless, amorphous powder which is stored at −20°. The yields range from 19–23% based on ϵ_{max} as determined for 5-hydroxyuridine; $\lambda_{max}^{pH\,2}$ 280 nm., $\lambda_{max}^{pH\,12}$ 304 nm.; A_{280}/A_{260} 1.8 at pH 2, and 1.0 at pH 12.

The product gives a purple color with ferric chloride. The molar ratio of total phosphorus to acid-stable phosphorus is 3. Its electrophoretic mobility is the same as that of **1c**. The purity of **2c** may be verified by paper chromatography; it migrates as a single ultraviolet-absorbing area on a paper chromatogram developed with solvent A, and may be separated from the most probable contaminants, 5-bromouridine 5′-triphosphate or **1c**, in this way.

REFERENCES

(1) D. A. Smith and D. W. Visser, *J. Biol. Chem.*, **240**, 446 (1965).
(2) M. Grunberg-Manago and A. M. Michelson, *Biochim. Biophys. Acta*, **80**, 431 (1964).
(3) P. Roy-Burman, S. Roy-Burman, and D. W. Visser, *Federation Proc.*, **24**, 483 (1965).
(4) S. Roy-Burman, P. Roy-Burman, and D. W. Visser, *J. Biol. Chem.*, **241**, 781 (1966).
(5) D. Davidson and O. Baudisch, *J. Biol. Chem.*, **64**, 619 (1925).

[152] 3-Methyluridine 5′-Phosphate

Methylation of Uridine Derivatives with Diazomethane; Phosphorylation of Nucleosides with Pyrophosphoryl Chloride

N. K. KOCHETKOV, E. I. BUDOWSKY, and V. N. SHIBAEV

INSTITUTE FOR CHEMISTRY OF NATURAL PRODUCTS,
ACADEMY OF SCIENCES OF U.S.S.R., MOSCOW, U.S.S.R.

HN, O, N, HOCH$_2$, O, O, O, Ip — **1** (284.3)

$\xrightarrow{CH_2N_2}$

MeN, O, N, HOCH$_2$, O, O, O, Ip — **2** (298.3)

$\xrightarrow[\text{2. } H_2O,\ H^+]{\text{1. } P_2O_3Cl_4\ \mathbf{3}\ (251.8)}$

MeN, O, N, O, $(NaO)_2POCH_2$, O, HO, OH — **4** (382.2)

INTRODUCTION

3-Methyluridine 5′-phosphate (**4**) is a key intermediate in the synthesis of analogs of biologically important uridine derivatives, such as 3-methyluridine 5′-(D-glucosyl pyrophosphate),[1,2] 3-methyluridine 5′-pyrophosphate,[3,4] poly(3-methyluridylic acid)[3] and copolymers of **4** with uridylic acid,[3,4] cytidylic acid, and adenylic acid.[4] These analogs have been applied to study of (a) the relationship between the structure and biological functions of nucleoside glycosyl pyrophosphates,[5] (b) the influence of structure of polynucleotides on their secondary structure,[3,6] and (c) their ability to be split with nucleases[3] and to participate in protein biosynthesis.[4] The polynucleotide chain of transfer ribonucleic acid contains the 3-methyluridine moiety.[7]

The usual way of synthesis of **4** consists in methylation of 2′,3′-*O*-isopropylideneuridine (**1**) with diazomethane,[1–4] and phosphorylation of the resulting 2′,3′-*O*-isopropylidene-3-methyluridine (**2**). The second step may be accomplished by reaction with dimorpholinophosphinic chloride,[1,2] phosphorus pentaoxide–phosphoric acid,[3] or 2-cyanoethyl phosphate–dicyclohexylcarbodiimide.[4] The procedure described here employs pyrophosphoryl chloride[8] (**3**) for the phosphorylation of **2**.

PROCEDURE

2′,3′-*O*-Isopropylidine-3-methyluridine (2)

A solution of diazomethane in dry ether is prepared from 1-methyl-1-nitrosourea (8 g.), a 40% solution of potassium hydroxide (50 ml.), and ether (50 ml.). A solution of 2′,3′-*O*-isopropylideneuridine[9] (**1**) (Sec. III [132]) (1.50 g., 5.26 mmoles) in methanol (45 ml.) is treated at 0° with the dry, ethereal solution of diazomethane, added in 5-ml. portions until a permanent yellow color is formed; approximately 25 ml. of diazomethane solution is necessary for completion of the reaction.[a] After being kept at room temperature overnight, the reaction mixture is evaporated to dryness. The glassy residue of **2** (1.57 g.) may be used in the subsequent step without purification. If necessary, **2** may be purified by chromatography on alumina and recrystallization from tetrahydrofuran–cyclohexane. Pure **2** has m.p. 133.5–134°.

3-Methyluridine 5′-Phosphate (4)

The glassy **2** (1.57 g., 5.26 mmoles) is dried *in vacuo* at 100° for 1 hr.; it is then cooled in a Dry Ice–acetone bath, and treated with pyrophosphoryl chloride[10] (**3**) (2.3 ml., 4.5 g., 17.8 mmoles). The mixture is allowed to warm up to room temperature, and is stirred for 4 hr.; it is then cooled in Dry Ice–acetone, and water (5 ml.) is added dropwise. The mixture is warmed to room temperature with intense stirring, diluted to 150 ml. with water, and heated at 100° for 15 min. After being cooled, the solution is treated with ammonium magnesium chloride hexahydrate (7 g.) in water (15 ml.), and aqueous ammonia is added to pH 9. After 1 hr. at 0°, the precipitate is filtered off, and extracted with three 5-ml. portions of water containing ammonia (pH 9). The solution and washings (A_{260} 42,000) are combined and diluted with water to 5 l., and the solution is applied to a column (6 × 16.5 cm.) of Dowex-1 X4 (HCO_3^-, 100–200 mesh). The column is washed with water (1 l.), giving an eluate containing 3-methyluridine (A_{260} 2,300). Elution of **4** is now achieved by washing the column with 0.2 *M* triethylammonium hydrogen carbonate buffer of pH 9.0. Ultraviolet-absorbing fractions (2–7 l. of eluate, A_{260} 35,000) are combined and evaporated *in vacuo*, and water (3 × 200 ml.) is added to and evaporated from the residue. The product is dissolved in water (50 ml.), and the solution is passed through a column (4 × 10 cm.) of Dowex-50 (Na^+); the effluent contains the disodium salt of **4** (A_{260} 33,800, corresponding to a 74% yield). After lyophilization of the eluate, a white powder of the pentahydrate of the disodium salt of **4** is obtained; yield 1.84 g.; λ_{max}

[a] The reaction may be followed by thin-layer chromatography on alumina, with 30:1 chloroform–ethanol as the solvent; **2** has R_f 0.4, and **1** has R_f 0.1.

262 nm., ϵ_{max} 8,700; λ_{min} 234 nm., ϵ_{min} 3,100 (water). The product is homogeneous by paper chromatography, R_f 0.39 in 5:2 ethyl alcohol–0.5 *M* ammonium acetate of pH 7.5. During paper electrophoresis in triethylammonium hydrogen carbonate buffers, **4** shows a mobility equal to that of uridine 5′-phosphate (UMP) at pH 7.5 (R_{UMP} 1.0), but lower mobility at pH 10.3 (R_{UMP} 0.80).

REFERENCES

(1) N. K. Kochetkov, E. I. Budowsky, and V. N. Shibaev, *Izv. Akad. Nauk SSSR, Otd. Khim. Nauk*, **1962**, 1035.
(2) N. K. Kochetkov, E. I. Budowsky, V. N. Shibaev, G. I. Eliseeva, M. A. Grachev, and V. P. Demushkin, *Tetrahedron*, **19**, 1207 (1963).
(3) W. Szer and D. Shugar, *Acta Biochim. Polon.*, **8**, 235 (1961).
(4) A. M. Michelson and M. Grunberg-Manago, *Biochim. Biophys. Acta*, **91**, 92 (1964).
(5) E. I. Budowsky, T. N. Drushinina, G. I. Eliseeva, N. D. Gabrieljan, N. K. Kochetkov, M. A. Novikova, V. N. Shibaev, and G. L. Zhdanov, *Biochim. Biophys. Acta*, **122**, 213 (1966).
(6) W. Szer, M. Swierkowski, and D. Shugar, *Acta Biochim. Polon.*, **10**, 87 (1963).
(7) R. H. Hall, *Biochem. Biophys. Res. Commun.*, **12**, 361 (1963); *Biochemistry*, **4**, 661 (1965).
(8) W. Koransky, H. Grunze, and G. Münch, *Z. Naturforsch.*, **17b**, 291 (1962).
(9) P. A. Levene and R. S. Tipson, *J. Biol. Chem.*, **106**, 113 (1934).
(10) H. Grunze, *Z. Anorg. Allgem. Chem.*, **296**, 63 (1958).

[153] 2-Thiouridine 5′-Phosphate

Reaction of Anhydronucleosides with Hydrogen Sulfide; Phosphorylation of Nucleosides with 2-Cyanoethyl Phosphate

N. K. KOCHETKOV, E. I. BUDOWSKY, and V. N. SHIBAEV

INSTITUTE FOR CHEMISTRY OF NATURAL PRODUCTS,
ACADEMY OF SCIENCES OF U.S.S.R., MOSCOW, U.S.S.R.

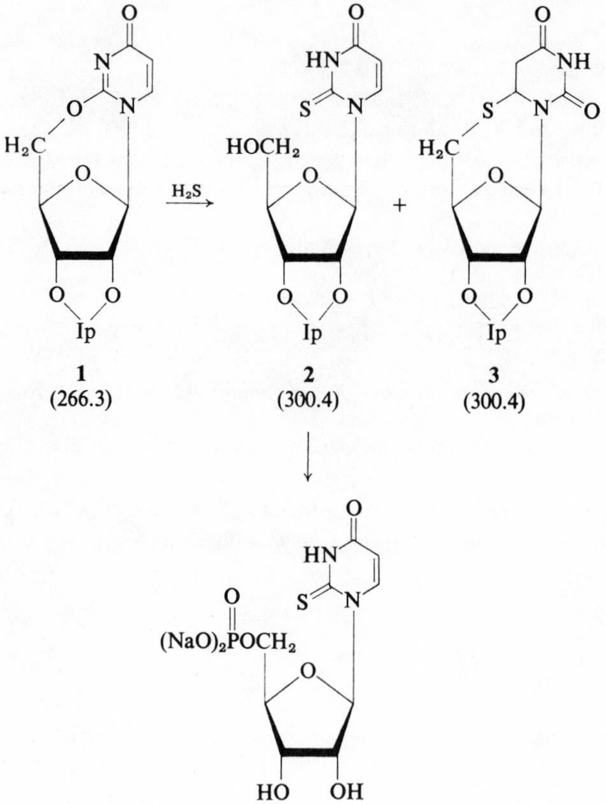

1 (266.3) $\xrightarrow{H_2S}$ 2 (300.4) + 3 (300.4)

2 → 4 (384.3)

INTRODUCTION

2-Thiourine 5′-phosphate (**4**) has been used as the starting material for the synthesis of 2-thiouridine 5′-(α-D-glucopyranosyl pyrophosphate),[1,2] 2-thiouridine 5′-pyrophosphate,[3] and poly(2-thiouridylic acid).[3] The first compound has been applied to study of the specificity of carbohydrate-metabolizing enzymes.[4] The polynucleotide chain of transfer ribonucleic acid contains the 2-thiouridine residue.[5] 2′,3′-*O*-Isopropylidene-2-thiouridine (**2**) is prepared by the reaction of 2,5′-anhydro-(2′,3′-*O*-isopropylideneuridine) (**1**) with hydrogen sulfide[6]; another product of this reaction is 5′-deoxy-5′,6-epithio-5,6-dihydro-2′,3′-*O*-isopropylideneuridine[7] (**3**), which was originally referred to[6] as the disulfide of **2**. 2-Cyanoethyl phosphate (hydracrylonitrile phosphate) is used for the phosphorylation of **2**.

PROCEDURE

2′,3′-*O*-Isopropylidene-2-thiouridine (2)

A solution of 2,5′-anhydro-(2′,3′-*O*-isopropylideneuridine)[8] (**1**) (Sec. III [99]) (475 mg., 1.72 mmoles) in anhydrous pyridine (20 ml.) is saturated with dry hydrogen sulfide.[a] The flask is stoppered (drying tube), and the mixture is allowed to warm up to room temperature. After evaporation of the excess of hydrogen sulfide is complete, the flask is tightly stoppered, and kept at room temperature for 6 days. Nitrogen is bubbled through the resulting, yellow solution (to remove hydrogen sulfide), and the solution is evaporated to dryness *in vacuo*. The residue is treated with absolute ethanol (10 ml.), and the resulting, white precipitate of compound **3** (271 mg.) is filtered off. The filtrate is evaporated to dryness, the residue is dissolved in acetone (3 ml.), and the solution is applied to a column (2 × 18 cm.) of neutral alumina (Brockman, activity grade II). The column is washed successively with acetone (400 ml.) and 10:1 acetone–methanol (250 ml.). Evaporation of the acetone eluates yields 75 mg. of **3**. Glassy **2** is obtained by evaporating the fractions eluted with acetone–methanol; yield 124 mg. (24%). The resulting preparation of **2** may be used for phosphorylation without further purification.

2-Thiouridine 5′-Phosphate (4)

A mixture of **2** (96.1 mg., 0.32 mmole) and *M* 2-cyanoethyl phosphate (Sec. V [159]) in pyridine[9] (1.48 ml.) is dried by co-evaporation with anhydrous pyridine[b] (3 × 10 ml.), and then evaporated to dryness. To a solution of the residue in anhydrous pyridine (10 ml.) is added dicyclohexylcarbodiimide (656 mg., 3.2 mmoles), and the mixture is kept at 60° for 4 hr. Water (20 ml.) is added, and the mixture is kept at room temperature for 30 min. The precipitate of dicyclohexylurea is filtered off, the

[a] Hydrogen sulfide is obtained by heating a mixture of paraffin and sulfur, and is condensed in a trap cooled with Dry Ice–acetone; the condensate (about 5 ml.) is then distilled into the reaction flask cooled to −30°.

[b] Pyridine, distilled from barium oxide and stored over Linde Molecular Sieves 4A, is used.

filtrate is evaporated to dryness, and the residue is co-evaporated with water (3 × 5 ml.). The residue is dissolved in 70% acetic acid (20 ml.) and heated at 100° for 45 min. The solution is cooled, and evaporated to dryness *in vacuo*. The residue is co-evaporated with water (3 × 5 ml.) and dissolved in a mixture of 10 ml. of *N* potassium hydroxide and 0.5 ml. of 2-mercaptoethanol. The solution is heated at 100° for 15 min., cooled, and filtered. The filtrate is passed through a column (4 × 6 cm.) of Dowex-50 (H^+), the pH of the effluent is adjusted to 8 with alkali, and the solution is applied to a column (2 × 14 cm.) of Dowex-1 X8 (HCO_3^-) (100–200 mesh). The column is washed successively with water (650 ml.), 0.05 *M* triethylammonium hydrogen carbonate buffer of pH 7.5 (1 l.), and 0.3 *M* buffer (2 l.; uridine 5'-phosphate, A_{260} 300, is eluted). Elution with 0.5 *M* buffer (800 ml.) yields **4** (A_{268} 2,100; 0.204 mmole). The fractions containing **4** are combined, and evaporated to dryness *in vacuo*, and the residue is co-evaporated with water (3 × 100 ml.) and then dissolved in water (20 ml.). The solution is passed through a column of Dowex-50 (Na^+); the effluent contains the disodium salt of **4** (A_{268} 2,000, corresponding to a yield of 61%). After lyophilization of the solution, a white powder of the trihydrate of the disodium salt of **4** is obtained; yield 87.5 mg.; λ_{max} 273 and 219 nm., ϵ_{max} 11,100 and 13,800; λ_{min} 243 nm., ϵ_{min} 4,950 (0.01 *N* HCl and water); λ_{max} 270 and 240 nm., ϵ_{max} 10,500 and 14,900, λ_{min} 259 nm., ϵ_{min} 9,800 (0.01 *N* KOH); the isosbestic point is at 268 nm. (ϵ 10,400). The preparation of **4** is homogeneous by paper chromatography; R_f 0.25 (5:2 ethanol–0.5 *M* ammonium acetate of pH 7.5).

REFERENCES

(1) N. K. Kochetkov, E. I. Budowsky, V. N. Shibaev, G. I. Eliseeva, M. A. Grachev, and V. P. Demushkin, *Tetrahedron*, **19**, 1207 (1963).

(2) N. K. Kochetkov, E. I. Budowsky, and V. N. Shibaev, *Khim. Prirodn. Soedin.*, **1965**, 409.

(3) P. Lengyel and R. W. Chambers, *J. Am. Chem. Soc.*, **82**, 452 (1960).

(4) E. I. Budowsky, T. N. Drushinina, G. I. Eliseeva, N. D. Gabrieljan, N. K. Kochetkov, M. A. Novikova, V. N. Shibaev, and G. L. Zhdanov, *Biochim. Biophys. Acta*, **122**, 213 (1966).

(5) J. A. Carbon, L. Hung, and D. S. Jones, *Proc. Natl. Acad. Sci. U. S.*, **53**, 979 (1965).

(6) D. M. Brown, D. B. Parihar, A. R. Todd, and S. Varadarajan, *J. Chem. Soc.*, **1958**, 3028.

(7) R. W. Chambers and V. Kurkov, *J. Am. Chem. Soc.*, **85**, 2160 (1963).

(8) D. M. Brown, A. R. Todd, and S. Varadarajan, *J. Chem. Soc.*, **1957**, 868.

(9) G. Tener, *J. Am. Chem. Soc.*, **83**, 159 (1961).

[154] 4-Thiouridine 5′-Phosphate

Acetonation of Nucleosides; Phosphorylation of Nucleosides with 2-Cyanoethyl Phosphate (Hydracrylonitrile Phosphate)

V. N. SHIBAEV, M. A. GRACHEV, and S. M. SPIRIDONOVA

INSTITUTE FOR CHEMISTRY OF NATURAL PRODUCTS,
ACADEMY OF SCIENCES OF U.S.S.R., MOSCOW, U.S.S.R.

$BzOCH_2$ … BzO OBz **1** (572.7) → $HOCH_2$ … HO OH **2** (260.3) → $HOCH_2$ … O O Ip **3** (300.4) → $(NaO)_2P(O)OCH_2$ … HO OH **4** (384.3)

where R = [4-thiouracil-1-yl: S, HN, O, N] and Bz is benzoyl.

INTRODUCTION

4-Thiouridine 5′-phosphate (**4**) has been used as an intermediate in the synthesis of 4-thiouridine 5′-(D-glucopyranosyl pyrophosphate),[1] the biologically active analog of the naturally occurring uridine derivative. The polynucleotide chain of transfer ribonucleic acid contains the 4-thiouridine moiety.[2] The synthesis of 4-thiouridine (**2**) has been described by Fox *et al.*[3]; however, the pure substance had not been isolated.

PROCEDURE

4-Thiouridine (2)

A solution of 2′,3′,5′-tri-*O*-benzoyl-4-thiouridine[3] (**1**) (Sec. III [135]) (5.4 g., 9.4 mmoles) in dry methanol (130 ml.) is treated with 132 ml. of 0.1 *M* sodium methoxide in methanol, and refluxed for 5 hr. The hot solution is treated with acetic acid (1 ml.), cooled, and evaporated to dryness. The residue is dissolved in water (5 ml.), the solution is washed with chloroform (3 × 5 ml.), the combined chloroform extracts are extracted with water (5 ml.), and the water layers are combined and evaporated to dryness. The residue is dissolved in water (2 ml.), isopentyl alcohol (6 ml.) and acetone (4 ml.) are added, and the mixture is shaken until it becomes homogeneous. The resulting solution is applied to a column (5 × 32 cm.) of cellulose, and the column is washed with 3:2:1 (v/v) isopentyl alcohol–acetone–water.[a] The fractions containing **2** are combined, and evaporated *in vacuo*, with gradual addition of water to facilitate the evaporation of isopentyl alcohol. The residue is dissolved in ethanol (5 ml.), and the solution is treated with ether (50 ml.), and rapidly filtered from a gummy precipitate that results. The filtrate is cooled to 0°, yielding yellow needles of **2** (1.12 g.), m.p. 135–138° (dec.). Processing of the gummy precipitate and of the mother liquor gives an additional quantity of **2**; total yield, 1.70 g. (69%); λ_{max} 331 and 245 nm., ϵ_{max} 21,200 and 4,000; λ_{min} 274 and 225 nm., ϵ_{min} 2,400 and 1,600 (water); λ_{max} 316 nm., ϵ_{max} 19,700; λ_{min} 268 nm., ϵ_{min} 2,400 (0.1 *N* NaOH).

2′,3′-*O*-Isopropylidene-4-thiouridine (3)

A solution of **2** (2.44 g., 10 mmoles) in dry *N*,*N*-dimethylformamide[b] (20 ml.) is treated with 2,2-diethoxypropane (2.5 ml.) and 2–3 drops of a saturated solution of hydrogen chloride in dry *p*-dioxane. The mixture is kept at room temperature for 4 hr., and is then poured into a solution of 2 ml. of ammonia in 50% aqueous methanol (200 ml.). The resulting solution is applied to a column (3 × 23 cm.) of Dowex-1 X8 (HCO_3^-, 100–200 mesh). The column is washed with 50% aqueous methanol (50 ml.) and then with 0.2 *M* triethylammonium hydrogen carbonate (pH 7.5) in 50% methanol. The separation is controlled by measuring the absorbance of eluates at 330 and 260 nm. Fractions containing **3** are evaporated, and the residue is co-evaporated with water (3 × 100 ml.), giving 2.03 g. (72%) of **3**, which is homogeneous by paper chromatography and may be used for phosphorylation without further purification. Pure **3** may be obtained by crystallization from ethanol–heptane, m.p. 170–172°.

4-Thiouridine 5′-Phosphate (4)

A mixture of **3** (1.54 g., 5.13 mmoles) and 22.5 ml. of *M* 2-cyanoethyl phosphate (Sec. V [159]) in pyridine[4] is evaporated; the residue is dried by co-evaporation with dry pyridine (3 × 30 ml.), dissolved in dry pyridine

[a] The process of purification may be monitored by observing the movement of a yellow zone through the column, or by paper chromatography of fractions in 43:7 (v/v) butyl alcohol–water; compound **2** has R_f 0.14 and shows a characteristic, blue fluorescence under ultraviolet light.

[b] *N*,*N*-Dimethylformamide is distilled from phosphorus pentaoxide immediately before the reaction.

(220 ml.), and treated with dicyclohexylcarbodiimide (10.7 g., 51.5 mmoles). The flask containing the mixture is tightly stoppered, and kept at 60° for 5 hr. Water (3.5 ml.) is then added, the mixture is kept at room temperature for 30 min., and filtered, and the filtrate is evaporated *in vacuo*. The residue is co-evaporated with water (3 × 30 ml.), and treated with 70% acetic acid (230 ml.). The mixture is heated at 100° for 30 min., and evaporated to dryness, and the residue is co-evaporated with water (2 × 30 ml.). The residue is dissolved in *M* potassium hydroxide (230 ml.), and the solution is heated at 100° for 15 min., cooled, and filtered. The filtrate is diluted with water to 600 ml. and applied to a column (2.5 × 25 cm.) of DEAE-Sephadex A-25 (HCO_3^-). The column is successively washed with water (500 ml.), 0.05 *M* triethylammonium hydrogen carbonate buffer of pH 7.5 (2.5 l.; compound **2**, A_{330} 8,050, is eluted), 0.2 *M* buffer (2.5 l.; uridine 5′-phosphate, A_{260} 5,250), and 0.3 *M* solution of buffer (this elutes **4**, A_{330} 79,500). The fractions containing **4** are combined and evaporated, and the residue is co-evaporated with water (3 × 300 ml.), and dissolved in water (500 ml.). The solution is passed through a column of Dowex-50 (Na^+), giving a solution of the disodium salt of **4**, A_{330} 66,300, corresponding to a yield of 61%. Lyophilization of the solution gives 1.2 g. of the disodium salt of **4**. The ultraviolet spectrum of the compound is similar to that of **2**. The preparation of **4** is homogeneous by paper chromatography, R_f 0.26 (5:2 ethanol–0.5 *M* ammonium acetate of pH 7.5), and by paper electrophoresis, R_{UMP} 1.25 (0.05 *M* triethylammonium hydrogen carbonate of pH 7.5).

REFERENCES

(1) N. K. Kochetkov, E. I. Budowsky, V. N. Shibaev, and M. A. Grachev, *Izv. Akad. Nauk SSSR, Ser. Khim.*, **1963**, 1592.
(2) M. N. Lipsett, *J. Biol. Chem.*, **240**, 3975 (1965).
(3) J. J. Fox, D. Van Praag, I. Wempen, I. L. Doerr, L. Cheong, J. E. Knoll, M. L. Eidinoff, and A. Bendich, *J. Am. Chem. Soc.*, **81**, 178 (1959).
(4) G. Tener, *J. Am. Chem. Soc.*, **83**, 159 (1961).

[155] Uridylyl-(3′ → 5′)-cytidylyl-(3′ → 5′)-uridine

Stepwise Synthesis of an Oligoribonucleotide Chain

A. HOLÝ

INSTITUTE OF ORGANIC CHEMISTRY AND BIOCHEMISTRY,
CZECHOSLOVAK ACADEMY OF SCIENCES, PRAGUE, CZECHOSLOVAKIA

INTRODUCTION

The present method for the synthesis of uridylyl-(3′ → 5′)-cytidylyl-(3′ → 5′)-uridine (**7**) exemplifies the stepwise synthesis of an oligoribonucleotide chain by use of a 2′,3′-*O*-(ethoxymethylene) group to protect selectively the 2′,3′-*cis*-diol system of the ribonucleoside,[1] an acid-labile 1-ethoxyethyl group to protect the 2′-hydroxyl group of the nucleotide,[2] an acetyl group to protect the 5′-hydroxyl group, and finally, a (dimethylamino)methylene group (by reaction with a *N*,*N*-dimethylformamide acetal) to protect specifically the amino groups of basic nucleosides.[3] This combination of protecting groups makes possible the preparation, in every reaction step, of an oligoribonucleotide derivative that has a single, unprotected hydroxyl group at C-5′ and that is suitable for further use in the stepwise synthesis of oligoribonucleotides.[4]

PROCEDURE

2′-*O*-(1-Ethoxyethyl)cytidylyl-(3′ → 5′)-2′,3′-*O*-(ethoxymethylene)uridine (3)

The silver salt[a] (1.4 g., 2 mmoles) of *N*-acetyl-5′-*O*-acetyl-2′-*O*-(1-ethoxyethyl)cytidine 3′-phosphate (**1**) (Sec. IV [143]) is dissolved, with ice-cooling, in 10 ml. of 50% aqueous pyridine, the solution is placed on a column (1.5 × 15 cm.) of Dowex-50 (pyridinium$^+$) (100–200 mesh) ion-exchange resin[b] (prewashed with 100 ml. of 50% aqueous pyridine), and the column is eluted at 2° with 200 ml. of 50% aqueous pyridine. The eluate is introduced, in portions, into an evacuated, 250-ml., round-bottomed distillation flask immersed in a water bath maintained at 25°. The flask is equipped with a dropping funnel and a capillary tube, and is connected to a receiver cooled in Dry Ice–ethanol. The eluate is concentrated at 25°/0.1 mm. Hg to 10 ml., and to this solution 50 ml. of pyridine is added and evaporated off. A solution of 1.21 g. (4 mmoles) of 2′,3′-*O*-(ethoxymethylene)uridine (**2**) (Sec. III [129]) in 25 ml. of pyridine is then added, and the mixture is concentrated, as above, to about 5 ml. The concentrate is carefully co-evaporated (not to dryness) with five 50-ml.

[a] For the preparation of this silver salt, see Sec. IV [143].

[b] For the packing of the ion-exchange resin column, see Sec. IV [148].

portions of anhydrous pyridine,[c] and the residue is shaken with 5 ml. of anhydrous pyridine[c] and 2.5 g. of dicyclohexylcarbodiimide, in a tightly stoppered flask at room temperature for 3 days. To the mixture is then added 20 ml. of 50% aqueous pyridine containing 5% of triethylamine, and the resulting mixture is shaken at room temperature for 2 hr. Water (100 ml.) is added, and the resulting suspension is washed with three 50-ml. portions of ether. The aqueous layer is concentrated at 30°/15 mm. Hg to about 10 ml., to the concentrate is added 20 ml. of 30% methanolic ammonia, and the resulting mixture is heated at 50° for 1 hr., and evaporated to dryness at 30°/15 mm. Hg on a rotary evaporator. The residue is dissolved in 50 ml. of 20% aqueous pyridine, the suspension is filtered through a thin layer of Hyflo Super Cel, and the material on the filter is washed with 10 ml. of water. The filtrates are combined, and placed on a column (4 × 100 cm.) of *O*-[2-(diethylamino)ethyl]cellulose (HCO_3^-).[d] The column is washed with approximately 2 l. of water to remove the neutral fraction (the optical absorbance of the eluate being determined by continuous measurement of the ultraviolet absorption), and is then eluted with a linear gradient obtained with 2 l. of 0.2 *M* triethylammonium hydrogen carbonate (pH 7.5) in the reservoir and 2 l. of water in the mixing chamber, at a flow-rate of 3 ml. per min. The first elution peak (buffer concentration, approximately 0.05–0.10 *M*) contains the reaction product. The corresponding fractions are evaporated to dryness at 35°/15 mm. Hg, and the residue is successively co-evaporated with two 50-ml. portions of water and two 50-ml. portions of 96% ethanol. After being dried over phosphorus pentaoxide at 0.1 mm. Hg, the foamy product is dissolved in 20 ml. of absolute methanol[e] containing a single drop of triethylamine, and the solution is added dropwise, with stirring, to 200 ml. of absolute ether.[f] The precipitate is collected with suction, washed with 50 ml. of absolute ether,[f] and dried over phosphorus pentaoxide at 0.1 mm. Hg. The chromatographically and electrophoretically homogeneous[g] triethylammonium salt of compound **3** weighs 780 mg. (50%); purity, 95–98%,

[c] Commercial pyridine is dried over potassium hydroxide at room temperature, distilled, and stored over molecular sieves.

[d] For packing of the column with DEAE-cellulose, see Sec. IV [140].

[e] Dried by the magnesium procedure.

[f] Dried with sodium.

[g] R_f values on Whatman No. 1 paper in 7:1:2 isopropyl alcohol–concentrated aqueous ammonia–water: 0.72 (**2**), 0.55 (**3**), 0.15 (**4**), and 0.04 (**7**). R_f values on Whatman No. 1 paper in 5:2:3 butyl alcohol–acetic acid–water: 0.15 (**4**), 0.20 [uridine 2′(3′)-phosphate], and 0.10 (**7**). Mobilities on Whatman No. 1 paper at 40 volt/cm. for 45 min. in 0.05 *M* sodium hydrogen citrate (pH 3.5): 0.25 (**3**), 0.25 (**4**), 0.55 (**5**), and 0.70 (**7**); in 0.05 *M* triethylammonium hydrogen carbonate (pH 7.5): 0.28 (**3**), 0.28 (**4**), and 0.52 (**7**). The mobilities are relative to that of uridine 3′(2′)-phosphate, 1.00.

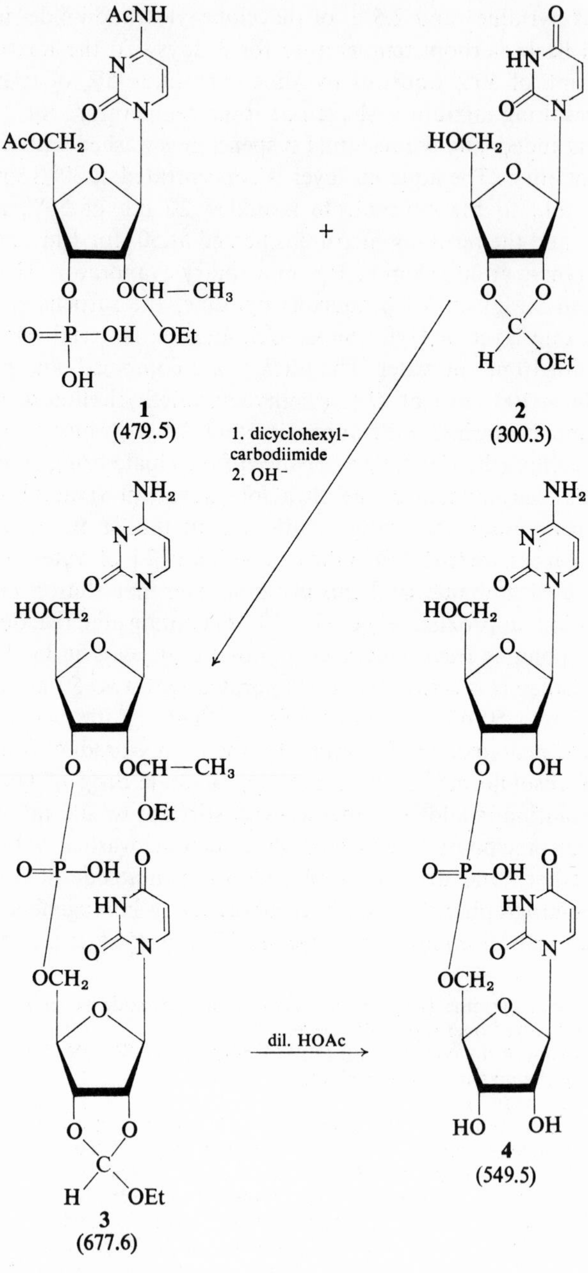

AcNH N O N AcOCH$_2$ O O OCH—CH$_3$ O=P—OH OEt OH

1
(479.5)

\+

O HN O N HOCH$_2$ O O O C H OEt

2
(300.3)

1. dicyclohexyl-
carbodiimide
2. OH$^-$

NH$_2$ N O N HOCH$_2$ O O OCH—CH$_3$ OEt O=P—OH O HN O N OCH$_2$ O O O C H OEt

3
(677.6)

dil. HOAc →

NH$_2$ N O N HOCH$_2$ O O OH O=P—OH O HN O N OCH$_2$ O HO OH

4
(549.5)

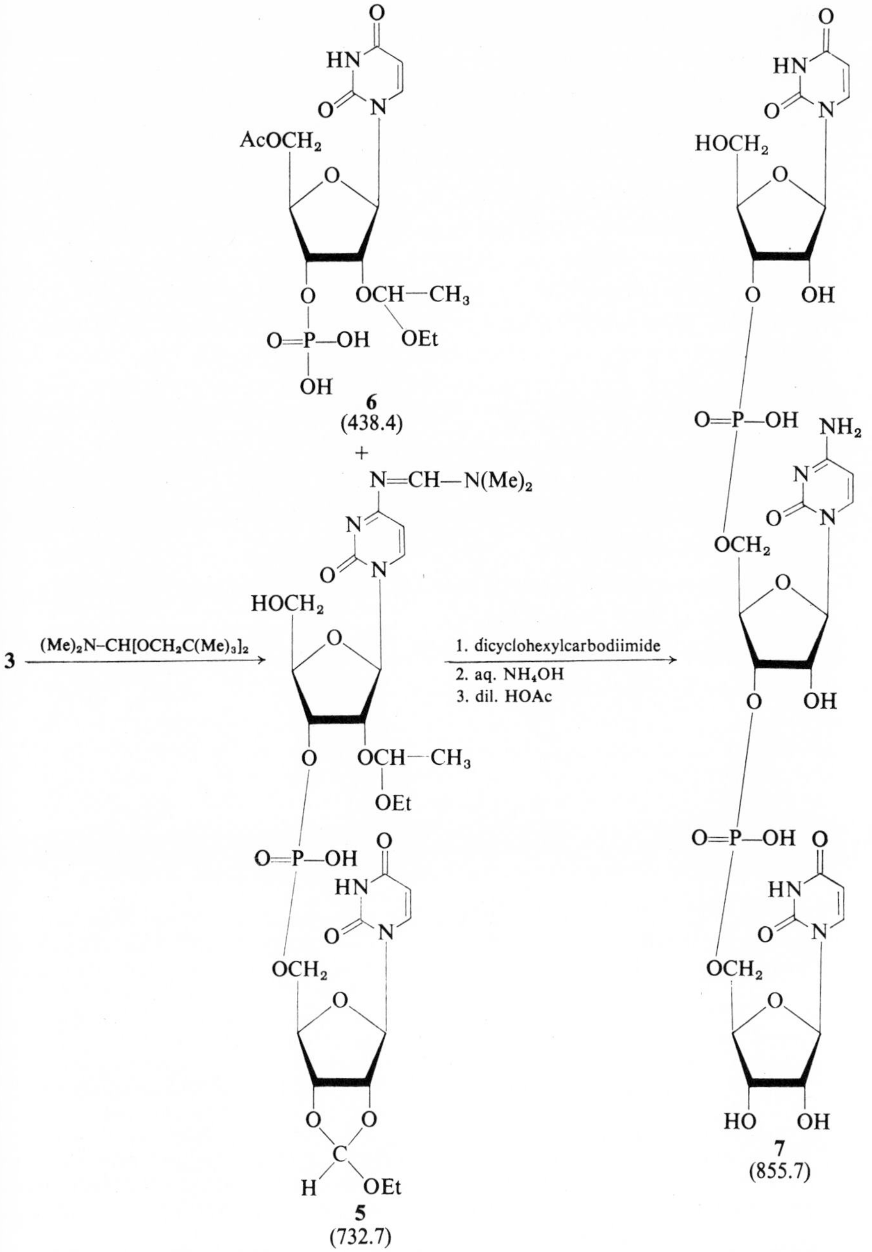

O
HN
O N
$AcOCH_2$
O
O OCH—CH_3
O=P—OH OEt
OH
6
(438.4)
+
N=CH—$N(Me)_2$
N
O N
$HOCH_2$
O

3 $\xrightarrow{(Me)_2N\text{–}CH[OCH_2C(Me)_3]_2}$

O OCH—CH_3
OEt
O=P—OH O
HN
O N
OCH_2
O
O O
C
H OEt
5
(732.7)

1. dicyclohexylcarbodiimide
2. aq. NH_4OH
3. dil. HOAc

O
HN
O N
$HOCH_2$
O
O OH
O=P—OH NH_2
N
O N
OCH_2
O
O OH
O=P—OH O
HN
O N
OCH_2
O
HO OH
7
(855.7)

as shown by spectroscopic, molecular-weight determination[h]; $\lambda_{max}^{pH\,2}$ 269 nm., $\lambda_{min}^{pH\,2}$ 235 nm.

Cytidylyl-(3′ → 5′)-uridine (4)

A solution of 156 mg. (0.2 mmole) of the triethylammonium salt of compound **3** in 1.0 ml. of 10% aqueous acetic acid in a 100-ml., round-bottomed flask is kept at room temperature for 5 hr. Water (5 ml.) is then added, and the solution is freeze-dried. The residue is dissolved in 2 ml. of water, and the solution is applied to 4 sheets of preparative Whatman 3MM paper; chromatography is performed with 7:1:2 isopropyl alcohol–concentrated ammonium hydroxide–water. The zones of the product (R_f value 0.18) are eluted[i] with water (total about 20 ml.) previously made alkaline with ammonium hydroxide to pH 8.5. The eluates are combined and freeze-dried, and the residue is dried at room temperature under diminished pressure (diffusion pump). The chromatographically and electrophoretically homogeneous[g] ammonium salt of cytidylyl-(3′ → 5′)-uridine (**4**) weighs 110 mg.; purity, 80–85%; content of the (2′ → 5′)-isomer[j] 1.5–2.5% (as determined by spectrophotometric, molecular-weight determination, using $\epsilon_{260}^{pH\,2} = 16.8 \times 10^3$).

***N*-[(Dimethylamino)methylene]-2′-*O*-(1-ethoxyethyl)cytidylyl-(3′ → 5′)-2′,3′-*O*-(ethoxymethylene)uridine (5)**

A solution of 623 mg. (0.8 mmole) of the triethylammonium salt of compound **3**,[k] 6 ml. of *N,N*-dimethylformamide, and 1.2 ml. of *N,N*-

[h] The molecular weight is determined spectroscopically by using the value 16.8×10^3 for $\epsilon_{260}^{pH\,2}$. In some preparations, the product (**3**) is contaminated with triethylammonium acetate. The presence of this salt does not interfere with the preparation of compound **4**. The contaminant is then removed in the subsequent step, *i.e.*, in the course of preparation of compound **5**.

[i] Elution of the bands may be conducted simultaneously from, for example, four strips of paper, according to the method described in Sec. IV [140], footnote *o*.

[j] The content of the (2′ → 5′)-isomer is determined by dissolving 1 mg. of the substance in 0.1 ml. of a solution containing 2 mg. of pancreatic ribonuclease (five times recrystallized) per ml. of 0.05 *M* 2-amino-2-(hydroxymethyl)-1,3-propanediol ("Tris") hydrochloride buffer (pH 8), incubating at 37 for 4 hr., and applying the digest to a 4-cm. wide strip of Whatman No. 1 paper. The chromatography is performed in 7:1:2 isopropyl alcohol–concentrated aqueous ammonia–water alongside authentic specimens of uridine, cytidine 2′(3′)-phosphate, and cytidylyl-(3′ → 5′)-uridine (**4**). The spots are detected under ultraviolet light, cut out, and eluted in test tubes, each containing 10 ml. of 0.01 *N* hydrochloric acid. Elution of a corresponding area of a blank chromatogram is performed simultaneously. It is recommended that the place on the chromatogram corresponding to the R_f value of cytidylyl-(3′ → 5′)-uridine be eluted, even when the spot is not visible under ultraviolet light. After 2 hr., with occasional shaking, the optical absorbance (at 260 nm.) of the spot eluates is measured, along with the above-mentioned blank. The isomerization is expressed (in %) as the ratio of the absorbance of cytidylyl-(2′ → 5′)-uridine to the sum of the absorbances of the uridine, cytidine 3′-phosphate, and cytidylyl-(2′ → 5′)-uridine eluates.

[k] When the salt of compound **3** is less pure, a larger quantity has to be used, determined by the result of the determination of the molecular weight.

dimethylformamide bis(2,2-dimethylpropyl) acetal[5] is kept in a 100-ml., round-bottomed flask, under exclusion of atmospheric moisture, at room temperature overnight. The mixture is then evaporated to dryness at 35°/0.1 mm. Hg, and the residue is co-evaporated with two 25-ml. portions of pyridine (to remove the last traces of *N,N*-dimethylformamide), and dissolved in 10 ml. of 50% aqueous pyridine. The solution is placed on a column (1.5 × 15 cm.) of Dowex-50 (pyridinium$^+$) (100–200 mesh) ion-exchange resin,[b] prewashed with 100 ml. of 50% aqueous pyridine. The column is eluted at 2° with 100 ml. of 50% aqueous pyridine, the eluate is evaporated to dryness, and the residue is co-evaporated with pyridine at 25°/0.1 mm. Hg, and then dissolved in 50 ml. of pyridine. To the solution is added 1 ml. of triethylamine, the mixture is evaporated almost to dryness under the above conditions, and the residue is co-evaporated with three 50-ml. portions of a 10% solution of triethylamine in 1:1 *p*-dioxane–toluene. The residue is dissolved in 20 ml. of methanol, and the solution is added dropwise, with stirring, to 200 ml. of ether. The resulting precipitate is collected with suction, washed with 50 ml. of ether, and dried over phosphorus pentaoxide at 0.1 mm. Hg. The electrophoretically homogeneous[g] triethylammonium salt of compound **5** weighs 600 mg. (90%); purity 95–98% according to molecular-weight determination[l] (calc. M.W. 833.9); λ_{max} 262 and 317 nm., λ_{min} 235 nm., in 96% ethanol.

Uridylyl-(3′ → 5′)-cytidylyl-(3′ → 5′)-uridine (7)

A solution of 166.7 mg. (0.2 mmole) of the triethylammonium salt of compound **5** and 0.6 mmole of the barium salt of compound **6**[m] in 10 ml. of 50% aqueous pyridine is placed on a column (1.5 × 10 cm.) of Dowex-50 (pyridinium$^+$) (100–200 mesh) ion-exchange resin.[b] The elution is performed at 2° with 100 ml. of 50% aqueous pyridine, and the eluate is co-evaporated with a total of 250 ml. of pyridine (in the apparatus and under the conditions given for the preparation of compound **3**). The final residue is dissolved in 5 ml. of anhydrous pyridine,[c] and the solution is shaken with 1.0 g. of dicyclohexylcarbodiimide at room temperature for

[l] Prior to the measurement, the solution of the triethylammonium salt (**5**) in 0.01 *N* hydrochloric acid is kept at room temperature for 15 hr. For calculation of the molecular weight, the value $\epsilon_{260}^{pH\,2} = 16.8 \times 10^3$ is used.

[m] The barium salt of compound **6** is prepared by the reaction of 5′-*O*-acetyluridine 3′-phosphate (see Sec. IV [149]) with ethyl vinyl ether, in analogy to the preparation of *N*-acetyl-5′-*O*-acetyl-2′-*O*-(1-ethoxyethyl)cytidine 3′-phosphate (**1**), except for the final precipitation, which, in the present case, is performed with barium trifluoroacetate (instead of silver trifluoroacetate). For 10 mmoles of 5′-*O*-acetyluridine 3′-phosphate, a solution of 6.0 g. of anhydrous barium trifluoroacetate in 100 ml. of methanol and 500 ml. of ether is used. The precipitate is dissolved in methanol, and the compound is reprecipitated with five volumes of ether. The product thus purified is collected, washed with ether, and dried over phosphorus pentaoxide at 0.1 mm. Hg. The molecular weight is determined spectrophotometrically, using $\epsilon_{260}^{pH\,2}$ 10.0 × 10^3 (content 85–90%).

5 days. To the mixture is then added 10 ml. of a 10% solution of triethylamine in 50% aqueous pyridine, and the mixture is shaken at room temperature for 2 hr.; 10 ml. of concentrated ammonium hydroxide and 30 ml. of 30% methanolic ammonia are added, and the mixture is heated at 50° for 3 hr. Water (100 ml.) is then added to the cooled mixture, and the resulting mixture is washed with three 50-ml. portions of ether. The aqueous layer is diluted with 50 ml. of pyridine, and concentrated at 30°/15 mm. Hg to about 10 ml., and the concentrate is diluted with 20 ml. of 50% aqueous pyridine, and filtered through a thin layer of Hyflo Super Cel. The material on the filter is washed with 10 ml. of water, and the filtrates are combined and placed on a column (4.5 × 100 cm.) of *O*-[2-(diethylamino)ethyl]cellulose (HCO_3^-).[d] Elution is performed with a linear gradient of triethylammonium hydrogen carbonate (pH 7.5) (2 l. of water in the mixing chamber and 2 l. of 0.2 *M* buffer solution in the reservoir) at a flow-rate of 3 ml./min. The optical absorbance of the eluate is estimated by continuous measurement of the ultraviolet absorption. The composition of the eluates is determined by paper chromatography in 7:1:2 isopropyl alcohol–concentrated ammonium hydroxide–water, and electrophoresis in 0.05 *M* triethylammonium hydrogen carbonate.[g] The eluate should separate into three fractions showing three main peaks. That containing peak 1 is not processed. Those containing peaks 2 and 3 are separately evaporated at 35°/15 mm. Hg on a rotary evaporator. Each residue is co-evaporated with two 50-ml. portions of water at 35°/15 mm. Hg, and then dissolved in methanol and made up to 50 ml. with methanol. A sample of each solution is diluted 1:100 with 0.01 *N* hydrochloric acid and the optical absorbance at 260 nm. is measured. Peak 2 contains 2,180 A_{260} units (0.13 mmole, 65%) of compound **3**, and peak 3 (buffer solution concentration, 0.12–0.17 *M*) contains 2,500 A_{260} units[n] of an absorbing material. An aliquot (50 A_{260} units) of the stock solution of peak-3 material is evaporated to dryness at 35°/15 mm. Hg on a rotary evaporator, the residue is heated with 0.5 ml. of 80% aqueous acetic acid at 50° for 2 hr., and the resulting solution is applied to a 10-cm. wide strip of Whatman 3MM paper, and chromatographed in 5:2:3 butyl alcohol–acetic acid–water, alongside authentic samples of uridine 2′(3′)-phosphate and cytidylyl-(3′ → 5′)-uridine (**4**). The presence of the product **7** is shown[o] by a spot having R_f 0.10. The zones are eluted with 15 ml. of 0.01 *N* hydrochloric acid, and the optical absorbance of the eluates is measured

[n] The A_{260} unit is defined as that amount of material per milliliter of solution that produces an absorbance of 1.00 in a cell of 1-cm. light-path at 260 nm.

[o] This identification of the presence of a trinucleoside diphosphate is of a general character. In the present solvent-system, the R_f values of a trinucleoside diphosphate are always lower than those of the corresponding nucleotide.

at 260 nm., with a blank prepared similarly by elution of an equal area of the same chromatogram. The yield of the product is expressed as a percentage of the optical absorbance of peak 3, and converted into μmoles by using the value $\epsilon_{260}^{pH\,2} = 26.8 \times 10^3$ [1 μmole of UpCpU (**7**) = 26.8 A_{260} units]. The spectroscopically determined yield of compound **7** is 25% (based on starting compound **3**) or 71% (based on the weight of compound **3** that reacted); total 1,330 A_{260} units, *i.e.*, 50 μmole.

Isolation of Uridylyl-(3′ → 5′)-cytidylyl-(3′ → 5′)-uridine (7) from Peak 3

Eluates corresponding to peak 3 are placed in a 250-ml., round-bottomed flask, and evaporated to dryness at 35°/15 mm. Hg on a rotary evaporator. The residue is dissolved in 5 ml. of aqueous acetic acid (20%), and the solution is kept for 5 hr., and then freeze-dried. The residue is dissolved in 2 ml. of water, applied quantitatively to two sheets of Whatman 3MM paper, and chromatographed in 5:2:3 butyl alcohol–acetic acid–water for 24 hr. The bands corresponding to product **7** are eluted with water that has previously been made alkaline (pH 8–9) with ammonium hydroxide.[i] The eluate (50 ml.) is freeze-dried, the residue is dissolved in 2 ml. of 50% aqueous ethanol, the solution is treated with 3 ml. of a 5% solution of calcium chloride in 96% ethanol, and the solution is added dropwise to 100 ml. of acetone. The resulting precipitate (the calcium salt of compound **7**) is collected by centrifugation (3,000 r.p.m.), successively washed under the same conditions with 10 ml. of acetone and 10 ml. of ether, and dried over phosphorus pentaoxide at 0.1 mm. Hg. The chromatographically and electrophoretically homogeneous[g] calcium salt of compound **7** weighs 40 mg.; content 80%, as determined by spectroscopic, molecular-weight determination, using $\epsilon_{260}^{pH\,2} = 26.8 \times 10^3$.

REFERENCES

(1) J. Žemlička, *Chem. Ind.* (London), **1964**, 581.
(2) J. Smrt and S. Chládek, *Collection Czech. Chem. Commun.*, **31**, 2978 (1966).
(3) J. Žemlička, S. Chládek, A. Holý, and J. Smrt, *Collection Czech. Chem. Commun.*, **31**, 3198 (1966).
(4) A. Holý and J. Smrt, *Collection Czech. Chem. Commun.*, **31**, 3800 (1966).

Section V

Reagents, Intermediates, and Miscellaneous Compounds

[156] 1-Bromo-3-methyl-2-butene (3,3-Dimethylallyl Bromide)

ROSS H. HALL and M. H. FLEYSHER

DEPARTMENT OF EXPERIMENTAL THERAPEUTICS, ROSWELL PARK MEMORIAL INSTITUTE, BUFFALO, NEW YORK 14203

$$(Me)_2C{=}CHC({=}O)OH \xrightarrow{LiAlH_4} (Me)_2C{=}CHCH_2OH \xrightarrow{PBr_3} (Me)_2C{=}CHCH_2Br$$

1 (100.1) **2** (86.1) **3** (149.0)

INTRODUCTION

The title compound (**3**) has utility in the synthesis of *N*-(3-methyl-2-butenyl)adenine (Sec. I [4]) and *N*-(3-methyl-2-butenyl)adenosine (Sec. III [68]), from adenine and adenosine, respectively.

PROCEDURE

3-Methyl-2-buten-1-ol[1] (2)

A solution of 33 g. (0.33 mole) of senecioic acid (3,3-dimethylacrylic acid,[a] **1**) in 500 ml. of absolute ether is added dropwise during 5 hr. to a slurry of 11.4 g. (0.30 mole) of lithium aluminum hydride in 400 ml. of absolute ether contained in a 3-l., three-necked flask equipped with a stirrer and a reflux condenser. Stirring is continued for 4 hr., and then 300 ml. of water is added slowly, while the mixture is cooled in an ice bath. Sulfuric acid (0.1 *N*) is added until the precipitate dissolves (*ca.* 650 ml.), the layers are separated, and the ether layer is washed with three 200-ml. portions of water, and dried with sodium sulfate. The ether is evaporated off, and the product (**2**) is distilled at 84–86°/80 mm. Hg; yield, 17 g. (65.5%).

1-Bromo-3-methyl-2-butene[2] (3)

A solution of 34 g. (0.39 mole) of compound **2** in 35 ml. of petroleum ether (b.p. 39–60°) is added during 80 min. to a cold (−15°), strongly agitated solution of 39.1 g. (0.14 mole) of phosphorus tribromide in 40 ml. of petroleum ether containing 1.7 ml. of anhydrous pyridine. The mixture is kept at room temperature overnight, and the layers are separated. The

[a] Columbia Organic Chemicals Co., Columbia, S.C.

sirupy, lower layer is washed once with 50 ml. of petroleum ether, and the petroleum ether solution and wash are combined, and evaporated to dryness to yield crude product (**3**) which is distilled at 85°/150 mm.Hg; yield 42 g. (71%).

REFERENCES

(1) J. Knights and E. S. Waight, *J. Chem. Soc.*, **1955**, 2830.
(2) W. Kuhn and H. Schinz, *Helv. Chim. Acta*, **35**, 2008 (1952).

[157] 1,3,5-Tri-*O*-acetyl-2-deoxy-D-*erythro*-pentose (1,3,5-Tri-*O*-acetyl-2-deoxy-D-ribose)

Acetolysis of a Natural 2′-Deoxynucleoside as a Source of a Fully Acylated 2-Deoxy-D-ribofuranose

MORRIS J. ROBINS and ROLAND K. ROBINS

DEPARTMENT OF CHEMISTRY, UNIVERSITY OF UTAH, SALT LAKE CITY, UTAH 84112

$\xrightarrow[C_5H_5N]{Ac_2O}$ ⟶ OAc +

1 (251.2) **2** (377.4) **3** (260.2) **4** (177.2)

INTRODUCTION

1,3,5-Tri-*O*-acetyl-2-deoxy-D-*erythro*-pentose (**3**) is an excellent derivative for use in the fusion synthesis[1] of 2′-deoxynucleosides. The direct acetylation of 2-deoxy-D-*erythro*-pentose gives the crystalline β-D anomer of **3** in 2% yield[2]; the present method[3] gives an 80% yield of the distillable triacetate **3** (probably, the pure α-D anomer), together with a quantitatve yield of *N*-acetyladenine (**4**) from commercial 2′-deoxyadenosine (**1**).

PROCEDURE

1,3,5-Tri-*O*-acetyl-2-deoxy-D-*erythro*-pentose (3)

2′-Deoxyadenosine monohydrate[a] (**1**) (269 g., 1 mole) is added to 570 ml. of pyridine and 566 ml. of acetic anhydride in a 2-l., round-bottomed flask. The suspension is stirred mechanically, and the temperature is kept below 85° by occasional cooling in an ice bath. The solid dissolves completely, and the temperature falls after *ca.* 20 min. The clear solution is

[a] Purchasable from International Chemical and Nuclear Corporation, City of Industry, California.

kept at room temperature overnight, and then the solvents are removed under diminished pressure (oil pump) at 60° (bath temperature) on a rotary evaporator. The colorless, amorphous residue is dissolved in 200 ml. of hot 5:1 (v/v) ethanol–toluene, and the solution is re-evaporated to dryness.

The crude *N*-acetyl-3′,5′-di-*O*-acetyl-2′-deoxyadenosine (**2**) is dissolved in 600 ml. of glacial acetic acid and 150 ml. of acetic anhydride, and the solution is heated at 100° (inner temperature[b]) in an oil bath, with mechanical stirring, for 4.5 hr. The mixture is cooled in ice, and 173 g. (98%) of colorless crystals[c] of *N*-acetyladenine (**4**) is filtered off, and washed with 500 ml. of chloroform. The filtrates are combined and evaporated to dryness under diminished pressure (oil pump) at 60°, giving a brown sirup that is dissolved in chloroform (500 ml.); 2 g. of compound **4** is removed by filtration. The filtrate is successively washed with ice-cold 3 *N* sulfuric acid, ice-water, ice-cold aqueous sodium hydrogen carbonate, and ice-water (to pH 7), dried (anhydrous sodium sulfate), filtered through a thin bed of carbon–Celite,[d] and evaporated to dryness under diminished pressure, giving a yellow, fairly viscous sirup. Vacuum distillation through a 5-in. Vigreux column gives a forerun of 8 g., followed by 208 g. (80%) of compound **3**, b.p. 124–125°/0.08 mm. Hg, $[\alpha]^{25}_{D}$ +25.0° (*c* 0.66, methanol).

REFERENCES

(1) T. Sato, T. Simadate, and Y. Ishido, *Nippon Kagaku Zasshi*, **81**, 1440 (1960).

(2) H. Venner and H. Zinner, *Chem. Ber.*, **93**, 137 (1960).

(3) M. J. Robins and R. K. Robins, *J. Am. Chem. Soc.*, **87**, 4934 (1965).

[b] During this period, the inner temperature should be carefully regulated, to effect complete acetolysis, but to avoid decomposition of the rather sensitive sugar.

[c] These crystals begin separating from the hot acetolysis solution after heating for about 20 min.

[d] Johns–Manville brand of diatomaceous earth.

[158] 2-Deoxy-3,5-di-*O*-*p*-toluoyl-D-*erythro*-pentosyl Chloride

Preparation of a Crystalline O-*Acyl-2-deoxypentofuranosyl Halide*

CLARITA C. BHAT

DEPARTMENT OF CHEMISTRY, GEORGETOWN UNIVERSITY,
WASHINGTON, D.C. 20007

HCl–MeOH; p-MeC_6H_4CCl (O); HCl–AcOH

1 (134.1) **2** (148.1) **3** (384.3) **4** (388.8)

INTRODUCTION

The title compound (**4**) is a relatively stable, crystalline *O*-acylglycosyl halide, and has utility in the direct synthesis of 2-deoxy-D-*erythro*-pentofuranosyl nucleosides (2′-deoxyribonucleosides).

PROCEDURE

Methyl 2-Deoxy-3,5-di-*O*-*p*-toluoyl-D-*erythro*-pentoside[1] (3)

To a solution of 13.6 g. (0.1 mole) of 2-deoxy-D-*erythro*-pentose (**1**) in 243 ml. of methanol is added 27 ml. of a 1% solution of hydrogen chloride in methanol. The mixture is kept in a stoppered flask for 12–15 min.,[a]

[a] The optimal yield of methyl 2-deoxy-D-*erythro*-pentofuranoside (**2**) is obtained in this time interval.

after which the reaction is stopped by adding, with vigorous stirring, 5 g. of silver carbonate. After filtration, the clear solution is evaporated to a sirup under diminished pressure.

Residual methanol is removed by repeated evaporations under diminished pressure with small volumes of dry pyridine, the sirupy **2** is dissolved in 80 ml. of pyridine, and the solution is cooled in an ice bath. To this solution is quickly added 34 g. (0.22 mole) of *p*-toluoyl chloride. The reaction mixture is stirred at 0° for 1 hr., gradually warmed to 40–50° and kept either at this temperature for 2 hr., or at room temperature overnight. The mixture is poured, with stirring, onto 300 ml. of crushed ice, and when the ice has melted, the mixture is extracted with three 150-ml. portions of ether. The ether extracts are combined and successively washed with water, dilute sulfuric acid, and aqueous sodium hydrogen carbonate. Evaporation of the extract under diminished pressure yields a yellowish sirup, which may be crystallized at this point to give a mixture of anomers of methyl 2-deoxy-3,5-di-*O*-*p*-toluoyl-D-*erythro*-pentoside (**3**); yield 26.9 g. (70%).

2-Deoxy-3,5-di-*O*-*p*-toluoyl-D-*erythro*-pentosyl Chloride (4)

For the direct preparation of the chloride, the above sirup (**3**) is dissolved in 40 ml. of glacial acetic acid, and to this solution is added 80 ml. of glacial acetic acid presaturated with dry hydrogen chloride at 10°. Hydrogen chloride is passed into the solution for 10 min., whereupon the chloride (**2**) solidifies, forming a thick, crystalline mass. After 30 min., the crystals are filtered off by suction, and thoroughly washed with dry ether. The crystals are then suspended in dry ether, filtered off, and stored in a vacuum desiccator containing soda-lime and phosphorus pentaoxide; yield, 27.5 g. (70%, based on **1**), m.p. 109° (dec.), $[\alpha]_D^{25}$ $+108 \rightarrow +65°$ (90 min., *c* 1.0, *N*,*N*-dimethylformamide). Compound **2** is stable for weeks when stored over soda-lime and phosphorus pentaoxide, but decomposes in a few hours if exposed to moisture. Small amounts of the chloride (**4**) may be recrystallized from toluene or carbon tetrachloride.

REFERENCE

(1) M. Hoffer, *Chem. Ber.*, **93**, 2777 (1960).

[159] Hydracrylonitrile (Barium Phosphate) [2-Cyanoethyl Barium Phosphate]

Preparation of an Intermediate in the Synthesis of Phosphoric Esters

SARAH J. CLAYTON, JERRY D. ROSE, and WILLIAM E. FITZGIBBON, Jr.

KETTERING-MEYER LABORATORY, SOUTHERN RESEARCH INSTITUTE, BIRMINGHAM, ALABAMA 35205

$$\underset{\substack{\mathbf{1}\\(71.1)}}{HOCH_2CH_2CN} \xrightarrow[\substack{(153.4)}]{\substack{POCl_3\\ \mathbf{2}}} \xrightarrow[\substack{(255.5)}]{\substack{Ba(OAc)_2\\ \mathbf{3}}} \underset{\substack{\mathbf{4}\\(322.4)}}{Ba\langle O_2\rangle P(\rightarrow O)-OCH_2CH_2CN \cdot 2H_2O}$$

INTRODUCTION

The barium salt[1] (**4**) of 2-cyanoethyl phosphate is frequently used in the synthesis of phosphoric esters of nucleosides.

PROCEDURE

2-Cyanoethyl (Barium Phosphate)[1] (4)

A solution of 70 g. (0.46 mole) of freshly distilled phosphoryl chloride (**2**) in 460 ml. of anhydrous ether is placed in a 1-l., 3-necked flask fitted with a thermometer, a sealed stirrer, and a 100-ml. pressure-equalizing funnel closed with a drying tube. The flask is placed in an ice–salt bath, the solution is cooled to $-10°$, and a solution of 32.6 g. (0.46 mole) of hydracrylonitrile (**1**) in 41.9 g. (0.46 mole) of anhydrous pyridine is slowly added to the vigorously stirred mixture at such a rate that the temperature is maintained at -10 to $-15°$. After the addition, which requires about 1.5 hr., the mixture is stirred for 1 hr. at the same temperature, and the suspension is slowly poured, with efficient stirring, into a mixture of 1.72 l. of water, 184 ml. of pyridine, and 700 g. of ice. A solution of 229 g. of barium acetate (**3**) in 690 ml. of water is then added, and the resulting suspension is kept at room temperature for 2 hr. (to permit evaporation of much of the ether and allow aggregation of the barium phosphate formed) and filtered. Ethanol (6 l.) is slowly added, with stirring, to the clear filtrate, the white suspension is refrigerated overnight, and the product is collected by filtration (which is very slow) and well washed with 50% ethanol followed by pure ethanol. The material is air-dried and then dried for 16 hr. at room temperature *in vacuo* over phosphorus pentaoxide. The lustrous, white platelets of **4** weigh 87 g. (59%).

For use in nucleotide syntheses, the barium salt **4** (21.2 g.) is dissolved in boiling water (5.5 l.), and 141 ml. of *N* sulfuric acid is added to the cold solution (5°). After removal of the precipitated barium sulfate by filtration, the resulting aqueous solution of 2-cyanoethyl phosphate is evaporated to dryness *in vacuo*. The residual sirup, dried by addition and distillation of three 250-ml. portions of dry pyridine, is dissolved in 475 ml. of dry pyridine.

REFERENCE

(1) G. M. Tener, *J. Am. Chem. Soc.*, **83**, 159 (1961).

[160] Chlorobis(*p*-methoxyphenyl)phenylmethane

A Reagent for the Selective Etherification of the 5′-Hydroxyl Group of Furanoid Nucleosides

A. HOLÝ

INSTITUTE OF ORGANIC CHEMISTRY AND BIOCHEMISTRY,
CZECHOSLOVAK ACADEMY OF SCIENCES, PRAGUE, CZECHOSLOVAKIA

MeO—C₆H₄—Br (**1**, 187.0) —Mg→ [MeO—C₆H₄—MgBr] (**2**) —Ph—C(=O)OMe (**3**, 136.1)→ [(MeOC₆H₄)₂C(Ph)—OH] (**4**) —AcCl→ (MeOC₆H₄)₂C(Ph)—Cl (**5**, 348.8)

where Ac is acetyl.

INTRODUCTION

Because of the relative ease of removal of the bis(*p*-methoxyphenyl)-phenylmethyl group as compared with the trityl group, chlorobis(*p*-methoxyphenyl)phenylmethane (**5**) has been used instead of chlorotri-phenylmethane for the selective etherification of primary hydroxyl groups; for example, the 5′-hydroxyl group of furanoid nucleosides. In particular, it has been employed to advantage in the preparation of 2′-deoxy-*N*-[(di-methylamino)methylene]-5′-*O*-[bis(*p*-methoxyphenyl)phenylmethyl]ade-nosine (Sec. III [57]).

PROCEDURE

Chlorobis(*p*-methoxyphenyl)phenylmethane (5)

In a 2-l., three-necked flask, fitted with a mechanical stirrer, a 500-ml. dropping funnel, and a reflux condenser protected from atmospheric moisture (calcium chloride tube), is placed 21.8 g. (0.9 g.-atoms) of magnesium turnings prewashed with ether and dried at about 15 mm. Hg. A small amount (10–20 mg.) of iodine is added, and the flask is warmed briefly with a free flame until iodine vapor is visible, and allowed to cool. The magnesium turnings are covered with 200 ml. of ether, and 2–3 drops of *p*-bromoanisole (**1**) and 1 ml. of methyl iodide are added, with stirring. As soon as the reaction begins, a mixture of 165 g. (0.88 mole) of *p*-bromoanisole (**1**) and 300 ml. of dry ether (predried by distillation over phosphorus pentaoxide and stored over sodium) is gradually added, with constant stirring, at such a rate as to cause gentle refluxing. When the addition is complete, the mixture is refluxed, with stirring, on a steam bath for 1 hr. The bath is removed, and 60 g. of freshly redistilled (b.p. 198–199°) methyl benzoate (**3**) is added dropwise, with stirring, at such a rate as to keep the mixture refluxing. When the addition is complete, the mixture is refluxed, with stirring, on a steam bath for 1 hr., and kept at room temperature overnight. After the flask has been cooled by immersion in ice–water, a solution of 115 g. of ammonium chloride in 300 ml. of water is added at such a rate as to cause gentle refluxing. The mixture is then steam-distilled until the distillate is clear. The distillation residue is cooled, and extracted with two 200-ml. portions of ether, and the extracts are combined, dried with magnesium sulfate, filtered, and evaporated to dryness. The residue is dissolved in 400 ml. of cyclohexane, the solution is treated with 100 ml. of acetyl chloride, and the latter solution is refluxed, with exclusion of atmospheric moisture (calcium chloride tube), on a steam bath for 15 min. The solution is evaporated to dryness, and the residue is dissolved in a mixture of 100 ml. of cyclohexane and 10 ml. of acetyl chloride by boiling briefly. The resulting solution is kept in a refrigerator (+2°) overnight, to deposit the product (**5**), which is rapidly filtered off with suction, washed with cyclohexane, and dried in a vacuum desiccator over potassium hydroxide; yield 74.5 g. (25%), m.p. 114° (lit.[1] 114°).

REFERENCE

(1) M. Smith, D. H. Rammler, I. H. Goldberg, and H. G. Khorana, *J. Am. Chem. Soc.*, **84**, 430 (1962).

[161] 5-Nitro-1-(2,3,4,6-tetra-*O*-acetyl-β-D-glucosyl)-2(1*H*)-pyridone

O → N-*Glycosyl Rearrangement*

DAVID THACKER AND T. L. V. ULBRICHT

TWYFORD LABORATORIES, ELVEDEN ROAD, LONDON, N.W. 10, ENGLAND

O_2N … AcOCH$_2$ … OAc … AcO … OAc
$\xrightarrow{SnCl_4}$
O_2N … AcOCH$_2$ … OAc … AcO … OAc

1 (470.4) **2** (470.4)

INTRODUCTION

Numerous *O*- → *N*-glycosyl rearrangements have been described in which mercuric bromide is the catalyst,[1] and it has been shown that the reaction is subject to general Lewis acid catalysis.[2] β-*O*- → α-*O*-Glycosyl anomerization may accompany the rearrangement, but, under the conditions described, rearrangement is essentially the only reaction.

PROCEDURE

5-Nitro-1-(2,3,4,6-tetra-*O*-acetyl-β-D-glucosyl)-2(1*H*)-pyridone (2)

To a solution of 5-nitro-2-(2,3,4,6-tetra-*O*-acetyl-β-D-glucosyloxy)-pyridine[3] (**1**) (2.0 g., recrystallized from methanol) in benzene (100 ml.) is added stannic chloride (4.52 g.). The solution is stirred at 22° for 24 hr., and poured into saturated, aqueous sodium hydrogen carbonate (50 ml.), and the organic layer is separated, washed successively with aqueous sodium hydrogen carbonate and water, dried (anhydrous magnesium sulfate), and evaporated to dryness under diminished pressure at 40°. The resulting sirup is purified by preparative thin-layer chromatography on silica gel GF_{254} (Merck, Darmstadt) with ethyl acetate, followed by crystallization from methanol–isopropyl alcohol; yield, 0.96 g. (48%), m.p. 99–100°, $[\alpha]_D^{22}$ +30° (*c* 1, dichloromethane), λ_{max}^{MeOH} 305 nm. (ϵ 10,300).

REFERENCES

(1) T. L. V. Ulbricht, *Proc. Chem. Soc.*, **1962,** 298; T. L. V. Ulbricht and G. T. Rogers, *J. Chem. Soc.*, **1965,** 6123, 6130; G. Wagner, *Z. Chem.*, **6**, 367 (1966).
(2) D. Thacker and T. L. V. Ulbricht, *Chem. Commun.*, **1967**, 122.
(3) G. Wagner and E. Fickweiler, *Arch. Pharm.*, **298**, 62 (1965).

[162] Salts of 7,8,9,10-Tetrahydro-7,7-dimethyl-3*H*-pyrimido[2,1-*i*]purine and of its 8-Bromo and 8-Iodo Derivatives

Cyclization of N-*(3-Methyl-2-butenyl)adenine at the 1-Position*

KERMIT L. CARRAWAY, MALCOLM RASMUSSEN, and JOHN P. HELGESON

DEPARTMENT OF CHEMISTRY AND CHEMICAL ENGINEERING,
UNIVERSITY OF ILLINOIS, URBANA, ILLINOIS 61801

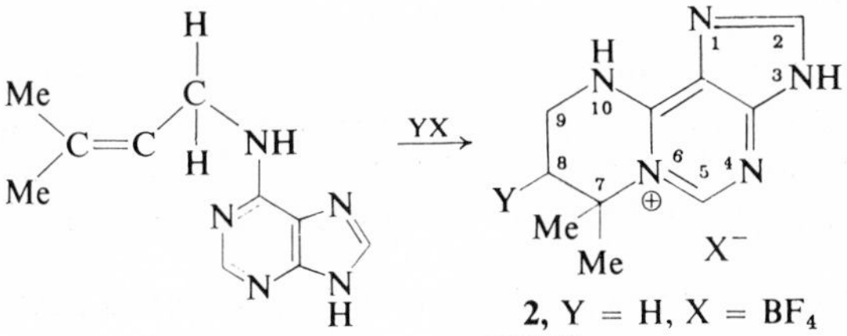

1
(203.3)

$\xrightarrow{YX}$

2, Y = H, X = BF_4
(291.1)
3, Y = H, X = CF_3COO
(317.3)
4, Y = I, X = I
(457.1)
5, Y = Br, X = Br
(363.1)

INTRODUCTION

Chemical investigation of isoprenoid purine derivatives has received considerable impetus from (a) the finding that *N*-(3-methyl-2-butenyl)-adenine[1] (**1**) is a highly active cytokinin[2] and (b) the isolation of **1** from *Corynebacterium fascians*, a plant pathogen,[3] and a D-ribofuranosyl derivative of **1** from serine transfer ribonucleic acid[4] and yeast and calf-liver soluble ribonucleic acid.[5] A characteristic of **1** is its acid-induced cyclization[5] to give[6] a tricyclic ring system as in **2** and **3**. Analogous cyclization of **1** to **4** and **5** may readily be effected under milder conditions by use of iodine or bromine.[7]

PROCEDURE

7,8,9,10-Tetrahydro-7,7-dimethyl-3*H*-pyrimido[2,1-*i*]purin-6-ium Fluoborate[6] (2)

To a suspension of 2.40 g. (11.8 mmoles) of *N*-(3-methyl-2-butenyl)-adenine (**1**) (Sec. I [3]) in 45 ml. of absolute ethanol is added 5.0 ml. of

50% aqueous fluoboric acid. The resulting solution is evaporated at 70° in a rotary evaporator during 10–15 min. The residue is shaken with 100 ml. of ethanol, the suspension is filtered, and the solid is dried, recrystallized from aqueous ethanol, and characterized as the fluoborate salt (**2**); yield 3.10 g. (91%), colorless needles, m.p. 278.5–279.5° (dec.).

The cyclization may also be effected by heating **1** in trifluoroacetic acid at 50° for 4 hr. Evaporation of the resulting solution of **3** to dryness, followed by treatment of the residue with an aqueous suspension of Dowex-1 (HCO_3^- or OH^-) and evaporation of the filtered solution, yields the free base. Recrystallization from ethanol gives colorless prisms, which decompose and melt indistinctly at about 285°.

7,8,9,10-Tetrahydro-8-iodo-7,7-dimethyl-3*H*-pyrimido[2,1-*i*]purin-6-ium Iodide[7] (4)

Iodine (0.148 g., 0.585 mmole) is added to a refluxing solution of 0.120 g. (0.59 mmole) of compound **1** in 11 ml. of absolute ethanol. After 30 min., the mixture is allowed to cool, and the product (**4**) is collected as yellow-orange platelets; yield 0.191 g. (71%), m.p. 240–241° (dec.).

8-Bromo-7,8,9,10-tetrahydro-7,7-dimethyl-3*H*-pyrimido[2,1-*i*]purin-6-ium Bromide[7] (5)

Bromine (4 drops) is added dropwise to a suspension of 0.038 g. (0.17 mmole) of compound **1** in 3 ml. of absolute ethanol. The yellow solution resulting is partially concentrated on a hot-water bath until crystallization begins. The reaction mixture is cooled and filtered, and the product is washed with acetone, affording almost colorless needles of **5**; yield 0.031 g. (47%), m.p. 261–262° (dec.).

This product (**5**) may also be prepared from **1** by treatment with *N*-bromosuccinimide in chloroform, followed by conversion of the resulting product into the hydrobromide. To a solution of 0.389 g. (2.18 mmoles) of *N*-bromosuccinimide in 25 ml. of dry chloroform is added dropwise a solution of 0.405 g. (2.00 mmoles) of compound **1** and 0.246 g. (2.02 mmoles) of benzoic acid in 50 ml. of dry chloroform. After 2 hr., the resulting yellow solution is concentrated under diminished pressure to about 15 ml., and the resulting solution is applied to a column of silica gel (100 g.) in chloroform. After the elution of benzoic acid and a small amount of a purine by-product, the free base is eluted with 3:7 ethanol–chloroform. Recrystallization from chloroform–ethanol gives 0.274 g. (49%) of colorless microcrystals, m.p. 192–197° (dec.). To a suspension of 0.027 g. (0.096 mmole) of this product in 2 ml. of absolute ethanol is added a solution of 0.009 g. (0.11 mmole) of hydrogen bromide in 0.8 ml. of ethanol. Dissolution is effected by boiling under reflux, and then 2 ml. of ether is added to the resulting hot, ethanolic solution. The resulting mixture is cooled and filtered, giving almost colorless, fine needles of **5**; yield 0.031 g. (89%), m.p. 262–263° (dec.).

REFERENCES

(1) A. Cavé, Dr. of Natl. Sci. Thesis, University of Paris, 1962; N. J. Leonard and T. Fujii, *Proc. Natl. Acad. Sci. U. S.*, **51**, 73 (1964).

(2) A. Cavé, J. A. Deyrup, R. Goutarel, N. J. Leonard, and X. G. Monseur, *Ann. Pharm. Franc.*, **20**, 285 (1962); J. H. Rogozinska, J. P. Helgeson, and F. Skoog, *Physiol. Plantarum*, **17**, 165 (1964); G. Beauchesne and R. Goutarel, *ibid.*, **16**, 630 (1963).

(3) D. Klämbt, G. Thies, and F. Skoog, *Proc. Natl. Acad. Sci. U. S.*, **56**, 52 (1966); J. P. Helgeson and N. J. Leonard, *ibid.*, **56**, 60 (1966).

(4) H. G. Zachau, D. Dütting, and H. Feldmann, *Angew. Chem.*, **78**, 392 (1966).

(5) R. H. Hall, M. J. Robins, L. Stasiuk, and R. Thedford, *J. Am. Chem. Soc.*, **88**, 2614 (1966).

(6) K. L. Carraway, Ph.D. Thesis, University of Illinois, 1966; N. J. Leonard, S. Achmatowicz, R. N. Loeppky, K. L. Carraway, W. A. H. Grimm, A. Szweykowska, H. Q. Hamzi, and F. Skoog, *Proc. Natl. Acad. Sci. U. S.*, **56**, 709 (1966).

(7) M. Rasmussen and N. J. Leonard, unpublished results.

[163] 2,3,5-Tri-*O*-benzoyl-D-ribosyl Bromide

An Intermediate in the Synthesis of D-*Ribofuranosyl Nucleosides*

JOHN D. STEVENS and HEWITT G. FLETCHER, JR.

NATIONAL INSTITUTE OF ARTHRITIS AND METABOLIC DISEASES,
NATIONAL INSTITUTES OF HEALTH, PUBLIC HEALTH SERVICE,
DEPARTMENT OF HEALTH, EDUCATION, AND WELFARE, BETHESDA,
MARYLAND 20014

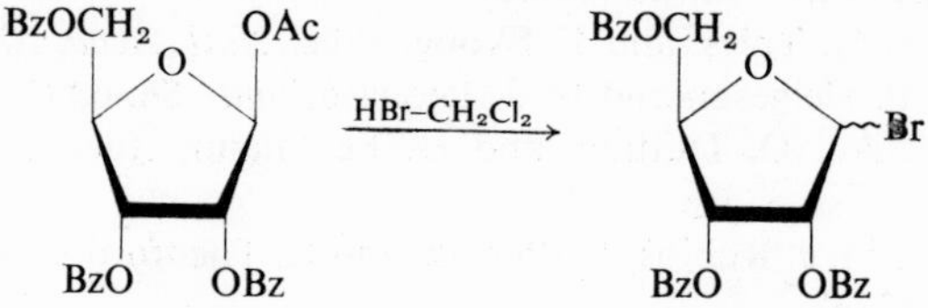

BzOCH$_2$ OAc → ($HBr–CH_2Cl_2$) → BzOCH$_2$ ~Br

BzO OBz BzO OBz

1 (504.5) **2** (525.4)

where Ac is acetyl and Bz is benzoyl.

INTRODUCTION

The title compound[1] (**2**) is readily prepared, in one step, from the commercially available 1-*O*-acetyl-2,3,5-tri-*O*-benzoyl-β-D-ribose (**1**), and has utility in the preparation of D-ribofuranosyl nucleosides.

PROCEDURE

2,3,5-Tri-*O*-benzoyl-D-ribosyl Bromide (2)

Hydrogen bromide is bubbled into an ice-cold solution of 1-*O*-acetyl-2,3,5-tri-*O*-benzoyl-β-D-ribose[a] (**1**) (25.2 g., 50 mmoles) in dichloromethane (150 ml.) for 15 min. After being kept at 0° for 1 hr. and at room temperature for 15 min., the solution is evaporated under diminished pressure to a thin sirup. Dry dichloromethane (25 ml.) and dry toluene (25 ml.) are successively added to, and distilled from the sirup, which may then be used immediately (without further purification) for condensation reactions with appropriate purine and pyrimidine derivatives. The n.m.r. spectrum of the sirup includes a singlet at τ 3.5 (H-1 of the β-D anomer of **2**) and a doublet centered at 3.10 (4.4 Hz, H-1 of

[a] Calbiochem, Box 54282, Los Angeles, California 90054.

the α-D anomer of **2**). By one method of integration, the relative intensities of these signals indicate an anomer ratio of $\alpha:\beta = 60:40$; another method gives $\alpha:\beta = 55:45$.

REFERENCE

(1) J. D. Stevens, R. K. Ness, and H. G. Fletcher, Jr., *J. Org. Chem.*, **33**, 1806 (1968).

Author Index

Numbers in **boldface** indicate authorship of a procedure description. Numbers in *italics* indicate the pages on which full references appear. Numbers in parentheses are reference numbers and show that an author's work is referred to although his name is not mentioned in the text.

C

D

E

F

I

L

M

N

S

W

Y

Z

Subject Index

Boldface numbers indicate the page giving the preparation of the compound.

A

B

E

F

G

H

O

P

R

S

T

U

X

Z

Index of Some General Reactions

A

B

D

I

M

O

R

T